Physical Chemistry
for Students of Pharmacy and Biology

Physical Chemistry
for Students of Pharmacy and Biology

Third Edition

S.C. Wallwork M.A., D.Phil., F.Inst.P.
and **D.J.W. Grant** M.A., D.Phil., C. Chem., F.R.I.C.

Longman
London and New York

Longman Group Limited London

*Published in the United States of America
by Longman Inc., New York*

*Associated companies, branches and representatives
throughout the world*

Second edition © S. C. Wallwork 1960
Third edition © Longman Group Ltd. 1977

First Published 1956
Second Impression (with minor corrections) 1959
Second Edition (revised) 1960
Second Impression 1961
Third Impression 1962
Fourth Impression 1965
Fifth Impression 1966
Sixth Impression 1968
Third Edition (revised) 1977
Second Impression 1978

Library of Congress Cataloging in Publication Data

Wallwork, Stephen Collier.
Physical chemistry for students of pharmacy and biology.

Includes index.
1. Chemistry, Physical and theoretical.
I. Grant, David James William, 1937–   joint author.
II. Title. [DNLM: 1. Chemistry, Physical. QD453 W214 p]
QD453.2.W35   1977   541'.3   73-94321

ISBN 0 582 44254 0

Printed in Great Britain by
Richard Clay (The Chaucer Press), Ltd,
Bungay, Suffolk

Preface to the First Edition

This book is intended to cover all the topics in physical chemistry which are required by students of pharmacy and biology taking courses leading to degrees and similar qualifications in universities and technical colleges. It is based on the recommended syllabus of the Pharmaceutical Society of Great Britain for the Pharmaceutical Chemist Qualifying Examination, and it is written as a result of the author's experience in delivering university lecture courses designed specifically for pharmacists and biologists.

Physical chemistry is concerned with explaining rather than describing the behaviour of chemical substances. The proper function of the teacher of physical chemistry should therefore be to emphasise the explanatory nature of the subject, and the attitude of the student should be one of seeking to understand rather than of learning the subject. It is inevitable, in a subject so closely allied to physics, that some of the processes of explaining and understanding should involve a rather mathematical approach. In this book the more mathematical aspects of physical chemistry have been reduced to a minimum, but mathematics cannot be dispensed with entirely, and the student is urged to make an attempt to understand the little that is essential. To this end, a mathematical introduction to the subject is provided in Chapter 1, and it is recommended that the principles set out there should be thoroughly mastered before proceeding with the rest of the book. The student will be able to judge whether he has understood the material in this and subsequent chapters if he attempts the numerical problems at the end of each chapter.

The student of pharmacy or biology should be encouraged to regard physical chemistry not as a separate study but as one which helps him to obtain a firmer and clearer appreciation of his main subject. For this reason, no attempt has been made to give a rigorous treatment of physical chemistry, but rather to concentrate on those aspects which are likely to prove of greatest value to the reader for whom the book is intended. Also, the opportunity has frequently been taken of giving examples of the application of physico-chemical principles to problems of biological or pharmaceutical importance.

Preface to the First Edition

To enlarge on such applications would be outside the scope of this book, but it is hoped that the material which has been given will serve as a link between this introductory account and the growing number of biological and pharmaceutical works which make use of the concepts of physical chemistry. Some of these works are included in the bibliography, and the author is glad to make acknowledgment of the debt which he owes to those which have formed the source of many of his examples.

The author records with pleasure his gratitude to those of his students and colleagues in the University of Nottingham who have helped in any way in the preparation of the book; and in particular to Dr M. W. Partridge, Reader in Pharmaceutical Chemistry, for many helpful discussions on the selection and presentation of the subject-matter, and to Dr W. E. Addison, Lecturer in Chemistry, for assistance in the preparation of the manuscript.

S. C. Wallwork The University of Nottingham *July* 1955.

Preface to the Third Edition

This edition has the same aim as the previous editions, namely, to cover all the physical chemistry required by students of pharmacy, biology and related sciences taking courses leading to a degree or similar qualifications in universities, polytechnics and technical colleges. It may also be of value to medical students and will form a useful introduction to physical chemistry for anyone with a background of biological sciences. Previous knowledge need be no more than G.C.E. 'A' level (or even 'O' level) in Chemistry.

The rapid developments in the pharmaceutical and biological sciences in recent years have emphasized the importance of physical chemistry. Since a degree in pharmacy has become a requirement for membership of the Pharmaceutical Society of Great Britain and of pharmaceutical societies in other countries, the physico-chemical understanding required by the practising pharmacist has increased in both depth and breadth. Pharmacists everywhere need to know the fundamental principles underlying the procedures which they follow so that their skills may be extended to solve new problems. Many of these principles are of a physico-chemical nature. In particular, pharmacists in hospitals and in industry are required to be familiar with those aspects of physical chemistry which are important in formulation and analysis. Similarly, the biological sciences which have for long been thought of as descriptive subjects, have become more and more quantitative. For a deeper understanding of their subjects biologists and biochemists turn not only to organic chemistry but to physical chemistry, physics and mathematics. All these trends are shown in the enormous increase in physical chemistry in the scientific literature in biology and pharmacy.

For these reasons it was felt that a new edition of this book should be produced and it was inevitable that the size should be increased considerably to accommodate all the necessary new material. The importance of solid dosage forms in pharmacy has necessitated expansion of the treatment of the solid state. The growing importance of spectroscopic methods of analysis is reflected in the more extended treatments of atomic and molecular spectra and in the introduction of sections on nuclear magnetic resonance and mass spectrometry. A quantitative treatment of enzyme inhibition has been included in the chapter on chemical reaction. Diffusion and rates of dissolution have been introduced into the chapter on solutions. A more extensive treatment of thermodynamics has formed the basis of a new chapter on energy and equilibrium. Electrolytes and conductivity have also been placed in a new chapter. Treatment of membranes has been expanded and some important electrochemical methods of analysis have been introduced. The physical chemistry of macromolecules, dispersions and other colloids have become so important in pharmacy and biology that a more extended treatment has become necessary. A descriptive chapter on rheology has also been included and gives molecular explanations of viscous flow, elasticity and viscoelasticity.

The bibliography in the earlier editions mentioned some of the biological and pharmaceutical books that made use of the concepts of physical chemistry. These concepts are now applied so widely, however, that it seems inappropriate to draw attention to just a few examples, so the bibliography has now been omitted.

As in previous editions, the mathematical treatment has been kept to a minimum, and the first chapter still contains all the necessary preparation. The IUPAC recommendations for the physical quantities and their symbols have in most cases been followed. SI units have been used throughout and other common units have been mentioned.

A few more simple experiments which illustrate physico-chemical principles have also been added. As before, these experiments do not require any special apparatus or instruments. It is recognised that most courses in pharmacy now include practical work on analysis using electronic instruments and the basic physico-chemical principles underlying analysis are discussed in the book. Textbooks of practical physical chemistry and equipment manuals should be consulted for details of these techniques.

We should like to express our gratitude to the following: Professor James Crossland, Dr. I. W. Kellaway and Dr. C. J. Timmons for discussions on membranes, membrane potentials and spectroscopy, the late Professor M. W. Partridge, Professor R. B. Cŭndall, Dr. Harold Booth, Dr. T. F. Palmer, Professor R. E. Grigg, Dr. J. M. Newton and Dr. Christopher Marriott for reading and commenting on some sections of the manuscript. Figures 4.22 and 4.23 have been adapted from Figures 2.9 and 2.8 in 'Cell Biology' by John Paul and we are grateful to the author and to the publishers, Messrs. Heinemann for permission to do this.

S. C. Wallwork   D. J. W. Grant   The University of Nottingham
May 1976

Contents

	Page
Preface to the First Edition	v
Preface to the Third Edition	vii
List of Experiments	xiii

Chapter 1 Mathematical Preparation   1

Introduction — Graphs — The Equation of a Straight Line — Indices — Logarithms — Exponentials — Growth and Decay — Rates of Change, Calculus Notation — Some Standard Differentials — Successive Differentiation — Partial and Complete Differentials — Integration — Integration between Limits — Problems.

Chapter 2 The Properties and Constitution of Matter   21

Gases, the Gas Law — Diffusion of Gases — Kinetic Theory of Gases — Partial Pressures — Deviations from the Gas Laws — Van der Waals Equation — Andrews' Isothermals for Carbon Dioxide — Joule-Thomson Effect — Liquefaction of Gases — Intermolecular Forces — The Liquid State — Liquid Crystals — The Solid State — X-ray Crystal Structure Determination — Equilibrium Diagram for a Single Component — Sublimation — Polymorphism — Solid and Liquid Phase Diagram — Experiment 2.1: The construction of the phase diagram for the system naphthalene/1-naphthol by the cooling curve method — Crystal Structure — Crystallisation — Problems.

Chapter 3 The Structure of Atoms   57

Atomic Theory—Atomic Weights — The Periodic Table — The Electron — Radioactivity — Radioactive Series — Positive Rays — X-ray Frequencies — The Atomic Nucleus — Isotopes — Nuclear Stability — Nuclear Reactions — Tracers — Measurement of Radioactivity — Radioactive Decay — Biological Effects of Radiation — Atomic Spectra — Atomic Structure — Atomic Structure and the Periodic Table — The Electronic Theory of Valency — Directional Bonds — Problems.

Chapter 4 The Structure of Molecules   96

The Basis of Structural Chemistry — Wave-mechanical Explanation of Organic Structures — Dipole Moments — Refractive Index and Molar Refraction — Optical Rotatory Dispersion and Circular Dichroism — Molecular Spectra — Fluorescence — Nuclear Magnetic Resonance — Electron Spin Resonance — Mass Spectrometry — Biological Structure — Problems.

x   Contents

Page

Chapter 5   Chemical Reaction   145

Introduction — Order of Reaction and Molecularity — Experimental Study of the Progress and Rate of a Reaction — First-order Reactions — Experiment 5.1: Determination of k_1 for the hydrolysis of methyl acetate in the presence of an acid catalyst — Second-order Reactions — Zero-order Reactions — Determination of Order of Reaction — Temperature Dependence of Rate Constants — Collision Theory of Reaction Rates — Theory of Absolute Reaction Rates — Unimolecular Reactions — Photochemical Reactions — Chain Reactions — Mechanisms of Organic Reactions — Homogeneous Catalysis — Heterogeneous Catalysis — Adsorption Isotherms — Enzyme Reactions — Inhibition of Enzyme Reactions — Competitive Inhibition — Non-competitive Inhibition — Inhibition by Covalent Bonding — Problems.

Chapter 6   Properties of Mixtures and Solutions   199

Mixtures and Solutions — Concentration — Solubility of Solids in Liquids — Solubility of Liquids in Liquids — Experiment 6.1: To determine the critical solution temperature for the phenol-water system — Experiment 6.2: To investigate the limits of miscibility of the system toluene—water—ethanol — Solubility of Gases in Liquids — Vapour Pressure of Liquid Mixtures: (a) Miscible liquids; (b) Immiscible liquids — Experiment 6.3: To determine the molecular weight of chlorobenzene by steam distillation — Relationships between Activities, Solubilities and Drug Action — Distribution (or Partition) between Immiscible Solvents — Experiment 6.4: To determine the partition coefficient of iodine between 1,1,1-trichloroethane and water — Partition Chromatography — Experiment 6.5: To illustrate the relationship between partition coefficient and R_F value in thin-layer partition chromatography using aromatic acids — Effect of Non-volatile Solutes on Vapour Pressure and Related Properties — Experiment 6.6: To determine the molecular weight of p-toluidine by Rast's method — Osmotic Pressure — Ionisation of Solutes and Colligative Properties — Biological Importance of Colligative Properties — Diffusion — Diffusion by the Free Boundary Method — Diffusion by the Porous Disc Method — Experiment 6.7: Determination of the diffusion coefficient of hydrochloric acid (0.4 mol dm^{-3}) at ambient temperature by means of the porous disc method — Dissolution Rates of Solids — Problems.

Chapter 7   Energy and Equilibrium   268

Heat and Work — Conservation of Energy, the First Law of Thermodynamics — Enthalpy — Extensive and Intensive Quantities — Heat Capacity — Thermochemistry — Bond Energies — Resonance Energy — Reversible and Irreversible Processes — Entropy and the Second Law of Thermodynamics — Statistical Thermodynamics and the Third Law — Entropy of Mixing — Entropy Changes from Calorimetry — Entropy and Enthalpy as Criteria for Spontaneity — Gibbs Free Energy — Effects of External Conditions on Phase Changes — Partial Molar Quantities of Substances in Mixtures — Chemical Potential and Activities — Activities and Activity Coefficients — Gibbs Free Energy Changes in Chemical Reactions — Dynamic Equilibrium in Chemical Reactions — Equilibrium Constants — Problems.

Contents xi

Page

Chapter 8 Electrolytes 326

Strong Electrolytes — Equilibria in Ionic Systems — Solubility Product — Experiment 8.1: Determination of the solubility product of potassium hydrogen tartrate — Electrolysis — Transport Numbers — Measurement of Transport Numbers — Some Factors Influencing Transport Numbers — Deviations of Strong Electrolytes from Ideal Behaviour — Weak Electrolytes — Resistance and Conductance — Resistivity and Conductivity — Measurement of Conductance and Conductivity of Electrolytes — Molar Conductivity of Electrolytes — Molar Conductivity and Electric Mobility of Ions — Conductimetric Titrations — Problems.

Chapter 9 Acids and Bases 356

The Definitions of Acid and Base — The Strengths of Acids and Bases — Buffer Solutions — Buffers in Biological Systems — Hydrolysis of Salts — Neutralisation Curves — Enthalpy of Neutralisation — Experiment 9.1: Determination of the enthalpy of neutralisation of various acid-base pairs — Amphoteric Electrolytes or Ampholytes — Indicators — Experiment 9.2: To determine the hydrolysis constant of ammonium chloride (i.e. the acid dissociation constant of the ammonium ion) by an indicator method — pH and Drug Action — Acid-Base Catalysis — Problems.

Chapter 10 Electrochemical Cells 392

Introduction — Transformation of Chemical into Electrical Energy — Reversible Electrode Processes — Electrode Potentials — Energy Changes from Measurements on Cells — Sign Convention — Standard Electrode (or Oxidation-Reduction) Potentials — Formal Electrode Potentials — Measurement of Electrode Potentials — The Significance of Electrode Potentials — Potentiometric Titrations — Glass Electrode — Oxidation-Reduction Indicators — Oxidation Reduction Processes in Biological Systems — Concentration Cells and Junction Potential — Membrane Potentials — Irreversible Electrodes, Overvoltage and Polarisation — Polarography — Clark Oxygen Cell — Amperometric (or Polarometric) Titrations — Dead-Stop End-Point Titrations — Experiment 10.1: The quantitative determination of sulphonamide drugs using a dead-stop end-point titration — Problems.

Chapter 11 Surface Chemistry 442

Introduction — Gas-Solid Interface — Liquid-Solid Interface — Experiment 11.1: The adsorption of oxalic acid from aqueous solutions by charcoal — Adsorption Chromatography — Ion Exchange — Gas–Liquid and Liquid–Liquid Interfaces — Surface Tension and Interfacial Tension — Adsorption at Liquid Surfaces — Experiment 11.2: To show the effect of increasing number of CH_2 groups on the surface tension of aqueous solutions of alcohols and the determination of surface excess — Soluble and Insoluble Surface Films — The Study of Insoluble Surface Films — Types of Insoluble Surface Films — (a) Gaseous or vapour films — (b) Condensed films — (c) Expanded films — Reactions in Surface Films — Problems.

Page

Chapter 12 Colloids 475

The Nature of Colloidal Particles — Classification of Colloidal Systems — Colloidal Dispersions — Solutions of Macromolecules — Association Colloids — Lyophilic and Lyophobic Colloids — Preparation of Colloidal Solutions — The Size and Molecular Weight of Colloidal Particles — Osmotic Pressure of Lyophilic Colloids — Light Scattering by Colloidal Particles — Sedimentation by Ultracentrifugation — Viscosity of Colloids — Experiment 12.1:Determination of the volume of the hydrated glycerol molecule — The Electrical Properties of Colloids — Electrophoresis — Isoelectric Point of Proteins — Stability of Lyophobic Colloids — Precipitation of Lyophilic Colloids — Interactions between Colloidal Systems — Suspensions — Emulsions — Problems.

Chapter 13 Rheology 531

Introduction — Deformation, Elasticity and Flow — Shear — Newtonian Flow — Experiment 13.1: The determination of the kinematic viscosity of light liquid paraffin at room temperature — Rotational Viscometers — Deviations from Newtonian Flow — Thixotropy — Elastomers — Viscoelasticity — Creep — Other Methods of Studying Viscoelasticity — The Weissenberg effect — Problems.

Answers to Problems 558

Appendices

Physical Quantities and Units — Periodic Table — Atomic Weights 567

Tables of Logarithms and Antilogarithms 576

Index 581

List of Experiments

	Page

Experiment 2.1 The construction of the phase diagram for the system naphthalene/1-naphthol by the cooling curve method — 52

Experiment 5.1 Determination of k_1 for the hydrolysis of methyl acetate in the presence of an acid catalyst — 152

Experiment 6.1 To determine the critical solution temperature for the phenol–water system — 205

Experiment 6.2 To investigate the limits of miscibility of the system toluene–water–ethanol — 209

Experiment 6.3 To determine the molecular weight of chlorobenzene by steam distillation — 221

Experiment 6.4 To determine the partition coefficient of iodine between 1,1,1-trichloroethane and water — 227

Experiment 6.5 To illustrate the relationship between partition coefficient and R_F value in thin-layer partition chromatography using aromatic acids — 234

Experiment 6.6 To determine the molecular weight of p-toluidine by Rast's method — 242

Experiment 6.7 Determination of the diffusion coefficient of hydrochloric acid (0.4 mol dm^{-3}) at ambient temperature by the porous disc method — 258

Experiment 8.1 Determination of the solubility product of potassium hydrogen tartrate — 328

Experiment 9.1 Determination of the enthalpy of neutralisation of various acid-base pairs — 375

xiv List of Experiments

Page

Experiment 9.2 To determine the hydrolysis constant of ammonium chloride (i.e. the acid dissociation constant of the ammonium ion) by an indicator method 381

Experiment 10.1 The quantitative determination of sulphonamide drugs using a dead-stop end-point titration 438

Experiment 11.1 The adsorption of oxalic acid from aqueous solutions by charcoal 452

Experiment 11.2 To show the effect of increasing number of CH_2 groups on the surface tension of aqueous solutions of alcohols and the determination of surface excess. 466

Experiment 12.1 Determination of the volume of the hydrated glycerol molecule 503

Experiment 13.1 The determination of the kinematic viscosity of light liquid paraffin at room temperature 539

1 Mathematical Preparation

Introduction

The development of the sciences of pharmacy and biology has, until recently, called for the application of only the simpler mathematical concepts. Indeed, it is sometimes the case that such sciences are deliberately chosen for study because of their generally non-mathematical nature. However, the importance of physical chemistry in helping to establish the fundamental principles on which pharmacy and biology are based is growing rapidly, and a proper understanding of physical chemistry is not possible without a certain minimum mathematical background. The student will not wish to approach certain aspects of physical chemistry, therefore, before receiving further guidance in the mathematical principles which are involved.

The present chapter summarises all the mathematical concepts which are necessary for the understanding of the rest of the book. They have been kept to a minimum, so that they hardly form an adequate basis for a more serious study of the whole of physical chemistry (though they would still form a suitable introduction to a further study of mathematics which this might involve). However, the student who ensures that he is familiar with the contents of this chapter before proceeding further should find that no section in the later chapters will present any mathematical difficulty.

Graphs

It is often very convenient to represent scientific relationships in a graphical manner. This applies particularly to the representation of experimental data, where errors in individual observations may tend to obscure the relationship connecting them all. By graphical representation, such errors would merely be apparent in the amounts by which the individual points are displaced from the mean line through them. Moreover, this line will often be the most accurate way of expressing the relationship inherent in the experimental data.

Suppose we wish to illustrate the relationship connecting two variable quantities x and y. We imagine that these are the only two variables, so that y depends only on x. This is expressed mathematically by saying that y is a function of x, and it is written symbolically as $y = f(x)$. To draw the graph, we first take two lines at right angles, and call the horizontal one the x axis and the vertical one the y axis. The x axis is marked out in successively increasing values of x, with negative values to the left and positive to the right. The y axis is similarly marked out in successively increasing values of y, not necessarily on the same scale,

Fig. 1.1 A straight-line graph

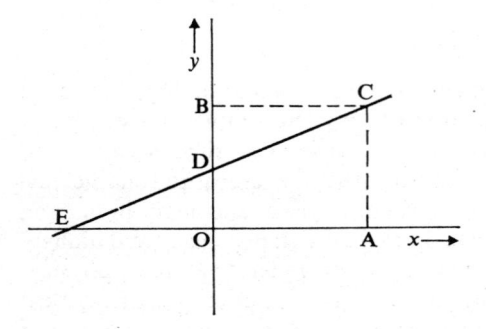

with the positive values upwards and the negative values downwards. Then, from the table of values of x and y, each point is plotted out by reading along the x axis until the x value is reached, and then along the line from this position parallel to the y axis until the y value is reached (or vice versa). The x value at each point is referred to as the abscissa (OA or BC for the point C in Fig. 1.1), and the y value is referred to as the ordinate (OB or AC for the same point). The points are then connected by a smooth curve, or, if there are evidently errors in some or all of the points, the best smooth curve is drawn passing as closely as possible to the maximum number of points.

The Equation of a Straight Line

If it is known beforehand that the results of a certain experiment should follow some theoretical relationship, the graphical presentation of the results is always arranged, wherever possible, to take the form of a straight line. This involves studying the theoretical relationship, and choosing the variables to be plotted in such a way that they obey the general equation for a straight line. The reason for this is that the form of the graph can be drawn easily with a ruler, and, if it is necessary to extrapolate the line beyond the range of the experimental points, there is no doubt as to its direction. The general equation of a straight line is:

$$y = mx + c,$$

where m and c are constants. The constant c is the value of y when $x = 0$ (OD in Fig. 1.1), and it is called the intercept on the y axis. The constant m is the slope, or gradient, of the line, defined as the tangent of the angle which the line makes with the x axis. (In Fig. 1.1 this is the angle DEO, and its tangent is DO/OE, or CA/AE, i.e. the difference in the ordinates divided by the corresponding difference in the abscissae.) This can be seen by substituting in the equation above (x_1, y_1) and (x_2, y_2) for two points on the line. Thus,

$$y_1 = mx_1 + c$$

and

$$y_2 = mx_2 + c.$$

Subtracting,

$$y_2 - y_1 = m(x_2 - x_1).$$

The gradient is given by $(y_2 - y_1)/(x_2 - x_1)$, and this is therefore equal to m.

Many relationships which at first sight appear to be more complicated can be put into a form suitable for plotting as a straight-line graph. Thus, in equation [5.5], a somewhat complicated expression is quoted, which connects the concentration $(a-x)$ of a reactant remaining in a reaction mixture, with the time t which has elapsed since the beginning of a first-order reaction. However, this equation can be put in the form shown in equation [5.8], which is seen to be a straight-line relationship if y is written in place of $\log(a-x)$, m in place of $-k/2.303$, x in place of t, and c in place of $\log a$.

Indices

If, in an expression, a number occurs multiplied by itself several times, it is usually written in index form. In this form, the number, which is called the base, is written once, with an index as a superscript indicating the 'power' to which the number is raised. The 'power' is the number of times the base would have occurred if the expression had been written out in full. Thus, $2 \times 2 \times 2 \times 2 \times 2$ is two raised to the power five, and it is written 2^5, i.e. two with index five. A number with a negative index is the reciprocal of the corresponding quantity with a positive index. Thus, $2^{-5} \equiv 1/2^5$. A number with a fractional index indicates that the corresponding root of the number should be taken. Thus,

$$4^{\frac{1}{2}} \equiv \sqrt{4}; \quad 4^{\frac{1}{4}} \equiv \sqrt[4]{4}; \quad 4^{-\frac{1}{4}} \equiv 1/4^{\frac{1}{4}} \equiv 1/\sqrt[4]{4}.$$

Any number raised to the power 1 is the number itself, i.e. $2^1 \equiv 2$.

When two numbers in index form, and having the same base, are multiplied together, the result can be expressed as the base raised to the power of the sum of the separate indices. Thus,

$$2^3 \times 2^2 = (2 \times 2 \times 2) \times (2 \times 2) = 2^5$$

(the index 5 being obtained more simply by: $3+2 = 5$).

When a number in index form is divided by another having the same base, the result is obtained by subtracting the index of the denominator from the index of the numerator. Thus,

$$\frac{2^3}{2^2} = \frac{2 \times 2 \times 2}{2 \times 2} = 2^1$$

(the index 1 being obtained more simply by: $3-2 = 1$).

It follows from this that any number raised to the power 0 is equal to unity. Thus, 3^0 can be regarded as, say, $3^2/3^2$ (since $2-2 = 0$), and such a fraction must always be equal to unity.

When a number in index form is itself raised to a power, it may also be expressed by writing the base raised to the power of the product of the two indices. Thus,

$$(2^3)^2 = 2^3 \times 2^3 = 2^6$$

(the index 6 being obtained more simply by: $3 \times 2 = 6$).

In expressing very large or very small quantities, it is common practice to write them as numbers having one figure before the decimal point, multiplied by ten raised to the appropriate power.

Thus,   4 650 000   is written as 4.65×10^6,

and   0.000 017 3   is written as 1.73×10^{-5}.

Logarithms

It has already been seen how multiplication and division of numbers in index form involves simply addition and subtraction of indices. This is the basis of the conversion of multiplication and division processes into additions and subtractions by means of logarithms. Each number is expressed as 10 raised to the appropriate power, and this power is called the logarithm of the number. It is then only necessary to add and subtract the logarithms in the correct manner to obtain the result, still in the form of a logarithm or power of 10. To convert this back to an ordinary number, the antilogarithm must be taken. A number expressed as 10 to a certain power is called a logarithm to the base 10, and is usually written '$\log_{10}$', or simply 'log'. Most tables provided to aid calculations (such as those at the end of this book) give values of logarithms and antilogarithms to this base. In the derivation of theoretical expressions, however, it is often found that logarithms are employed having as base a number written symbolically as e (actually $e = 2.718$), and multiplication and division can be carried out in the same way using this base. A logarithm to the base e is sometimes called a 'natural' logarithm, and it is indicated by writing 'ln', or sometimes '$\log_e$'.

Thus,   if $y = 10^x$, $x = \log_{10} y$ or $\log y$,
and   if $y = e^x$, $x = \ln y$ or $\log_e y$.

In this book 'log' will be used to imply a logarithm to the base 10, and 'ln' will imply a logarithm to the base e.

The various relationships for logarithms of products and quotients follow directly from a consideration of the logarithm as an index.

Thus, $\log(a \times b) = \log a + \log b$,
          $\log(a/b) \;\; = \log a - \log b$,
          $\log(a^2) \;\;\;\; = \log a + \log a = 2 \log a$,

or, in general,

          $\log(a^n) \;\;\; = n \log a$,
          $\log(a^{-n}) = \log(1/a^n) = \log 1 - \log a^n = -n \log a$
                             (since $\log 1 = 0$),
          $\log(a^{1/n}) = (1/n) \log a$.

(The same relationships hold exactly for ln.)

Because tables of logarithms to the base e are not always available, it is necessary to be able to convert readily from one form of logarithm to the other. This may be done as follows:

If $x = \ln y$, $y = e^x$.

Now, writing e as 10^μ, where $\mu = \log e$,

    y becomes $(10^\mu)^x = 10^{\mu x}$,

$\therefore$ $\log y = \mu x = \mu \ln y = \log e \times \ln y$,

$\therefore$ $\ln y = \log y / \log e$.

Now, $\log e = \log 2.718 = 0.4343$,

$\therefore$ $\ln y = \log y / 0.4343$,

or, $\ln y = 2.303 \log y$.

The tables normally give logarithms only of numbers between 1 and 10. Any other number can be brought within this range by the method, mentioned above, of expressing it in terms of a number with one figure before the decimal point multiplied by ten raised to the appropriate power. This power then appears before the decimal point in the logarithm, and is called the characteristic, the rest of the logarithm being known as the mantissa. Thus,

6 Mathematical Preparation

$\log 247.6 = \log(2.476 \times 10^2)$ and $\log 0.002\,476 = \log(2.476 \times 10^{-3})$

$\qquad = \log 2.476 + \log 10^2 \qquad\qquad = \log 2.476 + \log 10^{-3}$

$\qquad = 0.3988 + 2 \qquad\qquad\qquad\quad = 0.3938 - 3$

$\qquad = 2.3938. \qquad\qquad\qquad\qquad = \bar{3}.3938.$

Note that $\bar{3}$ (bar three) means that the characteristic is minus three, but the mantissa is positive, i.e. $+0.3938$. The characteristic is found by inspection. If it is positive, it is one less than the number of digits before the decimal point; if negative, it is numerically greater by one than the number of zeros after the decimal point.

If it is desired to take, say, the square root of a number, it can be seen from the previous examples that the logarithm must be divided by two. If the original number were less than unity, the logarithm would be negative. To proceed, the negative logarithm may first be written as an entirely negative number, and then reconverted to the 'bar' form after dividing. Thus, to find $\sqrt{0.002\,476}$,

$$\log 0.002\,476 = \bar{3}.3938 = -2.6062$$

$$\tfrac{1}{2} \log 0.002\,476 = -1.3031 = \bar{2}.6969$$

$$\text{Antilog } \bar{2}.6969 = 10^{-2} \times 4.976 = 0.049\,76,$$

which is the square root required.

An alternative, and rather shorter procedure, is to write the logarithm in such a way that the characteristic is divisible by the appropriate number, in this case 2. Thus we may write $\bar{3}.3938$ as $(-4+1.3938)$. Then, dividing by two, the logarithm for the square root becomes $(-2+0.6969)$, which is $\bar{2}.6969$, the logarithm obtained above.

Exponentials

A quantity which is written in the form of a power of a number such as e is called an exponential. For instance, y is an exponential function in the equation $y = e^x$. Such an exponential quantity increases more and more rapidly as the value of x increases, as shown in Fig. 1.2(a). Many relationships which are encountered in physical chemistry are found to be of exponential form, but usually of the type $y = e^{-x}$. This is the reciprocal of the relationship shown in Fig. 1.2(a) and has the form shown in Fig. 1.2(b). As an example may be quoted the relationship found on p. 75 for the number of atoms N remaining out of an original number N_0 of a disintegrating radioactive substance, after time t from the beginning of the disintegration. This expression is $N = N_0 . e^{-\lambda t}$, where λ is the disintegration constant. A graph showing the number of atoms remaining, plotted against time, would have the form of Fig. 1.2(b).

Fig. 1.2 Graphs of the functions (a) $y = e^x$ (b) $y = e^{-x}$

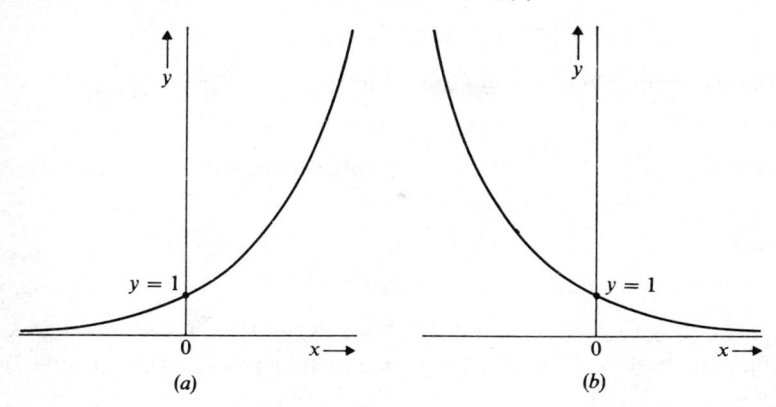

To reduce an exponential expression to one which does not involve indices, it is simply necessary to take the natural logarithm of the expression. Thus, if $y = e^x$, then $\ln y = x$. Similarly, taking logarithms of the expression in the example quoted above, $\ln N = \ln N_0 - \lambda t$.

Growth and Decay

It often happens when discussing biological systems that it is necessary to consider a population that increases or decays by a constant factor for each equal interval in time. For example, if a bacterial culture increases by each cell dividing into two so that on the average the number of cells doubles every thirty minutes the population will increase as follows:

time (min.)	0	30	60	90	120	150	...
population	1000	2000	4000	8000	16 000	32 000	...

In general, if the population initially of size y_0 increases by a factor r in unit time, the growth will be as follows:

time	0	1	2	3	... t ...
population	y_0	$y_0 r$	$(y_0 r)r = y_0 r^2$	$(y_0 r^2)r = y_0 r^3$	... $y_0 r^t$ ...

This form of series in which each term is derived from the previous one by multiplication by a constant factor is called a geometric series or geometrical progression. The growth law can therefore be expressed by the equation

$$y = y_0 r^t.$$

This equation is clearly of exponential form with r as the base and t the exponent so a graph of y against t would resemble the graph of y against x for the function $y = e^x$ in Fig. 1.2(a). In fact the growth law

Mathematical Preparation

is sometimes expressed as an exponential with the base e as follows,

$$(y/y_0) = r^t.$$

Take the natural logarithm of each side of the equation:

$$\ln(y/y_0) = t \ln r.$$

Putting $\ln r = l$ this becomes

$$\ln(y/y_0) = lt$$

hence
$$y/y_0 = e^{lt}$$

or
$$y = y_0 e^{lt}.$$

A more direct route for obtaining the last equation from $y = y_0 r^t$ is simply to replace r by e^l (which is equivalent to putting $\ln r = l$) when

$$y = y_0(e^l)^t = y_0 e^{lt}.$$

In plotting an experimental growth law to find the factor l or r, a straight-line graph is obtained by plotting $\ln y$ against t. This is because

$$\ln(y/y_0) = lt \text{ can be written}$$
$$\ln y - \ln y_0 = lt$$

or
$$\ln y = lt + \ln y_0.$$

This is of the form $y = mx + c$ where $\ln y$ takes the place of y, t takes the place of x, l is the slope m and $\ln y_0$ is the constant c (the intercept at $t = 0$). The constant l which determines the slope of such a graph is known as the growth rate constant.

Returning to the question of bacterial growth: the time taken for the population to double in size is known as the 'generation time'. If this is given the symbol T, then $t = T$ when $y = 2y_0$. The generation time is related to the growth rate constant as follows:

since
$$\ln(y/y_0) = lt$$

putting $y = 2y_0$ and $t = T$
$$\ln(2y_0/y_0) = lT$$
$$\therefore \ln 2 = lT$$

or
$$T = \ln 2 / l.$$

In this way the generation time can be found by measuring the slope of the linear, logarithmic growth plot.

If the factor r by which the population is multiplied after time t is less than unity then the population actually decays. The logarithm of a number between 0 and 1 is a negative quantity so the constant l, which is $\ln r$, is negative. So in the expression $y = y_0 e^{lt}$, l is negative and a graph of this function is similar to that of $y = e^{-x}$ in Fig. 1.2(b). The best known example of spontaneous decay is that of radioactive decay mentioned in the previous section and discussed in more detail on p. 75. Here we have the expression $N = N_0 . e^{-\lambda t}$ or $\ln N = \ln N_0 - \lambda t$. Comparing with the growth law in its logarithmic form $\ln y = \ln y_0 + lt$ we see that it is identical if $l = -\lambda$. We have seen that l is negative for decay and the $-\lambda$ emphasises this with λ as a positive constant known as the radioactive disintegration constant. The time required for the number of atoms remaining to drop to half the initial number is called the 'half life period'. It is the counterpart of the generation time in growth situations and is usually given the symbol $t_{\frac{1}{2}}$. It is shown on p. 75 that $t_{\frac{1}{2}} = \ln 2/\lambda$.

Rates of Change, Calculus Notation

Suppose a graph of a function of y in terms of x takes the form of a curve as shown in Fig. 1.3. If we wish to find the average rate of change of y with respect to x over a certain range of values of x, we find the increase in the value of y over this range and divide it by the increase in

Fig. 1.3 The rate of change of y with respect to x

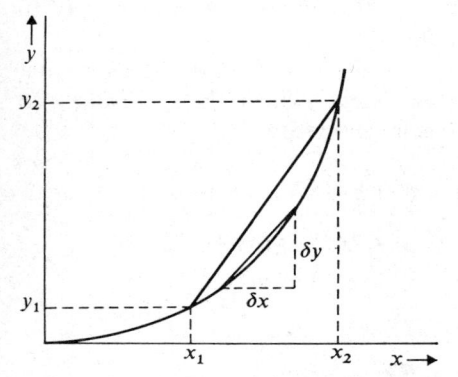

the value of x. As an example, suppose y represents the number of cm travelled by a falling particle after a time x seconds since its release. The average speed of the particle over a few seconds of its motion at any point (the rate of change of distance with time) is obtained by finding the increase in distance $(y_2 - y_1)$ and dividing it by the increase in time $(x_2 - x_1)$. The value of $(y_2 - y_1)/(x_2 - x_1)$ gives the average gradient of the graph over this range. The speed of the particle does not remain constant over this range at the calculated average value, but it varies continually. By choosing smaller and smaller intervals of time, however, over which the average is taken, a value is obtained which approximates more and more closely to the actual speed at one point. Suppose δy represents a small increase in y which takes place while x is increasing by a small amount δx. The average speed over this small range, or the average rate of change of y with respect to x, is given by δy/δx, and this is the gradient of the graph over this small range. Now, we may take the small increases δy and δx to be as small as we choose, and in the limit, as δx tends towards zero, the line showing the gradient will become a tangent to the curve at a point. This gradient will represent the actual speed or rate of change of y with respect to x at that point. This limiting value of δy/δx is called dy/dx, the differential coefficient of y with respect to x. Expressing the process mathematically, we have:

$$\underset{\delta x \to 0}{\text{Limit}} \left(\frac{\delta y}{\delta x} \right) = \frac{dy}{dx}.$$

This method of representation of rates of change, by means of the differential coefficient, is the calculus notation. It must be emphasised that dy is a single symbol representing the differential of y, just as δy is a single symbol representing a small increase in y.

The process of finding the limit dy/dx is called differentiation. It may be carried out by calculation from the equation connecting x and y, as shown in the example below, when it is known as differentiation from first principles. However, this process is used merely to derive general rules for differentiation.

Suppose the expression $y = x^2$ is to be differentiated from first principles. Let y change from y to $y+\delta y$, while x changes from x to $x+\delta x$. Then, when the change has taken place:

$$(y+\delta y) = (x+\delta x)^2$$
$$= x^2 + 2x.\delta x + (\delta x)^2.$$

But before the change:

$$y = x^2.$$

∴ Subtracting, $\delta y = 2x.\delta x + (\delta x)^2,$

∴ $\dfrac{\delta y}{\delta x} = 2x + \delta x.$

∴ $\underset{\delta x \to 0}{\text{Limit}} \left(\dfrac{\delta y}{\delta x} \right) = 2x,$

∴ $\dfrac{dy}{dx} = 2x.$

Hence, the rate of change of y with respect to x at any given value of x is equal to $2x$.

Some Standard Differentials

By methods such as in the example quoted above we may prove that, in general, if $y = x^n$, then $\dfrac{dy}{dx} = nx^{n-1}$, and if $y = ax^n$ (where a is a constant), $\dfrac{dy}{dx} = anx^{n-1}$. The same rule applies in such expressions as $y = \dfrac{1}{x^n}$. If it is written as $y = x^{-n}$, then from the rule:

$$\dfrac{dy}{dx} = -nx^{-n-1} = -nx^{-(n+1)} = \dfrac{-n}{x^{n+1}}.$$

When y is expressed as the sum of a number of functions of x, then $\dfrac{dy}{dx}$ is the sum of the differentials of each of the functions with respect to x. This is shown as follows:

Let $y = u+v+w$, where u, v and w are each functions of x. If y increases by an amount δy as u increases by δu and v increases by δv and w increases by δw, then

$$y + \delta y = (u + \delta u) + (v + \delta v) + (w + \delta w).$$

Subtracting the equation for y

$$\delta y = \delta u + \delta v + \delta w.$$

Dividing through by δx

$$\dfrac{\delta y}{\delta x} = \dfrac{\delta u}{\delta x} + \dfrac{\delta v}{\delta x} + \dfrac{\delta w}{\delta x}.$$

In the limit as all the increments approach zero

$$\dfrac{dy}{dx} = \dfrac{du}{dx} + \dfrac{dv}{dx} + \dfrac{dw}{dx}.$$

If $y = e^x$, then $\dfrac{dy}{dx} = e^x$. This follows because e^x can be expressed as an infinite series:

$$e^x = 1 + \dfrac{x}{1} + \dfrac{x^2}{2 \times 1} + \dfrac{x^3}{3 \times 2 \times 1} + \dfrac{x^4}{4 \times 3 \times 2 \times 1} + \cdots$$

Each term of this series can be differentiated by the general rule given above as follows

$$\dfrac{d(e^x)}{dx} = 0 + \dfrac{1}{1} + \dfrac{2x}{2 \times 1} + \dfrac{3x^2}{3 \times 2 \times 1} + \dfrac{4x^3}{4 \times 3 \times 2 \times 1} + \cdots$$

$$= 1 + \dfrac{x}{1} + \dfrac{x^2}{2 \times 1} + \dfrac{x^3}{3 \times 2 \times 1} + \cdots$$

$$= e^x.$$

When $y = \ln x$, $\dfrac{dy}{dx} = \dfrac{1}{x}$. This is proved as follows:

$$x = e^y$$

$$\therefore \dfrac{dx}{dy} = e^y = x$$

$$\therefore \dfrac{dy}{dx} = \dfrac{1}{x}.$$

The last step is true because for small finite increments in x and y, namely δx and δy, $1 \Big/ \dfrac{\delta x}{\delta y} = \dfrac{\delta y}{\delta x}$. So in the limit as δx and δy tend to zero $1 \Big/ \dfrac{dx}{dy} = \dfrac{dy}{dx}$.

If y is a product of two different functions of x, namely u and v, this is written

$$y = uv.$$

Let y increase by δy as u increases by δu and v increases by δv, then the new values are connected by the equation

$$y + \delta y = (u + \delta u)(v + \delta v)$$
$$= uv + u\,\delta v + v\,\delta u + \delta u\,\delta v.$$

Subtracting the equation for y

$$\delta y = u\,\delta v + v\,\delta u + \delta u\,\delta v.$$

Dividing throughout by δx:

$$\dfrac{\delta y}{\delta x} = u \cdot \dfrac{\delta v}{\delta x} + v \cdot \dfrac{\delta u}{\delta x} + \dfrac{\delta u\,\delta v}{\delta x}.$$

In the limit, as all the increments tend to zero the ratios of each pair of increments remain finite, and are replaced by the corresponding differentials, while the last term can be regarded as a finite ratio of two increments multiplied by a third increment which tends to zero. The whole term therefore tends to zero and the equation becomes:

$$\dfrac{dy}{dx} = u \cdot \dfrac{dv}{dx} + v \cdot \dfrac{du}{dx}.$$

A very useful technique for the differentiation of complicated expressions is the chain rule which shows how to differentiate a function of a function. Thus, if y is a function of u, and u in turn is a function of x then

$$\dfrac{dy}{dx} = \dfrac{dy}{du} \cdot \dfrac{du}{dx}.$$

This follows because it is true by multiplication of fractions for the corresponding finite increments:

$$\frac{\delta y}{\delta x} = \frac{\delta y}{\delta u} \cdot \frac{\delta u}{\delta x}$$

and in the limit as all the increments tend to zero the ratios of the pairs of increments are replaced by the corresponding differentials. A typical situation where this chain rule has to be used is in differentiating an expression such as

$$y = \sqrt{x^2+3}.$$

Here the basic function of x is x^2+3, and y is a function (the square root) of this function. The method of differentiating is to put the basic function of x equal to some other variable u and to differentiate y as a function of u. Multiply this by the differential of u with respect to x and then substitute for u in terms of x. Thus:

$$u = x^2+3$$

$$y = \sqrt{u} = u^{\frac{1}{2}}$$

$$\frac{dy}{du} = \tfrac{1}{2}u^{-\frac{1}{2}}$$

$$\frac{du}{dx} = 2x$$

$$\therefore \frac{dy}{dx} = \frac{dy}{du} \cdot \frac{du}{dx} = \tfrac{1}{2}u^{-\frac{1}{2}} \cdot 2x$$

$$= \frac{x}{u^{\frac{1}{2}}} = \frac{x}{\sqrt{x^2+3}}$$

The method can be extended to any number of variables, e.g.:

$$\frac{dy}{dx} = \frac{dy}{du} \cdot \frac{du}{dv} \cdot \frac{dv}{dx}.$$

Successive Differentiation

If an expression for y in terms of x has been differentiated with respect to x and then the result is differentiated with respect to x again, this is called double differentiation and the symbol for the double differential is written $\dfrac{d^2y}{dx^2}$; that is $\dfrac{d}{dx}\left(\dfrac{dy}{dx}\right) = \dfrac{d^2y}{dx^2}$

As an example, suppose $y = 3x^3 + 2x^2 + x$

then
$$\frac{dy}{dx} = 9x^2 + 4x + 1$$

$$\frac{d^2y}{dx^2} = 18x + 4.$$

Higher derivatives can be found by successively differentiating the result of the previous differentiation. For example, the third derivative is $\frac{d}{dx}\left(\frac{d^2y}{dx^2}\right)$ and is written $\frac{d^3y}{dx^3}$. The numbers written as indices are not powers but merely indicate the number of times the differentiation process has been carried out. Higher derivatives than the second are rarely found in scientific expressions.

Partial and Complete Differentials

If a variable is expressed as a function of two independent variables its rate of change must be expressed in a more complex way. The simplest rate of change to calculate is that with respect to just one of the variables while the other remains constant. This is known as a partial differential with respect to this variable. As an example consider the equation

$$T = \frac{1}{R} \cdot pV$$

which relates the pressure p and volume V of 1 mole of an ideal gas to its temperature T and the gas constant R. This equation expresses one variable T as a function of the two variables p and V. The rate of change of temperature with pressure at constant volume is the partial differential of T with respect to p, written with the symbol $\frac{\partial T}{\partial p}$. If it is necessary to emphasise that the volume is constant, V is added as a subscript thus $\left(\frac{\partial T}{\partial p}\right)_V$. It is found by the ordinary process of differentiation with R and V as constants, i.e.

$$\frac{\partial T}{\partial p} \quad \text{or} \quad \left(\frac{\partial T}{\partial p}\right)_V = \frac{1}{R} \cdot V.$$

Similarly the rate of change of temperature with volume at constant pressure is the partial differential of T with respect to V. Thus

$$\frac{\partial T}{\partial V} \quad \text{or} \quad \left(\frac{\partial T}{\partial V}\right)_p = \frac{1}{R} \cdot p.$$

The same principles apply to more complicated functions and where there are more variables, e.g.

if
$$y = u^3 + 2u^2v + uv^2w + w^3$$
$$\frac{\partial y}{\partial u} = 3u^2 + 4uv + v^2w$$
$$\frac{\partial y}{\partial v} = 2u^2 + 2uvw$$
$$\frac{\partial y}{\partial w} = uv^2 + 3w^2.$$

The complete differential of a quantity which is a function of two or more variables is the sum of the increments in this quantity due to increments in all the variables. Now the partial differential with respect to any one of these variables is the rate of change with respect to this variable or the slope of the graph of the quantity plotted against this variable, with all the other variables kept constant. Thus in the above expressions $\frac{\partial y}{\partial u}$ is the slope of the graph of y against u when v and w are both constant. The increment in y due to an increment du in u is therefore the slope multiplied by this increment in u, or $\left(\frac{\partial y}{\partial u}\right)$ du. The total increment in y due to increments du in u, dv in v and dw in w is therefore given by:

$$dy = \left(\frac{\partial y}{\partial u}\right) du + \left(\frac{\partial y}{\partial v}\right) dv + \left(\frac{\partial y}{\partial w}\right) dw.$$

This is an expression for the complete differential of y. If u, v and w are themselves each functions of x then the complete differential of y with respect to x is given by:

$$\frac{dy}{dx} = \left(\frac{\partial y}{\partial u}\right)\frac{du}{dx} + \left(\frac{\partial y}{\partial v}\right)\frac{dv}{dx} + \left(\frac{\partial y}{\partial w}\right)\frac{dw}{dx}.$$

Notice that $\frac{du}{dx}$, etc., are ordinary differential coefficients and not partial differentials because u is a function of only the one variable x, and similarly for v and w.

Integration

The process of integration is the reverse of differentiation. It is the process of finding y in terms of x if the expression for dy/dx is known. Suppose dy/dx is some function of x, written $f(x)$, then y is the integral of $f(x)$ with respect to x. Written symbolically, this statement becomes:

$$y = \int f(x)\,dx,$$

where the sign $\int$ is the integral sign.

The rules for integration are derived most readily from the corresponding rules for differentiation. However, an arbitrary constant C must be added after every process of integration. This is because, on differentiating back to prove the rule for integration, the differential coefficient of the constant is zero since a constant has no rate of change with respect to any variable. C is called the integration constant.

Hence, it follows from the rules for differentiation that, if $\dfrac{dy}{dx} = x^n$ (where n may have any value except -1),

then
$$y = \int x^n\,dx$$
$$= \frac{x^{n+1}}{n+1} + C.$$

This may be proved by differentiating again, i.e.:

$$\frac{dy}{dx} = (n+1)\cdot\frac{x^{(n+1)-1}}{(n+1)} + 0$$
$$= x^n.$$

Similarly, if $\dfrac{dy}{dx} = \dfrac{1}{x^n}$,

then
$$y = \int \frac{1}{x^n}\,dx$$
$$= \int x^{-n}\,dx$$
$$= \frac{x^{-n+1}}{-n+1} + C.$$

But if $\dfrac{dy}{dx} = \dfrac{1}{x} = x^{-1}$ (the particular case for which the rule above does not hold),

then,
$$y = \int \frac{1}{x}\,dx$$
$$= \ln x + C.$$

A constant factor in the expression for dy/dx may always be written before the integral sign.

Thus, if
$$\frac{dy}{dx} = a \cdot x^n$$

then
$$y = \int a \cdot x^n \, dx$$
$$= a \int x^n \, dx$$
$$= \frac{a}{n+1} \cdot x^{n+1} + C.$$

In most cases, the constant C represents the value of y when $x = 0$, and in practice this can often be evaluated.

Integration between Limits

Instead of finding a general expression for an integrated function such as y in terms of a variable x and an integration constant C it is sometimes necessary to know the change in y as x changes from a lower limit a to an upper limit b. In symbols we may say that if $\frac{dy}{dx} = f(x)$ then:

$$y_b - y_a = \int_a^b f(x) \, dx.$$

This is called a definite integral and the process of evaluating it is called definite integration or integration between limits. The integration is carried out in the same way as usual to give the integrated function $F(x)$ and then the value of $F(x)$ is found when $x = b$ and when $x = a$ and the second value is subtracted from the first. Because any integration constant would cancel out in the subtraction process it is not included in $F(x)$. To indicate that the change in the integrated function is required on going from the lower limit to the upper limit it is usual to place the integrated function inside square brackets and to indicate the limits to the right of the closing bracket, thus: $[F(x)]_a^b$.

As an example of the application of this technique we will take the definite integration of the expression on p. 164 for the variation of the logarithm of a rate constant for a reaction k with temperature T:

$$\frac{d \ln k}{dT} = \frac{E}{RT^2}.$$

The energy of activation E and the gas constant R can both be taken as constants. Comparing with the general expression given above, ln k is taking the place of y and T is taking the place of x. Definite integration between a lower limit T_1 and an upper limit T_2 therefore gives an expression for $\ln k_2 - \ln k_1$ where k_1 is the value of k at temperature T_1 and k_2 is its value at T_2. From this expression it is possible to calculate the activation energy E if the rate constant k is measured at two temperatures. The definite integral is written:

$$\ln k_2 - \ln k_1 = \int_{T_1}^{T_2} \frac{E}{RT^2} \, dT.$$

Since E and R are constants they may be placed outside the integration sign. Also the difference between the two logarithms may be written as the logarithm of the fraction.

$$\therefore \ln \frac{k_2}{k_1} = \frac{E}{R} \int_{T_1}^{T_2} \frac{1}{T^2} \, dT.$$

Now $1/T^2 = T^{-2}$ and $\int T^{-2} \, dT = -1 \cdot T^{-1} = -1/T$,

$$\therefore \ln \frac{k_2}{k_1} = \frac{E}{R} \left[-\frac{1}{T} \right]_{T_1}^{T_2}$$

$$= \frac{E}{R} \left[-\frac{1}{T_2} - \left(-\frac{1}{T_1} \right) \right]$$

$$= \frac{E}{R} \left[\frac{1}{T_1} - \frac{1}{T_2} \right].$$

This may also be written:

$$\ln \frac{k_2}{k_1} = \frac{E}{R} \left[\frac{T_2 - T_1}{T_1 T_2} \right].$$

Problems

1 The variation with temperature of the volume of 1 mol of a gas at one atmosphere (101 325 Pa) pressure is given by the equation:

$$V = 0.082t + 22.4$$

where V is the volume in dm³ and t is the temperature in degrees Celsius (centigrade). Plot a graph showing this variation from values of V calculated at $-150, -50, +50, +150$, and $+250$ °C. Extrapolate the graph to find the temperature at which $V = 0$. Also find the value of V when $t = 0$, and measure the gradient of the graph.

2 In a volume containing N molecules of a gas, a number N_1 have energies greater than a fixed quantity E at temperature T, given by the equation:

$$N_1 = N.e^{-E/RT},$$

where R is the gas constant. Rewrite this equation in such a form that, by plotting a function of N_1 versus a simple (non-exponential) function of temperature, a straight-line graph should be obtained.

3 The degree of dissociation (α) of an acetic acid solution is approximately related to the dissociation constant K_c, and to the concentration c, by the equation $K_c = \alpha^2 c$. Given that $K_c = 1.80 \times 10^{-5}$ mol dm^{-3}, calculate α for a solution of concentration 0.865 mol dm^{-3} using tables of logarithms.

4 Using the equation given in problem 2, calculate the value of N_1/N for the following conditions: $E = 80\,000$ J mol^{-1}, $R = 8.314$ J K^{-1} mol^{-1}, and $T = 300$ K.

5 Using the example on p. 7 of the variation with time of the average population of a bacterial colony, plot an appropriate graph to find the growth rate constant. By inspection of the figures for the population growth determine the generation time and use the relationship between generation time and growth rate constant to confirm the value already obtained for the latter.

6 Given that the rate constant k of a reaction varies with temperature T according to the equation:

$$\ln k = \text{constant} - E/RT,$$

deduce an expression for the rate of change of $\ln k$ with respect to temperature, assuming that E and R are constants, and evaluate the expression for the conditions quoted in problem 4.

7 Use the chain rule to find dy/dx for the following expressions which are similar to those relating the amount of reactant decomposed to the duration of chemical reactions (a is a constant):

(a) $y = \ln(a-x)$
(b) $y = 1/(a-x)$.

8 From the definition of enthalpy, H:

$$H = U + pV$$

find expressions for the partial differential of H with respect to (a) the pressure, p, at constant volume, V, and (b) the volume at constant pressure (remembering that the internal energy, U, may also vary with pressure and volume).

9 When colloid particles reach an equilibrium distribution during sedimentation in an ultracentrifuge the variation of concentration,

c, of colloid at a distance x from the axis of rotation is given by the expression:

$$Mx.dx = \text{constant} \cdot \frac{1}{c} \cdot dc,$$

where M is the molecular weight of the colloid. If the concentration is c_1 at distance x_1 from the axis and c_2 at distance x_2, integrate each side of this expression between the appropriate limits to obtain an expression for M in terms of the constant, c_1, c_2, x_1 and x_2.

2 The Properties and Constitution of Matter

Gases, the Gas Law

In the gaseous state, molecules of a substance are free to move about independently in space. Three variables affect the conditions of a gas – pressure, volume and temperature. Boyle in 1662 found that at constant temperature the volume of a given mass of gas was inversely proportional to the pressure exerted on it, i.e.

$$V \propto 1/p \quad \text{or} \quad pV = \text{constant.}$$

Charles in 1787 and Gay-Lussac in 1802 found that at constant pressure the volume of a given mass of gas increases by the same fraction for every degree rise in temperature. This fraction was established later as approximately 1/273 of the volume at 0° C. Thus, if V_0 is the volume at 0° C and V is the volume at $t°$ C,

$$V = V_0(1 + t/273).$$

From the form of this relationship it follows that if $t = -273°$ C

$$V = V_0(1 - 273/273) = 0.$$

In this way there arose the concept of an absolute zero of temperature – the temperature at which gases would have zero volume if they could exist at very low temperatures. Temperatures on the absolute scale are obtained approximately by adding 273 to the temperature in degrees Celsius (centigrade) and the unit is the kelvin expressed by the symbol K. Thus $0°$ C ≈ 273 K. (More accurately, following the definition of K on p. 40, $0°$ C $= 273.15$ K). Another way of expressing this is to say that the absolute temperature T is related to the Celsius (centigrade) temperature t by

$$T = T_0 + t \quad \text{where} \quad T_0 \approx 273.$$

If a given mass of gas occupies volume V_1 at temperature t_1 and V_2

at temperature t_2, we may write

$$V_1 = V_0(1+t_1/273)$$
$$V_2 = V_0(1+t_2/273)$$
$$\therefore V_1/V_2 = (273+t_1)/(273+t_2).$$

Putting the temperatures on to the absolute scale we have

$$V_1/V_2 = T_1/T_2,$$

so that $\qquad V_1/T_1 = V_2/T_2.$

Thus, in general, $V/T =$ constant, or volume at constant pressure is proportional to the absolute temperature.

Combining the pressure and temperature relationships we obtain the general gas law or equation of state for gases

$$pV/T = \text{constant}.$$

It is readily seen that at constant temperature this equation reduces to Boyle's law and at constant pressure it reduces to Charles' (or Gay-Lussac's) law.

Now, according to Avogadro's hypothesis, equal volumes of all gases, under the same conditions of temperature and pressure, contain equal numbers of molecules. This leads to the conclusion that the constant in the gas equation should be independent of the nature of the gas, provided that it refers to the same number of molecules in each case. For 1 mole of any gas, we write,

$$\frac{pV}{T} = R,$$

and R is called the gas constant. In practice, this equation represents the ideal behaviour, to which ordinary gases only approximate, so it is called the ideal gas equation.

R may be calculated on the basis of Avogadro's hypothesis because 1 mole of an ideal gas is known to occupy 22.414 dm³ at an atmospheric pressure of 101 325 Pa[†] and a temperature of 273.15 K. The gas constant is therefore given by:

101 325 (kg m s⁻² m⁻²) × 22.414 × 10⁻³ (m³ mol⁻¹)/273.15 (K)

= 8.3143 (kg m² s⁻² K⁻¹ mol⁻¹) = 8.3143 J K⁻¹ mol⁻¹.[‡]

For n moles of an ideal gas, the volume occupied at a given temperature and pressure is n times the volume occupied by one mole. The gas equation may theferore be written more generally (for n moles) as:

$$pV = nRT.$$

[†] A pascal (Pa) is a newton per square metre and a newton (N) is 1 kg m s⁻².
[‡] A joule (J) is a newton-metre, i.e. kg m² s⁻².

Diffusion of Gases

Gases have the property of filling uniformly any space into which they are introduced (a behaviour contrary to that of solids and liquids). They do this by a process of diffusion, moving outwards from the source. The velocities of diffusion of gases were first measured by Graham, who determined the rate of diffusion through a porous material. He enunciated the *law of diffusion*, that the rate of diffusion of a gas is inversely proportional to the square root of its density. Since the density of a gas is proportional to its molecular weight, it follows that the rate of diffusion is also inversely proportional to the square root of its molecular weight; e.g.

$$\frac{\text{rate of diffusion of } O_2}{\text{rate of diffusion of } CO_2} = \frac{\sqrt{\text{density of } CO_2}}{\sqrt{\text{density of } O_2}} = \frac{\sqrt{44}}{\sqrt{32}} = \sqrt{\frac{11}{8}} = 1.17.$$

Thus, oxygen diffuses 1.17 times as quickly as carbon dioxide. Such effects as these have to be taken into consideration in connection with diffusion into the tissue of the lung in a study of respiration. Similarly, in the administration of gaseous anaesthetics, the lower the molecular weight of the vapour, the more quickly will it diffuse into the system from the lung.

Kinetic Theory of Gases

Phenomena such as the diffusion of gases suggest that matter in the gaseous state consists of molecules in rapid motion. This view is supported by the fact that all the properties of gases can be explained both qualitatively and quantitatively by its aid. Thus, the pressure of a gas is due to the bombardment of the walls of the vessel by the molecules, and, if the volume of the vessel is increased, the number of bombardments per second on a given area of wall will decrease correspondingly. So the pressure decreases in agreement with the gas law. The increase of pressure at constant volume caused by a rise in temperature is explained by saying that the higher the temperature of the gas, the more rapid is the motion of the molecules, so that the number of bombardments of the wall per second is increased. Thus, a rise in temperature merely involves an increase in the kinetic energy (energy of motion) of the molecules. This view of the gaseous state is called the kinetic theory of gases.

In the course of their motion the molecules collide, not only with the walls of the vessel but also with each other. These collisions must be considered to be perfectly elastic, so that they involve no loss of momentum. If this were not so, the rate of bombardment, and hence the pressure, would continually decrease, and this is contrary to experience. In deriving quantitative relationships on the basis of the kinetic theory of gases we also assume that the molecules are infinitely small compared with the space that they occupy and that there is no force of cohesion between them. The fact that these two assumptions are not strictly correct accounts for the deviations from ideal behaviour found with all real gases. A hypothetical gas for which the assumptions are true would be an ideal gas, and it can be proved that it would obey the ideal gas law.

At any particular temperature, the molecules of a gas do not all move with the same speed. Because of the random collisions with each other some will temporarily gain momentum at the expense of others and at any moment in time there is a distribution of velocities with most molecules having velocities near the average but some having very low velocities and some very high. The theoretical form of the distribution was first worked out by Maxwell and it is illustrated in Fig. 2.1. As the temperature is raised the average velocity increases

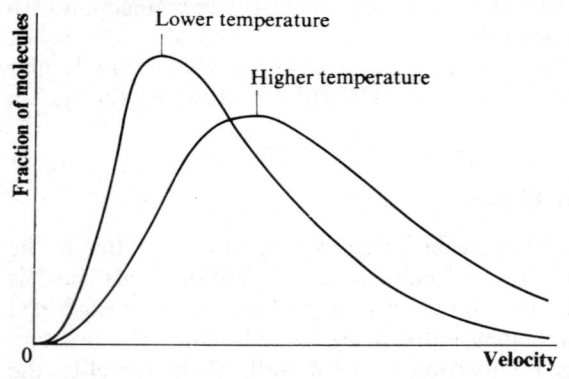

Fig. 2.1 Distribution of molecular velocities

and also the fraction of molecules travelling with higher velocities is increased, i.e. the velocities become more spread out in the direction of higher velocities. Since the energy of a gas molecule of mass m and velocity u is $\frac{1}{2}mu^2$ the distribution of molecular velocities is closely related to the distribution of molecular energies derived by Boltzmann. In particular it can be deduced that the fraction of molecules having a relative velocity along a line joining them such that their combined kinetic energy in this direction is greater than or equal to E is given by the exponential term $e^{-E/RT}$. This Boltzmann factor is used in the collision theory of reaction kinetics (see p. 166).

Partial Pressures

Where there is a mixture of gases occupying the same volume, each gas can be regarded as exerting a partial pressure which is independent of the presence of the other gases. The partial pressure is defined as the pressure which that gas would exert if it alone occupied the volume of the mixture. Dalton's law of partial pressures states that the total pressure of a mixture of gases is equal to the sum of the partial pressures of the constituent gases. This law is clearly in accord with the kinetic theory of gases and, like the other laws of gas behaviour so far discussed, it applies strictly only to ideal gases. It may be proved from the general gas law as follows:

The pressure p_A exerted by n_A moles of a gas A which occupies a volume V at temperature T is given by

$$p_A V = n_A RT.$$

Similarly, the pressure p_B exerted by n_B moles of a gas B occupying an equal volume at the same temperature is given by

$$p_B V = n_B RT.$$

If the two gases are caused to occupy the same volume V at this temperature and they behave ideally then the gas law will apply to the mixture in the following form:

$$p_{total} V = n_{total} RT.$$

Now
$$n_{total} = n_A + n_B$$
$$\therefore\ p_{total} V = n_A RT + n_B RT.$$

Replacing $n_A RT$ by $p_A V$ and $n_B RT$ by $p_B V$ we have

$$p_{total} V = p_A V + p_B V.$$

Dividing through by V

$$p_{total} = p_A + p_B,$$

which is Dalton's law. The proof can clearly be generalised to any number of gases.

If the gas equations for A and B are divided in turn by the gas equation for the total mixture we have

$$\frac{p_A}{p_{total}} = \frac{n_A}{n_{total}} \quad \text{and} \quad \frac{p_B}{p_{total}} = \frac{n_B}{n_{total}}.$$

The expression n_A/n_{total} is called the mole fraction of A and is usually given the symbol x_A, so we may write

$$p_A = \frac{n_A}{n_{total}} \cdot p_{total} = x_A \cdot p_{total}$$

and similarly for p_B. In this way the contribution of each gas to the pressure of a gas mixture may be calculated if the proportional composition is known.

As an example of the application of the concept of partial pressures, it is instructive to calculate the partial pressures of the main constituents of air in the lung during respiration. We may take the composition of alveolar air to be, nitrogen 80 per cent, oxygen 14 per cent, carbon dioxide 6 per cent, by volume. If the total pressure due to these constituents is 96 kPa (the rest of the pressure being due to water vapour), the partial pressures will be:

$$N_2 : \frac{80}{100} \times 96 = 76.8 \text{ kPa}$$

$$O_2 : \frac{14}{100} \times 96 = 13.4 \text{ kPa}$$

$$CO_2 : \frac{6}{100} \times 96 = 5.8 \text{ kPa}.$$

Such information as this is necessary when calculating the amount of each of these constituents dissolved in blood plasma (see p. 212).

Deviations from the Gas Laws

From the gas law $pV = nRT$ it can be seen that, for a fixed number of moles of gas at constant temperature, the product pV should be constant even if the pressure p is varied. But this is an ideal law and real gases only obey it at low pressures when the molecules are too far apart for appreciable molecular interaction. If a graph is plotted of pV versus p, an ideal gas would give a horizontal straight line but real gases give graphs of the type shown in Fig. 2.2. The deviation from the ideal line is greater the nearer a gas is to the conditions which produce liquefaction, as is seen in the case of carbon dioxide.

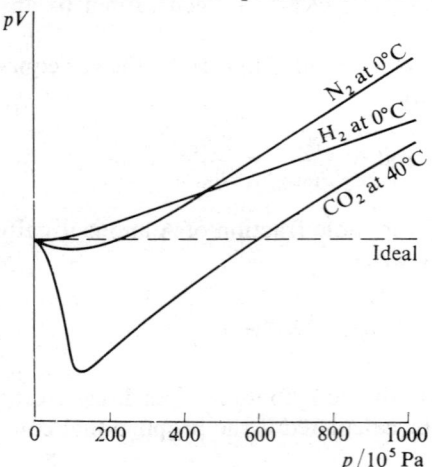

Fig. 2.2 Variation of pressure × volume with pressure

Van der Waals Equation

The deviations from the gas laws are due partly to the existence of cohesive forces between molecules, and partly to the fact that gas molecules occupy a finite volume. To allow for these effects, van der Waals suggested a modified form of the gas equation which should apply to real gases. This equation (for one mole of gas) is:

$$\left(p + \frac{a}{V^2}\right)(V-b) = RT,$$

where a and b are constants for any one gas. The term a/V^2 represents the attractive force tending to pull a molecule which is near the wall of the vessel back into the bulk of the gas. This force will reduce the pressure relative to that which the gas would exert if it were ideal, so that, for a real gas, $(p+a/V^2)$ replaces p for an ideal gas. The finite volume of the molecules decreases the volume of free space available for motion, so the volume term is reduced to $(V-b)$.

Andrews' Isothermals for Carbon Dioxide

Carbon dioxide may be taken as a typical example of a gas which can easily be liquefied. Andrews measured the variation of volume with pressure for a fixed mass of gas at a series of constant temperatures, and expressed his results graphically, in the form of curves shown in Fig. 2.3.

Fig. 2.3 Andrews' isothermals for carbon dioxide

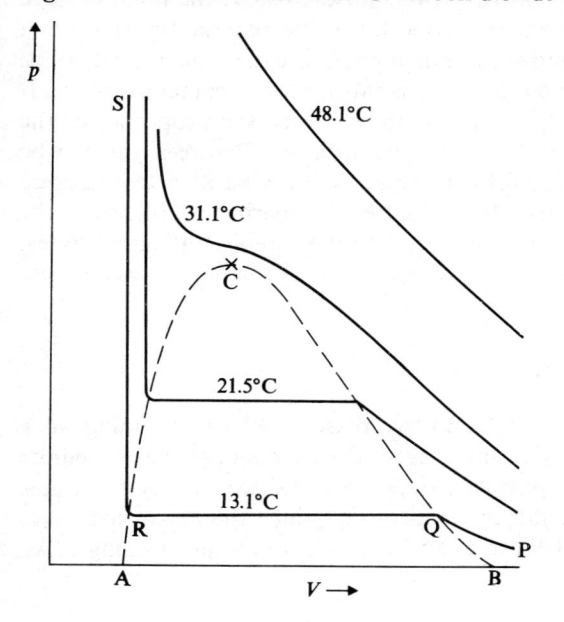

The curve PQRS at 13.1° C may be considered first. Starting with low pressure and high volume at P, the pressure is increased by decreasing the volume. As far as Q, the curve follows approximately the relationship $p \propto 1/V$, which should hold for an ideal gas. After Q, however, it is found that there is no further increase in pressure over quite a considerable range of decrease in volume. The explanation of this is that at Q the gas is beginning to liquefy under the influence of pressure, and that further reduction in volume simply converts more gas into liquid at the same pressure. Over the range, RQ, gas and liquid exist together at the same temperature, and the pressure corresponding to this state of affairs is the saturated vapour pressure of the liquid at that temperature. At R, all the gas has been converted into liquid, and further reduction in volume produces a very rapid increase in pressure along RS, because liquids are not very compressible.

The curve PQRS is one of a family of similar curves which all exhibit horizontal portions within the area marked out by the broken curve ACB. However, within this family of curves, it is found that increasing the temperature shortens the range of the horizontal portion RQ, so that the volume of the material when it is entirely gaseous approaches more closely the volume which it occupies when it is entirely liquid. This must mean that the densities of liquid and gas are becoming more equal at higher temperatures. It is also found that the higher the temperature, the further from the vertical is the portion RS, so that the compressibility of the liquid is increasing. These and other observations show that, on increasing the temperature, the properties of the gaseous and liquid states are becoming more and more alike, and suggest that eventually the two will coincide. Above the point C there is, in fact, no abrupt change in properties as the material transfers from the gaseous to the liquid state. The point C is called the critical point, and the corresponding temperature is called the critical temperature. It is best defined as the temperature above which the properties of the liquid and gaseous states are indistinguishable. Before a gas can be converted into a distinguishable liquid it must be at a temperature below the critical temperature. Andrews' experiments illustrate the principle that there is continuity between the gaseous and liquid states, or, in other words, there is no fundamental distinction between these two states of matter.

Joule-Thomson Effect

When a gas expands, it cools, for two reasons: Firstly, in doing work on its surroundings, by pushing away the existing gases at a definite pressure, there is a compensating decrease in the heat energy of the gas. This must result in a fall in temperature, since the expansion takes place too rapidly for heat to enter the gas from its surroundings. (An expansion in which heat is not allowed to enter or leave the system is called an adiabatic expansion.) This cooling effect would be observed for an ideal gas, since it results merely as a consequence of the gas doing work. The second effect is due to the gas not being ideal, and arises from the energy required to overcome the forces of cohesion between the molecules in moving them further apart. This energy must also come from heat energy of the gas itself, which is therefore cooled. It is this second effect which is known as the Joule–Thomson effect. For some gases at ordinary temperatures, notably hydrogen and helium, the Joule–Thomson effect brings about a rise, and not a fall, in the temperature of the gas. This is related to pV increasing with p for such gases at ordinary temperatures and pressures, as shown in Fig. 2.2. If these gases are first cooled below their 'inversion temperatures', they behave normally in that the Joule–Thomson effect then produces a fall in temperature on expansion.

In handling anaesthetic gases, the possibility of a change in temperature when the gases are released from the cylinders must always be borne in mind.

Liquefaction of Gases

The cooling effects mentioned above, namely, that due to doing work on expansion, and the Joule–Thomson effect, are both employed in the liquefaction of gases. In the method devised by Claude for the liquefaction of oxygen, the gas was first compressed to about 40 atmospheres (atm) pressure, cooled to remove the heat generated on compression, and then allowed to expand to 1 atm pressure in an engine where the gas does work by pushing back a piston. The cold, expanded gas is used to cool the incoming compressed gas. Eventually the incoming gas is cooled sufficiently by the expanded gas for liquefaction to occur.

In applying the Joule–Thomson effect, a gas at about 200 atm pressure is allowed to stream through a valve or throttle, and continues to flow at a considerably lower pressure. Under these conditions, hardly any external work is done by the gas, but, provided the gas is below its inversion temperature, it will cool by doing work against the cohesive forces of its molecules. As before, the expanded gas is used to help to cool the incoming gases, and eventually liquefaction occurs.

Another effect which has been used in the liquefaction of gases is the cooling which occurs when a liquid is caused to boil under reduced pressure. The liquid must obtain its latent heat of evaporation from its surroundings, which therefore become colder. This method may be applied as a cascade process, each boiling liquid being used to liquefy a gas with a somewhat lower boiling-point. However, unlike the expansion methods, this process cannot be used to produce liquid hydrogen, because there is no gas liquefiable by this method which boils at a temperature below the critical temperature for hydrogen.

Intermolecular Forces

The non-ideal behaviour of gases and the possibility of liquefying them shows that there are attractive forces between gas molecules. Such forces exert an even greater effect in the liquid and solid states where the molecules are closer together. They arise chiefly from the polar nature of the molecules. As is described on p. 100, many molecules possess permanent dipole moments due to a separation between their positive and negative electrical centres. Such dipolar molecules attract one another by the electrostatic attraction between the positive end of one molecule and the negative end of an adjacent molecule. A related but simpler situation arises when a dipolar molecule is attracted by an ion. This type of interaction is responsible for the hydration of ions in aqueous solution. The oxygen atom of a water molecule is negative relative to the hydrogen atoms so positive ions in solution attract the oxygen atoms and negative ions attract the hydrogen atoms.

Ions and dipolar molecules also have the ability to induce dipoles in otherwise non-polar molecules. Such induced dipoles are always in such direction as to give attractive interactions with the ions or dipoles causing them. In this way non-polar solvent molecules are attracted to polar solutes.

Even when there are no permanent dipoles in molecules there are still forces of attraction between them. An extreme case is the interaction between inert gas atoms. London explained these forces in terms of correlation between temporary dipoles set up because of the unsymmetrical distribution of electrons at any instant in time. Electrons are in constant motion round atomic nuclei and although their probability distributions averaged over quite short intervals of time are symmetrical about the nucleus, at any instant it is quite possible for there to be more charge on one side of an atom than the other. The resulting instantaneous dipole induces an attractive dipole in a nearby atom and as the instantaneous dipole changes in direction the induced dipole keeps in step with it so that the interaction is always attractive.

The London dispersion forces just described are non directional in character, that is, they are equally strong in all directions. Dipoles, on the other hand, attract most strongly along the line of the dipole. This directional effect is lost in gases and, to a large extent, in liquids also due to the rapid thermal motion of the molecules, though there is still an overall attraction between two rotating dipoles because the inductive effect will tend to correlate their orientations. The combined effect of London dispersion forces, dipole–dipole and dipole-induced dipole forces is spoken of as the van der Waals attraction since it is these that give rise to the a/V^2 term in the van der Waals equation. Besides affecting the properties of gases these attractive forces are also responsible for holding molecules together in the liquid and solid states and for causing the processes of solution and adsorption.

A particularly strong and therefore particularly important type of intermolecular force is the hydrogen bond. This occurs when two highly electronegative atoms such as F, O or N are linked by a hydrogen atom bearing a partial positive charge. It may be represented as

$$\overset{\delta-}{A}\text{—}\overset{\delta+}{H}\ldots\overset{\delta-}{B}$$

though in general the two electronegative atoms A and B are linked by covalent bonds to further atoms which comprise the molecules linked by the hydrogen bond. Except in the very strongest hydrogen bonds, the hydrogen atom is strongly bonded to only one of the two atoms A and B (indicated above by the bond between A and H) and the interaction between H and the other atom B is weaker. To a first approximation, then, the hydrogen bond may be described in terms of the electrostatic attraction between the dipole A—H and the negative charge on B. This resembles the interaction between a dipole and an ion mentioned above, or possibly dipole–dipole interaction because B is almost certainly the negative end of another dipole. The extra strength of a hydrogen bond can be explained in terms of the small size of the hydrogen atom, especially when some of its electron cloud has been transferred to the atom A, thus allowing the dipole A—H to come very close to the atom B. This electrostatic picture has been found to be insufficient to explain all the experimental facts about hydrogen bonds and there is now acceptance of the fact that there is some covalent contribution to the interaction H....B. Hydrogen bonds vary in strength from about 10 to 40 kJ mol^{-1} depending upon the nature and charge of the atoms A and B. As the strength increases it is found that the bond A—H becomes progressively weaker and longer and the interaction H....B becomes increasingly stronger and shorter. It is thought that the electron density between H and B increases so that there is more covalent character in this part of the link. In special cases, such as the [F..H..F]$^-$ ion, the limit is reached in which the hydrogen atom is equally linked to both atoms A and B by about a 'half-bond' on each side.

Hydrogen bonding is the most important cause of association of molecules. This reveals itself in abnormally high apparent values when molecular weights are determined in the vapour state or in solution. It also causes high boiling points (compare H_2O, which is a liquid at room temperature, with H_2S, which is a gas in spite of its higher molecular weight), low vapour pressures and high latent heats of vaporisation (p. 295). Hydrogen bonding between solute and solvent molecules also favours solubility. These effects only occur, of course, where the hydrogen bonding is *between* molecules (intermolecular). It is also possible for hydrogen bonding to occur *within* a molecule (intramolecular) as, for example, in ortho-nitrophenol (I) or salicylaldehyde (II).

Such intramolecular hydrogen bonding cannot occur with the meta and para isomers because the two oxygen atoms are too far apart in each case and they therefore form intermolecular hydrogen bonds. Because of this the ortho isomers are much more volatile than the meta and para isomers. Salicylic acid (III), having an additional OH group, can form both inter- and intramolecular hydrogen bonds. These affect the OH stretching frequencies in the infra-red spectrum (see p. 115).

In the solid state, intermolecular hydrogen bonding always has a profound effect on the crystal structure. The molecules nearly always arrange themselves in such a way that every hydrogen atom that can take part in intermolecular hydrogen bonding does so. Because each hydrogen bond is strongest when the atom B is along the extension of the line A—H, such intermolecular hydrogen bonding has an orientating effect on the molecules that are linked together. This may give rise to abnormally open structures. This explains why ice, which is a fully hydrogen-bonded structure, is less dense than water. Typical hydrogen bond distances (between the atoms A and B) are 0.30 nm for N—H...O interactions and 0.27 nm for O—H...O interactions. Biological materials are all highly hydrogen-bonded, e.g. by N—H...O hydrogen bonds in proteins (see p. 139) and by O—H...O in carbohydrates. The large number of such interactions for each molecule results in a large total intermolecular binding force and such molecules decompose on heating before they separate.

The Liquid State

Although, above the critical temperature, there is no sharp distinction between the gaseous and liquid states, under normal conditions there are obvious distinctions. A liquid, like a gas, takes up the shape of the vessel into which it is put, but it does not fill the whole of the vessel. It is bounded by a definite surface above which the material only exists in the form of vapour. It is more dense than the corresponding vapour, so the molecules must be closer together on the average. The forces of attraction between molecules, which give rise to the a/V^2 term in the van der Waals equation (and which are consequently known as van der Waals forces), increase rapidly as the distance between the molecules decreases. The molecules are therefore held together far more strongly in a liquid than in a gas. These attractive forces are counteracted by the thermal motion of the molecules, so that a liquid expands on heating (unless special effects are operating) until the balance is again restored. The finite size of the molecules is also of greater importance in a liquid, since the molecules are much closer together, and when molecules are in contact, the attractive van der Waals force is balanced by a repulsive force due to compression of the molecules. This balance introduces a certain amount of structure into the arrangement of molecules in a liquid, by the temporary existence of clusters of molecules in contact. In this sense, a liquid is truly intermediate between a solid and a gas, and indeed, from a structural point of view, there is a continuous variation, with no sharp distinction throughout all three states of matter.

Because there is a distinct surface bounding a liquid, molecules in the surface will not be attracted by other molecules equally on all sides, as are molecules in the bulk of the liquid. Thus, a molecule will move about freely in the liquid until it reaches the surface, but it will not escape from the surface unless it arrives there with considerable kinetic energy. The unequal distribution of forces will also confer a special property on the surface called surface tension, which causes the surface to behave as though it were a skin under tension. This property will be considered in greater detail in Chapter 11.

The existence of a saturated vapour pressure of a liquid may be explained conveniently from a kinetic point of view. Since molecules are in motion in both the liquid and the vapour, there will always be a tendency for molecules to escape from the liquid into the vapour, and also to fall back from the vapour into the liquid. The higher the temperature, the more molecules of the liquid have sufficient energy to escape from the surface, in spite of the retarding forces which exist there. However, the more molecules of vapour there are above the liquid, the greater is the number of molecules which, in their random motion, will fall back into the liquid. So, at any one temperature, an equilibrium state will be reached, when the average number of molecules leaving the liquid per second is equal to the average number of molecules re-entering the liquid per second. The pressure in the vapour corresponding to this equilibrium state is the saturated vapour pressure at this temperature. As the temperature increases, we should expect the saturated vapour pressure to increase, and this is in agreement with what is observed (see Fig. 2.5). The energy required to make a molecule leave the liquid and enter the gas is obtained as heat energy from the surroundings. The heat required to convert 1 mole of liquid into vapour at the same temperture is the latent heat of evaporation per mole, or molar enthalpy of vaporisation.

Liquid Crystals

Some substances, when melted, do not form directly a normal liquid but an intermediate state which is usually turbid and which, on raising the temperature further, eventually clarifies at a sharp transition point to give a normal liquid phase. The substance in the intermediate state is known as a liquid crystal because it exhibits some properties of a liquid and some of a crystal. In particular it behaves like a crystal in giving interference colours when placed between a crossed polariser and analyser in a polarising microscope. The nature of polarised light is explained on p. 105 and it is sufficient to mention here that a polariser and an analyser transmit light vibrating in only one plane. When these planes are at right angles, the polariser and analyser are said to be crossed, and if the medium between them is isotropic (i.e. has the same structure in all directions, as in all normal liquids) then no light is transmitted by the combination of the two. Only a material possessing some order in such a way that its structure is anisotropic (i.e. different in different directions) is capable of so affecting the plane polarised light from the polariser that it has a component vibrating in the plane of the analyser which is then transmitted. The fact that light is transmitted when a liquid crystal is in position between the polariser and the analyser is evidence that there is at least partial order in the arrangement of molecules in the liquid crystal phase.

The ordering arises from the very anisotropic nature of the molecules that form liquid crystals. A typical example is p-azoxyanisole:

$$H_3C-O-\langle\bigcirc\rangle-\overset{\overset{O}{\uparrow}}{N}=N-\langle\bigcirc\rangle-O-CH_3$$

and the substances that form liquid crystals all consist of similar long thin molecules. It is clearly difficult for such molecules to undergo rotation about axes perpendicular to the long molecular axis so they tend to line up parallel to each other. With this tendency as a basis, two main types of liquid crystal phase may be distinguished. These are the nematic, or thread-like, and the smectic, or soap-like, phases. The nematic phase arises when rod-shaped molecules are all aligned approximately parallel to each other but otherwise they are not arranged in any sort of order (i.e. their relative positions in the direction of their lengths are random). Its name is derived from the appearance of mobile threads due to lines of optical discontinuity. The smectic phase is formed when the rod-shaped molecules pack together to form layers though there is still no long-range order within each layer. This phase tends to form visible strata which glide over one another as the liquid flows. Both types of phase are generally turbid and this is because the molecules form clusters within each of which the molecules have roughly the same orientation but the orientation varies from one cluster to another. The turbidity is due to scattering of light at the boundaries between the clusters.

Besides showing evidence of order by their appearance in polarised light, liquid crystals of the smectic type are also capable of diffracting X-rays in such a way as to show one-dimensional order. On the other hand liquid crystals flow like normal liquids, showing that the molecules still have considerable freedom of motion. The combination of layer formation and freedom of flow is important in lubrication and it is thought that many lubricants act in the form of liquid crystals. Many naturally occurring molecules, particularly proteins and lipids, can adopt a rod-like shape appropriate for liquid crystal formation so it is probable that they exist as liquid crystals in many biological environments. The combination of structure and fluidity is particularly well suited to biological functions. Liquid crystal systems are important in pharmacy because they occur in semi-solid emulsions, such as pharmaceutical creams.

The Solid State

The characteristic feature of the solid state is structure. There are a few solids which may be truly amorphous, but practically all have their component molecules or atoms arranged according to some regular structure. In the majority of these, the crystalline solids, the regularity is complete in three dimensions. A two-dimensional analogy is found in a wall-paper pattern, which is built up by regular repetition of a small two-dimensional unit of pattern. A crystal may be regarded as being built up by almost indefinite repetition in three dimensions of a single block of material. Such a block is called a unit cell of the structure, and it contains every element of the three-dimensional pattern. The repetition of this cell by adding identical blocks to it on every side will reproduce the whole crystal. In order to specify a structure completely, therefore, it is only necessary to describe the contents of one unit cell. The arrangement of atoms, molecules or ions in a three dimensional pattern can be deduced by the technique of X-ray crystal structure determination described in the next section.

The existence of regular structure in a solid does not mean that thermal motion is entirely lacking or that occasional defects in the pattern do not occur. The thermal motion is mainly restricted, however, to vibration of the atoms about their mean positions in the structure. The defects arise from a balance between the tendency of a group of particles to form an arrangement of lowest potential energy which is the perfectly regular crystal lattice and the tendency towards increasing disorder as the temperature is increased, expressed in the concept of entropy (p. 290). There are point defects in which some lattice sites are vacant and/or in which extra sites (interstitial sites) between the normal lattice sites are occupied. These account for the processes whereby atoms or molecules or ions can diffuse through a solid and reactions involving solids can occur. The diffusion occurs by the relatively easy movement of a particle from a normal lattice site into an adjacent vacant site leaving a vacancy where it was. Sometimes a complete layer of particles can suddenly terminate in the middle of the crystal structure. This is known as an edge dislocation and it accounts for the ease with which metals may be plastically deformed (strained beyond the point where they will return to their original shape when the stress is removed). This is due to the edge joining up with an adjacent layer of particles which is thereby broken to form a new edge in the next layer. A related defect is the screw dislocation in which a layer of particles is not completely flat but is like one turn of a spiral staircase. In this way the layer at any stage of its growth has a free edge and the rapid growth of crystals is accounted for by the ease with which newly arriving particles can add to the free edge.

As the temperature of a crystal is raised the vibrations of the particles become more intense and the number of defects increases until eventually the forces of attraction holding the components of the structure together are overcome, and at this point the solid melts. However, as mentioned previously, the liquid which is formed still contains clusters of molecules which remain stable over short time intervals and which retain the structure of the solid. The heat energy required to convert one mole of solid into a liquid at the same temperature is the latent heat of fusion per mole, or molar enthalpy of fusion.

X-ray Crystal Structure Determination

This technique has analogies with optical microscopy in which beams scattered by the object being viewed are recombined by the objective lens to form an image of the object. In order to see the atomic detail of crystal structures, radiation of wavelength of the same order of size as atomic dimensions (1 ångström unit or 0.1 nm) must be used but X-rays cannot be bent by passing through any substance so they cannot be refocussed optically. However the periodic repetition of the atomic pattern in the crystal structure means that a crystal acts as a three-dimensional diffraction grating for X-rays. Instead of recombining the diffracted beams optically, they are measured and recombined by calculation. The result of this calculation is the electron density as a function of position throughout the unit cell (the unit of the three-dimensional pattern which repeats indefinitely to produce the crystal). The reason why electron density is obtained from these calculations is that the electrons of the atoms scatter the X-rays in the first place. In order to recombine the diffracted beams by calculation it is necessary to know their intensities, their directions and their relative phases. Of these three quantities, the first two are observed in recording the diffraction pattern but there is no way of determining the phases experimentally. This is referred to as the phase problem in X-ray crystallography and it prevents the determination of crystal structures from being an automatic process.

A possible way of overcoming the phase problem is by postulating at least a partial crystal structure and calculating what would be the amplitude and phase of the scattered beam in each direction when X-rays are diffracted by this structure. Each amplitude should compare with the square-root of the intensity of the corresponding diffracted beam and, if the agreement is reasonably good, the postulated structure must be basically correct and the calculated phases must be approximately correct. An electron density distribution can then be calculated using the observed amplitudes but the calculated approximate phases. This is done by a process of Fourier synthesis in which a periodic function (the electron density) is built up by summing sine and cosine terms of amplitudes equal to the amplitudes of the diffracted beams in the correct phase relationship to each other. The highest points in the corresponding electron density function indicate the positions of atoms in the reconstructed image of the crystal structure. These should be more accurate positions than those originally postulated because the observed amplitudes were used with the calculated phases in the calculation of the image. The observed amplitudes correspond to the correct crystal structure even though the phases with which they are combined correspond to the postulated structure. The calculated image of the structure is therefore a compromise between the correct structure which gave rise to the amplitudes of the diffracted beams and the postulated structure used for the derivation of the phases.

The improved atomic positions are then used in the calculation of an improved set of phases and these are again combined with the observed amplitudes of diffraction in performing a second Fourier synthesis which should result in further improved atomic positions. This process of deriving improved phases and using them in a Fourier synthesis to obtain improved atomic positions is a cyclic process of refinement which can continue until there is no improvement from one cycle to the next. In practice an alternative method of refinement is usually employed in the later stages that is more suitable for automatic cycling on a computer. This is a least-squares method which calculates shifts in the atomic positions that will minimise the differences between the observed and calculated amplitudes of the diffracted beams.

If only a partial structure has been proposed in order to obtain the first set of calculated phases the first Fourier synthesis should show at least some of the remaining atoms. These are then added to the atoms used to calculate the phases and they bring about improvements to the phases additional to those brought about by refinement of the positions of the atoms previously included in the calculation. The second Fourier synthesis using these improved phases should then show more of the remaining atoms and more should be revealed at each successive cycle. In this way all the rest of the structure is eventually revealed and the atomic positions are then refined.

It is obvious that the main difficulty in applying this method is in postulating the initial trial structure. If the structures of the molecules are known their arrangement in the unit cell may sometimes be sufficiently restricted by the size, shape and symmetry of the cell to allow the crystal structure to be postulated to a sufficiently close approximation. If the structure of the molecules is unknown further information is required and this can be provided by the intensities of the diffracted X-ray beam. If a modified form of Fourier synthesis is carried out using the squares of the observed amplitudes in place of the amplitudes themselves and omitting the phase terms the resulting function has peaks which reveal interatomic vectors in the crystal structure. That is to say that the distances and directions of peaks from the origin of the cell correspond to distances and directions between pairs of atoms in the structure. Such a synthesis is called a Patterson synthesis and it is one of the commonest ways of starting a crystal structure determination. Unfortunately, if there are N atoms in the unit cell, there are $N(N-1)$ vectors between them and the task of deducing the structure from the vectors may well prove impossible for larger structures. For this reason the 'heavy atom' method is often employed.

In this method it is arranged that one atom in each molecule dominates the X-ray scattering by having a much larger number of electrons (i.e. a much larger atomic number) than the rest. In a Patterson synthesis calculated from the diffraction pattern from such a structure only the vectors between the heavy atoms show up as strong peaks and, since these are relatively few in number, it is not usually too difficult to deduce the positions of the heavy atoms. These are then used to calculate the phases for the first Fourier synthesis. Besides showing the heavy atoms used in its calculation this electron density function usually shows many, if not all, the light atoms in the structure. These are then refined by further phase calculations and Fourier syntheses in the usual way. The heavy atom method is a common way of finding crystallographically an unknown molecular structure and the technique is now being applied for this purpose by many organic chemists. An outstanding example of the application of this technique was in the solution of the crystal structure of vitamin B_{12} (cyanocobalamine).

An entirely different way of solving the phase problem in crystallography has recently gained popularity. This depends upon mathematical relationships between phases of X-ray diffracted beams that have a higher probability of being true the stronger are the beams being related. This method is complementary to the heavy atom method because it works best when all atoms are equal in weight. It is, however, too complex to describe further in this book.

Equilibrium Diagram for a Single Component

Most substances, under appropriate conditions, can exist as a solid, a liquid or a gas. The transformations from one phase to another can be brought about by changes in pressure or temperature or both. The conditions under which the separate phases exist and the conditions under which each phase transforms into another are summarised in an equilibrium diagram or phase diagram. As a typical example, that for water is shown in Fig. 2.4. This shows that ice exists at low temperatures and liquid water at higher temperatures. Water vapour is seen to be capable of existence over a range of temperatures but the pressure must be low enough.

Fig. 2.4 Phase diagram for water

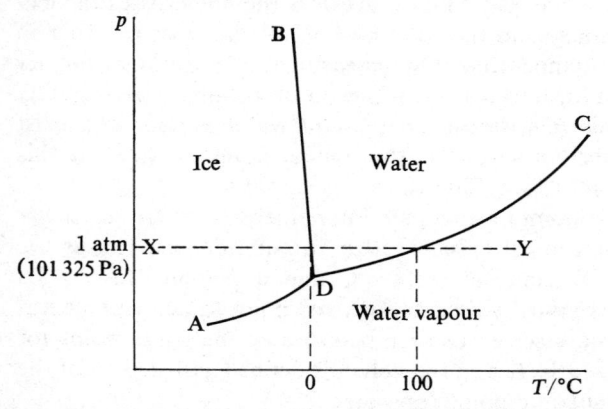

The line BD shows the variation with pressure of the melting point of ice. Since ice contracts slightly on melting an increase in pressure is favourable to melting (an illustration of Le Chatelier's principle – see p. 316), so the line slopes slightly to the left as it rises from D to B. Thus, a vertical line representing an increase in pressure at constant temperature will pass from a region where ice is the stable phase to a region where water is the stable phase. Most solids expand slightly on melting so the melting curve corresponding to BD usually slopes slightly to the right as it extends towards higher pressures.

The line AD shows how the temperature of sublimation of ice

varies with pressure and the line DC illustrates the variation with pressure of the boiling point of water. In both these phase changes there is a large increase in volume on passing into the vapour phase so the transformations are inhibited by an increase in pressure. This causes the phase boundary lines AD and DC to slope upwards to the right and to show a large variation of the temperature of transformation for a small change in pressure. All three lines AD, DB and DC representing phase changes are governed by the Clapeyron–Clausius equation (see p. 303) which shows that the slope dp/dT is proportional to the enthalpy change (latent heat) for the transformation and inversely proportional to the increase in volume. This shows that the steep slope of DB is due to a large enthalpy change on melting and a small volume change and the negative slope arises from the negative volume change (volume of liquid less than the volume of solid). The shallower slopes of AD and DC are due to the larger volume changes but AD is steeper than DC because of the larger enthalpy change for sublimation (the volume changes for sublimation and for boiling being very similar).

A horizontal line in the diagram shows the changes that take place as the temperature is altered at constant pressure. Such a line is XY at a pressure of 1 atmosphere (101 325 Pa). Below 0° C it passes through an area where ice is the stable phase. At 0° C the line crosses the ice/water phase boundary and the solid and liquid can exist together in equilibrium at this temperature. On increasing the temperature further all the ice melts to form water and this is the stable phase until 100° C is reached. At this temperature water and water vapour can exist together in equilibrium so this is the boiling point of water at this pressure. Above 100° C only the vapour is stable.

There is just one point on the diagram, namely D, where ice, water and water vapour can all exist together in equilibrium. This is the triple point and it occurs at 0.0075° C and a pressure of 587 Pa (4.4 mm Hg).[†] It is actually used as the fixed point to define temperatures on the absolute scale. The temperature of the triple point for water is taken as 273.16 K and the kelvin is defined as 1/273.16 of the temperature of the triple point for water.

The line showing how the boiling point of water varies with pressure is DC. Unlike DA and DB, which continue indefinitely (except that there are complications with DB due to the existence of high-pressure forms of ice), DC comes to a definite end at C, the critical point for water. It will be remembered (p. 28) that this is the point above which liquid and vapour become indistinguishable. The liquid can therefore only be obtained as a stable state at temperatures and pressures below C.

---

† 760 mm Hg = 1 atm = 101 325 Pa

Sublimation

If a solid is heated at pressures below that of the triple point (i.e. below the point D in Fig. 2.4) the solid is converted directly to vapour on crossing the phase boundary AD without passing through an intermediate liquid state; that is, it sublimes. For most substances this only occurs at pressures below normal atmospheric pressure. In the case of ice, for example, the pressure must be below 587 Pa. The possibility of removing water from a solid without producing liquid water is exploited in the process of freeze drying which results in finely divided dry solids and avoids decomposition of materials which would be adversely affected by drying at higher temperatures. There are some materials, however, which have triple points above atmospheric pressure. Examples are grey arsenic and violet phosphorus. For such substances, the solid sublimes on heating at atmospheric pressure and melting can only be brought about by heating at high pressures. The temperature of sublimation is the temperature at which the vapour pressure of the solid is equal to the pressure of the surrounding atmosphere. A typical example of this type of behaviour is afforded by carbon dioxide. Solid carbon dioxide is readily available as the refrigerant 'dry ice'. This material sublimes at atmospheric pressure at $-78°$ C. However, at room temperature, if the gas is compressed to 50 atmospheres, it liquefies, and is stored in this form in cylinders. This is only possible because room temperature lies below the critical point for carbon dioxide, as described on p. 28.

Polymorphism

Polymorphism is the existence of at least two different crystal structures of the same chemical substance. Sometimes the phenomenon is known as allotropy but this term is usually restricted to polymorphism among elements. For some well-known substances the polymorphs or allotropes have acquired different names, e.g. diamond and graphite, or descriptions, e.g. grey and white tin, but for most organic substances the polymorphs are known by Greek prefixes, e.g α-, β- and γ-quinol, by capital letters after the name, e.g. chloramphenicol palmitate A, B and C, or by Roman numerals in parentheses following the name, e.g. picric acid (I) and (II).

The importance of polymorphism is that the different forms have different physical properties such as density, refractive index, thermal and electrical conductivity, solubility and sometimes colour and they may also have different chemical stabilities. These differences in properties are highly relevant in pharmacy. For example, when preparing drugs in solid form it is important to ensure that the most appropriate polymorphic form is produced. If transformation from one polymorphic form to another takes place in a pharmaceutical preparation, the new form may have a different solubility affecting the rate of availability of a drug or it may grow as larger crystals during the transformation giving rise to unacceptable particle sizes affecting the rate of solubility, the consistency of creams, etc. An aspect which is important in the preparation of suppositories is that different polymorphic forms have different melting points so it is important that a polymorphic change to a form that melts above the body temperature should not take place.

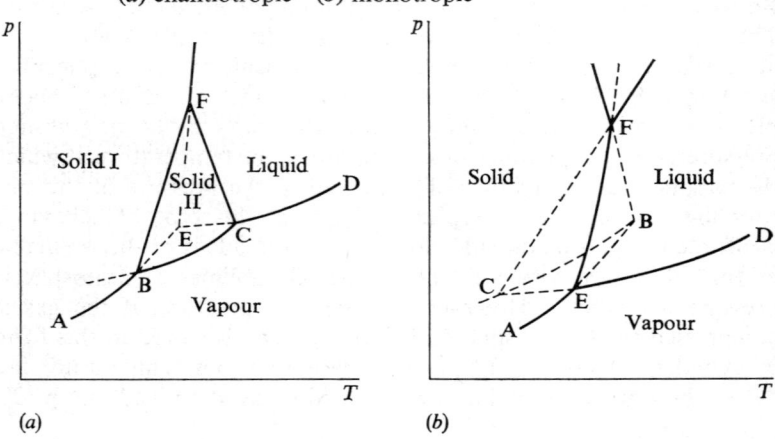

Fig. 2.5 Phase diagram for polymorphic systems
(*a*) enantiotropic (*b*) monotropic

Information about melting points and stabilities of polymorphs may be deduced from a phase diagram. The contrasting behaviours are illustrated in Fig. 2.5. In both parts of the Figure, the curve AB represents the variation of vapour pressure with temperature of polymorphic form I of the solid and the curve BC shows the corresponding variation for polymorphic form II. Where a curve is dotted it represents the behaviour of a metastable species. Thus in Fig. 2.5(*a*), at temperatures higher than that of the point B the vapour pressure curve AB for form I continues as the dotted curve BE because, having a higher vapour pressure than form II (curve BC) form I will have a tendency to transform spontaneously into form II. In Fig. 2.5(*b*) the curve BC is always above the curve AB until the point B is well inside the liquid range so the solid form II is always unstable relative to the solid form I at normal pressures. In both parts of the Figure curve EF represents the variation with pressure of the melting point of the solid form I and CF is the melting point curve for the solid form II. However, in Fig. 2.5(*a*) form II is stable at higher temperatures, so EF represents the melting points of a metastable species whereas, in Fig. 2.5(*b*), form II is the metastable species and EF represents the melting points of a stable species. On both diagrams curve BF shows the variation with pressure of the transition point where form I is converted into form II. For the monotropic system in Fig. 2.5(b) this transition point cannot be attained in practice except at very high pressures (beyond F) because the line BF is above the melting points of both forms.

The difference between the two systems represented by Fig. 2.5(a) and (b) is best appreciated by considering a horizontal line representing an increase in temperature at a constant pressure a little below that of the point D. In Fig. 2.5(a) such a line will pass through a region where solid form I is stable and then through a region where solid form II is stable and finally through the liquid and vapour regions. In Fig. 2.5(b) a similar horizontal line passes through the solid region always where form I is stable and then passes through the liquid and vapour regions. Where a system has two or more stable solid forms it is said to be enantiotropic and where it has only one stable form it is said to be monotropic. It is, however, possible to crystallise the unstable form of a monotropic system from a supercooled liquid or vapour (crystallising at temperatures to the left of curve CF or the continuation of curve BC in Fig. 2.5(b). It is also possible to crystallise the high-temperature form II of an enantiotropic system along the line FC of Fig. 2.5(a) and then cool sufficiently rapidly through the transition temperature BF that the transition to the low-temperature form I does not take place. In either case an unstable form of the solid is obtained and this may transform slowly on keeping to the stable form. As mentioned above, such transformations can be a serious disadvantage if they take place in pharmaceutical preparations. On slowly heating the metastable form of an enantiotropic polymorph it will change at temperatures below those of the line BF to the stable form, then it will transform again at the transition temperature to the high-temperature form and this will eventually melt. In a monotropic system, however, the metastable form first melts as the temperature crosses curve CF and the substance solidifies again and finally the stable form melts at the curve EF. Benzophenone provides an example of this behaviour – the metastable form melts at 26.5° C and then solidifies and the stable form melts at 48.5° C.

Solid and Liquid Phase Diagrams

We have seen how each solid phase of a single substance has its own melting point curve showing the variation of melting point with pressure. At constant (e.g. atmospheric) pressure each pure solid has a fixed melting point. When two solids are mixed and melted together to form a miscible liquid the freezing point of the mixture depends upon the composition. Equilibrium diagrams can be drawn showing the regions of stability of the various phases that are formed and these diagrams vary in type according to whether or not the two solids form a compound and whether or not solid solutions are formed. Where solid solution formation does not occur, solidification of a miscible liquid mixture results in a physical mixture of the two solids. After separation of the excess component of one solid a mixture of constant composition, known as a eutectic mixture, crystallises. This eutectic mixture has the two solids initimately mixed in a finely divided form and it has found application in pharmacy as a method of preparing relatively insoluble drugs in a form which will dissolve more rapidly. A eutectic mixture is made of the drug with an easily soluble inert matrix. The matrix quickly dissolves, leaving the drug in a finely divided state so it is released into solution more rapidly than if small particles are produced by grinding or by other normal methods. An example is the enhancement of the dissolution rate of chloramphenicol by forming a eutectic mixture with urea.

To form a solid solution it must be possible for molecules of a second component to enter the crystal lattice of the first component without causing serious distortions of that lattice. This can happen in two ways: by substitution, in which molecules of component B take the place of molecules of component A, or by formation of an interstitial solid solution, in which small molecules B fit in between much larger molecules A. Substitution is the commonest method for forming solid solutions of organic substances. If the molecules are sufficiently similar it may be possible to have mutual solid solubility over the complete range of composition from pure A to pure B. This is the case with naphthalene and 2-naphthol. More often there is a limited composition range over which foreign molecules can be accommodated in a crystal lattice until the distortions become too great and a new crystal lattice is formed. If models of molecules A and B are superimposed and the overlapping volume is r and the non-overlapping volume is Δ, it is possible to calculate a factor e which is called the degree of molecular isomorphism:

$$e = 1 - \Delta/r.$$

For wide or complete solid solubility to be possible e should be at least 0.9. Interstitial solid solution formation is found mainly in hydrides and nitrides of metals where small atoms of hydrogen or nitrogen can fit into the interstices of the crystal lattice of larger metal atoms without changing the crystal structure. It is found that the solute particles should have a volume less than 20 per cent of that of the solvent particles and even then there is only a limited range of solid solubility. A parallel in the organic field is the incorporation of drug molecules in polyethylene glycol polymers of molecular weight 4000 and 6000. A solid solution is a molecular dispersion of one substance in another and, as such, it should be even better than a eutectic mixture at causing the fast release of a relatively insoluble drug. This has been found to be the case, for example, in solid solutions formed by sulphathiazole in urea.

The freezing point of a liquid is usually lowered by the addition of a second substance that dissolves in the liquid (see page 240). In an equilibrium diagram that shows the solid and liquid phases formed by the components at various temperatures, such as Fig. 2.6, the two

Fig. 2.6 A eutectic system in which no compound formation occurs

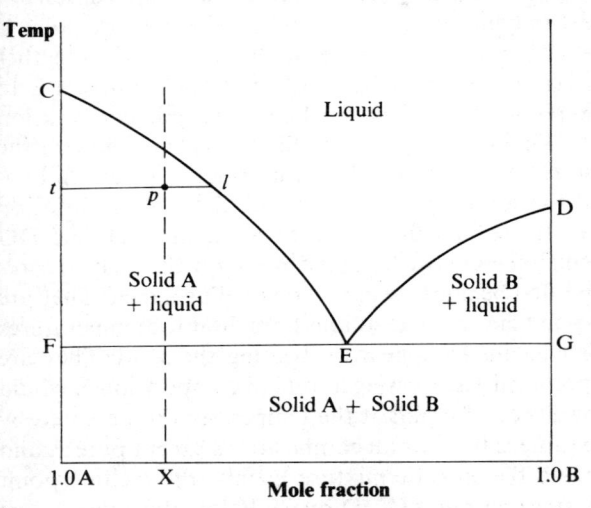

curves representing the freezing points starting at each of the pure components C and D therefore slope downwards towards lower temperature as the composition moves further from either end. For systems where no compound is formed between A and B the two curves meet at E, the eutectic point, which is the lowest freezing point for any mixture of A and B. If a liquid of composition X is suddenly cooled to a temperature t, represented by moving down the vertical line through X to the point p, the system is in a two-phase region where a mixture of solid and liquid can exist in equilibrium with each other. The compositions of the solid and liquid in equilibrium are found by imagining the compositions to move to the left and to the right along the horizontal line (known as a tie line) through p until phase boundaries are reached. To the left the phase boundary is the vertical line representing pure solid A. To the right the phase boundary is the liquid of composition represented by the point l. Cooling to a lower temperature would give a liquid of composition further to the right, i.e. a higher percentage of B. To cause the increase in the proportion of B in the liquid more solid A separates. If a liquid of composition X is slowly cooled, as soon as the vertical composition line crosses the liquid composition curve solid A would begin to separate. As the mixture cooled solid A would keep separating and the liquid composition would follow the curve CE until it reached the eutectic point E. On further cooling the liquid of composition represented by the eutectic point would crystallise as a heterogeneous mixture of A and B in the proportion represented by the composition at E. This is known as the eutectic mixture.

The line CE represents compositions of liquid that can exist in equilibrium with solid A and the line DE represents compositions of liquid that can exist in equilibrium with solid B. The eutectic point E is the only point at which solid A, solid B and liquid can exist together in equilibrium. A phase boundary line FEG is drawn through E to show that no liquid phase exists below this line. The liquid composition lines such as CE and DE are called liquidus lines. The lines showing the composition of the solid phases and their variation, if any, with temperature are called the solidus lines. In this case there is no variation with temperature so the solidus lines are the vertical lines CF and DG with their connecting line FG. The liquidus lines show the temperatures at which solid phase first begins to separate on cooling a melt. They are therefore freezing-point curves. The solidus lines show the temperatures at which liquid first begins to appear on heating the solid. They are therefore melting point curves. Heating a solid of composition X would first show the appearance of liquid at the temperature represented by the line FEG. The same is true for all compositions except pure A and pure B. For these and the eutectic mixture E only, the melting point is the same as the freezing point (C, D or E). Below the eutectic line FEG is a region where solid A and solid B can exist together over the whole composition range without the presence of liquid. Examples of this type of phase diagram are found in the systems $NaCl/H_2O$, naphthalene/p-toluidine and urea/paracetamol.

In systems where compound or complex formation can occur between the two components, the type of phase diagram depends upon whether or not the compound or complex can exist in equilibrium with a liquid of the same composition. If it can, the diagram is of the type shown in Fig. 2.7 and the compound or complex is said to have a congruent melting point. Where the components are molecular substances the 'compounds' formed between them are almost certainly molecular complexes but in this discussion they will be referred to as compounds. The appearance of the diagram is that of two simple eutectic systems combined in which there are three vertical solidus lines representing pure A, compound AB and pure B respectively. Pure A and compound AB form a eutectic mixture M, and pure B and compound AB form a eutectic mixture O. On cooling a liquid with composition within the range QM, pure A would first crystallise and the liquid would follow the composition curve LM until the eutectic

Fig. 2.7 Formation of a compound with a congruent melting point

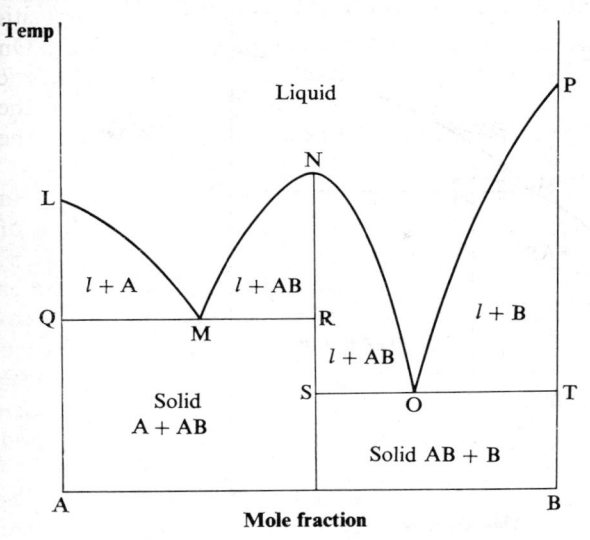

mixture of compound AB and pure A crystallised at M. However, cooling a liquid within the composition range MR would cause compound AB to crystallise and the liquid would follow the composition curve NM to the eutectic point M. Similarly, on the right-hand side of the diagram, whether solid B or solid compound AB first separates on cooling a melt depends on whether the composition of the melt lies between O and T or between S and O. The freezing points follow the liquidus curves LMNOP and the melting points lie on the solidus lines QMR and SOT. The vertical lines LQ, NS and PT complete the solidus. The pure compound AB will melt and freeze at the same temperature represented by the point N. Some compounds have a tendency to dissociate into their components when they are melted. It is found that the stability of the compound is related to the sharpness of the peak at N – the flatter the peak the greater the tendency to dissociate. An example of a system that gives this type of phase diagram is benzophenone/diphenylamine.

If a compound or complex is formed which, when melted, breaks down to form a different solid and a liquid the phase diagram has the appearance of Fig. 2.8. The compound is said to have an incongruent melting point. Thus on heating the compound AB represented by the composition O to temperatures above the line MOQ it decomposes and separates into solid B and a liquid whose composition lies on the curve MN. This decomposition is known as a meritectic reaction and the temperature at which it first occurs on heating, represented by the line MOQ, is called the meritectic point or the incongruent melting point. If this decomposition did not occur the compound would melt

Fig. 2.8 Formation of a compound with an incongruent melting point

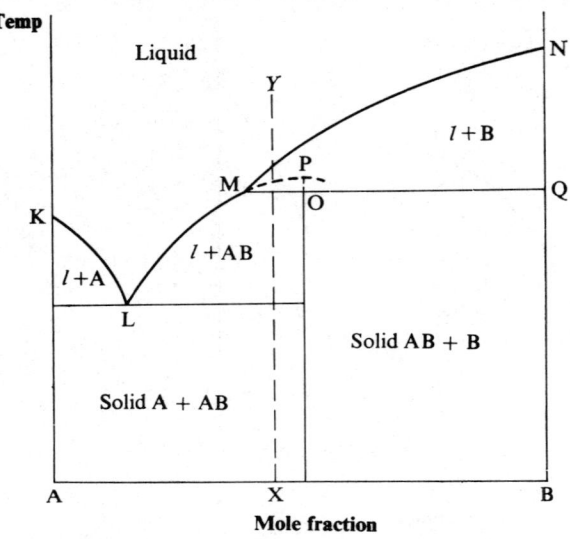

congruently at the point P. This point is not attainable in practice and it should be noticed that the compound cannot exist in equilibrium with a liquid of the same composition. As soon as the compound starts to melt at the point O the liquid that is in equilibrium with it has the composition corresponding to the point M. Cooling any liquid of composition within the range OQ results first in the separation of solid B while the liquid follows the liquidus NM. When the temperature reaches that of the line OQ the meritectic reaction takes place in reverse, some solid B reacting with the liquid M to give the solid compound O. On cooling a liquid of composition within the range MO, e.g. along the line YX, solid B first separates. On reaching the line MO the reverse meritectic reaction takes place, B reacting with liquid M to give solid O. However, this time when the reaction is complete, there is still some liquid of composition M left in equilibrium with the solid O and no B remains. This is seen by the fact that the line YX lies between the liquidus LM and the vertical solidus through O. On cooling further, the solid that separates is the compound AB and the liquid composition follows the line LM. When it reaches L a eutectic mixture of the compound and pure A crystallises. Within the composition range LM the compound AB is the solid which separates on cooling the liquid below the liquidus curve. Again the liquid composition follows the line ML and the eutectic mixture L finally crystallises. A similar behaviour is found when cooling liquid mixtures within the composition range KL except that the solid separating is pure A and the liquid composition follows the line KL to the eutectic L. An example of this type of phase diagram is found in the system benzene/picric acid.

Fig. 2.9 Formation of a continuous series of solid solutions

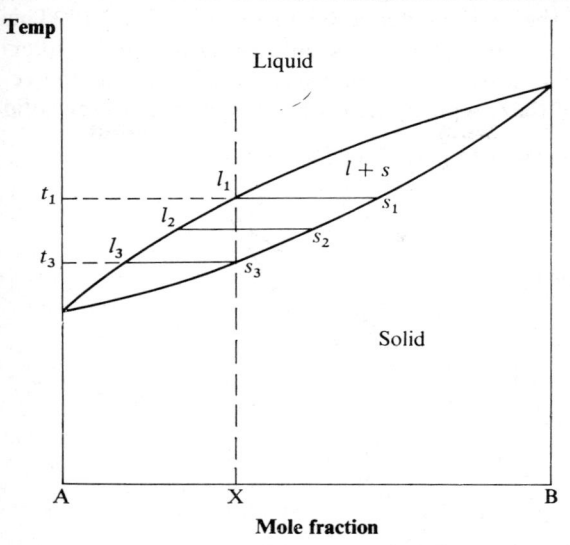

In cases where solid solution is possible over the whole composition range, a simple phase diagram of the type shown in Fig. 2.9 is found. If a liquid mixture of composition X is cooled, as soon as the temperature drops below that of the liquidus curve at this composition the mixture will separate out along a horizontal tie line to give a solid whose composition is represented by the point s_1 and a liquid represented by l_1. At a slightly lower temperature the mixture separates out along another tie line to give a solid s_2 and liquid l_2. As the temperature falls the compositions of the solid and liquid which are in equilibrium at each temperature are given by the points where the horizontal tie line at that temperature cuts the solidus and liquidus lines. The last trace of liquid has the composition given by l_3 and this is in equilibrium with a solid s_3. At lower temperatures than this the whole mass is solid. Conversely, if a solid solution of composition X is heated, the first trace of liquid appears at the temperature t_3 so this is the melting point of the solid. The freezing point is t_1 where the first trace of solid is formed on cooling. Over the whole range t_3 to t_1 there is a mixture of solid and liquid and there is no point during the melting process at which the temperature remains constant. In this respect, systems forming solid solutions differ from those forming eutectic mixtures. On heating the latter the temperature remains constant at the eutectic temperature while the solid is converted into a pure solid component or compound and a liquid of the eutectic composition. Besides the example given above of substances that form a continuous series of solid solutions further examples are p-dichlorobenzene and p-dibromobenzene, and anthracene and phenanthrene.

The commonest type of phase diagram where the solids are partially miscible with each other is the eutectic type shown in Fig. 2.10. This diagram resembles Fig. 2.6 for the simple eutectic system and is interpreted in a very similar way. The difference is the existence of the regions at either end of the composition range corresponding to the solid

Fig. 2.10 Eutectic system with partial solid solubility

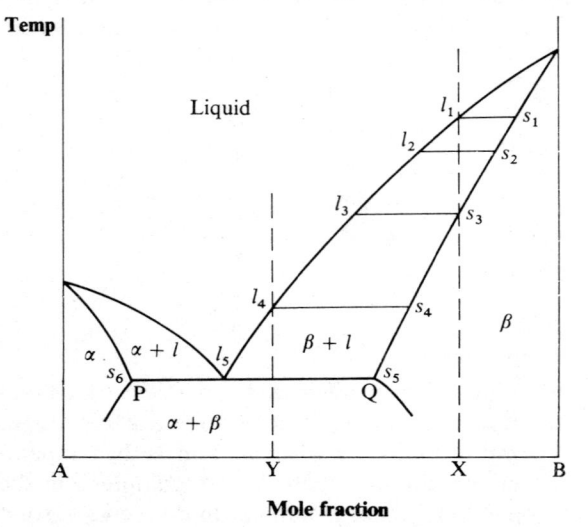

solution α of B in A and the solid solution β of A in B. Thus when a liquid of composition X is cooled below the liquidus line, e.g. to the tie line $l_2 s_2$, the solid s_2 which separates is a solid solution of A in B. In the range of temperatures over which there is a mixture of solid and liquid the solid varies in composition from s_1 to s_3 and the liquid varies in composition from l_1 to l_3. If a liquid of composition Y is cooled the behaviour is at first similar to that for the composition X. The first solid to separate is a solid solution of A in B of composition s_4 and on cooling to the eutectic temperature the solid composition varies from s_4 to s_5 and the liquid composition varies from l_4 to the eutectic composition l_5. Once the eutectic temperature is reached, however, it remains constant while the liquid at the eutectic composition all solidifies as a mixture of solid solutions α and β of compositions s_6 and s_5. They crystallise in the proportion which gives the solid separating the same overall composition as the liquid at the eutectic point l_5. It is usual for solid solubilities to decrease with decreasing temperature and this is shown by the outward slope of the solid phase boundaries between the regions representing the homogeneous solid solutions α and β and the region representing a heterogeneous mixture of the two (labelled $\alpha + \beta$ on the diagram). Transformations within the solid phase are slow at low temperatures so unless the heterogeneous solid is

Solid and Liquid Phase Diagrams 51

cooled very slowly the compositions of the solids separating at the eutectic temperature, namely s_5 and s_6, become frozen in. Examples of systems giving this type of phase diagram are naphthalene/1-naphthol and urea/sulphathiazole.

The positions of the phase boundaries between solid and liquid are most conveniently found by thermal analysis – by following the changes of temperature with time on heating or cooling specimens of known composition. The heating curves are usually followed by measuring with thermocouples the difference in temperature between the specimen and a similar quantity of an inert material both of which are heated together at a constant rate. This technique is known as differential thermal analysis (often abbreviated to DTA). Any process in the specimen such as melting which is accompanied by an absorption of heat causes the temperature of the specimen to be less than the temperature of the inert material. A negative temperature deviation will therefore mark the crossing of any phase boundary at which liquid is first formed and the negative deviation will persist until the next phase boundary is crossed at which the last trace of solid becomes liquid. In this way the temperatures of the solidus and liquidus can be found for specimens covering the whole composition range.

The cooling curve method will be described in more detail because it does not require any special apparatus. Mixtures of the two pure components covering the whole composition range are made up and each mixture is placed in a test tube. A thermometer is placed in the first mixture and the tube is heated until the sample is entirely liquid. The test tube is then supported in a boiling tube so that its rate of cooling is reduced and measurements of temperature are made at regular intervals, usually every minute or half-minute while the mixture is stirred (with the thermometer or with a separate stirrer). The cooling is followed until a few degrees below the point at which the mixture has become entirely solid. The mixture is heated to melting again to allow the withdrawal of the thermometer which is then cleaned before insertion in the next tube of mixture. The procedure is repeated for each mixture and graphs are drawn of temperature versus time.

For a single substance or compound or a eutectic mixture the cooling curve has the form shown in Fig. 2.11(a) with a region of constant temperature while all the liquid becomes solid of the same composition. Where the solid crystallising is a single type of solid solution (such as any composition in a system which forms a continuous series of solid solutions or a composition such as X in Fig. 2.10 for a system which forms a eutectic with solid solutions) a cooling curve of the type shown in Fig. 2.11(b) is found. As solid is formed the enthalpy of fusion is evolved and the rate of cooling is decreased. When all the liquid has solidified no more heat is evolved and the rate of cooling increases again but not to a rate as high as that for the cooling of the liquid. In cases where the liquid reaches the eutectic composition before solidification is complete as in Fig. 2.6 or for compositions lying between P and Q in Fig. 2.10 the cooling curve has the form shown in Fig. 2.11(c). There is a reduction in the rate of cooling as soon as the

Fig. 2.11 Cooling curves
(a) for a single substance or eutectic mixture
(b) for solid solution formation
(c) for eutectic formation

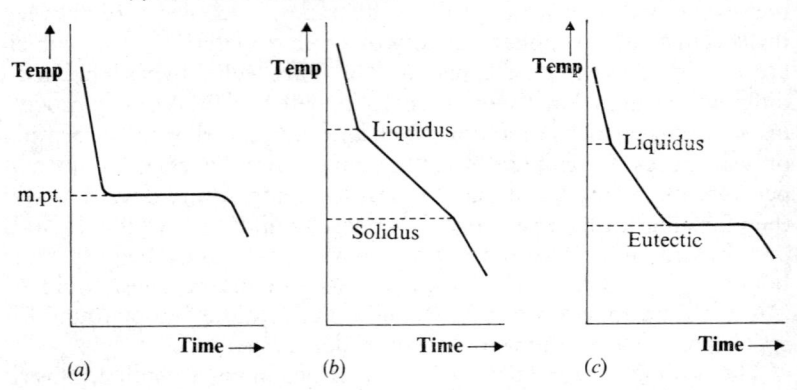

liquidus curve is crossed and solid starts to separate and this continues until the liquid has reached the eutectic composition. Then the temperature remains constant while all the liquid solidifies at the eutectic point. Finally the line falls due to the cooling of the solid.

Experiment 2.1 The construction of the phase diagram for the system naphthalene/1-naphthol by the cooling curve method.

Mixtures of naphthalene and 1-naphthol are made up in test-tubes as follows:

Tube no.	1	2	3	4	5	6	7	8	9	10	11
Naphthalene (g)	–	1	2	3	4	5	6	7	8	9	10
1–Naphthol (g)	10	9	8	7	6	5	4	3	2	1	–

A 110° C thermometer is placed in tube No. 1, which is heated gently in a bunsen flame until the contents have completely melted. The tube is then placed on a cork inside a boiling tube to serve as an air bath and it cools slowly. The contents of the tube are stirred continually and readings of temperature are taken every half-minute. A graph is plotted of temperature against time, preferably while the readings are being taken. The cooling and taking of readings should be continued until about 4° C below the point where the contents of the tube have become completely solid. The procedure is then repeated for each tube. The thermometer is removed from a solidified tube by warming it again gently with a bunsen flame to liquefy the contents. The thermometer may then be wiped fairly clean with a tissue or filter paper before it is inserted into the next tube. When all the cooling curves have been drawn they should be interpreted as in Fig. 2.11 to obtain the liquidus, solidus and eutectic points. Sometimes a dip is found in the curve at the liquidus point due to supercooling and this should be overcome by extrapolating back the fairly linear portion of the curve between the liquidus and solidus points until it meets the curve for the cooling of the liquid at the liquidus point. Also in cooling curves of the type of Fig. 2.11(b) the solidus point may be obscured by curvature. If so, the fairly linear portions on either side of the curve should be extrapolated and the point of intersection of the extrapolated lines taken as the solidus point. The phase diagram should then be drawn and all the areas and lines labelled.

Crystal Structure

The investigation of the structure of crystals by X-rays, described briefly above, has led to an understanding of the fundamental processes at work in the building-up of solids. The simplest structures of all are those of the metals, which depend on the closest packing of the metal atoms so as to fill space in the most economical way. The structures found are most frequently the same as those adopted when a number of spheres of the same size are shaken up together in a box. The structures of simple salts are based on the same principle, except that the spherical units which are packed together are now positive and negative ions of different sizes. These must be packed so that each positive ion is as close as possible to as many negative ions as can be accommodated round it, and vice versa, consistent with preserving the correct overall ratio of numbers of positive and negative ions to maintain electrical neutrality. Thus, in the structure of sodium chloride illustrated in Fig. 2.12(a), every Na^+ ion is surrounded by 6 Cl^- ions, and every Cl^- ion by 6 Na^+ ions, whereas in the fluorite structure (Fig. 2.12(b)) every Ca^{2+} is surrounded by 8 F^- ions, but each F^- ion is surrounded by 4 Ca^{2+} ions to maintain the ratio appropriate for CaF_2.

In the majority of organic compounds, and in some inorganic compounds, the structure is composed of covalently bound molecules held together by van der Waals forces. Again, the principle of closest packing applies, but the irregular shapes of the molecules which have to be packed together result in a very wide variety of structures. The main interest in the detailed determination of such structures is in the shapes and sizes of the molecules themselves. Most of our information about bond distances and bond angles in organic and inorganic compounds has come from such studies as these.

54 The Properties and Constitution of Matter

Fig. 2.12 The crystal structures of (a) NaCl, (b) CaF_2

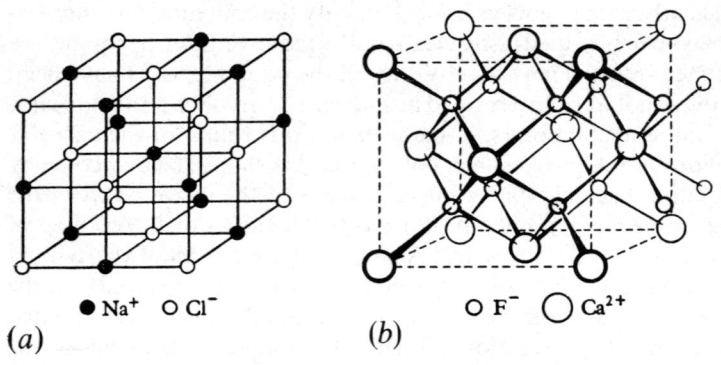

(a) ● Na^+ ○ Cl^-

(b) ○ F^- ◯ Ca^{2+}

When organic molecules are held together solely by van der Waals forces the carbon, nitrogen and oxygen atoms of different molecules are usually separated by about 0.35 nm. Separations appreciably less than this are usually evidence of some specific interaction between the molecules. The commonest example of this is hydrogen bonding, the effects of which on the structures of crystals have already been mentioned (p. 32). A further example is charge-transfer interaction which is found in complexes between electron donor and electron acceptor molecules. When this effect occurs between planar molecules the crystal structure consists of stacks of alternate donor and acceptor molecules with an interplanar separation of about 0.32 nm. By such disturbances of the normal packing arrangements and by reduction in the normal molecular separations molecular interactions may be detected and studied.

There are very few substances which are truly amorphous in the sense that they show no regularity in their structures. One example is provided by glass, which consists mostly of a silicon–oxygen framework. The grouping of oxygen atoms immediately round each silicon atom is much the same as in crystalline silicates, but the groups of atoms so formed are connected together in a random way in glass and in a regular way in the crystalline silicates. When a finely divided material is spoken of as an amorphous powder, the term usually implies simply that the substance does not look crystalline to the naked eye. Examination under a high-power microscope, or by X-rays, will usually establish the fact that the so-called amorphous powder is really micro-crystalline.

Crystallisation

When a substance can be prepared in a crystalline form it is usually in a reasonably pure state. Unless the substance forms mixed crystals or a molecular compound on crystallisation, the impurities will normally be left in the solution. The majority of solid materials are therefore purified by recrystallisation. A hot saturated solution is made of the crude material in a suitable solvent and, on cooling the solution, crystals are deposited. These are filtered off, washed with small quantities of the cold solvent to remove adsorbed mother liquor, and allowed to dry. The more slowly the solution cools the slower will be the crystallisation, and the crystals will be correspondingly larger and more perfectly formed. If very small crystals are required, as is sometimes the case for pharmaceutical purposes, the solution may be chilled suddenly, or rapid crystallisation may be assisted by subjecting the solution to supersonic vibration. The latter method is employed in the preparation of microcrystalline sulphathiazole.

It is a common experience that, when crystals of a wide variety of sizes are kept together for some time in contact with their mother liquor, the large crystals grow at the expense of the small ones, especially if the mixture is subjected to fluctuations in temperature. This effect is sometimes found in pharmaceutical creams. On keeping for some time they may become more granular due to disappearance of the smaller crystals and the growth of the larger ones.

Fractional crystallisation is the process of separating two substances by virtue of their different solubilities. If a hot saturated solution of the mixture is cooled, the less soluble component begins to crystallise first. Hence, when the solution has cooled completely, the solid deposited will be richer in the less soluble component, and the mother liquor will be richer in the more soluble component. This process is repeated, and, at each stage, the mother liquors from the fractions richer in the less soluble component are combined with crystals from the poorer fractions. This method is the most economical in the amount of material used. The crystalline fractions are gradually enriched, until pure crystals of the less soluble component are eventually obtained. The more soluble component is also obtained in the pure state by evaporation of the final mother liquor. The number of steps required to produce the pure materials depends on the closeness of the solubilites of the two components.

Problems

1 The pressure in a 3.4 m^3 cylinder of oxygen is 1.27×10^7 Pa at 293 K. What would be the pressure at 273 K? What volume would the gas occupy at atmospheric pressure (1.013×10^5 Pa) and 293 K? (Assume that oxygen behaves as an ideal gas.)

2 Calculate the relative rates of diffusion of oxygen and diethyl ether vapour into the tissue of the lung.

3 A sample of air at atmospheric pressure (1.013×10^5 Pa) and 273 K contains the following concentrations of gases: N_2, 0.0358 mol

dm^{-3}; O_2, 0.0063 mol dm^{-3}; CO_2, 0.0027 mol dm^{-3}. Calculate the volume percentage composition of the air and the partial pressure of each gas (in Pa).

4 Figure 2.13 is a phase diagram showing the melting and freezing points for the system acetylsalicylic acid/salicylic acid. Label each area to show the types of phase that can exist together within it, each liquidus and solidus line, and the eutectic point. Predict the type of cooling curve that would be obtained by cooling mixtures represented by the broken lines 1, 2 and 3 to beyond the point of complete solidification.

Fig. 2.13.

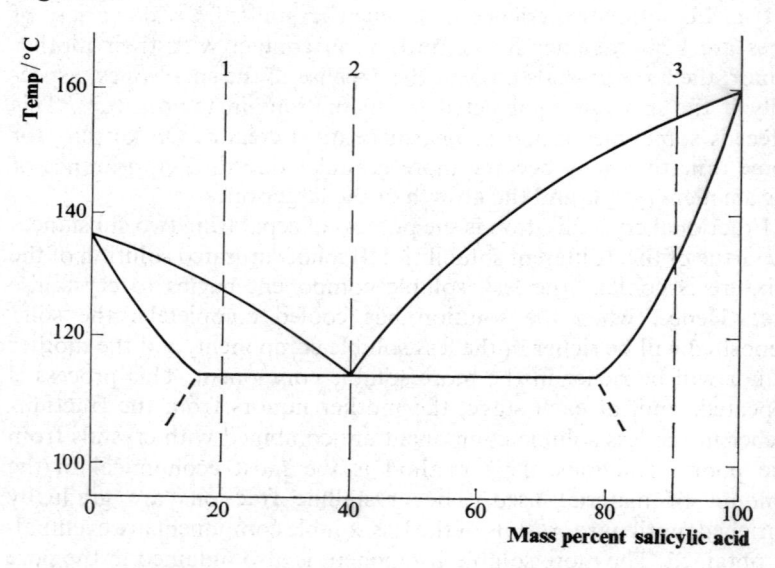

3 The Structure of Atoms

Atomic Theory

The view that matter is not homogeneous but must consist of ultimate particles, or atoms, dates from the time of the Greek philosophers Leucippus (*c.* 500 B.C.), Democritus (*c.* 450 B.C.) and Lucretius (57 B.C.). It was not until the laws of chemical combination had been established, however, that any quantitatively useful atomic theory could be put forward. These laws (the law of conservation of mass, the law of constant proportions, the law of multiple proportions and the law of reciprocal proportions) demonstrated that elements always combine in simple multiples of certain proportions by weight. Dalton's atomic theory showed how this fact could be interpreted in terms of each element consisting of atoms of the same weight and properties, with different elements having atoms of different weights and properties. An atom was therefore defined as the smallest particle of an element which can take part in a chemical change. A compound consists of elements combined in a fixed proportion by weight and this is interpreted in terms of molecules which consist of fixed numbers of atoms. A molecule is therefore the smallest particle of a compound that can exist in the free state.

Atomic Weights

The atomic weight of an atom was first defined as the number of times it is heavier than an atom of hydrogen. It was calculated from the weight of the element combining with or displacing 1 g of hydrogen. However, since many elements combine directly with oxygen and not with hydrogen the definition was later changed to be the number of times heavier the atom is than one-sixteenth of the weight of an atom of oxygen. In other words oxygen was defined as having an atomic weight of 16 exactly. This definition was eventually considered to be unsatisfactory when the full implications of the existence of isotopes was realised. Isotopes are discussed in more detail later (p. 65) but for present purposes it is sufficient to know that they are different forms of the same element (each called a nuclide) with slightly different weight but the same chemical properties. In view of this identity of chemical properties, the different chemical processes that an element can undergo do not change the relative abundance of the isotopes of the element. Over the enormous periods of geological time therefore the abundances for nearly all elements have reached a constant distribution for all specimens of each element throughout the world. The average atomic weight for each type of atom should therefore be a constant in whatever specimen it is measured. However, now that extremely accurate methods are available for measuring atomic weights it has been found that some atomic weights show variations larger than can be accounted for by experimental error. These are due to slight differences in the isotopic abundances. Oxygen itself consists of a mixture of isotopes. Besides the major nuclide of isotopic mass about 16, there is 0.037 per cent of a nuclide of isotopic mass about 17 and 0.204 per cent of another nuclide of isotopic mass about 18. The possible variations in isotopic abundance make the average atomic weight of oxygen an unsatisfactory standard against which to measure all other atomic weights. It is clearly more satisfactory to use a single isotope as the standard and that chosen is the major isotope of carbon of mass number 12. This is defined as having an isotopic mass equal to exactly 12.

The unit in terms of which the masses of atoms of nuclides are often expressed is the unified atomic mass constant, usually given the symbol m_u. It is defined as one-twelfth of the mass of an atom of carbon-12 and it has the value $(1.660\ 53 \pm 0.000\ 01) \times 10^{-27}$ kg. The isotopic mass of a nuclide is the mass of an atom of the nuclide divided by the unified atomic mass constant. Ratios of isotopic masses can be measured with great accuracy in a mass spectrometer (see p. 134) so derivation of the ratio of the mass of a nuclide to that of the carbon-12 nuclide is a relatively straightforward matter. Multiplication of this ratio by 12 gives the isotopic mass. The atomic weight of an element is the weighted mean of the masses of the naturally occurring isotopes divided by the unified atomic mass constant. In practice this means adding together the isotopic masses each multiplied by one-hundredth of its percentage abundance. Thus nitrogen consists of 99.63 per cent of nitrogen-14 of isotopic mass 14.003 07 and 0.37 per cent of nitrogen-15 of isotopic mass 15.000 11. The atomic weight is therefore $14.003\ 07 \times 0.9963 + 15.000\ 11 \times 0.0037 = 14.0067$. An alternative, but equivalent, definition of atomic weight is the ratio of the average mass per atom of the natural nuclidic composition of an element to one-twelfth of the mass of an atom of the nuclide ^{12}C. In books on SI units this is sometimes described as the relative atomic mass of an element and this should be regarded as an alternative name for atomic weight. The symbol recommended for it is A_r. In the same way the molecular weight or the relative molecular mass of a compound M_r is the ratio of the average mass per formula unit of the natural nuclidic composition of the substance to one-twelfth of the mass of an atom of the nuclide ^{12}C.

This should not be confused with the molar mass of a substance, given the symbol M, which is the mass of substance divided by the amount of substance in moles. Now the mole is the amount of substance which contains as many elementary entities as there are carbon atoms in 0.012 kg of carbon-12. So for carbon-12 one mole has a mass of 0.012 kg. Its molar mass is therefore 0.012 kg mol^{-1} or one-thousandth of its atomic weight. This factor of a thousand holds for the ratio of each atomic or molecular weight to its molar mass expressed in kg mol^{-1}.

The Periodic Table

If the elements are arranged in order of increasing atomic weight, it is seen that there is a marked periodicity in properties. This fact led Mendeléeff to propose his Periodic Table, in which elements of similar chemical properties appear in the same group or column. In constructing the table, it was apparent that, in order to preserve the correct periodicity of properties, certain positions had to be left unfilled. It was realised that these must correspond to elements not yet discovered, and the search for them was thereby stimulated. It has since been found that if the strict order of increasing atomic weight is adhered to, there are certain anomalies in the table. Thus, argon has the properties of an inert gas, and potassium those of an alkali metal; yet, to place these elements in their correct groups, potassium with the smaller atomic weight, 39.1, must follow argon of atomic weight 39.9. Similarly, tellurium has properties similar to selenium, whereas iodine is a member of the halogen group; yet, according to their atomic weights, their positions should be reversed. These anomalies could only be explained when the study of atomic structure had led to the discovery of isotopes (see p. 65). When the elements have been placed in position in the table, allowing for missing elements and for the few anomalies, they are numbered in order of increasing atomic weight, starting with hydrogen, the lightest element, as No. 1. The number which each element receives in this way is called the atomic number, and, as will be seen later, it is in some respects more important than the atomic weight.

At certain places in the periodic table there occur series of elements showing some similarities in properties. These are the transition series, and the most familiar is comprised of the elements from scandium to zinc. There is also a series of elements which are so similar in properties that they are all included in the same position in the periodic table.

These elements are the rare earths (or lanthanides), and they constitute an inner transition series. Again, the understanding of these series depends upon a knowledge of the structure of atoms.

The realisation that Dalton's postulate of the indivisibility of atoms was incorrect followed a study of the phenomena of conduction of electricity through gases at low pressure. Research on this topic by J. J. Thomson led to the discovery of a particle more fundamental than the atom, namely, the electron.

The Electron

If a tube containing two electrodes is evacuated to a pressure of about 1.33 Pa (1/100 mm of mercury), and the electrodes are connected to a high-voltage source, faintly luminous rays are seen, apparently emanating from the negative electrode (or cathode). They travel in straight lines from the cathode irrespective of the position of the anode, and they are known as cathode rays. The rays are capable of rotating a light paddle-wheel placed in their path, and they are deflected by electric or magnetic fields. It was therefore suggested that the rays might consist of streams of negatively charged particles emitted by the cathode. J. J. Thomson carried out quantitative experiments on the deflection of the rays by electrostatic and magnetic fields, and was able to measure the ratio of charge to mass, Q/m, for the particles. The value of Q/m was found to be independent of the material of the electrodes and of the nature of the gas in the tube, so it was realised that these particles are even more fundamental in nature than atoms.

Previous experiments on electrolysis of solutions (see p. 331) had suggested that there was a fundamental unit of electric charge associated with ionic substances, and this unit had been termed the 'electron'. The term was therefore applied to the fundamental unit of charge found in the gas discharge tubes. The value of Q determined first by electrolysis experiments and then confirmed by other methods, combined with the ratio Q/m, enabled the mass m of the electron to be determined. It was found to be about 1/1840 of that of the hydrogen atom. Since electrons of identical mass are emitted by all types of cathode materials in a discharge tube, Thomson considered that all atoms are built up by embedding negative electrons in a uniform sphere of positive electricity. He considered that all the mass of an atom was contained in its electrons, and the positive sphere merely provided sufficient positive charge to produce an atom bearing no resultant charge. This theory assumed that even the lightest atom, hydrogen, must contain about 1840 electrons, but, as a result of further discoveries now to be considered, the theory was soon discredited.

Radioactivity

Some of the elements of high atomic weight are found to be undergoing continuous, spontaneous disintegration by giving out rays which will affect a photographic plate. If these rays are passed through an electric or a magnetic field, they are found to consist of three types of ray, which are deflected in different ways by the field. One type, called alpha-rays, is deflected as though it were a stream of positively charged particles; the second type, called beta-rays, is deflected in the opposite direction, corresponding to a stream of negatively charged particles; the third type, called gamma-rays, is not deflected at all, even by powerful fields. This type of ray is not a stream of particles but an electromagnetic radiation like light, though of much shorter wavelength and hence of higher energy.

By studying the deviations of α- and β-rays when acted upon by both electric and magnetic fields, it is possible to calculate Q/m for their respective particles, as was done for electrons by J. J. Thomson. From these experiments, it is found that β-particles are identical with electrons, and α-particles have a value of Q/m such that the units of mass on the atomic weight scale are twice the number of units of charge on the electronic scale. Measurements of the charge of α-particles show that they must be helium nuclei of mass $4\ m_u$ and charge $+2e$, where e is the electronic or protonic charge. This is confirmed by proving spectroscopically the existence of helium in a tube through the thin walls of which α-particles have passed.

The α-, β-, and γ-rays are also distinguishable by their penetrating powers. α-Particles are easily absorbed on passing through material, and have a range of only a few centimetres in air. β-Particles have a higher penetrating power, but, because they are very light in weight, they are easily deflected from their course by collison with other particles, and so they do not travel far in a straight line. γ-Rays are the most penetrating of all, and use is made of this property in the medical application of radiotherapy.

Important information about the nature of atoms was obtained by Rutherford, in an experiment in which he bombarded thin metal foils with α-particles. Most of the particles passed through the foil with little or no deflection, whereas a few were scattered through large angles. He found that he could only explain these results if he assumed that the mass of the atoms of the foil was concentrated in heavy nuclei, small in comparison with the size of the atoms. The α-particles hitting, or passing close to, the nuclei would be scattered through large angles. A somewhat analogous effect would be observed if one were to shoot peas at a rather coarse wire netting – only those hitting the wires would be deflected, and the rest would pass through. Rutherford therefore concluded that the atom consisted of a positively charged nucleus containing the mass of the atom surrounded by sufficient negatively charged electrons to neutralise the positive charge.

Radioactive Series

Each type of radioactive decay, producing one substance from another, involves the expulsion from the atom of either an α- or β-particle. Where γ-rays are emitted, they are simply an indication of the release of energy as radiation at the same time as the expulsion of a particle. When an α-particle is expelled, it must originate from the nucleus of the radioactive atom, because it has relatively large mass (4 on the atomic weight scale). The new atom which is formed will therefore have an atomic weight 4 less than the parent atom. But the α-particle also has a charge of $+2e$, so, when it is expelled, the nucleus which remains will bear a resultant positive charge two electronic units less than the parent atom. This change in the nucleus of the atom is found to produce an entirely different chemical element, and, from its properties, it is found that this element must be placed in the Periodic Table in a position two places to the left of the original element, i.e. the atomic number and the positive charge on the nucleus have both decreased by two units. This, and other evidence, shows that the positive charge in electronic units on the nucleus of an atom is the same as the atomic number.

When a β-particle is ejected from an atom it is possible that it might have been one of the electrons surrounding the nucleus. But, whereas the loss of an electron from outside the nucleus of an atom merely converts it into the corresponding positive ion, β-particle emission produces a completely different element. The new element has properties consistent with its being the element one place to the right in the periodic table from the parent substance. This is in agreement with the β-particle being emitted from the nucleus, since this would reduce the number of negative particles in the nucleus by one, and hence increase the resultant positive charge on the nucleus by one electronic unit. The atomic number is therefore also increased by one.

By studying the sequence of elements in a radioactive series, and the types of particle emission by which each substance is produced, it is possible to follow the changes in atomic weight and atomic number taking place from one element to the next throughout the series.

If expulsion of an α-particle is followed by two further radioactive transformations each involving ejection of a β-particle, the atomic number first decreases by two and then increases by two, i.e. it returns to its original value. It has, however, lost four atomic mass units of mass so it is an isotope of the original element. Such recurrence of atomic numbers after loss of mass is a common feature in radioactive series – the series of elements produced by the successive radioactive disintegrations of a parent element – and was the way in which the existence of isotopes was first recognised. The radioactive series starting with the most abundant isotope of uranium is illustrated in Fig. 3.1. The symbol for each isotope is preceded by two numbers, the lower one being the atomic number and the upper one being the mass

Fig. 3.1 The uranium radioactive series

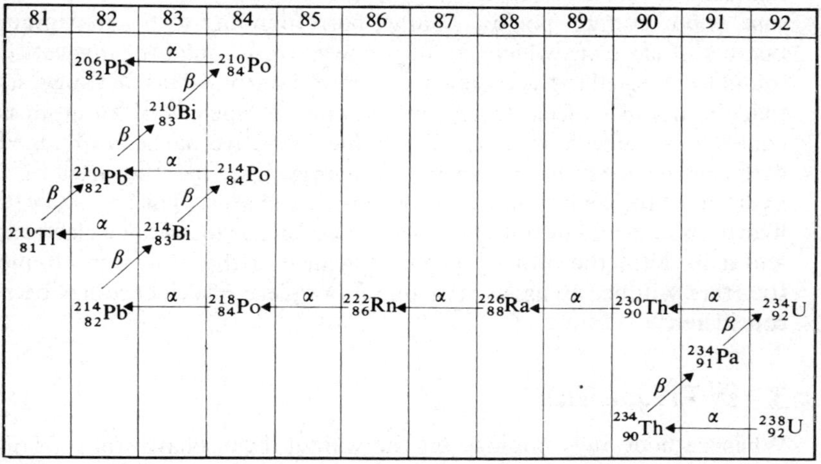

81	82	83	84	85	86	87	88	89	90	91	92
	$^{206}_{82}$Pb ←α—	$^{210}_{83}$Bi —β→	$^{210}_{84}$Po								
	←β										
	$^{210}_{82}$Pb ←α—	$^{214}_{83}$Bi —β→	$^{214}_{84}$Po								
$^{210}_{81}$Tl ←α—	←β										
	$^{214}_{82}$Pb ←α—		$^{218}_{84}$Po ←α—		$^{222}_{86}$Rn ←α—		$^{226}_{88}$Ra ←α—		$^{230}_{90}$Th ←α—		$^{234}_{92}$U
										$^{234}_{91}$Pa —β→	←β
									$^{234}_{90}$Th ←α—		$^{238}_{92}$U

number, i.e. the whole number nearest to its isotopic mass. The values of these numbers for each isotope can be deduced from those of the preceding isotope in the sequence of disintegrations by taking account of the mass and charge of the particle expelled in generating it. In this way the positions of all the members of the family can be plotted in a periodic table. In Fig. 3.1 the isotopes are arranged in columns according to the group in the periodic table and each column is headed by the atomic number which applies to all the isotopes in that column.

Positive Rays

Further valuable evidence about the structure of atoms has come from the investigation of other rays found in discharge tubes, called positive rays. These are rays of particles which are attracted to the cathode and, if the cathode has a hole cut in it, they are found streaming through the hole in a direction away from the anode. The fact that the particles travel away from the anode, and are accelerated towards the cathode before they pass through it, shows that they are positively charged. Their deflection in electrostatic and magnetic fields shows that they are very heavy compared with the particles of cathode rays. The value of Q/m for these particles is found to vary according to the nature of the gas present at low pressure in the tube. In fact, the positive particles are found to be atoms, or small groups of atoms, from the molecules of the gas, which have become positively charged by losing some of the electrons surrounding their nuclei. For each of the light atoms, it is possible to obtain positive particles of similar mass, but bearing different positive charges, corresponding to different numbers of electrons having been removed from the atom. In each case, the maximum positive charge, corresponding to the maximum number of electrons which can be removed from round the nucleus, is found to be equal to the atomic number of the atom. This fact suggests that the number of electrons surrounding the nucleus of an atom is equal to the atomic number. The lightest positive particle, obtained with hydrogen as the gas in the discharge tube, has the mass of a hydrogen atom with one unit of positive charge. This must be a hydrogen nucleus, i.e. a hydrogen atom which has lost its one electron, and it is called the proton. The significance of the proton in atomic structure will be discussed after one more piece of evidence has been considered.

X-ray Frequencies

When cathode rays impinge on the wall of the discharge tube, or on the anode, if this is placed in their path, X-rays are emitted. These are electromagnetic waves intermediate in energy between light and γ-rays, but resembling the latter in their ability to penetrate matter. In tubes designed to produce X-rays, the stream of electrons from the cathode is made to strike the anode, which is cooled by water, or by cooling-fins, to dissipate the large quantity of energy released as heat. Just as light has a wavelength or a frequency which is associated with a specific colour, so X-rays of various frequencies can be observed. For each element used as an anode of an X-ray tube, there is a characteristic frequency of X-rays produced. The X-rays are 'harder', i.e. have higher frequency and higher energy (and therefore higher penetrating power) the higher is the atomic number of the element constituting the material of the anode. There is no simple correlation between frequency and atomic weight, but if the square root of the characteristic frequency is plotted against the atomic number of the element, a straight-line graph is obtained.

This work, which was carried out by Moseley, was the first to show that the atomic number of an element is a more fundamental property than the atomic weight. Moseley concluded that the atomic number of an atom, or some fundamental property closely related to it, probably represented the positive charge on the nucleus of an atom. As we have seen, this view is also supported by a study of the changes taking place during radioactive decay. It has also been confirmed by experiments on the scattering of α-particles carried out by Chadwick, in which the charge on the nucleus could be estimated.

The Atomic Nucleus

Since many atomic weights are approximately whole numbers, and since there exists a positive particle – the proton – with A_r equal to unity, it might be considered that the nuclei of atoms consist of protons. This view introduces a serious difficulty, however. For an atom of atomic weight A, it would mean that the nucleus would have to bear a positive charge of A electronic units. But if the atom is to remain neutral, the positive charge on the nucleus must be equal to the total negative charge of the surrounding electrons. However, the number of surrounding electrons is equal to the atomic number of the element, Z and Z is not equal to A, but is about $A/2$ for light elements. To overcome this difficulty, we must assume that the nucleus contains, besides A protons $(A-Z)$, electrons. The resultant positive charge on the nucleus then has the correct value of Ze.

Rutherford suggested in 1920 that the electrons in the nuclei might be combined with some of the protons to form neutral particles of unit mass, which he called neutrons. Such neutral particles were later found experimentally as products of the bombardment of certain light atoms with α-particles and they were recognised to be separate particles in their own right rather than combinations of protons and electrons. Thus the nucleus will contain Z protons and $(A-Z)$ neutrons, so that the total mass is A and the positive charge is Ze.

Isotopes

The studies in the foregoing sections show that it is the charge on the nucleus that determines the position of an atom in the Periodic Table and establishes its chemical identity. The nuclear mass can vary by the existence in the nucleus of different numbers of neutrons but providing the number of protons remains the same the nuclides will be the same element. As has been mentioned already, nuclides having the same atomic number or nuclear charge but different masses are known as isotopes. Most naturally occurring elements have two or more isotopes and this accounts for the fact that most atomic weights differ appreciably from whole numbers even though all isotopic masses are all close to whole numbers on the atomic weight scale.

The existence of isotopes in normal elements was first demonstrated by Aston from a study of the deflection of positive rays by magnetic and electrostatic fields, as in the experiments for determining Q/m. Isotopes of different values of m are deflected to different extents. This method has been developed by Aston and others into the technique of mass spectrometry referred to above and described on p. 134. Isotopes behave chemically in the same way, so the relative abundance of the isotopes of an element is unchanged by chemical processes and the average atomic weight remains the same. Physical processes which depend upon mass (such as gaseous and thermal diffusion, centrifugation, fractional distillation and electrolysis) take place at different rates for different isotopes, and so do rates of reaction, so it is possible to effect a separation of isotopes.

Nuclear Stability

When the exact masses of atoms are compared with the exact masses of the protons, electrons and neutrons of which they are composed it is found that there is a mass deficiency (the true mass defect) in the atom. Because of the equivalence of mass and energy enunciated by Einstein in the form

$$E = mc^2$$

(where E is the energy equivalent to the mass m and c is the velocity of light) the mass defect corresponds to the release of energy in the hypothetical process whereby the nucleus of the atom is synthesised from its constituent protons and neutrons. Nuclear energies are usually quoted in million electronvolts (MeV). The electronvolt (eV) is the energy of an electron when accelerated through a potential difference of 1 volt. It is therefore equal to the work done in transferring the electronic charge (see p. 331) across a potential difference of 1 volt. This is 1.6022×10^{-19} (C) $\times 1$ (V) $= 1.6022 \times 10^{-19}$ J. So 1 MeV is equivalent to 1.6022×10^{-13} J. Mass defects are quoted in terms of the unified atomic mass constant m_u equivalent to $1.660\ 53 \times 10^{-27}$ kg. Thus the energy equivalent to m_u is

$$E = 1.660\ 53 \times 10^{-27}\ (\text{kg}) \times (2.998 \times 10^8)^2\ (\text{m s}^{-1})^2$$
$$= 1.4925 \times 10^{-10}\ \text{J}$$
$$= 1.4925 \times 10^{-10}/1.6022 \times 10^{-13}\ \text{MeV}$$
$$= 931.5\ \text{MeV}$$

Thus for an atom of ^{4}He of mass $4.002\ 60\ m_u$ composed of two hydrogen atoms of mass $1.007\ 825\ m_u$ and two neutrons of mass $1.008\ 665\ m_u$ the mass defect is

$$2 \times 1.007\ 825 + 2 \times 1.008\ 665 - 4.002\ 60 = 0.030\ 38\ m_u.$$

This is therefore equivalent to $0.030\ 38 \times 931.5 = 28.30$ MeV. This amount of energy (ΔE) can be regarded as the binding energy of the particles constituting the nucleus. Some idea of the stability of nuclei can be obtained by calculating the binding energy per nuclear particle or nucleon. The number of nucleons (protons plus neutrons) is equal to the mass number A. The binding energy per nucleon is therefore equal to $\Delta E/A$. When this quantity is plotted against A a curve of the form shown in Fig. 3.2 is obtained. This has a broad maximum for elements having A about 40 to 120 and decreases rapidly for lighter elements and slowly for heavier elements. In nuclear fusion reactions there is a tendency for light nuclei low on this curve to coalesce to form heavier nuclei higher on the curve. Similarly in nuclear fission reac-

Fig. 3.2 Binding energy per nucleon plotted against mass number

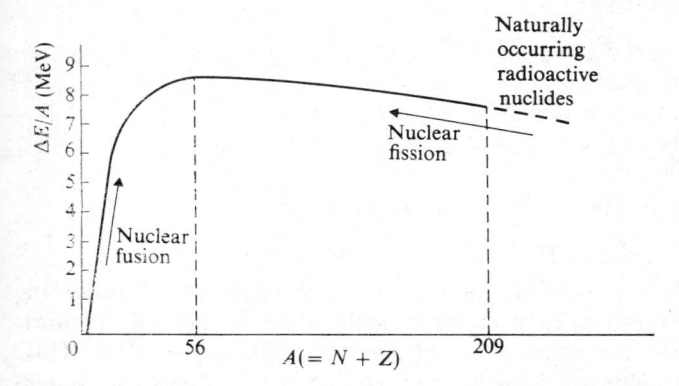

tions, very heavy nuclei split up to give lighter nuclei high on the curve. Some isotopes, notably ^{4}He, ^{12}C and ^{16}O, are more stable than this smooth curve would suggest and have points (not shown in Fig. 3.2) lying above the curve. The most abundant element in the composition of the earth, ^{56}Fe, lies at about the highest point on the curve.

The tendency for a nucleus to disintegrate by radioactive decay does not depend mainly on its position on this curve. It seems to depend much more on the relative numbers of protons and neutrons in the nucleus. The number of protons is given by the atomic number Z and the number of neutrons N is given by $A-Z$. If Z is plotted against N for stable nuclides a narrow band is found which follows the line shown in Fig. 3.3. Nuclides which lie above or below this band are radioactive and disintegrate in such a way as to come within the band. If they lie below the curve (excess neutrons) they disintegrate by giving off negatrons, i.e. negatively charged β particles, and for each disintegration of this type one neutron is converted into a proton.

e.g. $^{32}_{15}\text{P} = {}^{32}_{16}\text{S} + {}^{\ \ 0}_{-1}\beta$.

Compare this behaviour with the stability of $^{31}_{15}\text{P}$. If they lie above the curve (excess protons) they may suffer the alternative form of β decay in which positrons (particles having the same mass and magnitude of charge as electrons but positively charged) are given off, each of

Fig. 3.3 Number of protons plotted against number of neutrons for stable nuclides

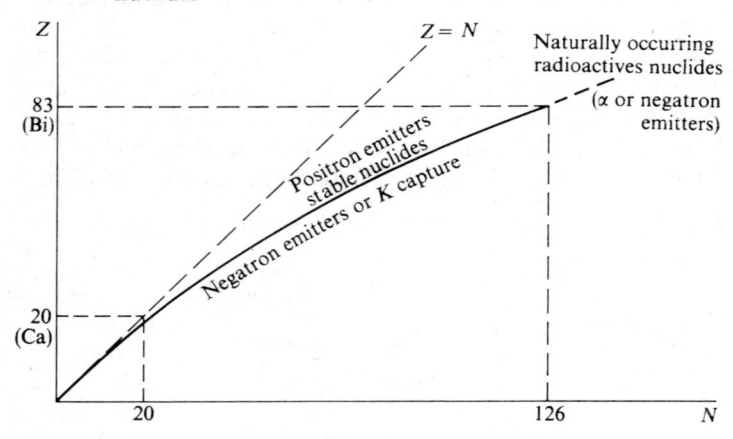

[Graph: Z axis vs N axis. Line Z = N shown dashed. Curve of stable nuclides labelled "Positron emitters stable nuclides" above and "Negatron emitters or K capture" below. Dashed lines at Z = 83 (Bi) and Z = 20 (Ca), and N = 20 and N = 126. Region beyond bismuth labelled "Naturally occurring radioactives nuclides (α or negatron emitters)".]

which converts a proton into a neutron,

e.g. $^{30}_{15}P = {}^{30}_{14}Si + {}^{0}_{+1}\beta$.

Alternatively it is possible for a proton-rich nucleus to capture an orbital electron (usually from the K shell and hence called K capture) and so convert a proton into a neutron, e.g. $^{125}_{53}I + {}^{0}_{-1}e = {}^{125}_{52}Te$. This process is accompanied by the emission of X-rays due to an outer electron falling into the K shell to replace the captured electron. The naturally occurring radioactive nuclei lie above bismuth ($A = 209$) in the periodic table and have neutron-rich nuclei. Many of these disintegrate by α-particle emission which leaves the ratio of protons to neutrons almost unchanged (Z and N both decrease by 2). However this is followed by negatron emission which causes the nuclide to approach the stability line. Often α or β decay results in a nucleus in an excited energy state and it relaxes to a lower energy state by giving out electromagnetic radiation in the form of γ-rays. This process does not, of course, alter Z or N. For the very heaviest elements, above uranium ($A = 238$) in the Periodic Table, disintegration may take place by nuclear fission in which the nucleus splits into two smaller nuclei of approximately equal mass. This process is accompanied by an appreciable loss of mass which is converted into enormous quantities of energy and is the source of atomic energy.

Most of the common elements have even atomic numbers (even values of Z) and their most abundant isotopes have even mass numbers so that they must also have even values of N. The combination of even Z and even N seems, then, to be particularly stable. Conversely the combination of odd Z and odd N seems particularly unfavourable since it is found in only 1.5 per cent of stable naturally occurring nuclides.

Nuclear Reactions

Most atomic nuclei, when bombarded with light particles, such as protons, deuterons (^{2}H nuclei), α-particles, or neutrons, undergo nuclear reactions in which the mass or charge, or both, of the nucleus is changed. In the course of the reaction another light particle is often liberated, though the nucleus may lose its excitation energy merely as γ radiation. If only the mass of the nucleus is changed during a nuclear reaction the product is an isotope of the original substance. If the charge is altered a different element is formed but often as a different isotope from the common naturally occurring isotope. In both cases, the new isotope may be radioactive, even if it is a light element, so this is an important method of producing 'tracer' atoms (see p. 70).

In nuclear reactions both the charges and the mass numbers must balance on the two sides of the equation thus:

$$^{14}_{7}N + ^{1}_{0}n = ^{14}_{6}C + ^{1}_{1}H$$

where the mass numbers total 15 on both sides and the charges total 7 on both sides. Here ^1_0n stands for the neutron, which is commonly used in nuclear reactions because of the absence of charge which enables it to enter the highly charged positive nuclei of atoms and because of its ready availability in nuclear reactors. This particular nuclear reaction also occurs naturally to a small extent due to the action of neutrons associated with cosmic rays. This provides the source of radioactive carbon dioxide in the atmosphere which exchanges with carbon in living material but not in inanimate matter and so gives rise to the technique of radiocarbon dating. The radioactive carbon decays by negatron emission to revert to the normal isotope of nitrogen:

$$^{14}_{6}C = ^{14}_{7}N + ^{0}_{-1}\beta.$$

There is a concise method of indicating nuclear reactions. The bombarding and emitted particles are written as symbols inside brackets and separated by a comma; to the left and right of the brackets are written the chemical symbols of the target and product nuclei. Thus the reaction above for the production of radiocarbon would be indicated as N(n,p)C, or more explicitly as ^{14}N(n,p)^{14}C. Similarly the reaction

$$^{14}_{7}N + ^{4}_{2}He = ^{17}_{8}O + ^{1}_{1}H$$

would be written ^{14}N(α,p)^{17}O.

Tracers

One of the most important applications of radioactive nuclides is in tracing the distribution of chemically similar substances with which they are mixed as they take part in physical or chemical changes or in biological processes. Usually the radioactive nuclide is an isotope of a non-radioactive atom in the molecule of the substance whose distribution is to be followed. Thus, if the efficiency of a precipitation is to be studied the material to be precipitated can be mixed with a small proportion of the same substance 'labelled' by the fact that it contains a radioactive isotope of one of the atoms. The supernatant liquid is separated from the precipitate after the precipitation and its activity is compared with that of a known concentration of the tracer. From this and the known proportion of the tracer in the mixture with the substance under investigation, the total concentration of the latter in the supernatant liquid can be found. From this the solubility can be derived and the method works well even where the solubility is so low that ordinary analytical methods are inapplicable.

Radioactive tracers are important in diagnostic medicine. For example, the efficiency of a particular organ in taking up or secreting a particular fluid can be measured by arranging for the fluid to be labelled with a radioactive tracer and then measuring the activity. Thus $Na^{131}I$ solution is used to study the uptake of iodide ion by the thyroid gland. The radioactive iodine is made by the nuclear reactions:

$$^{130}_{52}Te + ^{1}_{0}n = ^{131}_{52}Te + \gamma$$

$$^{131}_{52}Te = ^{131}_{53}I + _{-1}^{0}\beta$$

and it is a β^- and γ emitter. Similarly $Na_2H^{32}PO_4$ solutions are used in blood studies where it is detected by its β^- emission. The radioactive phosphorus is made by the nuclear reaction:

$$^{32}_{16}S + ^{1}_{0}n = ^{32}_{15}P + ^{1}_{1}H.$$

Organic compounds labelled with ^{14}C are widely used in studies of metabolic processes and chemical reactions.

Tracers need not necessarily be radioactive, though these are most convenient because of the ease of detection. Non-radioactive tracers are usually detected by mass spectrometry (p. 134). In particular oxygen does not have a radioactive isotope of half-life long enough to be a suitable tracer and ^{18}O is often used in tracer studies. Thus reaction of an ester with $H_2^{18}O$ solves the problem of whether it is the C—O or the O—R bond which breaks in the ester:

$$R-C\begin{array}{c}\diagup O \\ \diagdown O-R'\end{array} + H-^{18}O-H \rightarrow R-C\begin{array}{c}\diagup O \\ \diagdown ^{18}O-H\end{array} + R'OH$$

or

$$R-C\overset{O}{\underset{O-R'}{\diagup}} + H-{}^{18}O\overset{H}{\diagdown} \rightarrow R-C\overset{O}{\underset{O-H}{\diagup}} + R'\,{}^{18}OH$$

After the reaction the ^{18}O atoms are found exclusively in the acid, showing that the first mechanism is the correct one.

Measurement of Radioactivity

When α-, β- or γ-rays pass through a gas between two electrodes they ionise some of the gas molecules, creating electrons and cations. These move in opposite directions, the electrons towards the anode and the cations towards the cathode. The variation of number of ions collected with applied potential difference has the form shown in Fig. 3.4. In region 1 some recombination of electrons and cations occurs before

Fig. 3.4 Effect of increasing voltage on counter operation

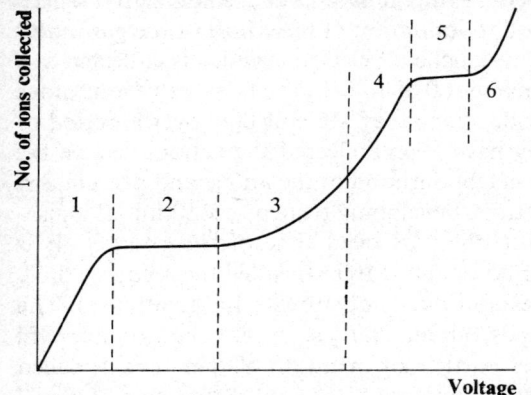

they can reach the electrodes. In region 2 there is saturation collection, i.e. all the ions and electrons formed by the radiation are collected and this situation is not affected by increasing the applied voltage. Eventually further increase in voltage increases the kinetic energy of the electrons so much that they can now ionise further gas molecules and 'gas multiplication' occurs. At first (region 3) the charge collected at any one voltage is still proportional to the number of ionisations first produced by the radiation. The proportionality constant increases with increasing voltage. Eventually a second plateau region 5 is reached in which all the gas is ionised by further collisions with electrons whatever the applied voltage. Beyond this the gas ionises by the action of the high applied voltage alone irrespective of incoming radiation (region 6), so continuous discharge occurs.

Instruments that work in region 2 are called ionisation chambers and they are used for accurate work where absolute measurements of the activity are required. Their disadvantage is that the ion currents are very small and highly sensitive and stable equipment is needed to measure them. Instruments operating in region 3 are called proportional counters. They are set up not only to count the pulses of ion current, one of which is produced by each α- or β-particle or γ-ray quantum, but also to measure the energy of the radiation from the size of the pulses. The number of ions and electrons produced in the initial ionisation of the gas depends upon the energy of the particle or quantum, and the size of the pulse formed after gas multiplication is proportional to the initial ionisation and hence also proportional to the energy of the radiation. The number of pulses per second is a measure of the intensity of the radiation. One of the advantages of proportional counters is that they operate at atmospheric pressure so they are often made to be demountable and samples that emit α- or low energy β-particles can be placed actually inside the counter.

The most popular type of instrument is the Geiger–Müller (or G–M) counter which operates in region 5. Even a single initial ionisation would eventually lead to ionisation of the whole of the gas in this region, thus causing an 'avalanche' of electrons which is collected as a pulse at the anode within about 0.05 to 0.3 μs. The positive ions move more slowly to the cathode where they are collected over a period of about 100 μs. Until they have been collected the cations reduce the potential gradient in the neighbourhood of the anode and prevent any further avalanche of electrons developing from a second initial ionisation. This period is called the true dead time of the counter. It is followed by a further period of about 100 μs, called the recovery time, during which the counter still does not provide full-size pulses. The total of these two periods, about 200 μs, before the counter will respond fully to a further particle or quantum of radiation, is called the paralysis time. Observed counting rates may be corrected for this paralysis time by multiplying by the factor $1/(1 - Nt)$ where N is the observed counting rate and t is the paralysis time. When the cations strike the cathode electrons may be displaced which can cause a further avalanche of electrons and the whole process might be repeated to give multiple pulses if it were not quenched. This is done by adding an organic vapour or a halogen gas to the gas in the counter. Such quenching agents dissipate the excess energy by dissociation. The high electric field strengths required for proportional or Geiger–Müller counters are achieved by having the anode in the form of a wire supported centrally in a cylindrical tube. In the commonest form of the counter the side walls of the tube act as the cathode and a thin end window of mica or aluminium permits the entry of α- or β-particles or γ-rays. Filters may be placed in front of the window to exclude α- or α- and

β-particles so that the activity due to each type of particles may be deduced.

Scintillation counters operate on the completely different principle that certain materials, known as phosphors, emit flashes of light when exposed to radiation. The phosphor is in contact with a photomultiplier tube where the light causes emission of electrons from the photosensitive cathode. These electrons are accelerated to the first of a series of anodes, called dynodes, each more positive than the previous one. The accelerated electrons on collision with the first dynode release 3 or 4 times as many electrons which are accelerated towards the second dynode and so on through often 11 stages. The resultant pulse is then electronically amplified and counted. The size of the output pulse is proportional to the energy expended by the incident radioactive particle or γ-ray quantum on passing through the phosphor, so the energy of the incident radiation can be measured. Phosphors for detecting α-particles are usually zinc sulphide films activated with silver. For β-particles anthracene is usually used as a solid phosphor but a solution of p-terphenyl in toluene is often used. Single crystals of sodium iodide activated with about 0.1 per cent thallium iodide are the most commonly employed phosphors for γ-rays. Low energy β-particle emitters can best be counted by dissolving the sample in a suitable phosphor solvent or by adding a phosphor to the sample solution.

Radioactive Decay

When the radioactivity of freshly isolated materials was studied quantitatively it was found that the activity gradually decreased with time. Moreover, the rate of decrease was proportional to the activity remaining. This can be explained as follows. The radioactivity is due to the disintegration of the atomic nuclei of the original substance and if the probability of a nucleus disintegrating in unit time is λ then the number of nuclei disintegrating in unit time is λN, where N is the number of nuclei remaining at any time t. The change in the number of nuclei $-dN$ (the negative sign indicating a decrease in N) in time dt is therefore λN. dt. Hence the rate of change of N with time is given by

$$\frac{-dN}{dt} = \lambda N.$$

The activity (a) measured in counts per unit time by a counter is proportional to the number of nuclei disintegrating in unit time (λN). That is

$$a = k\lambda N.$$

This proportionality constant k depends on the geometry of the counter and allows for the fact that not all the disintegrations are counted but a constant fraction of them. The activity decreases with time because the number of nuclei remaining (N) decreases with time. The rate of decrease of a with time is $-da/dt$ and this is obtained by differentiating with respect to time the negative of the expression for a, i.e.

$$\frac{-da}{dt} = -k\lambda \frac{dN}{dt}.$$

Putting $-dN/dt = \lambda N$ we have

$$\frac{-da}{dt} = k\lambda(\lambda N).$$

But $k\lambda N = a$

so
$$\frac{-da}{dt} = \lambda a.$$

Thus the rate of decrease of activity at any time is proportional to the activity at that time, as was observed.

It is instructive to integrate the equation for $-dN/dt$ to obtain an expression for the number of nuclei (N) remaining at time t in terms of the number present initially (N_0). This is done by first rearranging the equation thus:

$$\frac{dt}{dN} = -\frac{1}{\lambda N}.$$

Integrating (by the rule on p. 16):

$$t = -(1/\lambda) \ln N + C.$$

To evaluate the integration constant C, we must substitute the value of N when $t = 0$; this is the number of atoms which were present when the disintegration began, namely N_0. Then:

$$0 = -(1/\lambda) \ln N_0 + C,$$
or,
$$C = (1/\lambda) \ln N_0.$$

The integrated equation then becomes:

$$t = -(1/\lambda) \ln N + (1/\lambda) \ln N_0,$$
or,
$$t = -(1/\lambda)(\ln N - \ln N_0).$$

Multiplying throughout by $-\lambda$, and simplifying the logarithmic term:

$$-\lambda t = \ln \frac{N}{N_0}.$$

Hence:
$$\frac{N}{N_0} = e^{-\lambda t}$$

or,
$$N = N_0 \cdot e^{-\lambda t}.$$

λ, the constant which represents the probability of a nucleus disintegrating in unit time, is called the radioactive disintegration constant. It is obviously a measure of the rate of decay but two other constants that measure the rate of decay are more often quoted. One is the 'half life period' $t_{\frac{1}{2}}$ which is the time taken for the number of nuclei (N) to be reduced to half of the original number (N_0). It is related to λ thus:

$$\tfrac{1}{2} N_0 = N_0 e^{-\lambda t_{\frac{1}{2}}}$$

Dividing both sides of this equation by N_0 we have

$$\tfrac{1}{2} = e^{-\lambda t_{\frac{1}{2}}}$$

or
$$2 = e^{\lambda t_{\frac{1}{2}}}.$$

Hence
$$\ln 2 = \lambda t_{\frac{1}{2}}$$

or
$$t_{\frac{1}{2}} = \ln 2 / \lambda.$$

Since $\ln 2$ has the value 0.693, we have

$$t_{\frac{1}{2}} = 0.693/\lambda.$$

Now $a = k\lambda N$ and $a_0 = k\lambda N_0$, where a_0 is the initial activity,

so
$$\frac{a}{a_0} = \frac{N}{N_0}$$

and $t_{\frac{1}{2}}$ is therefore also the time for the activity to drop to half its original value.

The second alternative measure of the rate of decay is called the 'mean life time' and it is the time taken for the number of nuclei to fall to $1/e$ of the original number (or for the activity to decrease by this factor). It is often given the symbol τ. Thus, when $t = \tau$, $N = N_0/e = N_0 e^{-1}$.

Hence
$$N_0 e^{-1} = N_0 e^{-\lambda \tau}.$$

To make both sides of this expression equal, the powers of e must be equal.

$$\therefore \ -1 = -\lambda \tau$$

or
$$\tau = 1/\lambda.$$

The product formed by decay of a radioactive substance is usually itself radioactive. So the rate of decay of the original material represents the rate of formation of the radioactive product. This product, in turn, may disintegrate at a definite rate to form a third substance, and so on. This process continues throughout a whole radioactive series, each member being produced by disintegration of the preceding one. When such a series has been in existence for a long time, it will have reached a position of equilibrium, such that each member is disintegrating at the same rate as that at which it is being formed. Expressed mathematically, if N_1, N_2, N_3, etc., represent the numbers of atoms of elements 1, 2, 3, etc., at equilibrium, then:

$$-\frac{dN_1}{dt} = -\frac{dN_2}{dt} = -\frac{dN_3}{dt} = \text{etc.}$$

But, for each element:

$$\frac{dN}{dt} = -\lambda N,$$

$$\therefore \lambda_1 N_1 = \lambda_2 N_2 = \lambda_3 N_3 = \text{etc.}$$

That is to say, the amounts of the elements present at equilibrium are inversely proportional to their disintegration constants. This fact provides a method of finding the disintegration constants for elements which disintegrate too slowly for the constant to be measured directly.

Biological Effects of Radiation

In view of the growing application of radioactive materials in atomic energy projects and in industry and research, a short digression from the main topic of atomic structure is justified to consider the effects which radiation may have on biological materials and on the human body in particular. The word 'radiation' is used in a general sense here to include the effect of α- and β-particles and neutrons as well as X-rays and γ-rays. Three factors must be taken into consideration in this connection: the intensity of the radiation, the rate at which it decreases due to both radioactive decay and elimination from the body, and the penetrating power of the radiation.

The amounts of radioactive substances present in a given sample are expressed in terms of the curie (Ci). This was originally defined as the amount of a substance which disintegrates at the same rate as 1 g of pure radium. As a result of more accurate experimental work, the definition has now been modified, and is the amount of substance in which 3.7×10^{10} atoms disintegrate per second. This is such a high rate of disintegration that amounts are expressed in mCi or μCi.

The actual amount of radiation emanating from a radioactive substance in the form of γ-rays is measured in terms of the roentgen (R). This is the amount of radiation which will ionise 1 kg of dry air at

N.T.P.† to such an extent that 2.58×10^{-4} coulombs of charge are produced. It corresponds to the absorption of about 8.77×10^{-3} J per kg of dry air at N.T.P. The intensity of radiation is therefore quoted in roentgens per second. Besides applying to γ-radiation, this unit is also used in quoting the intensity of X-rays.

The roentgen strictly applies only to X-rays and γ-rays, but α- and β-particles and neutrons also exert mainly an ionising effect on a medium through which they pass, so it has been suggested that a similar unit should be used to describe their effect. This is the *roentgen equivalent physical* or Rep which is the amount of radiation of any type whose absorption produces the same amount of energy as 1 R of X-rays or γ-rays. In biological tissue this amount of energy is 9.3×10^{-6} J per cm³; larger than the amount of energy produced by the same amount of ionisation in air because of the different constitution of the material.

In view of the dependence of the definition of the Rep on the somewhat uncertain knowledge of the relationship between the amount of ionisation and the amount of energy absorbed in any given material, a more convenient unit has been defined. This is called *radiation absorbed dose* or rad and it is the amount of radiation whose absorption causes the liberation of 10^{-2} J of energy per kg of material. It can be seen that the Rep and the rad are nearly the same for soft tissue, but they differ considerably for bone.

The Rep and the rad are both defined in terms of the physical effect which the radiations have on the material which absorbs them. Another unit which is related directly to the effect of greatest interest – biological damage – is the *roentgen equivalent man* or Rem. It was originally defined as the quantity of radiation producing the same biological damage in man as that produced by 1 Rep of X-rays or γ-rays. In view of the uncertainties involved in the definition of the Rep, the Rem is now defined in a way which avoids dependence on the Rep. It uses instead the concept of *relative biological effectiveness* (or RBE) of radiations. This is the ratio of the number of rads of X-rays or γ-rays which must be absorbed to the number of rads of the radiation in question which must be absorbed to produce the same biological effect. For example, fast neutrons have a RBE of about 10 for cataract formation, which means that about 10 times as many rads of X-rays or γ-rays must be absorbed as of fast neutrons to produce the same average probability of cataract formation. The Rem is then defined such that a dose in Rems is given by the product of the dose in rads and the RBE for the particular radiation.

To summarise this rather confusing state of affairs with regard to all these definitions of amounts of radiation, it may be said that the

† N.T.P. = normal temperature and pressure = 101 325 Pa and 373.15 K.

rad is the most important definition in terms of the physical effect (energy absorption) of radiation and the Rem is the definition which compares the biological effect of the radiation with that of 1 rad of X-rays or γ-rays.

The definitions which have just been discussed are very useful in assessing the effect of a dose of radiation of constant intensity. A more likely situation is that of exposure to or ingestion of radioactive material which disintegrates and emits progressively more feeble radiation. A given number of atoms of radioactive material produces a greater amount of radiation initially the shorter is its half life but this radiation decays correspondingly more rapidly. The extent of radiation damage which is caused depends therefore on how freshly produced the material was at the start of the exposure to its radiations and on how extended the exposure was. The total integrated dose from the beginning to the end of the exposure can be assessed if the constitution of the radioactive material is known together with the ages and half lives of all the constituents. When the radiating material is very heterogeneous in character it may be that an average decay law is known even if the details for the separate constituents are not known. It has been found, for instance, that the radioactive fall-out from a nuclear explosion normally decays such that the radiation at a time t after the explosion is proportional to $t^{-1.2}$.

In cases where radioactive materials are ingested the intensity of radiation from these materials decreases partly because of radioactive decay and partly through the normal biological processes which cause the materials to pass out of the body. The latter process is characterised by the *biological half life* which is the time taken for the body to eliminate one-half of the substance in question. The combined effect of the radioactive decay and the biological elimination is sometimes stated in terms of the *effective half life* which is the time required for the amount of a particular radioactive element to fall to half of the original value as a result of both processes. The elements which are the most dangerous when ingested are those which have relatively short radioactive half lives but fairly long biological half lives. They give a large amount of radiation initially and are retained by the body for at least as long as the initial active period lasts. Particularly dangerous are the two isotopes of strontium with mass numbers 89 and 90 which are among the abundant products of nuclear explosions involving uranium or plutonium. Being similar in chemical properties to calcium, these elements become incorporated in the bones where their radiations act directly on the bone marrow and destroy its ability to produce blood cells.

It is in connection with problems of ingestion of radioactive substances that the factor of the penetrating power of their radiations becomes most important. The general order of penetrating power of the α-, β-

and γ-radiations has already been described on p. 61. From this it is clear that materials giving rise to only α- or β-particles can only cause appreciable damage to the body if they are ingested. (Care must also be taken to avoid allowing them to come into contact with the skin as this can cause cancer of the skin.) Again, inside the body, their effect becomes more serious if they tend to concentrate in regions particularly sensitive to damage. The case of strontium has already been mentioned but another is that of radioactive caesium of mass number 137. This tends to concentrate in the gonads where it is likely to cause undesirable mutations. It is therefore a serious hazard from the genetic point of view.

A great deal remains to be ascertained about the biological effects of radiation, e.g. the precise manner in which the damage to cells is caused, and whether or not there is a threshold dose below which there is no measurable biological effect. However, there is general agreement as to the maximum radiation dose which personnel should be permitted to receive if some exposure is unavoidable in their occupation. This is 0.1 Rem per week if the parts of the body exposed include the particularly sensitive organs such as the gonads, the blood-forming organs and the lenses of the eyes. Rough figures have also been arrived at for the maximum permissible body contents of various radioactive materials. These are expressed in terms of the curie, the unit mentioned on p. 76. For example, the maximum safe body content of radium is considered to be 0.1 microcurie, i.e. approximately 0.1 microgram or 10^{-7} g. Full details of permissible doses and recommendations for protection against radiation are given in the official reports of the International Commission on Radiological Protection.

Atomic Spectra

The evidence so far considered suggests that an atom consists of a nucleus surrounded by a number of electrons equal to the atomic number, but it does not indicate how these electrons are arranged. Rutherford suggested that they revolved round the nucleus like planets round the sun – the attraction between the unlike charges of electrons and nucleus being balanced by the centrifugal force of rotation. This view was criticised, however, on the grounds that the electrons would gradually lose energy by radiation and fall in towards the nucleus. The study of atomic spectra, on the other hand, showed that the criticism could be met if the views of the energies of electrons in atoms were modified.

If an atom were to lose energy continuously by radiation, it should be possible to observe electromagnetic radiation of continuously varying frequency emanating from the atom. In practice, atoms do not radiate energy at all until they are excited. Then, the spectrum of the radiation emitted is found to consist not of a continuous range of frequencies but of a few sharply defined frequencies.

The nature of light energy and of its absorption and emission by matter was studied by Planck at the beginning of the present century. It led him to a concept of far-reaching consequence, namely, that energy is not continuous but is exchanged in 'packets' or quanta of a definite size. For energy in the form of electromagnetic radiation, the size of the quantum is proportional to the frequency of the radiation v, and is equal to hv joules, where h is Planck's constant (6.6262×10^{-34} J s).

These ideas were applied to the problem of the energies of electrons in atoms by Bohr. He interpreted the sharp lines of definite frequency in the spectra of atoms as arising from changes in energy of the electrons as they moved from one state of constant energy to another. If an electron moves from a state of high energy to a state of low energy, the energy lost by the electron appears as a quantum of light which is seen in the spectroscope as a line of frequency v, related to the difference in energy ΔE between the two energy states of the electron by the Planck equation:

$$\Delta E = hv.$$

A change in the electron's energy between the same two energy states, but in the reverse direction, would take place if the atom absorbed light of this frequency. This would give an absorption line at the same place in the spectrum.

Spectra are obtained for any form of electromagnetic radiation by separating out the components of different wavelength or frequency. The most familiar example is the spectrum of visible light obtained when the path of a beam of white light is bent by a prism. The amount by which the path is bent is greater the higher the frequency of the light, so that the blue and violet rays are bent most and the red rays are bent least. This method of separating the components of different wavelengths depends upon the phenomenon of refraction but an alternative method is to use the phenomenon of diffraction. This consists of passing the light through a glass plate on which are ruled equally spaced parallel lines very close together or reflecting light from a similarly ruled mirror. These are known as diffraction gratings. The light waves emanating from the spaces between the lines interfere with each other producing strong diffracted beams only in certain specific directions, depending on the wavelength of the light and the spacing of the lines on the grating. In this way also the components of different wavelength are seen in different directions.

The spectrum may be observed visually in a spectroscope, recorded photographically in a spectrograph or detected and scanned by some light-sensitive device in a spectrometer. In each case the information which is obtained is a record of the variation of intensity over a range of wavelengths. Outside the visible range the photographic method can still be applied but the use of some device which converts the radiant intensity into an electric current is more common. The current is then amplified and converted into a plot of intensity against wavelength by a recorder.

Spectra may be studied by causing the material under observation either to emit or to absorb radiation. Emission spectra of atoms are produced by raising the appropriate materials to a high temperature in a flame or by passing an electric discharge through suitable gases. The spectrum is then seen as a number of bright lines occurring at sharply defined frequencies or wavelengths. For absorption spectroscopy in the visible range the most convenient source of radiation is a tungsten filament electric lamp which gives a continuous spectrum of smoothly varying intensity. The presence of atoms between the lamp and the spectroscope then produces the corresponding atomic spectrum as dark lines on the bright background due to the lamp. This is because radiation of particular wavelengths is absorbed by the atoms and so reaches the spectroscope with diminished intensity. The dark lines of an absorption spectrum appear at the same wavelengths as the bright lines of an emission spectrum from the same substance and these wavelengths are characteristic of the particular type of atom. In this connection it is interesting to note that much information about the chemical composition of the sun and stars can be obtained by observation of the dark lines in their spectra. These characterise the atomic species which absorb radiation in the cooler outer layers of the stars.

An analytical technique based on atomic emission spectra is flame photometry. It is applied mainly to the alkali and alkaline earth elements. The sample to be analysed is in the form of a solution and this is converted into a mist by means of an 'atomiser'. The mist is fed into a non-luminous gas flame where the temperature is sufficient to excite the atomic emission spectra of the components. Ideally the emitted light should be split up into a spectrum by a prism or diffraction grating but instead a narrow spectral range is often isolated from the total light output by an appropriate filter (e.g. yellow for sodium). The intensity of light passed by the filter is measured by a photoelectric cell and galvanometer. For each element the instrument is calibrated by finding the difference in galvanometer reading between each of a series of standard solutions and a blank. The blank should contain the same substances as are present in the test solution in approximately the same concentrations but with the element being analysed missing. The standard solutions should be made up like the blank but in addition they should contain increasing known amounts of the element to be analysed. A graph should be plotted of the difference in galvanometer reading between the standard solution and the blank versus the concentration of the element to be analysed in the standard solution. The galvanometer reading for the solution under test is then found and the difference between it and the blank is used to read off the concentration of the required element on the calibration curve. The process may be repeated for other elements by inserting different filters and making separate calibration curves by varying the concentration of each of the other elements in making up the standard solutions.

A related analytical technique using absorption spectra is atomic absorption spectroscopy. This technique is more sensitive than flame photometry and can be used to determine most elements in the periodic table. This time, the technique of spraying a solution to be analysed into a flame is merely a method of producing a cloud of vapour of fixed dimensions which will absorb incident light. Even though some light is emitted by this cloud, as in flame photometry, much more is absorbed because, even at the temperature of the flame, the great majority of the atoms are still in their unexcited states. The light source is a monochromator or a hollow cathode lamp that contains the element which is being determined and therefore emits light of the characteristic frequency of that element. The absorption of this light by the element sprayed into the flame is measured by a photomultiplier by comparing the intensity transmitted through the flame before and during the spraying of the element being measured. Before measuring the unknown concentration the apparatus is calibrated by making similar measurements on standard solutions of the same element as in the case of flame photometry. Measurements are made with a different light source for each element to be determined. This technique is highly selective so it is useful where an element is to be determined in the presence of a large amount of another which might interfere in flame photometry.

Atomic Structure

In picturing the energy states of electrons in atoms Bohr used the same concept as Rutherford, namely, that of electrons revolving in orbits round the nucleus. The new idea which Bohr was able to introduce on the grounds of the evidence from atomic spectra was that of quantised energy levels for electrons. An electron could revolve round a nucleus for ever in a state of constant potential energy. It could not gradually lose energy as a result of this rotation because only by losing a whole quantum of energy at one time could its potential energy change at all. This is contrary to the classical view, and it would seem that, for particles of atomic dimensions, energy changes must follow quantum rules rather than classical rules.

In order to interpret all the details of atomic spectra in terms of fixed energy levels it was found necessary to define the energy of each electron by means of four quantum numbers, which together indicated the total energy. Although the ideas which these quantum numbers were intended to express have been modified by the subsequent development of wave mechanics, the nomenclature has been retained. The principal quantum number n gives the approximate energy and originally indicated the 'shell' to which the orbit belonged – the numbers 1, 2, 3, etc., corresponding to the K, L, M, etc., shells respectively. The subsidiary quantum number l, besides indicating which subdivision of the energy level the electron belongs to, also defines the shape of the orbit. In the Bohr theory it shows whether the electron moves round the nucleus in a circular or an elliptical orbit and, if the latter, how elliptical the orbit is. The magnetic quantum number m indicates the extent to which the electron's energy is modified if the atom is in a magnetic field, the modification being seen in atomic spectra as a splitting of the lines. Finally, there is the spin quantum number s which was originally intended to show whether the electron was spinning forwards or backwards relative to its motion round the nucleus.

The Bohr theory has two fundamental weaknesses. The first, which was realised from the beginning, was that there was no theoretical justification for the assumption which had to be made that energy levels of electrons in atoms are quantised. The second was only realised when various difficulties in explaining the properties of particles of sub-atomic dimensions led Heisenberg to propose his 'uncertainty principle'. In qualitative terms this states that, for such particles, it is not possible to know precisely both their position and their energy. The Bohr theory contradicts this principle because it defines both an exact orbit and an exact energy for each electron.

Both these weaknesses were overcome simultaneously when the ideas of wave mechanics were applied to electrons in atoms. These ideas are not easy to picture, and indeed the function which is in the form of a 'standing wave' and which describes mathematically the behaviour of an electron has no physical significance. However, the square of the amplitude of the wave at any given point near an atom (the amplitude being the height or depth of the wave measured from the mean level) represents the probability of finding the electron in a small element of volume at that point. This means that the position of an electron is never precisely specified since one can never do more than find regions where there is a greater probability than average of the electron being there.

The fact that only certain definite energy levels can occur in an atom arises automatically in wave mechanics because there are only certain wavelengths possible for a standing wave existing between specified limits. A useful analogy here is that of a violin string which

84   The Structure of Atoms

can only vibrate so that its length corresponds to half a wave, or a whole wave, or one and a half waves, etc., corresponding to the fundamental note and the various harmonics. In the case of electron waves in an atom, each mode of vibration corresponds to a particular energy. Thus, the restriction of wavelengths to certain definite values implies the existence of only certain definite energy levels.

It is found that three quantum numbers are necessary to define the energy of an electron in an atom according to the wave-mechanical theory. Since these correspond in their function to the quantum numbers n, l and m described above, the same nomenclature has been retained. Their significance in indicating the way in which the electron is distributed about the atomic nucleus is rather different, however. The principal quantum number n fixes the approximate magnitude of the energy of the electron as before. The subsidiary quantum number l mainly serves the purpose of defining the shapes of the regions which the electron has a high probability of occupying, but it also fixes the energy more closely. The magnetic quantum number m defines the orientation of these shapes in space. It only gives their orientations relative to each other, unless there is a magnetic field present to provide a reference direction against which to measure these orientations (in much the same way that the earth's magnetic field provides the reference direction of magnetic North on a map). The 'spin' quantum number enters into wave mechanics in a rather different way since the electron is no longer thought of as a spinning particle. It has to be introduced in connection with the symmetry of the wave function for an electron, but this will not be considered further here. It suffices to say that this quantum number can only have two values, as in the Bohr theory.

The quantum numbers are restricted in the values they can take by rules first deduced experimentally from the study of atomic spectra and then confirmed theoretically by wave mechanics. The rules are:

n may be any positive whole number
l may have positive whole number values from 0 to $(n-1)$. These values are often represented by letter symbols. For $l = 0, 1, 2, 3, 4$, etc., these are respectively s, p, d, f, g, etc., The s here should not be confused with the spin quantum number s mentioned below.
m may have negative and positive whole number values from $-l$, increasing by steps of 1, through 0, to $+l$
s can have only two values $+\frac{1}{2}$ and $-\frac{1}{2}$

The way in which electrons occupy the energy levels in atoms defined by the four quantum numbers is governed by three further rules:

(1) Except as determined by the subsequent rules, electrons in the normal state of the atom go into the lowest energy level available.

(2) Where there is a group of orbitals available of the same energy the electrons occupy them singly until all the orbitals are singly filled. This is a statement of Hund's rule.

(3) No two electrons have all four quantum numbers the same. This is Pauli's exclusion principle.

With these rules it is possible to deduce the electronic structures of successive elements of the Periodic Table in their free gaseous states.

Atomic Structure and the Periodic Table

When the principal quantum number, n, is 1, l can only be 0 and m can only be 0. So the one electron of the hydrogen atom goes into a $1s$ orbital. In helium the second electron can also go into the same orbital (same n, l, and m) provided it has a different value of the spin quantum number. This sub-shell and shell is then full, so in the third element, lithium, the third electron must go into a new shell with $n = 2$. This has two sub-shells, that with $l = 0$ and that with $l = 1$. The s sub-shell with $l = 0$ has lower energy so the third electron is to be found in this orbital. The electron configuration of lithium can therefore be written $1s^2 2s^1$. The fourth electron in beryllium also goes into the $2s$ orbital, giving the configuration $1s^2 2s^2$. The fifth electron in boron must go into a $2p$ orbital with $l = 1$. This time there are three $2p$ orbitals with $m = -1$, 0 and $+1$ respectively but in the absence of a magnetic field all three have the same energy. Thus in the next two elements, carbon and nitrogen, the two further electrons continue to fill singly the orbitals of different m values. This process is illustrated in Fig. 3.5. At oxygen with 8 electrons the electrons start to double up in the $2p$ orbital and fluorine has two doubly filled $2p$ orbitals. By the time neon is reached with 10 electrons all the $2p$ orbitals are doubly filled and the shell with $n = 2$ is also complete. This point represents the completion of the first short period of the periodic table.

The second short period commences by putting one further electron in the $3s$ orbital to form a sodium atom. The next element, magnesium, has a second electron in the $3s$ orbital. Then from aluminium to argon the electrons are fed into the $3p$ orbitals, half-filling the sub-shell at phosphorus and filling it at argon.

For $n = 3$, l can have the value 2, making it possible for the first time to have d orbitals giving rise to the first long period of elements. However, the $4s$ orbitals are lower in energy than the $3d$ orbitals so potassium, the next element after argon, does not have any electrons in the $3d$ orbitals but has one in the $4s$ orbital. Calcium then has two electrons in the $4s$ orbital but the next element, scandium, has its one further electron in a $3d$ orbital. There are five d orbitals that can hold 10 electrons so from scandium to zinc the successive electrons fill

Fig. 3.5 Electron configuration of the ground states of neutral gaseous atoms

Atom	1s	2s	2p	3s	3p	4s	3d
H	↑						
He	↑↓						
Li	↑↓	↑					
Be	↑↓	↑↓					
B	↑↓	↑↓	↑				
C	↑↓	↑↓	↑ ↑				
N	↑↓	↑↓	↑ ↑ ↑				
O	↑↓	↑↓	↑↓ ↑ ↑				
F	↑↓	↑↓	↑↓ ↑↓ ↑				
Ne	↑↓	↑↓	↑↓ ↑↓ ↑↓				
Na	As neon			↑			
Mg				↑↓			
Al				↑↓	↑		
Si				↑↓	↑ ↑		
P				↑↓	↑ ↑ ↑		
S				↑↓	↑↓ ↑ ↑		
Cl				↑↓	↑↓ ↑↓ ↑		
Ar				↑↓	↑↓ ↑↓ ↑↓		
K	As argon					↑	
Ca						↑↓	
Sc						↑↓	↑
Ti						↑↓	↑ ↑
V						↑↓	↑ ↑ ↑
Cr						↑	↑ ↑ ↑ ↑ ↑
Mn						↑↓	↑ ↑ ↑ ↑ ↑
Fe						↑↓	↑↓ ↑ ↑ ↑ ↑
Co						↑↓	↑↓ ↑↓ ↑ ↑ ↑
Ni						↑↓	↑↓ ↑↓ ↑↓ ↑ ↑
Cu						↑	↑↓ ↑↓ ↑↓ ↑↓ ↑↓
Zn						↑↓	↑↓ ↑↓ ↑↓ ↑↓ ↑↓

up the d orbitals, first singly as required by Hund's rule and then pairing up with opposite spin. The filling is not quite regular because the $3d$ and $4s$ orbitals are very similar in energy, so an electron is transferred at chromium and copper from the $4s$ to the $3d$ shell to give the particularly stable configuration of a half-filled or completely filled sub-shell. This filling of the $3d$ sub-shell while the $4s$ sub-shell remains approximately unchanged constitutes the first transition series of elements. After zinc the filling of the N shell is resumed by putting electrons in the $4p$ sub-shell, so there follows the sequence of normal group elements from gallium to krypton where the first long period is complete.

The second long period is built up very much like the first with electrons going first into the $5s$ orbital in the case of rubidium and strontium and then into the $4d$ orbitals from yttrium to cadmium before filling the $5p$ orbitals from indium to xenon. The elements from yttrium to cadmium therefore represent a second transition series.

At xenon the N shell is incomplete because the $4f$ orbitals have not yet been used and only the $5s$ and $5p$ orbitals of the O shell have been filled. Nevertheless, the next orbital in order of increasing energy is the $6s$ and this is filled with first one and then two electrons in caesium and barium. At lanthanum the filling of the $5d$ sub-shell is commenced but immediately afterwards, at cerium, electrons begin to enter the $4f$ shell and in the series of very similar elements, known as the lanthanides, which follow, the $4f$ orbitals are successively filled. Again there are slight irregularities due to the similarity in energy between the $4f$ and $5d$ levels, but this detail may be ignored. At hafnium the filling of the $5d$ orbitals is resumed and there follows another series of transition elements until mercury is reached when the $5d$ sub-shell is full. Thereafter, the $6p$ orbitals are filled from thallium to radon.

A similar sequence is followed in the last row of the Periodic Table. First the $7s$ orbital is filled, then at actinium an electron goes into the $6d$ orbital. The actinides follow in which the $5f$ sub-shell is being filled, with again some irregularities in which the $6d$ occupancy increases to 2 or falls to zero. After the completion of the $5f$ sub-shell the last few elements so far known in the Periodic Table continue to fill the $6d$ sub-shell.

The Electronic Theory of Valency

One of the most important consequences of knowledge of electron energy levels and configurations is that it provides an explanation of the valencies exhibited by the elements. In combining to form compounds, elements can form basically two types of chemical bond –

ionic and covalent. In forming the ionic bond one atom donates one or more electrons to another. The donating atom reduces the number of electrons round its nucleus, thereby forming an ion with a resultant positive charge. The accepting atom increases the number of electrons surrounding its nucleus so that it becomes an ion with a resultant negative charge. These positive and negative ions are then attracted to each other by electrostatic forces. In the solid state this attraction is strong, with the result that ionic substances have high melting points. Also, ionic substances will only dissolve in polar solvents, such as water, capable of diminishing the ionic attraction, and thereby forming discrete ions which move about independently in the solution. If it is necessary to emphasise that a compound is ionic, it is represented by indicating the charges on the ions, thus: Na^+Cl^-.

The covalent bond is formed by atoms sharing electrons via the overlap of atomic orbitals to form molecular orbitals. In many compounds, bonding molecular orbitals are localised in the region between pairs of atoms and one bonding molecular orbital is formed by the overlap of two orbitals, one from each atom. As in the case of atomic orbitals, each molecular orbital can contain up to two electrons of opposite spin. A chemical bond is therefore visualised as a pair of electrons in a localised molecular orbital. The attraction of the pair of electrons by two nuclei instead of one and the smearing of electron density over an increased volume of space both give rise to a reduction in energy of the system. A bond will therefore always tend to form when there are two electrons in two atomic orbitals that can overlap. Usually the two electrons come one from each atom but it is also possible to form a covalent bond from an electron pair in one atomic orbital and an empty orbital of low enough energy on another atom. The bond is then known as a dative covalent bond or a coordinate bond because the electron pair is donated by one atom. In forming a dative covalent bond the donor atom decreases in electron density and becomes partially positively charged, whereas the acceptor atom gains electron density and becomes partially negatively charged. A dative covalent bond is therefore polar in nature but the charges are not as large as those developed in forming ionic compounds.

With these principles in mind and with the aid of Fig. 3.5 we can now understand the characteristic valencies exhibited by atoms. Hydrogen has one electron in a $1s$ orbital. In forming covalent compounds, this $1s$ orbital can overlap with another half-filled orbital on another atom to form a molecular orbital into which is placed the pair of electrons – one electron from the hydrogen atom and one from the other atom. All the other possible orbitals of hydrogen are too high in energy to be involved in further bonding so the covalency of hydrogen is one. The $1s$ electron can be donated to another atom in forming ionic compounds but the proton H^+ so formed has such a small size that it exerts powerful electric fields which cause dative covalent bonding from solvent molecules. The proton is therefore never found alone except in gaseous ion beams but is associated in solution with at least one solvent molecule. However, by donating only one electron the ionic valency of hydrogen is also one.

Helium has two electrons in the $1s$ orbital so it could only form a covalent bond by first promoting one of its electrons to the $2s$ orbital, but this would involve too much energy. Similarly, the $1s$ orbital is too stable for ionisation to take place except with the expenditure of a great deal of energy so helium does not form ionic compounds. It is therefore chemically inert and is a member of the group of noble gases.

Lithium has a completed $1s$ orbital and one electron in a $2s$ orbital. Due to the partial shielding of the positive nuclear charge by the completed $1s$ orbital the electron in the $2s$ orbital is not tightly bound and is readily ionised off. Lithium therefore readily forms a monovalent cation. The same is true for all the alkali metals because they each have one electron in an s orbital of higher principal quantum number that the completed sub-shell of the proceeding element. In each case, then, this one s electron is well shielded by the completed shells and sub-shells of lower principal quantum number and is readily ionised. The alkali metals therefore typically form ionic compounds in which they bear a single negative charge. Covalent bonding, though uncommon, is possible, e.g. in the diatomic gaseous molecules Na_2, etc., and takes place by the overlap of the $1s$ orbital with one other orbital, again making the alkali metal monovalent.

Beryllium has the electron configuration $1s^2 2s^2$, and since the $2s$ orbital is full it might be expected that beryllium would be inert like helium. In this case, however, one of the $2s$ electrons can readily be promoted to a $2p$ orbital only slightly higher in energy and there are then two half-filled orbitals ready to form two covalent bonds. The reduction in energy on forming the two bonds more than repays the energy needed for the promotion of the electron to make the bonding possible. Beryllium is therefore divalent. The alternative way of forming compounds by ionisation does not take place because of the rather large ionisation energy for beryllium. However, the heavier elements in the same alkaline earth group have lower ionisation energies and tend to be more ionic in their compounds than covalent. They are still divalent because both $2s$ electrons are ionised off.

The additional electron in boron goes into a $2p$ orbital but, as in the case of beryllium, not much energy is required to raise one of the $2s$ electrons into a second $2p$ orbital to give three singly occupied orbitals that can overlap to form three bonds. Boron is therefore trivalent. It is not readily ionised and therefore forms covalent compounds. The other elements in group III which also have $s^2 p^1$ ground states ionise more readily and form mainly ionic trivalent compounds. Another interesting feature, however, is that on descending the group, promotion or ionisation of the s electron becomes more difficult and thallium in particular forms both univalent and trivalent compounds.

Carbon is like beryllium and boron in obtaining its normal valency state by promotion of a $2s$ electron into the $2p$ sub-shell. This gives three unpaired p electrons (in agreement with Hund's rule) and one unpaired s electron so the valency is four. Again ionisation is difficult, so covalent compounds are formed. Throughout the whole of group IV covalency is dominant but on descending the group the promotion energy for the s electron becomes comparable with the strengths of the two additional bonds that can be formed thereby, so the tendency to form divalent compounds increases.

The ground state of nitrogen has the electron configuration $2s^22p^3$, giving three singly occupied p orbitals available for bond formation. To increase the valency above 3 would require the promotion of a $2s$ electron to the next available empty orbital which is $3s$. This is too high in energy for this process to take place, so nitrogen is trivalent. Loss of electrons to form positive ions is not energetically feasible but it is possible for nitrogen to gain electrons to attain the favourable state of a completed shell. This can happen when the atom is alone, to form the nitride ion N^{3-}, or when it is has already formed one or two electron pair bonds, as in the ions NH_2^- and NH^{2-}. Of the other elements in group V, phosphorus resembles nitrogen in forming mainly covalent compounds but arsenic, antimony and bismuth are progressively more metallic in nature with some tendency to form ionic compounds containing the $3+$ ion obtained by removing the outermost p electrons.

Oxygen has four electrons in the $2p$ sub-shell so two of them must be paired. This leaves two unpaired p electrons available for covalent bond formation making oxygen divalent. However, in many oxides oxygen exists as the O^{2-} ion in which a stable completed shell structure has been achieved by accepting two electrons. The completed shell may also be achieved partly by covalent bond formation and partly by electron acceptance as in the hydroxyl ion OH^-. The rest of the group VI elements, S, Se, Te and Po, form mainly covalent compounds though the ions such as S^{2-} can exist in compounds with highly electropositive metals, and ions such as HS^- are known.

Fluorine is only one electron short of the completed shell configuration $1s^22s^22p^6$, and since this structure is a particularly stable one the chemistry of fluorine is dominated by its high electron affinity. It readily forms the monovalent F^- ion. Covalent compounds of fluorine are also common in which the completed shell structure is achieved by overlap of its singly occupied p orbital with another singly occupied orbital. The halogens generally show these two characteristics with the electron affinity decreasing on descending the group.

At neon the stable completed shell structure has already been achieved in the isolated atom so the element is chemically inert. To achieve a state where a singly filled orbital is available for overlap with another singly filled orbital in covalent bond formation the promotion of one of the $2p$ electrons to a $3s$ orbital would be necessary. This would require more energy than would be repaid in bond formation so it does not take place. In the same way, all the elements in the noble gas group have their s and p sub-shells filled with electrons and the promotion energy to the next available orbital is too high for compound formation to take place except in special circumstances.

One of the most important properties of the transition elements is that they have variable valencies. These arise because of the similarity in energy between the d electron levels that are in the process of being filled in the series of elements in question and the s and p electrons of the next higher principal quantum number. So, in forming ionic compounds, electrons can be ionised off not only from the s sub-shell but also to varying extents from the d sub-shell. For forming covalent bonds, electrons can readily be promoted to the p orbitals so that singly filled d, s and p orbitals are available for overlap with singly filled orbitals on other atoms. In many compounds of transition elements known as coordination compounds, the bonding is best visualised as involving first the conversion of the metal to the appropriate oxidation state followed by the formation of co-ordinate bonds by donation of electron pairs from the surrounding anions or molecules to empty orbitals on the ion or atom of the transition element.

Directional Bonds

From the chemical point of view one of the most significant advantages of the wave-mechanical theory over the Bohr theory is that it automatically introduces the concept of directional properties of covalent bonds when atoms are joined to form molecules. This topic may be introduced conveniently here because it is intimately concerned with the shapes of atomic orbitals, but a more general discussion of molecular structure is to be found in the next chapter.

The shapes of s, p and d orbitals are shown in Fig. 3.6. It can be seen that, when $l = 0$, there is an s orbital which has a spherical probability distribution. The probability of finding the electron within this spherical region varies according to the distance from the atomic nucleus at the centre, and in a manner which depends upon the value of the principal quantum number n. The spherical symmetry of the s orbital means that the probability variation measured outwards from the centre is the same in any direction, so it can overlap

92 The Structure of Atoms

Fig. 3.6 Shapes of s, p and d orbitals

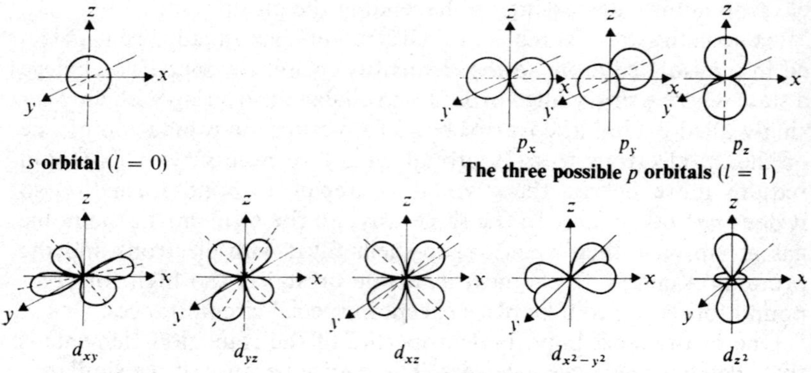

s **orbital** ($l = 0$)

p_x p_y p_z
The three possible p orbitals ($l = 1$)

d_{xy} d_{yz} d_{xz} $d_{x^2-y^2}$ d_{z^2}

The five d orbitals ($l = 2$)

with an atomic orbital on a neighbouring atom equally well in any direction. When $l = 1$ there are three p orbitals which have a dumb-bell shape, so the variation in probability is not the same in all directions but has high values only in a dumb-bell shaped region. When $l = 2$ the d orbitals have high electron probabilities, in all cases but one, along four directions at right angles. For all shapes of orbitals and for all modes of variation the probability of finding an electron in a small volume element at a particular point drops to zero as the point considered moves to large distances from the nucleus.

The three p orbitals point along three axes at right angles to each other. Hence, when bonds are formed by overlap of two or three p orbitals on the central atom with orbitals on two or three neighbouring atoms, strong overlap will only be achieved along the x, y, or z directions of the p orbitals. The bonds formed will therefore be at right angles to each other. This is approximately what happens when orbitals of the oxygen atom overlap with orbitals of two hydrogen atoms to form a water molecule. The oxygen atom has the configuration $1s^2 2s^2 2p^4$ so two of the p orbitals are singly occupied and one is doubly occupied. On the simple view of bonding described in the previous section, only the two singly occupied p orbitals are available for overlap with the hydrogen $1s$ orbitals to form O—H bonds. If these two p orbitals are taken as p_x and p_z the bonding may be pictured as in Fig. 3.7(a). The principle that the strongest bonding is obtained when there is maximum overlap of atomic orbitals thus leads to the prediction that the H—O—H bond angle in water should be 90°. (Note that it is not possible for two hydrogen $1s$ orbitals to overlap opposite lobes of the same $2p$ dumb-bell. Once one lobe has formed a bond the probability distribution of electrons changes markedly and the opposite lobe practically disappears.)

The H—O—H bond angle actually observed in structural studies is 105° and the discrepancy between this and the predicted 90° suggests that the scheme of overlap of Fig. 3.7(a) is an over-simplification of the problem. The oversimplification arises from considering the 2s and $2p_y$ orbitals, which contain a pair of electrons each in the isolated atom, as being excluded from the bonding scheme. They have similar energies to the $2p_x$ and $2p_z$ orbitals so it is more likely that all four orbitals will be involved in bonding. Provided all the electrons in the molecule can be accommodated suitably in the molecular orbitals that are formed by overlap of the atomic orbitals it does not matter whether they

Fig. 3.7 Two approximate representations of bonding in H_2O
(a) using pure p orbitals   (b) using sp^3 hybrid orbitals

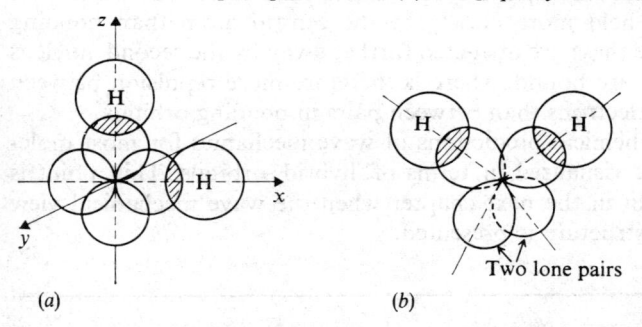

came from filled or partly filled orbitals in the isolated atoms. It is difficult to visualise where two hydrogen atoms should be placed for their orbitals to overlap most effectively with all three 2p orbitals and the 2s orbital. Pauling suggested that the way out of this difficulty is to visualise a new set of four oxygen orbitals equivalent to the original set but made by mixing or 'hybridising' the original orbitals in an appropriate way. When four equal sp^3 hybrid orbitals are produced by hybridizing one s and three p orbitals in this way they point towards the corners of a regular tetrahedron. Each consists mainly of a pear-shaped lobe which is more elongated than a p orbital and so gives better overlap with a hydrogen 1s orbital in bond formation. On this basis the water molecule would be pictured as in Fig. 3.7(b) with two lobes overlapping with hydrogen 1s orbitals and two filled with a lone pair of electrons each. According to this pictorial approximation the H—O—H bond angle would be expected to be 109° and this is much closer to the observed value than the 90° of Fig. 3.7(a). It is interesting to note that when a water molecule accepts two hydrogen bonds from other molecules (see p. 31) these hydrogen bonds make angles of about 109° with the covalent O—H bonds. This suggests that the attractive force of a hydrogen bond is dependent upon the electrons in the 'lone pair' orbitals in particular rather than on the electronegativity of the oxygen atom in general.

94   The Structure of Atoms

It is likely that the hybridisation of the oxygen orbitals in the water molecules is not quite the regular tetrahedral arrangement described above but is such as to give slight inequality between the lone pair and the bonded lobes. The proportions of s and p character in any one hybrid can be varied as the situation demands anywhere between pure s and pure p provided that the sum of the s contributions in all the hybrids totals one and the sum of the p contributions totals three. By appropriate choice of s and p character it is possible to obtain hybrids such that one pair of equal lobes occur at the required angle of 105° for forming bonds with the hydrogen atoms. The other two equal lobes that will contain the lone pairs of electrons must then have an angle of 115° between them. This is consistent with what is required on the grounds of repulsive forces between electrons. Lone pair electrons are held more closely to the central atom than bonding electrons since these are attracted further away by the second nucleus to which they are bound. There is therefore more repulsion between lone pairs of electrons than between pairs in bonding orbitals.

The stereochemical predictions of wave mechanics for most molecules must be visualised in terms of hybrid orbitals. This point is taken up again in the next chapter when the wave mechanical view of molecular structure is presented.

Problems

1 Thorium, of mass number 232 and atomic number 90, eventually forms a stable disintegration product after undergoing six alpha-ray changes and four beta-ray changes. Identify the nuclide formed (from its atomic number), and calculate its mass number.

2 Calculate the mass defect in forming ^{56}Fe of accurate isotopic mass 55.934 94 from hydrogen atoms of mass 1.007 825 m_u and neutrons of mass 1.008 665 m_u. Hence calculate the binding energy per nucleon in MeV.

3 Write out in full (showing mass numbers and nuclear charges) the equations corresponding to the following abbreviated forms for nuclear reactions, supplying by calculation the missing information indicated by x or y in each case:

$$^6\text{Li}(n, \alpha)x; \quad ^9\text{Be}(\alpha, y)\,^{12}\text{C};$$
$$^{130}\text{Te}(n, x)\,^{131}\text{Te} \rightarrow y + \beta^-;$$
$$^{59}\text{Co}(n, \gamma)x \rightarrow \,^{60}\text{Ni} + y + \gamma.$$

4 A specimen which originally contained 12×10^{20} atoms of ^{14}C undergoes radioactive decay, and the disintegration constant λ is 1.36×10^{-4} year^{-1}. Calculate the time which elapses before the number of atoms has dropped to 6×10^{20}, 4×10^{20}, 3×10^{20}, and

2×10^{20}, respectively. Hence, plot a graph of the number of atoms remaining versus time.

5 The nuclides $^{218}_{84}$Po, $^{214}_{82}$Pb, and $^{214}_{83}$Bi, which are among the elements produced by the radioactive disintegration of radium, have half life periods 3.05, 26.8, and 19.7 minutes, respectively. Calculate the disintegration constants for these three nuclides, and hence their relative abundances in a radium-bearing mineral.

6 Write the electronic structures in the form $1s^2 2s^2 2p\ldots$ for the following atoms: C, P, Ar, Fe, As.

4 The Structure of Molecules

The Basis of Structural Chemistry

We have seen already how the valency of an atom can be explained in terms of its electronic structure, and how this leads to an understanding of the formation of three types of chemical bond. Historically, this explanation of the situation was not forthoming until after the concept of valency had become highly developed, largely through its importance in interpreting the nature of organic compounds. Compared with the majority of inorganic compounds the molecular formulae of organic compounds are highly complex, and show an infinite variety of combining ratios. However, once the typical quadrivalency of carbon had been established, the way was opened for an interpretation of the molecular formulae in terms of the typical valencies of the elements involved. It was Kekulé who led the way in clarifying the situation by proposing the chain form of aliphatic compounds, and the ring form of aromatic compounds. He represented each covalent link between a pair of atoms as a chemical bond, drawn as a stroke in a diagram of each molecule. In this way he originated the method of drawing bond diagrams which was soon universally adopted. Kekulé made a further advance by assuming that, when carbon is linked to less than four other atoms, the quadrivalence is retained by the formation of multiple bonds. Thus ethylene (ethene) should be (I) with a double carbon-carbon bond and not (II), which would involve tervalent carbon.

$$(I)\ \begin{array}{c} H \\ H \end{array}\!\!\!>\!C\!=\!C\!<\!\!\!\begin{array}{c} H \\ H \end{array} \qquad (II)\ \begin{array}{c} H \\ H \end{array}\!\!\!>\!C\!-\!C\!<\!\!\!\begin{array}{c} H \\ H \end{array},$$

Kekulé's well-known explanation of the structure of benzene was that it could adopt either of the two forms shown in Fig. 4.1.

Further, to overcome the objection that either of these two forms should give four isomeric disubstitution products, whereas only three are known, he postulated that the molecule exists in a state of rapid oscillation between the two forms. This was the first exposition of the concept of resonance but it is now known that the idea of oscillation

Fig. 4.1 The Kekulé formulae for benzene

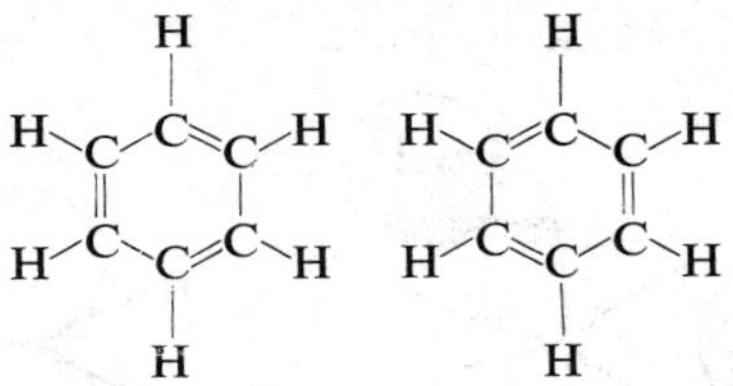

is erroneous. Each bond is permanently in a form intermediate between its extreme types and the properties of the molecule are to some extent intermediate between those of the extreme forms.

The criteria which must be satisfied before resonance can occur are that (a) the atomic nuclei in all the various extreme bond diagrams must occupy very nearly the same positions, and (b) the structures represented by the various extreme bond diagrams must all be of about the same energy. (Energies of molecules are discussed in a later section, p. 279). Both these criteria are satisfied if the same atoms are linked together in all the extreme structures and the only changes made are in the type of link.

Wave-mechanical Explanation of Organic Structures

It was mentioned on p. 90 that carbon attains its normal valency state by promotion of a $2s$ electron to a $2p$ orbital so that it has the configuration $2s^1 2p^3$ and a valency of four. The way the electrons distribute themselves about the carbon nucleus in forming four bonds depends very much on the environment. In order to form equal or almost equal bonds to four surrounding atoms the s and the three p orbitals are hybridised in an sp^3 manner, like those of the oxygen atom in Fig. 3.7(b). In this way four lobes are formed pointing towards the corners of a regular tetrahedron giving the characteristic arrangement of bonds found in all saturated carbon compounds.

In unsaturated compounds, such as ethylene, however, where the carbon atom is surrounded by three other atoms, a more appropriate form of hybridisation is sp^2. Since three atomic orbitals are mixed in forming the hybrids, three hybrid orbitals must result. In this arrangement the lobes of the three sp^2 hybrid orbitals point towards the corners of an equilateral triangle. In ethylene they will therefore overlap in these directions with the $1s$ orbitals of each of two hydrogen atoms and with another sp^2 lobe from the other carbon atom, as shown in Fig. 4.2(a). The sp^2 hybridisation still leaves one p electron on each carbon atom unhybridised. Each will be in a p orbital at right-angles to the two p orbitals that were mixed in forming the hybrids It must therefore be at right-angles to the plane of the three sp^2 hybrids. The two p orbitals on the two carbon atoms can now

98 The Structure of Molecules

Fig. 4.2 Overlap of orbitals in the ethylene molecule
(a) in the plane of the molecule
(b) at right angles to the plane

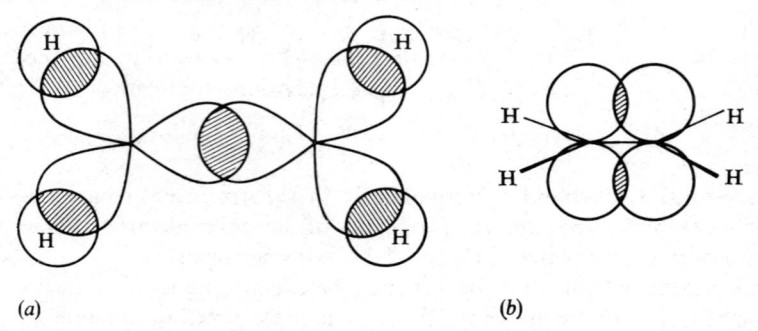

overlap sideways to form a further bond between the carbon atoms, as shown in Fig. 4.2(b). This is how a double bond is formed. There can only be strong overlap, however, if the two overlapping p orbitals are parallel. This is true when the two sp^2 hybrid planes are coplanar, i.e. when all the atoms of the molecule are coplanar. Twisting of one CH_2 plane relative to the other will be strongly resisted in contrast to the fairly free rotation about single bonds. A bond formed by sideways overlap of p orbitals is called a π (pi) bond whereas the end-on overlap of the sp^2 hybrids to give a molecular orbital which has cylindrical symmetry round the bond direction gives rise to what is called a σ (sigma) bond.

When a carbon atom bonds in a linear fashion to one atom on either side of it the appropriate hybrid orbitals are sp. The two combinations of s and p orbitals that make the pair of sp hybrids are $(s+p)$ and $(s-p)$ and these can be visualised readily as in Fig. 4.3. The $(s+p)$ orbital has the correct shape for good overlap with another orbital to the right of the carbon atom and $(s-p)$ overlaps well with an orbital to the left. This type of hybridisation leaves two p orbitals

Fig. 4.3 Formation of sp hybrid orbitals

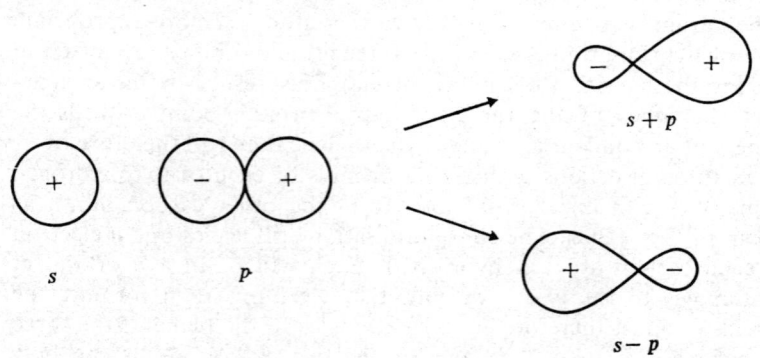

Wave-mechanical Explanation of Organic Structures 99

on the carbon atom unhybridised and orientated along axes at right angles to the axis of the sp hybrids. In acetylene (ethyne), for example, one sigma bond is formed between the carbon atoms by overlap of sp hybrid orbitals on each atom as in Fig. 4.4 and then two further π bonds are formed by the sideways overlap of the two p orbitals on each atom. This is how triple bonds are formed.

Fig. 4.4 Overlap of sp hybrid orbitals and hydrogen orbitals in acetylene

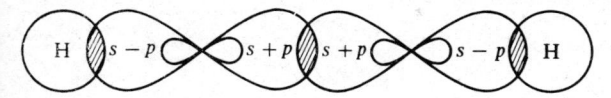

H ⬭ $s-p$ ⬭ $s+p$ ⬭ $s+p$ ⬭ $s-p$ ⬭ H

Compounds that have to be represented in bond diagrams in terms of resonance between alternative structures have to have a non-localised description of the bonding in wave-mechanical terms. Benzene provides a typical example. Since each carbon atom is surrounded by a regular arrangement of three other atoms the basic sigma bonds are formed, as in ethylene, by overlap of sp^2 hybrid orbitals. These leave a p orbital, orientated at right-angles to the molecular plane, at each carbon atom. Each of these p orbitals can overlap equally well with the p orbital of the atom on either side of it so localised bonds between pairs of carbon atoms are not formed. Instead, molecular orbitals which each stretch over the whole carbon skeleton are formed by taking sums and differences of the atomic p orbitals. Since we start with 6 p orbitals we must form 6 molecular orbitals, three of which will be bonding (lower energy than separated atoms) and three antibonding (higher energy than separated atoms). The 6 electrons from the atomic p orbitals all fit pairwise into the three bonding molecular orbitals which are illustrated schematically in Fig. 4.5. Even in

Fig. 4.5 The three bonding molecular orbitals in benzene

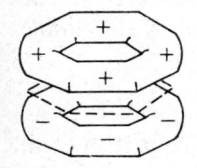

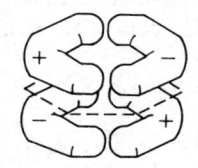

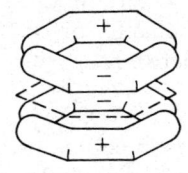

the second and third orbitals, which each have a break where the wave function passes through zero, it is the same orbital that covers the whole of the carbon skeleton and is therefore delocalised. This is why chemical modifications to one position in the molecule so readily have an effect also at remote positions in the molecule. Delocalisation of electrons occurs whenever there is conjugation, i.e. an alternation of single and multiple bonds. Illustrations of this are butadiene, CH_2=CH—CH=CH_2, where the central bond has considerable double-bond character, and vitamin A:

```
      CH₃ CH₃
        \ /
   CH₂—C
   /     \            CH₃              CH₃
  CH₂     C—CH=CH—C=CH—CH=CH—C=CH—CH₂OH
   \     //
   CH₂—C
        |
        CH₃
```

where the resonance arising from the conjugation over practically the whole length of the molecule accounts for the yellow colour of the compound.

Dipole Moments

Besides having resonance between single and double bonds, it is also possible for resonance to take place between covalent and ionic structures, thus:

$$\text{H—Cl} \leftrightarrow \text{H}^+\text{Cl}^-.$$

The effect of this is to give a certain amount of ionic character to the covalent bond, by causing a drift of electrons from the hydrogen atom towards the chlorine atom. The actual state of the molecule may be represented thus: $\overset{\delta+}{\text{H}}$—$\overset{\delta-}{\text{Cl}}$ or H–|→Cl where δ signifies a partial charge, of magnitude only a fraction of the charge of the fully ionised form, or –|→ indicates the direction of the electron drift. The molecule is therefore a dipole, i.e. from an electrical point of view it may be regarded as a pair of opposite charges $+Q$ and $-Q$ whose centres are separated by a certain distance l. A quantitative measure of the separation of charge, or of the electron drift, is given by the dipole moment μ where

$$\mu = Ql.$$

In SI units Q is in coulombs, C, and l is in metres, m, so μ is in C m. The former unit for dipole moments, still commonly quoted, is the Debye unit, D, given by

$$\mu(\text{in D}) = Q\ (\text{in e.s.u.}) \times l(\text{in cm}) \times 10^{18}.$$

Since 1 e.s.u. = 3.336×10^{-10} C and 1 cm = 10^{-2} m

$$1\ \text{D} = 3.336 \times 10^{-10} \times 10^{-2} \times 10^{-18} = 3.336 \times 10^{-30}\ \text{C m}.$$

Thus two opposite charges each equal to the charge on an electron (1.602×10^{-19} C) separated by 0.1 nm (1 Å) corresponds to a dipole moment of 1.602×10^{-29} C m or

$1.602 \times 10^{-29}/3.336 \times 10^{-30}$ D = 4.80 D.

It is possible to measure the dipole moment of a molecule by observing the effect of the substance on the strength of an electric field. The drift of electrons from one atom to another in a covalent bond always takes place when the two atoms are different. This arises because of their different electronegativities. The concept of electronegativity is most easily approached from a consideration of the structures of atoms. An inert gas structure of electrons has the effect of 'screening' the positive charge on the nucleus so that it exerts very little attraction on external negatively charged particles. This accounts for the unreactive nature of the inert gases themselves, but it also explains the ease with which an alkali metal atom loses the one extra electron to form a positive ion. The alkali metals are all, therefore, very electropositive in character. Moving from left to right across the Periodic Table, the balance is altered between two conflicting effects which determine the ease with which an atom loses or gains electrons. The screening effect of a completed shell is offset, as we move to the right, by the increasing charge on the atomic nuclei. Successive atoms therefore become progressively less and less electropositive, or more electronegative, until the halogens are reached, so that the most electronegative element immediately precedes the inert gases. Superimposed on this variation across the Periodic Table there is also a variation down each group. The more electrons there are around a nucleus, the smaller will be the attraction exerted by the nucleus on its external valency electrons. Therefore, we expect the electropositive nature to increase as we move down a group. This is found to be so, since the alkali metals become more electropositive and the halogens less electronegative with increasing atomic number. Combining the horizontal with the vertical variation, we can say that electronegativity will increase from the bottom left-hand corner of the Periodic Table to the top right-hand corner, excluding the inert gases.

Bearing in mind this general relationship, it is usually possible to predict the direction of electron drift in a bond joining any two atoms, and also to obtain some idea of its magnitude. The total effect of electron drifts in all the bonds of the molecule will combine to give the resultant dipole moment. Although it is not strictly justifiable to regard the effects in the separate bonds as acting independently of each other, it is found empirically that the dipole moment of a molecule can be predicted fairly accurately by combining bond moments for all the various bonds. This combination is carried out by vector addition – a process which can best be understood by reference to a drawing of a molecule with a diagram of the bond moments drawn to scale (Fig. 4.6).

The Structure of Molecules

Fig. 4.6 Vector addition of bond moments

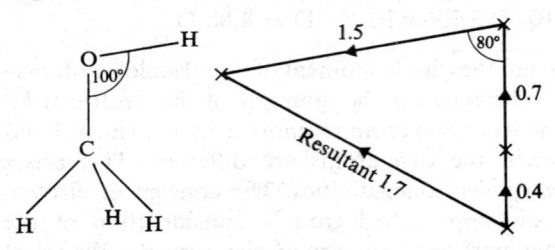

Starting from a point, successive lines are drawn proportional in magnitude to, and in the same direction as, the bond moments, taken in turn. Thus, the value of 0.4 is taken as the combined moment of the CH_3 group (in Debye units) acting in the direction of the C—O bond. Next comes the moment for the C—O bond itself, 0.7 in the same direction. The moment of the O—H bond acts in the direction from H to O, since O is the more electronegative. It is therefore represented by a line of length 1.5 units sloping diagonally from right to left at the appropriate angle to the C—O bond. The combined effect (or resultant) of all these contributions is given in magnitude and direction by the line joining the first point to the last. In this case, the total moment is found to be 1.7 D, in agreement with the experimentally determined value. The result may also be obtained by trigonometrical calculation, without recourse to scale drawing.

The concept of bond moments, and their vector addition, is important from the point of view of the determination of molecular structure, since it is possible to deduce an unknown bond angle from the known bond moments and the experimentally determined dipole moment. This process is essentially the same as that of vector addition illustrated above, but with the resultant total known and one of the angles unknown. However, there is usually some uncertainty in the experimental dipole moments, and this, combined with the approximation inherent in assuming bond moments, means that the bond angles so obtained cannot be regarded as being accurate. Indeed, the application of wave mechanics to the structure of molecules has shown that there are other effects contributing to the total dipole moment of a molecule besides the electron drifts in the bonds themselves. For example, in the water molecule a large part of the dipole moment is due to the lone pairs of electrons on the oxygen atom (see Fig. 3.7(b). There is therefore no theoretical justification for the empirically determined bond moments, and bond angles derived from them must be accepted with caution.

The simplest and most reliable deduction which can be made about molecular structure from dipole moment evidence is whether a molecule is completely symmetrical or is unsymmetrical, because in the former case the dipole moment must be zero. Thus we have, for two triatomic molecules:

$$H\diagdown O\diagup H \qquad O-C-O$$
$$\mu = 1.84\ D \qquad\qquad \mu = 0$$

A further important example is in *cis-trans* isomerism where the *cis*-isomer possesses a dipole moment but the *trans*-isomer does not. Other isomers which may be distinguished by dipole moments are *o-* *m-* and *p-* disubstituted benzenes:

$$\mu = 2.5\ D \qquad \mu = 1.7\ D \qquad \mu = 0$$

Certain atoms or groups, e.g. —Cl, —NO$_2$, when attached to a carbon atom withdraw electrons from it becoming the negative end of the dipole and are said to exert a negative inductive effect (—I effect). Other atoms or groups, e.g. —CH$_3$, attached to a carbon atom become the positive end of a dipole and are said to exert a positive inductive effect (+I effect). The sign and magnitude of the inductive effect is important in understanding chemical reactivity.

Refractive Index and Molar Refraction

The dipole moment is a measure of the permanent electron drift in a molecule. However, temporary electron drifts can be caused in any molecule, whether it has a permanent dipole moment or not, by induction from varying electric fields. These give rise to an induced moment, which is larger the more polarisable the molecule. The polarisability can be studied most conveniently by making use of the alternating electric field produced by light waves. The result of the interaction between the electric field of the light waves and the electrons in the molecules is to give rise to a certain refractive index n. Refractive index is given quantitatively by the extent to which a ray of light is bent on passing from a vacuum into the medium (see Fig. 4.7). If the angle between an incident ray and the normal to the surface where the ray enters the medium is i and the angle between the refracted ray inside the medium and the normal is r then the refractive index n is given by

$$n = \sin i/\sin r.$$

Fig. 4.7 Bending of light by refractive index effect

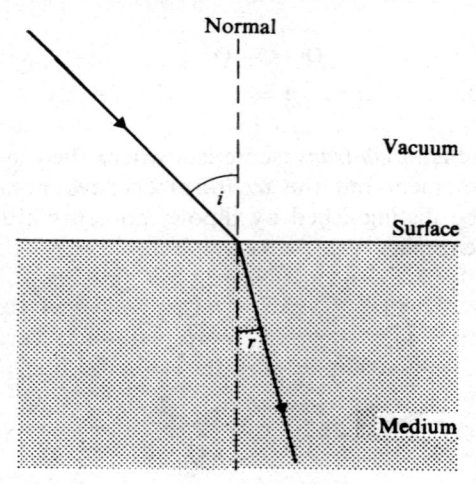

The interaction between the light and the medium also brings about a reduction in the velocity of the light as it travels through the medium. This is related to the refractive index by

n = velocity of light in the medium/velocity in vacuo.

The velocity of light in air is sufficiently close to its velocity in a vacuum to be able to write, when the medium is a liquid or a solid,

$n \approx$ velocity of light in the medium/velocity in air.

The polarisability of a molecule at the frequency of the light is given by the molar refraction R_m where

$$R_m = \frac{n^2-1}{n^2+2} \cdot V_m.$$

Here, V_m is the molar volume (the volume occupied by 1 mole of the substance) equal to the molar mass divided by the density. Since n is dimensionless R_m has the dimensions of volume divided by amount of substance and is measured in m^3 mol^{-1} or in cm^3 mol^{-1}. Both V_m and R_m are additive quantities and can be regarded as the sum of contributions from the separate parts of a molecule. In the case of R_m contributions from the separate bonds, known as bond refractions, are summed. However, when the molecule contains conjugated multiple bonds the extra electron mobility causes the molar refraction to be greater than the sum of the bond refractions.

Besides being related to molecular structure in the way just described, refractive index also has analytical applications. Since it is a property that can readily be measured for liquids with high precision (with a relatively simple optical instrument called a refractometer) it can be used to confirm the identity of a pure liquid compound. When used in this way it is advisable to measure the material under test, and a standard of the substance it is suspected of being, on the same refractometer under the same conditions. A more frequent application is as a criterion of purity because small quantities of impurities usually cause significant changes in refractive index. In the *British Pharmacopoeia* a range within which the refractive index must lie is often quoted for a liquid as a standard of purity. Liquid mixtures have refractive indices which lie between those of the components so, once a calibration graph has been prepared for a given pair of liquids, a measurement of the refractive index will allow the composition of any mixture of the two liquids to be determined. In all these applications the refractive indices must be measured at constant temperature by passing water from a thermostat tank through the refractometer, because refractive indices vary markedly with temperature.

Optical Rotatory Dispersion and Circular Dichroism

Optical activity of a molecule is its ability to rotate the plane of plane polarised light. Unpolarised light vibrates in all directions perpendicular to its direction of travel. After passing through a polariser (a piece of Polaroid or a Nicol prism) plane polarised light emerges in which vibrations only occur in one direction. In a polarimeter, the instrument used to measure the rotation of plane polarised light (Fig. 4.8), there is an analyser which also consists of polarising

Fig. 4.8 Components of a polarimeter and directions of vibration of light

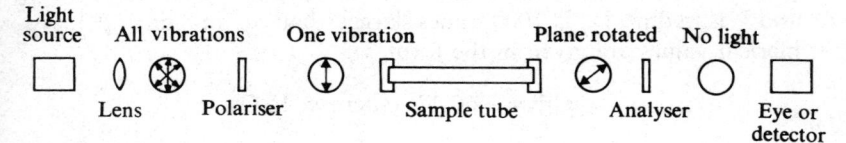

Light source	All vibrations	One vibration		Plane rotated	No light
☐	◊ ⊛	▯ ⊕	▭	⊘ ▯	○ ☐
	Lens Polariser		Sample tube	Analyser	Eye or detector

material which will permit the passage of light vibrating in one direction only. When there is no optically active material between the polariser and the analyser no light will emerge from the analyser when it is turned so that its plane of polarisation is at right angles to that of the polariser. If an optically active material is placed between the polariser and the analyser it rotates the plane of plane polarised light as it passes through the material. The angle of rotation α is measured by recording the angle through which the analyser must be turned to restore the situation where no light emerges. The optically active material is usually a solution in a sample tube. If the solution consists of a mass m of solute dissolved in a volume V of solvent and the tube is of length l, the specific optical rotatory power α_m of the substance is given by

$$\alpha_m = \alpha V/ml.$$

If α is measured in radians, m is in kg, V is in m³ and l is in m, then α_m has the SI units rad m² kg⁻¹. More often α is measured in degrees, m in g, V in dm³ and l in dm, so α_m is in ° dm² g⁻¹. Thus for sucrose in water at 20° C the specific optical rotatory power for sodium D light would be recorded as

$$\alpha_m \text{ (sucrose, H}_2\text{O, 293.15 K, 589.3 nm)} = +0.066\ 47\ °\ \text{dm}^2\ \text{g}^{-1}$$
$$(= 66.47\ °\ \text{dm}^2\ \text{kg}^{-1}).$$

In order to compare rotations for compounds of different molecular weights it is better to use molar optical rotatory power α_n given by

$$\alpha_n = \alpha V/nl$$

where n is the amount of substance (in moles). Since $n = m/M$ where m is the mass of substance (in kg) and M is the molar mass (in kg mol⁻¹), it is clear that $\alpha_n = \alpha_m \times M$ if the mass units for α_m are kg. Thus for sucrose, α_n (sucrose, H₂O, 293.15 K, 589.3 nm) $= +66.47 \times 0.3423$ ° dm² mol⁻¹ $= 22.753$ ° dm² mol⁻¹.

In the old notation which will be found in most books the quantity corresponding to specific optical rotatory power is called specific rotation and is given the symbol $[\alpha]$. It is defined as

$$[\alpha] = 100\alpha/lc,$$

where l is in decimeters and c is the concentration of the solution in g/100 cm³. Since c in these units is one tenth of m/V when m is in g and V is in dm³, $[\alpha]$ is 1000 times larger than α_m in ° dm² g⁻¹. Numerical values are given in the form

$$[\alpha]_D^{20} = +66.47° \text{ (sucrose, H}_2\text{O)},$$

where the subscript D refers to sodium D light and the superscript 20 indicates that the value is for 20° C.

The quantity in the old notation corresponding to the molar optical rotatory power is the molecular rotation [M] or [ϕ] defined by

$$[M] \text{ or } [\phi] = [\alpha].M_r/100,$$

where M_r is the relative molecular mass (or molecular weight) of the substance in question.

When rotations are measured at different wavelengths of light they are found to vary. This phenomenon is known as optical rotatory dispersion (abbreviated to ORD). If a plot of rotation against wavelength shows no maxima and minima the curve is known as a plain curve; if it shows maxima and minima it is known as a Cotton effect curve. A plain curve is known as positive or negative according to whether the rotation increases or decreases as the wavelength decreases. A Cotton effect curve is known as positive or negative according to whether the peak or the trough occurs at longer wavelength. These various types of curve are illustrated in Fig. 4.9.

Fig. 4.9 Different types of optical rotatory dispersion curve

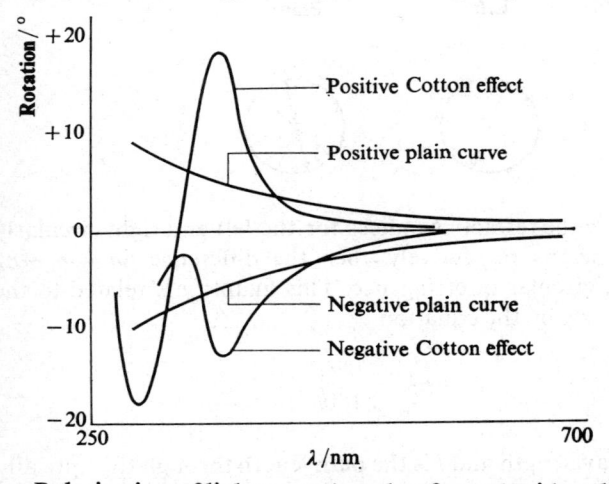

Polarisation of light can take other forms besides plane polarisation. Circular polarisation occurs when the direction of vibration of the light rotates with time and therefore with position along the direction of travel. If it rotates clockwise (looking along the direction of travel towards the source) the light is right circularly polarised and if it rotates anticlockwise the light is left circularly polarised. Plane polarised light can be regarded as the combination of equal beams of right and left circularly polarised light. If the right and left beams are equally tilted to the vertical their right and left components cancel out and give a resultant in the vertical plane. As the beams are propagated through a non-optically-active medium their directions of vibration turn in opposite directions at an equal rate so their left and right components always cancel out and the vertical resultant varies sinusoidally from maximum to minimum – the normal behaviour for plane polarised light,(Fig. 4.10(a)). In an optically active medium the refractive index is different for the right and left circularly polarised light so these two beams are propagated with unequal velocity and the rotations become progressively more different as they traverse the medium. They no longer combine to give a vibration in the vertical plane but in a plane which is rotated further from the vertical by an increasing angle as the beams progress through the medium (Fig. 4.10(b)).

Fig. 4.10 Combination of right and left circularly polarised light
(a) when the refractive indices for the two are equal
(b) when the refractive indices are unequal

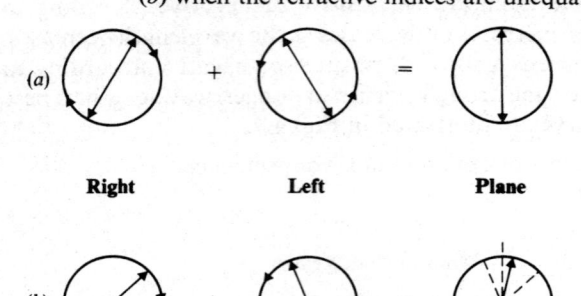

Right   Left   Plane

If n_L and n_R are the refractive indices for the left and right circularly polarised light beams respectively, then the difference $\Delta n = n_L - n_R$ is known as the circular birefringence. This quantity is related to the optical rotation, $\alpha°$, by the equation

$$n_L - n_R = \frac{\lambda \alpha}{180 l},$$

where λ is the wavelength and l is the path length through the optically active medium.

Besides experiencing a difference in refractive index, right and left circularly polarised light beams traversing a medium can experience a difference in absorption. If ε_L and ε_R are the molar absorption coefficients for the left and right circularly polarised light beams respectively then the difference $\Delta \varepsilon = \varepsilon_L - \varepsilon_R$ is known as the circular dichroism (usually abbreviated to CD). This difference in absorption for left and right beams is caused by chromophores that are themselves asymmetric (e.g. helical molecules) or become asymmetric by interaction with an asymmetric environment. It is this difference of absorption for left and right circularly polarised beams that causes the Cotton effect in optical rotatory dispersion. Each optically active absorption band contributes a 'partial rotation' to the total rotation at each wavelength considered. These partial rotations may be positive or negative and their algebraic sum gives the total rotation at any one wavelength. The effect of each optically active absorption band diminishes the further from the maximum of the absorption band is the wavelength under consideration, but never quite becomes zero. Even outside the region of absorption of the chromophore the effect is still quite appeciable (see Fig. 4.11). In principle then each structural feature of the molecule can influence the form of the total rotation curve over a wide range of wavelengths and every Cotton effect curve affects the others. Consequently, the correlation between structure and ORD is less direct than in spectroscopy.

Circular dichroism is easier to interpret than ORD because the CD peaks are discrete, the CD becoming zero at wavelengths outside an absorption band (see Fig. 4.11). The CD curves thus have higher resolution than ORD curves in cases where there are several optically active adsorption bands. CD and ORD are sufficiently closely related

Fig. 4.11 Circular dichroism (CD) and optical rotatory dispersion (ORD) at an absorption band ('A')

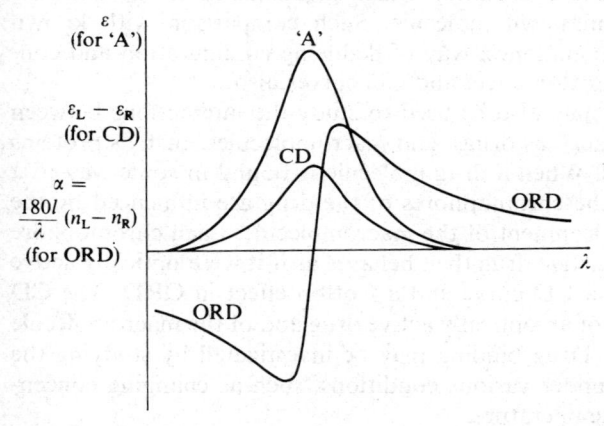

ε (for 'A')

$\varepsilon_L - \varepsilon_R$ (for CD)

$\alpha = \dfrac{180l}{\lambda}(n_L - n_R)$ (for ORD)

that each can be calculated, in principle, from the other. However, to calculate the ORD correctly it is necessary to know the CD over all wavelengths from 0 to ∞ and in practice CD measurements are only made over a limited wavelength range.

A good example of an asymmetric structure that gives rise to ORD and CD phenomena is the polypeptide chain as it occurs in simple polypeptides or in proteins. The three conformations α-helix, antiparallel β (see p. 138) and random coil give entirely different CD curves and by comparing the size of the α-helix peak with that for a similar polypeptide which is known to be entirely in the form of an α-helix the percentage of α-helix in any polypeptide or protein can be determined. It can also be deduced from a quantitative study of molar refraction per average residue weight at the maximum and minimum of the ORD curve. The chromophore whose ORD and CD curves are most frequently studied when it occurs in asymmetric environments is the $C=O$ group. In fact $C=O$ groups are often deliberately introduced at specific points in a molecule to examine the conformation near these points. This chromophore normally absorbs strongly just in the ultraviolet part of the spectrum and the circular dichroism associated with this absorption band varies in sign and amplitude according to the arrangement of the groups in the molecule near to the $C=O$ group and their position relative to the $C=O$ group. The sign and magnitude of the Cotton effect in the ORD curves varies correspondingly but here the effect may be complicated by the superposition of contributions to the ORD curve from further optically active absorptions in the molecule. Where circular dichroism or Cotton effect curves cannot be measured, either because they occur too far in the ultra-violet or because optical absorption is too strong, the plain curves in the observable region can still provide evidence of conformation. It is taken as being the same as that of the related molecule of known configuration and conformation which gives the curve most similar to that of the unknown molecule. Such comparison with known molecules is the commonest way of deducing configuration and conformation from Cotton effect and CD curves also.

CD and ORD may also be used to study the interactions between small molecules, such as drugs, and macromolecules, such as proteins and nucleic acids. When a drug molecule is bound in some way to a macromolecule, the chromophores of the drug are influenced by the dissymmetric environment of the macromolecule. Each chromophore of an optically inactive drug then behaves as if it were optically active and gives rise to a CD curve and a Cotton effect in ORD. The CD and ORD curves of an optically active drug and of the macromolecule are also affected. Drug binding may be investigated by studying the observed curves under various conditions, such as changing concentration, pH and temperature.

Molecular Spectra

All the evidence relating to molecular structure which we have discussed so far in this chapter has been somewhat indirect in nature. There are a few methods of investigating structure which are capable of giving more direct and more detailed information. Of these, the most important are molecular spectra, electron diffraction, and X-ray diffraction. The first two are applied mostly to gases and also to liquids, whereas the third finds its main application in the study of the solid state. X-ray diffraction has already been described and the main aspects of molecular spectra will now be considered briefly.

In atomic spectra, the absorption of light excites electrons to higher energy levels but, in molecular spectra, this is only one of three types of energy change that can occur. The molecule also possesses energy of rotation about its centre of gravity and energy due to vibration of the atoms about their mean positions. A change in either or both of these two further types of energy can also be detected spectroscopically. This is because both types are quantised, i.e. the molecule can only possess rotational and vibrational energies of certain fixed values, so to change either type of energy must involve absorption of fixed amounts of energy.

Table 4.1 shows the spectroscopic ranges from the far infra-red to the ultra-violet showing the energy changes corresponding to each region and the corresponding wavelengths, frequencies and wavenumbers. The relationship between frequency ν and wavelength λ for a wave-motion of velocity c is $c = \nu\lambda$. The SI unit for c is m s^{-1} and for light travelling in vacuo c has the value 2.998×10^8 m s^{-1}. The frequency ν is the number of waves per second passing a given point and has dimensions time^{-1}. The units can either be quoted as s^{-1} or by the special name Hertz with the symbol Hz. The wavelength λ has the SI unit m but for light it is usually expressed in nm. The wavenumber is usually given the symbol $\bar{\nu}$ though the recommended SI symbol is σ. It is the number of waves per unit length and so has the dimensions length^{-1} and is usually expressed in cm^{-1}. It is equal to $1/\lambda$ or ν/c and is the quantity commonly quoted for the positions of spectroscopic bands, especially in the infra-red region.

The difference in energy between successive levels of rotational energy of a molecule is very small, so, if it is desired to study this type of energy change alone, the absorption of long wavelength infra-red radiation must be observed. The frequency ν in this spectral region is so low that the size of the quantum absorbed, $h\nu$, is only sufficient to alter the rotational energy. From the spacing of successive absorption lines in this frequency range the moment of inertia of the molecule (i.e. its inertia towards a force causing it to rotate) can be calculated. The moment of inertia depends only on the masses of the atoms and their distances from the centre of gravity so, for simple molecules, accurate bond lengths can be derived from this information.

Successive levels of vibrational energy differ by energies of the order of 100 times greater than the separation of rotational energy levels. Hence, frequencies in the near infra-red region of the spectrum (i.e. not much beyond the red) give quanta $h\nu$ of the correct size to cause changes in the vibrational energy without changing the electronic energy of the molecule. Each vibrational energy change is usually accompanied by a change in rotational energy so the spectrum in this region consists of fine lines grouped into bands. The frequency at the centre of a band corresponds to a change in vibrational energy only and the lines distributed on both sides of the centre show how this energy is modified by simultaneous changes in rotational energy. From the spacings of these lines the moment of inertia of the molecule can be found in the same way as from the long-wave infra-red spectra. From the positions of the centres of the different bands the fundamental vibration frequencies of the molecule can be derived and these, in turn, provide information about the types or strengths of the bonds in in the molecule.

Table 4.1. Spectroscopic ranges. The energies quoted are based on $\varepsilon = h\nu$, $E = Lh\nu$ and $4.1840\ \text{J} = 1\ \text{cal}_{\text{th}}$

Frequency	300 GHz	6 THz		428 THz	749 THz	1500 THz
Wavenumber/cm^{-1}	10	200		14 286	25 000	50 000
Radiation		far infra-red	infra-red	visible		ultra-violet
Wavelength	0.1 cm	50 000 nm		700 nm	400 nm	200 nm
Energy						
ε/J per molecule	1.986×10^{-22}	3.973×10^{-21}		2.838×10^{-19}	4.966×10^{-19}	9.932×10^{-19}
E/kJ mol^{-1}	0.120	2.393		171	299	598
E/kcal mol^{-1}	0.0286	0.571		40.8	71.5	143

If the sample is a liquid or in solution, interactions between the molecules blur the lines of a vibrational band so much that only the general intensity distribution across the whole band can be observed. Even so, such spectra are very useful for identifying molecules and for recognising the presence of certain groups within the molecules. This is because each group gives rise to bands of characteristic frequencies and intensities in the near infra-red region of the spectrum.

Although each band actually corresponds to a vibration (or combination of vibrations) for the molecule as a whole, for most bands the vibration takes place mainly in the stretching of a particular bond or in the opening and closing of a particular bond angle (known as bending vibrations). The greater the stiffness of a stretching or bending vibration the higher the wavenumber or frequency at which the corresponding band occurs in the spectrum. Thus bands corresponding to the stretching of multiple bonds occur at higher frequencies that those for single bonds. Also the heavier are the atoms involved in a vibration the lower is the frequency. Bending vibrations occur at lower frequencies than stretching vibrations for comparable atoms. Thus bands at wavenumbers above 2500 cm^{-1} correspond to the stretching of bonds to hydrogen atoms. Stretching vibrations for triple bonds $C\equiv C$ and $C\equiv N$ occur in the region 2500–2000 cm^{-1}, whereas those for the double bonds $C=C$, $C=N$ and $C=O$ occur in the region 2000–1600 cm^{-1}. Single bond stretches are found below 1600 cm^{-1} and so are bending vibrations. In fact the region from about 1600 to 700 cm^{-1} is so specific to the particular structure of the molecule that it is sometimes spoken of as the fingerprint region. The ranges just quoted are for different multiplicities of bonds between carbon, nitrogen and oxygen atoms in general. Particular bonds have much more narrowly defined ranges. For example the $C=O$ stretching vibration occurs at 1725–1700 cm^{-1} in acyclic ketones, 1700–1680 cm^{-1} in aryl ketones, 1690–1655 cm^{-1} in quinones with two CO groups in the same ring and 1655–1635 cm^{-1} in quinones where there are two CO groups in two rings. In saturated aliphatic aldehydes the same vibration occurs at 1740–1720 cm^{-1}, in saturated aliphatic acids it is at 1725–1700 cm^{-1}, in saturated aliphatic esters it is at 1750–1735 cm^{-1}, and so on. Thus from the recognition of a band within a general range it may be possible first to detect the presence of a certain group within the molecule of a substance under investigation and then from the particular wavenumber to recognise the environment in which the group is found.

For example in the infra-red spectrum of solid salicylic acid shown in Fig. 4.12 the band E at 1650 cm^{-1} is in the general range for the stretching of a $C=O$ group in a carboxylic acid but the fact that it occurs within the narrower range 1680–1650 cm^{-1} shows that it is intramolecularly hydrogen-bonded by the OH group in the position ortho to it. The bands A, B, C and D are all due to vibrations of the

114　*The Structure of Molecules*

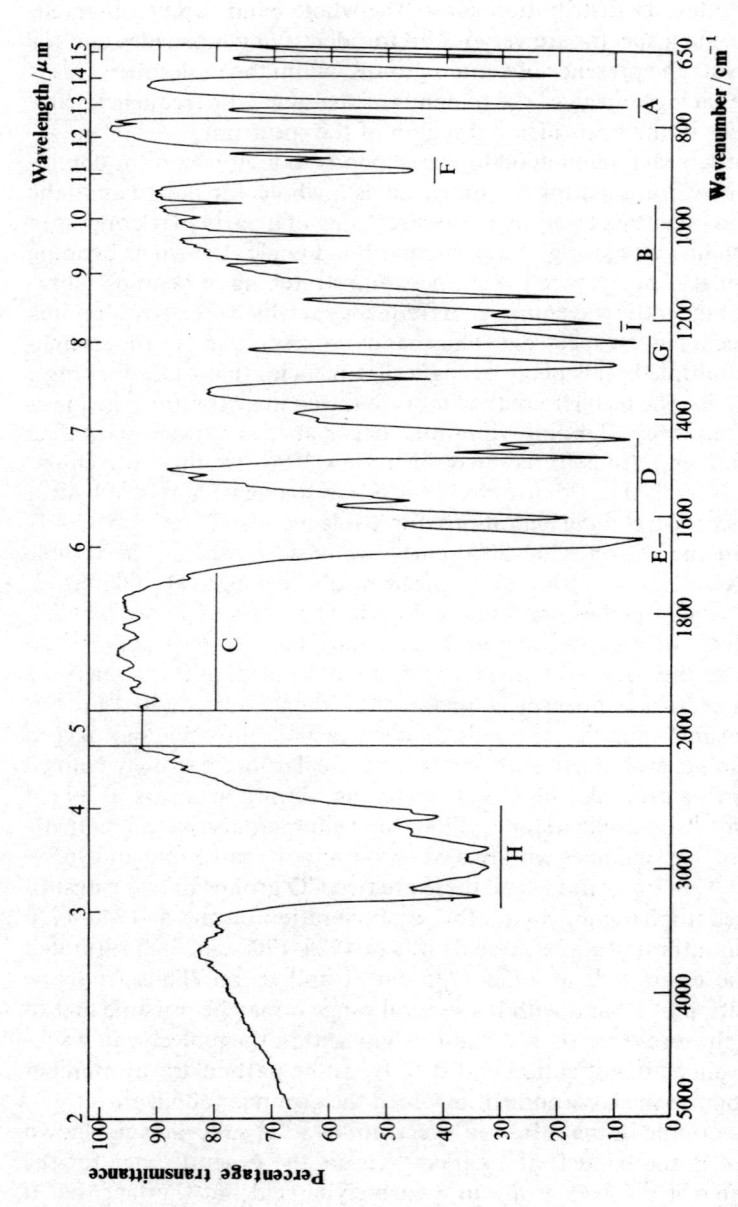

Fig. 4.12 The infra-red absorption spectrum of solid salicylic acid

phenyl ring. A is an out-of-plane vibration of the C—H groups, B are due to in-plane bending of the C—H groups and bands at C are combinations involving out-of-plane C—H vibrations. The arrangement of peaks in regions A, B and C is consistent with the ring being 1,2-disubstituted. The bands at D are due mainly to the stretching of the C—C bonds in the phenyl ring. Besides the C=O stretching band at E the bands at F, G and H are due mainly to the carboxyl group. F is an out-of-plane bending of the O—H group at a relatively high frequency due to intermolecular hydrogen bonding to the C=O group of a neighbouring molecule. The band at 1290 cm^{-1} results from coupled C—O and O—H in-plane deformation modes. Bands at H are due to O—H stretching of the COOH group and the phenolic OH group, lowered in frequency because of the hydrogen bonding. I is also due to the phenolic group.

If a molecule absorbs light in the visible or ultra-violet region of the spectrum, all three types of energy, rotational, vibrational and electronic, are altered. For any one change in electronic energy there are numerous bands corresponding to the various possible changes in vibrational energy which can also occur. Under high resolution the bands for gaseous spectra are again seen to be composed of lines due to the rotational energy changes also occurring. As might be expected, the analysis of such a spectrum is rather difficult but, if it can be done, a study of the spacings between successive vibrational bands leads to a value for the energy required to dissociate the molecule into fragments, known as the bond dissociation energy.

More often, spectra in the visible and ultra-violet range are observed for substances in solution. Again all the detail is lost and a smooth curve is obtained if the fraction of light absorbed is plotted versus frequency or wavelength. For such spectra, as also for infra-red spectra, the actual amount of light absorbed at a particular wavelength is obtained by comparing the intensity I of radiation which has passed through the spectrometer with the sample in position with the intensity I_0 in the absence of the sample. The factor $\log(I_0/I)$ is proportional to the amount of absorbing material present in the sample, providing the absorption of the substance is not modified by molecular interactions. The amount of absorbing material is proportional to the concentration, c, and the length of the light path (usually measured in cm). The relationship is often expressed:

$$\log(I_0/I) = A = \varepsilon cl,$$

in which A is called the absorbance and ε is called the molar absorption coefficient. When c is in mol dm^{-3} ε has the units dm^3 mol^{-1} cm^{-1}. In SI base units c is in mol m^{-3} and l is in m so ε has units m^2 mol^{-1}. (1 dm^3 mol^{-1}cm^{-1} = 10^{-1} m^2 mol^{-1}). The *British Pharmacopœia* uses the older term extinction for A and gives it the symbol E. Another old name for it is optical density which has the symbol D. When c is expressed as mass/volume percent (i.e. g per 100 cm³) the relationship is written:

$$\log(I_0/I) = A \quad \text{or} \quad E = A^{1 \text{ per cent}}_{1 \text{ cm}} cl \quad \text{or} \quad E(1 \text{ per cent, } 1 \text{ cm})cl$$

where $A^{1 \text{ per cent}}_{1 \text{ cm}}$ is called the absorptivity for a 1 per cent solution in a 1 cm cell or $E(1$ per cent, 1 cm$)$ is called the extinction coefficient for a 1 per cent solution in a 1 cm cell. The last two are the same quantity expressed by different names and symbols and they are useful when the molar mass is unknown or doubtful. It is for this reason that $E(1$ per cent, 1 cm$)$ is the quantity usually quoted in the *British Pharmacopæia*.

For a single, stable substance each of the quantities ε, $A^{1 \text{ per cent}}_{1 \text{ cm}}$ or $E(1$ per cent, 1 cm$)$ should have a constant value at a particular wavelength when calculated at a number of different concentrations. In other words A (or E) should be strictly proportional to concentration. If this is so the substance is said to be obeying the Beer–Lambert law and the equations above are algebraic expressions of that law. If the law is not obeyed, it is an indication that the absorbing species is taking part in an equilibrium and the position of equilibrium is varying with the overall concentration.

The law is based on the fundamental assumption that the fraction of the intensity of the incident light that is absorbed is proportional to the number of adsorbing molecules in the light path and the form of the law quoted above may be derived from this assumption in the following way. Consider a medium containing a total number n of absorbing molecules in the path of the light. In an infinitesimally thin section of the medium perpendicular to the light path let there be a number dx of absorbing molecules and let the decrease in intensity of the light in traversing this section be $-dy$. If the intensity of light entering the section is y the fraction of incident light absorbed is $-dy/y$. According to the fundamental assumption this is proportional to dx. Writing the constant of proportionality as k the relationship can be written

$$\frac{-dy}{y} = k\,dx.$$

To account for the absorption over the whole light path both sides of this expression must be integrated, the left hand side between the limits I_0 and I and the right hand side between 0 and n

$$-\int_{I_0}^{I} \frac{1}{y}\,dy = k\int_{0}^{n} dx$$

$$\therefore \quad -[\ln y]_{I_0}^{I} = k[x]_0^n$$

$$\therefore \quad -\ln I + \ln I_0 = kn - 0$$

$$\therefore \quad \ln(I_0/I) = kn$$

$$\therefore \quad 2.303 \log(I_0/I) = kn$$

or $\log(I_0/I) = k'n.$

Now n, the number of molecules in the light path, is proportional to c, their concentration, and l, the length of the light path, or

$$n = k''cl.$$

Combining this with the previous equation:

$$\log(I_0/I) = k'k''cl.$$

This is identical with the forms of the law quoted above if the product of the constants k' and k'' is replaced by ε when c is in mol dm^{-3} or by $A_{1\,\text{cm}}^{1\,\text{per cent}}$ or $E(1\text{ per cent}, 1\text{ cm})$ when c is in g per 100 cm^3.

For substances which obey the Beer–Lambert law absorption measurements clearly provide a useful method of quantitative analysis. The ratio I/I_0 which is known as the transmittance (T) is usually plotted by the spectrometer as a function of the wavelength or wave number. Log $(1/T)$, which is $\log(I_0/I)$ or A, is then calculated at a specific wavelength (usually a maximum absorbance or minimum transmittance) for which ε, or $A_{1\,\text{cm}}^{1\,\text{per cent}}$ or $E(1\text{ per cent}, 1\text{ cm})$ is known. These quantities, together with the cell length l, are then substituted into the appropriate Beer–Lambert law equation in order to obtain the concentration c. The method assumes that only the one species is present that absorbs light at the wavelength used and it is only really valid if the same solvent is used for the unknown solution as for the determination of ε, $A_{1\,\text{cm}}^{1\,\text{per cent}}$ or $E(1\text{ per cent}, 1\text{ cm})$. When more than one absorbing species is present it is necessary to measure the absorbance at as many points (preferably absorption maxima) as there are components in the solution and then solve this number of simultaneous equations for the concentrations.

Molecules containing electrons only in single bonds absorb light in the far ultraviolet region of the spectrum. In order to absorb in the near ultraviolet or visible region of the spectrum a molecule must contain a group in which the electrons are less tightly bound. Such a group is called a chromophore and is usually an unsaturated group such as C=C, C=O, N=N or N=O. There are other groups which can enhance the absorption, though incapable themselves of producing absorption in a readily accessible region of the spectrum. These are known as auxochromes and generally contain atoms with lone-pair electrons. Examples of such groups are OH, NH$_2$ and their alkyl and aryl derivatives. When two chromophores occur in the same molecule in such a way that they are not conjugated, the absorption spectrum is often the sum of the characteristic absorptions of the separate chromophores. In this way it may be possible to deduce something about the structure of a molecule from its visible and ultraviolet absorption spectrum. When two chromophores are conjugated, however, there is a shift of the absorption maxima to longer wavelengths. This is called a bathochromic effect. There is also an increase in the absorbance. Bathochromic shifts are also produced when an auxochromic group is substituted into a chromophore. The opposite effect of a shift to shorter wavelengths is called hypsochromic. This happens particularly when an extra group is inserted into a molecule in such a way as to interrupt a conjugation chain, e.g. on converting a phenyl to a benzyl derivative. A benzene ring is a chromophore but aromatic compounds in general have more complex absorption spectra than aliphatic compounds. Benzene has a high intensity absorption at 200 nm and a complex, lower-intensity band between 230 and 270 nm, which consists of four main peaks. This is due to a $\pi \to \pi^*$ transition which means that a delocalized π electron is promoted to an excited π state. As may be expected because of the greater resonance possibilities the effect of auxochromes substituted into an aromatic chromophore is more pronounced than in aliphatic compounds. As an example, the ultraviolet absorption spectrum of salicylic acid dissoved in ethanol is shown in Fig. 4.13. The complex aromatic peak system has now become one peak and is shifted to higher wavelenths (305 nm, a bathochromic shift) due to conjugation with the auxochromes OH and COOH. The band at 236 nm is another $\pi \to \pi^*$ transition called a 'conjugation band' because it is intensified by groups in conjugation with the delocalised π orbitals. The excited state in this case is a dipolar state due to electron transfer within the molecule:

The absorption at high wavenumbers (low wavelengths) is known as the 'end absorption' leading to a peak beyond the range of the spectrometer which is due to another $\pi \to \pi^*$ transition of the phenyl ring.

It is not usually possible to identify an unknown compound from its electronic spectrum because different chromophores can absorb at similar wavelengths and their absorption can be shifted by conjugation. Absence of absorption in a characteristic position is a better indication of the absence of a chromophore. For example, absence of absorption around 250 nm is fairly good evidence that the compound is not aromatic. The best form of identification is to compare the spectrum with a sample of the material that the compound is thought to be, paying attention to the intensities as well as the positions of the absorption bands. For substances that exhibit acid-base properties, the change in the spectrum brought about by a change in pH should also be characteristic.

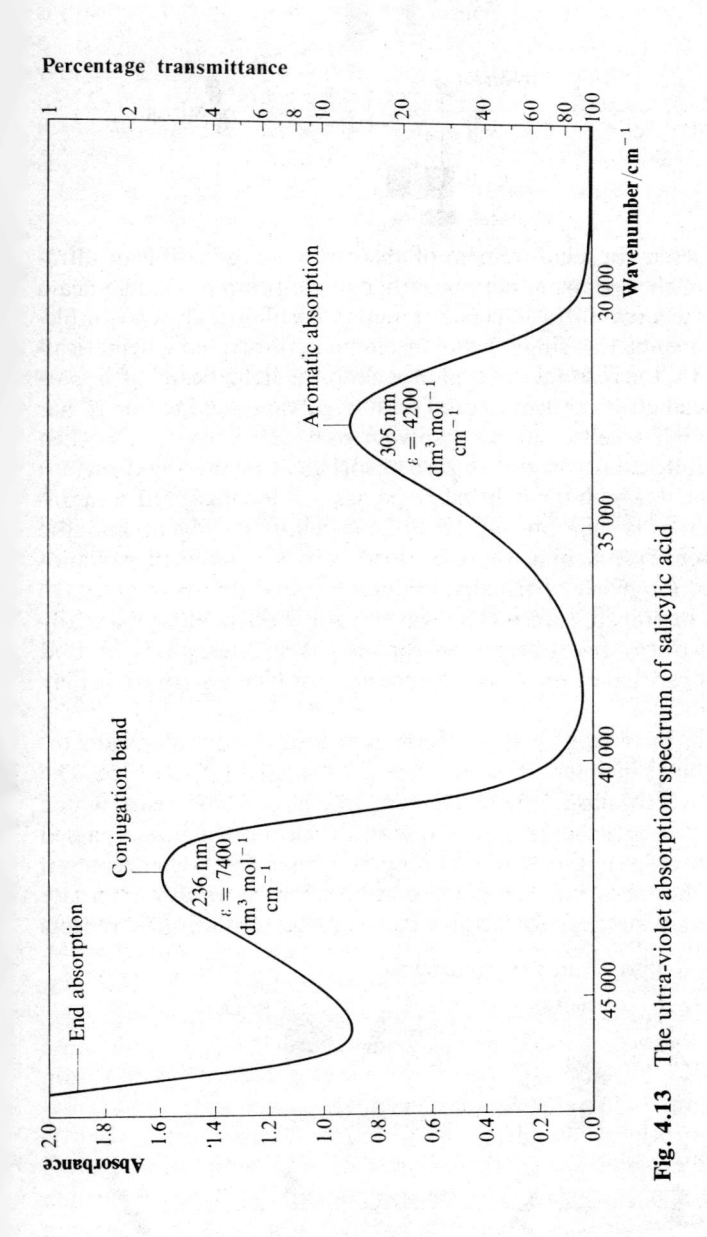

Fig. 4.13 The ultra-violet absorption spectrum of salicylic acid

120   The Structure of Molecules

Fig. 4.14 A single-beam spectrometer

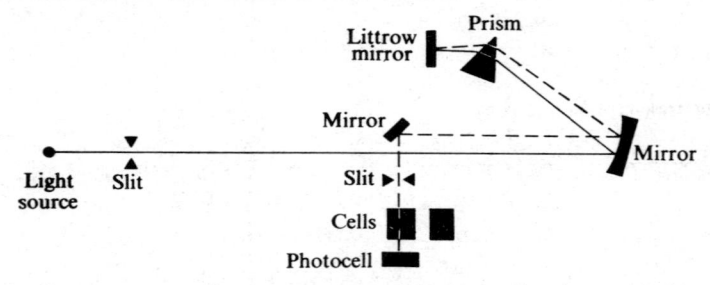

Spectrometers for measurement of absorption in the visible or ultraviolet part of the spectrum may be either single-beam or double-beam instruments whereas infra-red spectrometers are almost always double-beam instruments. In a single-beam instrument, illustrated schematically in Fig. 4.14, the response of a photocell to the light beam of a particular wavelength is first measured with a galvanometer after it has passed through a cell containing solvent only. This response is then adjusted to full-scale deflection or zero absorbance by altering the width of a slit through which the light beam passes. An identical cell containing the solution is then put in place of the cell with solvent and the reduced intensity reaching the photocell causes a reduced galvanometer deflection wich is usually calibrated directly in percentage absorbance or transmittance. The wavelength is selected by manually rotating the prism or a Littrow mirror behind it. A graph is plotted by hand of absorbance or transmittance against wavelength to display the spectrum.

Automatic recording of the spectrum requires a double-beam instrument, the principles of which are illustrated in Fig. 4.15. The mechanism which allows the instrument to compare the beam which has passed through the sample cell with the beam which has passed through the reference cell (containing solvent only) is the sectored mirror or chopper. This consists of a mirror disc with alternate sectors cut out so that for 50 per cent of each rotation it reflects

Fig. 4.15 A double-beam spectrometer

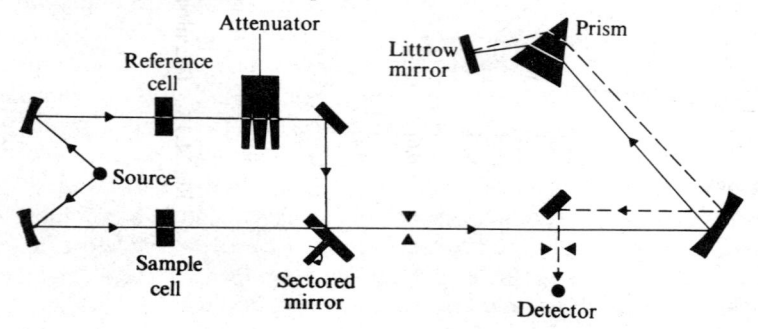

one beam and prevents the passage of the other and for the other 50 per cent it allows the first beam to pass out of the light path of the instrument to be absorbed and permits the passage of the second beam. As it rotates, therefore, an alternating intensity is received by the detector and converted into an alternating electric current. This current causes an attenuator to be moved into the reference beam until the amount it cuts down the reference beam is just equal to the amount of light absorbed by the sample in the other beam. The movement of the attenuator is linked with the movement of the pen of a recorder and represents per cent transmittance or per cent absorbance. The movement of the paper of the recorder is linked to the rotation of the Littrow mirror which causes the spectrum to be moved across the slit in front of the detector. The combination of the two movements causes the spectrum to be plotted out as transmittance or absorbance versus wavelength or wavenumber.

In the visible and ultra-violet regions of the spectrum the source is a tungsten lamp or a hydrogen discharge lamp respectively, the prism is made of silica and the detector is a photoelectric cell or sometimes a photomultiplier. In the infra-red region the source is an electrically heated glowing ceramic rod known as a Nernst filament, the prism is made of NaCl or KBr or is replaced by a diffraction grating and the detector is a thermocouple or a special pneumatic cell known as a Golay cell.

Fluorescence

After a molecule has absorbed light and has been raised to an electronically excited state, if the molecule does not undergo chemical reaction, the energy may be lost again by being converted into vibrational energy of the molecule or by being transferred to other molecules by collision or it may be re-radiated in the form of light. The two ways in which this re-radiation can take place are illustrated in Fig. 4.16 in which, for simplicity, only 7 vibrational energies ($v = 0$ to 6) are shown added to each electronic energy for the molecule. There can be fluorescence from the singlet excited state to the singlet ground state or there can be a radiationless transition to a triplet excited state followed by phosphorescence to the singlet ground state. The terms singlet and triplet refer to the multiplicity of the energy state which arises from the number of different ways in which the total electronic spin momentum S may be combined with the orbital angular momentum L. For a normal electron-paired molecule there are as many electrons with spin quantum number $+\frac{1}{2}$ as with spin quantum number $-\frac{1}{2}$ so the total spin quantum number S in the ground state is zero. There is only one way in which this may be combined with the orbital momentum quantum number (i.e. $L+0$) so only one total energy. Ab-

Fig. 4.16 Transitions between molecular energy levels

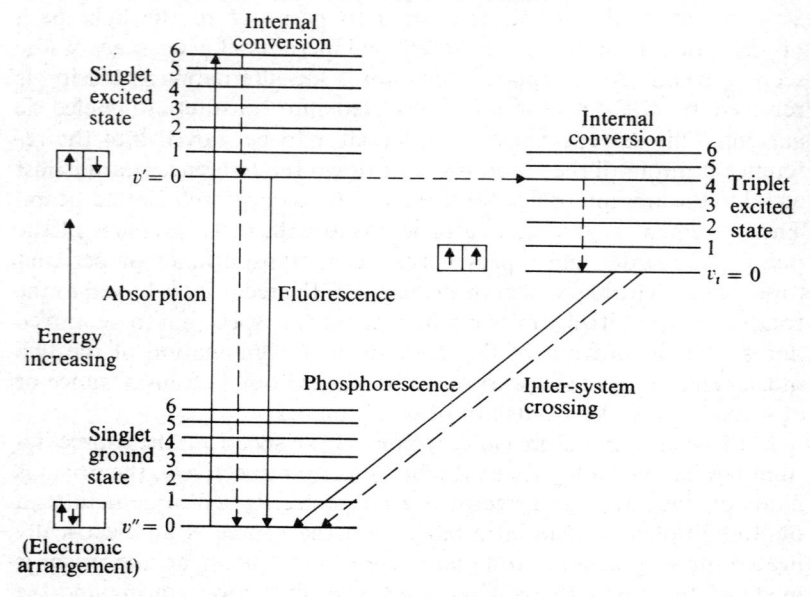

sorption of radiation promotes one electron to a higher energy level but leaves its spin unchanged, so S is still zero and the excited state is again a singlet. When transition to a triplet state takes place one electron has its spin reversed so that the two spins are now parallel (both have spin quantum numbers $+\frac{1}{2}$ or $-\frac{1}{2}$) so $S = +1$ or -1. The vector representing the total spin momentum may now be orientated in three ways relative to the orbital momentum vector, represented by the quantum number combinations $L+1$, $L+0$ and $L-1$, so there are three closely similar energies, all less than the energy of the excited singlet state in accordance with Hund's rule (p. 85).

Change of spin during transition from one electronic state to another is theoretically forbidden though in practice it takes place with low probability. This is why the singlet ground state usually gives rise to a singlet excited state on absorption of light. The reverse process, from singlet excited state to singlet ground state by fluorescence or radiationless deactivation processes is also an easy transition, so the lifetime of a molecule in an excited singlet state is very low (of the order of 10^{-8} seconds). The triplet state, however, may have lifetimes of between 10^{-4} and 30 seconds since energy losses by phosphorescence or by radiationless deactivation to the singlet ground state are processes of low probability. It should be noted that the initial absorption of radiation may be to give a high vibrational level in the excited singlet state but the vibrational energy is quickly removed before fluorescence by collision with other molecules, except for gases at low pressures. The fluorescence therefore generally takes place from the lowest vibrational level of the excited singlet state to various levels in the ground state. The energy and therefore the frequency (or wavenumber) of the light emitted is thus in general less than that of the light absorbed. This is why fluorescent dyes can absorb ultraviolet light and emit visible light. In solution the absorptions and fluorescences to the different vibrational levels are not resolved as separate bands but they do affect the shape of the overall band envelopes. It often happens that the intensity distributions in the absorption and fluorescence envelopes are approximate mirror images of each other.

Radiationless transfer from the excited singlet to the excited triplet state, even though it is theoretically forbidden, can take place, often rapidly, where the potential energy curves for the two states cross (i.e. where there is a combination of interatomic distances that gives rise to the same energy for both states, though this may correspond to low vibrational energy for the singlet and high vibrational energy for the triplet state). Such a transfer process is known as inter-system crossing and is indicated in Fig. 4.16 by the horizontal broken line. Once the transfer has taken place, loss of vibrational energy to the $v_t = 0$ level quickly takes place by collision with other molecules (internal conversion) followed by phosphorescence or some form of radiationless deactivation, which may again involve inter-system crossing. Phosphorescence is distinguished from fluorescence experimentally by studying the rate of decay of light emission after illumination of the sample has been cut off. Emission with a mean life for the decay greater than 10^{-4} seconds is likely to be phosphorescence though the situation is complicated by the fact that long life-time fluorescence can sometimes occur.

As implied above, fluorescence and phosphorescence always occur in competition with radiationless loss of energy by collision with other molecules. The latter is known as quenching and it can occur as self-quenching between molecules of the substance under investigation or it can be caused by the solvent or other solutes. When using the intensity of fluorescence or phosphorence to measure the amount of a substance in solution it must always be compared therefore with standard solutions made up in the same solvent and with the same concentration of any additional substances. The amount of light emitted as fluorescence is compared with the amount of light absorbed by the same molecules in a ratio known as the fluorescence efficiency defined as:

$$\text{fluorescence efficiency} = \frac{\text{number of quanta emitted by fluorescence}}{\text{number of quanta absorbed}}$$

The alternative methods of losing energy prevent the fluorescence efficiency from being unity, though it can approach this value, e.g.

for dilute solutions of quinine sulphate in 0.05 mol dm^{-3} sulphuric acid. Over a small concentration range the fluorescence efficiency may be taken as being constant and it may then be proved that the fluorescence intensity at low concentrations is proportional to the concentration of fluorescing material in solution. This is done as follows.

The intensity of light transmitted by the solution (I) is related to the incident intensity I_0 by the Beer–Lambert law (pp. 115–117):

$$\log(I_0/I) = \varepsilon cl,$$

where ε is the molar extinction coefficient, c is the concentration and l is the path length through the solution.
Hence

$$\ln(I_0/I) = 2.303\varepsilon cl$$

or

$$I_0/I = e^{2.303\varepsilon cl}$$
$$\therefore I = I_0 e^{-2.303\varepsilon cl}.$$

Put $\quad\quad\quad\quad -2.303\varepsilon cl = x.$

Then $\quad\quad\quad\quad I = I_0 e^x.$

The amount of light absorbed by the solution is $I_0 - I$.
This is equal to $I_0(1 - e^x)$.
The fluorescence intensity F is equal to the intensity absorbed multiplied by the fluorescence efficiency Q.

Hence $\quad\quad\quad\quad F = QI_0(1 - e^x).$

Expanding e^x as the series quoted on p. 11:

$$F = QI_0\left[1 - \left(1 + \frac{x}{1} + \frac{x^2}{2\times 1} + \cdots\right)\right].$$

When the concentrations (c) of fluorescing solute are small, x is small and terms in x^2 and higher powers may be neglected in the series.

Then
$$F = QI_0[1 - (1 + x)]$$
$$= -QI_0 x$$
$$= QI_0 \cdot 2.303\varepsilon cl$$
$$= \text{constant} \times c.$$

It is clear from the approximation made in this proof that the linear relationship between fluorescence intensity and concentration of fluorescing solute only holds for small concentrations. Other effects of increasing concentration are that the absorption of incident light becomes so strong that fluorescence is reduced in the parts of the cell further from the point of incidence of the light on the cell, and self-quenching mentioned above may also occur. Absorption and quenching may also be brought about by other solutes. In practice solutions used for fluorimetric analysis are typically 10^{-4} mol dm^{-3} to 10^{-9} mol dm^{-3} in the fluorescent material.

An instrument used for the quantitative study of fluorescence is called a fluorimeter if the range of exciting and fluorescence wavelengths is defined by filters and a spectrophotofluorimeter if prisms or grating monochromators are used to define very narrow ranges of exciting and fluorescent wavelengths and the intensity of the fluorescent light is measured by a photomultiplier. The fluorescent light is radiated in all directions from the sample but it is usually measured at right angles to the incident beam to reduce to a minimum the pick-up of stray incident light by the photomultiplier. Since fluorescence intensity is approximately proportional to the incident light intensity, very intense light sources – usually mercury or xenon arc lamps – are used. The fluorescence intensity is also approximately proportional to the extinction coefficient for absorption by the sample, so for analytical applications the incident monochromator is set to the maximum in the absorption spectrum of the sample. If, instead, the incident monochromator is caused to scan through the complete wavelength range while the fluorescence monochromator is kept at one wavelength (preferably the fluorescence maximum) the photomultiplier response, when corrected for the variation with wavelength of the intensity of the exciting source, corresponds to the absorption spectrum – in this context called the excitation spectrum – of the sample. Keeping the incident monochromator fixed at the maximum in the excitation (absorption) spectrum and causing the fluorescent monochromator to scan through the complete wavelength range the photomultiplier response is now the fluorescence spectrum of the sample. Such scans are usually carried out, when the sample has unknown spectroscopic characteristics, before using fluorescence intensity measurements for the quantitative analysis of the sample. With the monochromators set to the maxima in the excitation and the fluorescence spectra the maximum sensitivity is achieved.

To determine the amount of a fluorescent material present in a given solution a calibration curve of fluorescent intensity versus concentration is first prepared by making measurements on solutions of known concentration in the same solvent as the solution of unknown concentration and preferably, as mentioned above, in the presence of the same additional solutes as the sample to be studied and at the same pH. Any small fluorescence of the other components in the solution is allowed for by making measurements on a blank made up in the same way as the solutions already studied but without the fluorescent component to be estimated. The fluorescent intensity of the blank is subtracted from each total intensity of the calibration solutions and the unknown. When the fluorescent intensity has been measured for the sample the concentration of the fluorescent material can then be read off from the calibration graph. If there are two fluorescent components in the sample it may be possible to determine the concentrations of both by making measurements at two wavelengths, one corresponding to a peak in the fluorescence spectrum of one component and the other at the fluorescence peak of the other component. Alternatively the fluorescence of one component may be eliminated by adjusting the pH. For example, morphine and codeine both fluoresce at 355 nm but when the solution is made alkaline only the codeine fluorescence remains.

The main advantage of fluorimetry as an analytical technique is its sensitivity to small quantities of the material to be determined. In fact fluorescence spectrometry is generally more sensitive than absorption spectrometry by a factor of about 10^3. It is therefore often applied in determining the concentration of drugs and drug metabolites in blood, urine and other biological materials and in studying rates and mechanisms of drug absorption, metabolism and excretion. Many substances which are only weakly or not fluorescent can be converted into fluorescent materials by appropriate chemical reactions, the chemical group that is added is then known as a fluorophor. Phosphorescence spectrometry is more sensitive even than fluorescence spectrometry (often by a further factor of 10^3) because it is measured as a delayed effect after the exciting light has been cut off so there is no interference from stray incident light.

Nuclear Magnetic Resonance

A structural technique which is closely related to the spectroscopic techniques just described is that of nuclear magnetic resonance. It depends upon absorption and emission of energy associated with transition between energy levels but in this case it is the nuclei of atoms that undergo the transitions. Nuclei are always spinning on their axes and those that possess an odd number of protons or an odd number of neutrons in their nuclear structure such as ^1_1H, $^{19}_9F$, $^{13}_6C$ and $^{31}_{15}P$ exhibit a magnetic moment by virtue of their spin. If the nuclear spin quantum number is I the angular momentum due to spin is $\frac{h}{2\pi} \sqrt{I(I+1)}$ where h is Planck's constant, and the magnetic moment due to spin is $g_N \beta_N \sqrt{I(I+1)}$ where β_N is a fundamental unit of magnetic moment called the nuclear magneton and g_N is called the nuclear g factor. (It is a constant for each nucleus.) If placed in a magnetic field the spinning nucleus will tend to align itself with the field just as a compass needle tends to orientate itself parallel

Nuclear Magnetic Resonance

to the earth's magnetic field. The important way in which this analogy breaks down is that whereas a compass needle can take up any orientation of increasing energy away from the direction of the magnetic field a magnetic nucleus can only have orientations in which the energy is increased by quantised steps. The energy depends upon the component of the magnetic moment in the direction of the magnetic field and this is given by $\mu_z = m_N g_N \beta_N$ where m_N is the nuclear magnetic quantum number which can take the values I, $I-1$, $I-2$, ... $-I$. The energy of the nuclear magnet in an applied magnetic

Fig. 4.17 Quantised components of nuclear magnetic moments

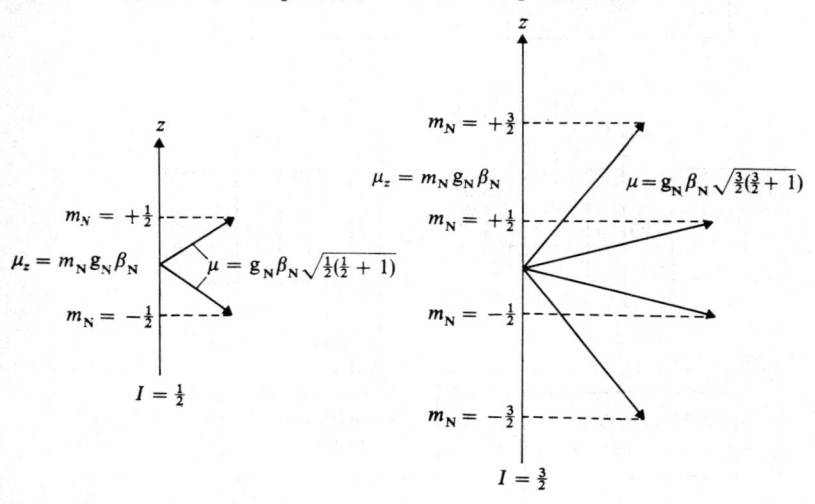

field of flux density B is given by $-\mu_z B$ and therefore equals $-m_N g_N \beta_N B$. The different values of m_N correspond to the different orientations of the spinning nucleus relative to the direction of the applied field as shown in Fig. 4.17 for the cases where $I = \frac{1}{2}$ as in protons and $I = 3/2$ as in $^{11}_5$B nuclei. When the orientation changes by energy being emitted or absorbed it does so by moving only to the adjacent position. That is to say, m_N changes by only one unit at a time. The change in energy ΔE is therefore $g_N \beta_N B$. If this is to be brought about in a spectroscopic process by the absorption of radiation the frequency of the radiation ν is given by

$$\Delta E = h\nu = g_N \beta_N B.$$

In a nuclear magnetic resonance experiment it is common to use magnetic flux densities B of the order of 1 tesla (10^4 gauss). The nuclear magneton, β_N, has the value 5.0504×10^{-27} J T^{-1} (where T is the symbol for tesla) and g_N values are of the order of unity so ΔE is of the order of 5×10^{-27} J. Now $h = 6.6262 \times 10^{-34}$ J s, so the frequency ν is of the order of 10^7 s^{-1}, i.e. 10 MHz. Instruments in common use employ frequencies of 60 or 100 MHz, in the v.h.f. radio range, and for protons (g_N = 5.584 90) these require fields of about 1.4 T and 2.3 T, respectively.

In the experimental technique for bringing about these transitions it is usual to have a fixed radio frequency and to vary the applied field until $g_N\beta_N B$ becomes equal to $h\nu$. A schematic view of the apparatus is shown in Fig. 4.18. The receiver coil is at right angles to the radio

Fig. 4.18 Nuclear magnetic resonance spectrometer (schematic)

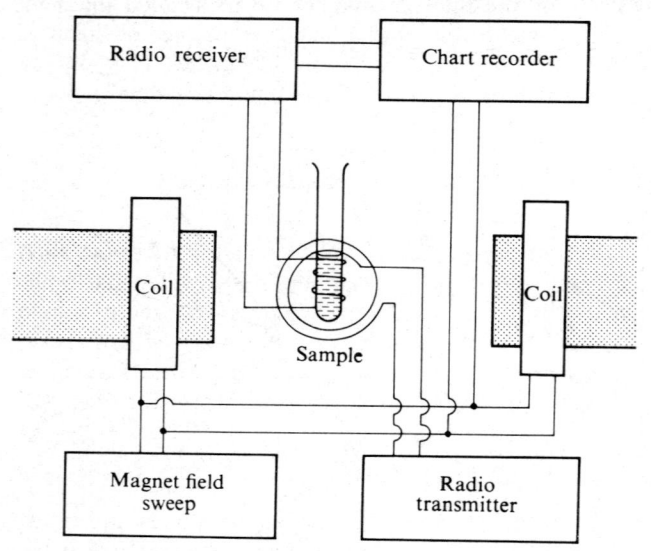

frequency transmitter coil so that there is no direct pick-up of the transmitted signal. The main magnetic field is applied at right angles to the axes of both the coils. When the magnetic field reaches the correct value for transitions to occur between the nuclear magnetic energy levels the transitions cause a field to be induced in the receiver coils and a signal is displayed on the chart recorder. This is the resonance condition.

The chemical significance of nuclear magnetic resonance arises from the fact that the field experienced by any one nucleus depends upon its chemical environment. In the first place, the nucleus is partly shielded from the external field by the electrons surrounding the nucleus and this effect differs according to the type of chemical group of which the nucleus is a part. This causes the resonance position for the nucleus to occur at slightly different applied fields according to its chemical environment and is called the chemical shift.

If the applied field gives a flux density B_0 the flux experienced by a shielded nucleus is reduced to $B = B_0(1-\sigma)$ where σ is called the screening constant for the nucleus in the particular environment. So the greater the shielding effect, i.e. the larger is σ, the higher must be the applied flux density B_0 to make the flux B experienced by the nucleus fit the equation $h\nu = g_N\beta_N B$. Alternatively, for a fixed applied flux density B_0, the greater is σ, the smaller is B and so the smaller is the frequency at which resonance takes place. Even when spectrometers are used under conditions of fixed ν the positions on the output chart at which resonances occur are usually calibrated in terms of the frequencies at which the resonances would occur if they were obtained under conditions of constant B_0.

The chemical shift is measured relative to the resonance position of some standard substance such as tetramethylsilane, $Si(CH_3)_4$, in the case of proton magnetic resonance. It is sometimes quoted as

$$\delta = \frac{\text{(frequency of sample resonance} - \text{frequency of standard resonance)}}{\text{frequency of oscillator}} \times 10^6.$$

The frequency differences are very small so the factor 10^6 is included to make δ values into convenient numbers. They are then parts per million and typical values range from 0 to 10 p.p.m. Tetramethylsilane has its hydrogen nuclei more highly shielded than hydrogen nuclei in most other environments so its resonance occurs at a lower frequency and δ values are positive numbers. Another way of expressing chemical shifts is in terms of τ where

$$\tau = 10.00 - \delta.$$

On this scale the more shielded nuclei have higher τ values and tetramethylsilane has $\tau = 10.00$. The protons in benzene have a δ value of 7.3. and a τ value of 2.7.

The second way in which the position of resonance depends upon the environment is that the signal is split into components by the presence of nearby magnetic nuclei with different resonance frequency and within the same molecule. For example, the high resolution nuclear magnetic resonance spectrum of protons in ethyl alcohol is shown in Fig. 4.19. The separate groups of peaks are recognisable from their chemical shifts as arising from the OH, CH_2 and CH_3 protons respectively and this is confirmed by the integrated areas of peaks in the three groups being in the ratio 1 : 2 : 3. It can be seen that the CH_2 peak is split into four lines and the CH_3 peak is split into three lines. Because this arises from the interaction of nuclear spins it is called spin–spin splitting. The analysis of this effect indicates the types of group adjacent to the one in question.

In the present example, a proton with $I = \frac{1}{2}$ can have two possible orientations relative to an external magnetic field. These can be represented by arrows pointing up and down respectively, representing the components of the magnetic moment along the direction of the applied field. Now, relative to a proton with a particular spin orientation in the methyl group the CH_2 protons can have their spins arranged in the four different ways shown in Fig. 4.20(a). Of these, the two central arrangements have the same total spin of 0 so they will have the same (zero) effect on the protons of the CH_3 group. The other

Fig. 4.19 High resolution NMR spectrum of protons in ethyl alcohol

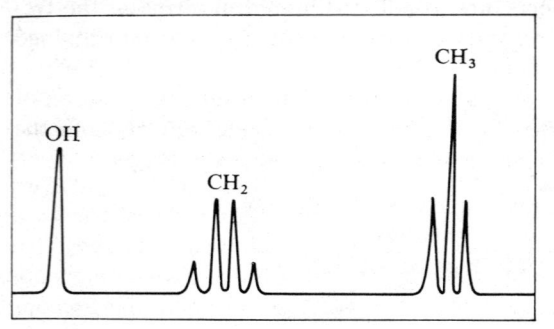

two arrangements will add and subtract a little to or from the field experienced by a CH_3 proton and so will shift the resonance position slightly. Because there are two arrangements with zero field effect and one each having a positive or negative effect, the probability of a CH_3 proton not having its resonance shifted is twice as great as the probability of experiencing either of the shifts. This is why the central peak of the CH_3 triplet is about twice the size of either of the two flanking peaks. In the same way we can work out, as in Fig. 4.20(b), that there are four different total spin arrangements of the three protons of a CH_3 group relative to any one proton on the CH_2 group. The CH_2 resonance is therefore split into four separate resonances, the central pair having three times the probability of the outer two. One

Fig. 4.20 The different possible spin arrangements of
(a) 2 protons (b) 3 protons

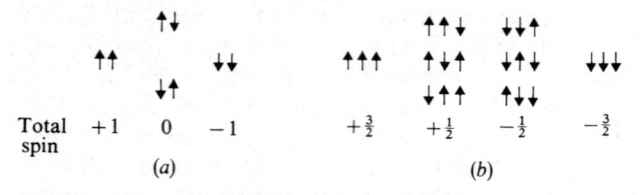

would expect the lines of the CH_2 resonance to be further split into pairs by the two possible orientations of the OH proton which is also adjacent to the CH_2 groups. However, in most samples of ethanol the OH protons undergo rapid exchange with one another, particularly when traces of acid or base are present, and they do not remain attached to the molecule long enough for their spins to influence the field experienced by the CH_2 protons. If the ethanol is very pure, eight resonance peaks may be resolved for CH_2 protons.

Although most n.m.r. work until recently has been done with protons, there has been an important extension of the technique over the last few years to the investigation of ^{13}C resonances. This has necessitated more sensitive methods of recording the spectra, partly because only 1 per cent of carbon atoms in a normal carbon compound contain ^{13}C nuclei. Enhanced sensitivity has been achieved in two main ways, which are often used in combination. One is to scan the spectrum several times and to average out the deflections over all the scans by means of a computer. This is the Computer Averaging of Transients (or CAT) method. Even where a genuine resonance is only of the same size as the background noise this will be positive on every trace and will average to a finite positive deflection whereas the noise will be randomly positive and negative and will average to zero. The difficulty with this method is to retain constant homogeneous magnetic fields over the long periods required for the multiple scans. The second method of increasing sensitivity is to apply a brief high-power pulse of the radio frequency which is broad enough to cover simultaneously the resonance frequencies of all the ^{13}C nuclei in their different environments. The signal received by the radio receiver is produced suddenly and then dies away in a manner that depends on the summed effects of all the different resonances. Fourier transformation of this decay signal by means of a computer gives the normal spectrum. This is therefore known as pulsed Fourier transform n.m.r. Accumulated repetition of the decay signal by the CAT method can further enhance such derived spectra.

The interpretation of ^{13}C n.m.r. spectra depends on the same principles of chemical shift and spin-spin splitting as in the case of proton n.m.r. but there are differences in detail. One big advantage of ^{13}C over ^{1}H n.m.r. is that the chemical shifts are much larger so the ^{13}C atoms in the different chemical environments in the molecule can readily be distinguished. Spin-spin splitting of ^{13}C resonance peaks is caused by the group of protons attached to the carbon atom and not by adjacent carbon atoms. This is because the abundance of ^{13}C atoms is so low that the chance of having two such nuclei at nearby positions in the same molecule is negligible. Thus a ^{13}C resonance peak will be a doublet if the ^{13}C atom is part of a CH group, a triplet if it is in CH_2 and a quartet if it is in CH_3, so the environment of each carbon atom can readily be recognised. However these principles are made more difficult to apply by the fact that samples are commonly irradiated by radio waves of ^{1}H resonance frequencies in addition to ^{13}C frequencies in order to enhance further the size of the ^{13}C signals. This double resonance technique has the effect of reducing more the extent to which the ^{13}C resonance is split by nearby ^{1}H nuclei the nearer is the applied radio frequency to the resonance frequency of these attached protons. A multiple peak becomes a single peak when these frequencies coincide. Indeed, a broad range of ^{1}H frequencies is often applied to reduce all the multiplets to singlets. Double resonance techniques also have the disadvantage of causing the integrated areas of peaks to be no longer proportional to the number of nuclei having the corresponding chemical shifts. In general, then, ^{13}C n.m.r. is usually applied as a diagnostic technique in conjunction with conventional proton n.m.r.

Electron Spin Resonance

Though less important for the investigation of molecular structure, the phenomenon of electron spin resonance (e.s.r.) is so similar to that of nuclear magnetic resonance that it deserves a mention here. It depends upon the fact that an unpaired electron possesses a magnetic moment, partly due to its spin and partly due to its orbital motion, and so its energy in a magnetic field depends on the orientation of the resultant spin vector relative to the direction of the field. If an atom has more than one unpaired electron the spin quantum numbers s are added to give a total spin quantum number S and the orbital momentum quantum numbers l are added to obtain a total orbital quantum number L. Finally S and L are combined to give the total angular momentum quantum number J corresponding to the total angular momentum $(h/2\pi) \sqrt{J(J+1)}$. This gives a magnetic moment $g\beta \sqrt{J(J+1)}$ where g is the splitting factor and β is the Bohr magneton. As in the case of nuclear magnetic moments, the magnetic moment of an atom can take up specific orientations relative to an applied magnetic field such that the component of the moment parallel to the field is $g\beta M$ where M, the magnetic quantum number, can have the values J, $J-1, J-2, \ldots -J$. The magnetic energy in each of these orientations is given by $g\beta MB$ where B is the magnetic flux density and transitions between these energy levels are restricted to those where M changes by 1. The energy which must be absorbed to increase M by 1 is equal to $g\beta B$ and for flux densities commonly used (0.3 tesla) this corresponds to radiation of frequency about 9000 MHz which is in the microwave region of the spectrum.

Microwaves are usually conducted along hollow tubes known as wave guides and they are caused to interact with the sample in a resonant cavity. This is a section of the wave guide with reflecting walls placed the correct distance apart to set up standing waves within the cavity. Because of this necessity to tune the cavity it is usual to observe electron spin resonance by maintaining a fixed microwave frequency and sweeping the magnetic field through the resonance value with the aid of auxillary coils on the electromagnet. The position of resonance depends upon the g value for the electrons and this in turn depends upon the relative contributions of the spin and orbital components of the total angular momentum. Where the orbital contribution is 'quenched', as it is for organic free radicals, the g value is close to 2.

The main application of e.s.r. is in the detection and estimation of the concentration of free radicals even when these are only present in small proportions. The area under the resonance peak is proportional to the number of free radicals and the apparatus can be calibrated with a substance known to be 100 per cent free radical. In this way it is possible to study, for example, the number of free radicals produced by exposure to radiation, and the production and decay of free radical intermediates in chemical reactions. In transition metal complexes the magnetic moment is due almost entirely to the total spin S of the unpaired electrons in the d shell. Since each unpaired electron contributes $\frac{1}{2}$ to S there are $2S$ unpaired electrons. The magnetic quantum number M can take values $S, S-1, S-2\ldots -S$, which totals $2S+1$ values, so in a magnetic field there are $2S+1$ energy levels. In the absence of the ligand field these levels would be equally spaced and the transitions between them would all involve the same energy difference and would therefore give only one peak. However the electric fields of the ligands cause the energy levels to be differently spaced and in general there are $2S$ different transitions between the $2S+1$ levels. This gives $2S$ peaks in the e.s.r. spectrum, equal to the number of unpaired electrons which can thus be measured. The number of unpaired electrons can be related, in turn, to the nature of the bonding to the ligands. The small amount of orbital contribution to the total angular momentum in transition metal complexes has a small effect on the g value and this is slightly different in different directions. This variation with direction will however reflect the symmetry of the ligand field, so this can be studied by measuring the variation of g with orientation in single crystals of the complex. In this way the orientations of the haem planes in heamoglobin crystals were first discovered by e.s.r. studies.

Hyperfine splitting of resonance signals due to the magnetic effects of nearby nuclei occurs with e.s.r. as well as with n.m.r. In the case of e.s.r. the orbital of the unpaired electron must encompass the magnetic nucleus if it is to cause any splitting and the extent of the splitting is related to the probability that the electron is in the vicinity of this particular nucleus. In general the orbital encompasses more than one magnetic nucleus and complex splitting patterns may result. The observed pattern is compared with calculated patterns until agreement is reached and in this way the probability distribution of the unpaired electron in a radical may be deduced.

Mass Spectrometry

As its name suggests, mass spectrometry is a technique for dispersing particles according to their masses or, more correctly, according to the ratio of the mass m to the electric charge Q on the particle. These particles are formed and given an electric charge usually by passing a stream of atoms or molecules through a beam of electrons (see Fig. 4.21). The electrons in the beam have high enough energy to ionise the atoms or molecules by knocking out electrons and so giving them a positive charge. In this process many molecules are also broken down into smaller fragments of varying sizes so the ionised beam contains particles of a wide variety of masses. Nearly all of them are

Fig. 4.21 Diagram of a mass spectrometer. The whole ion path is enclosed in an evacuated tube

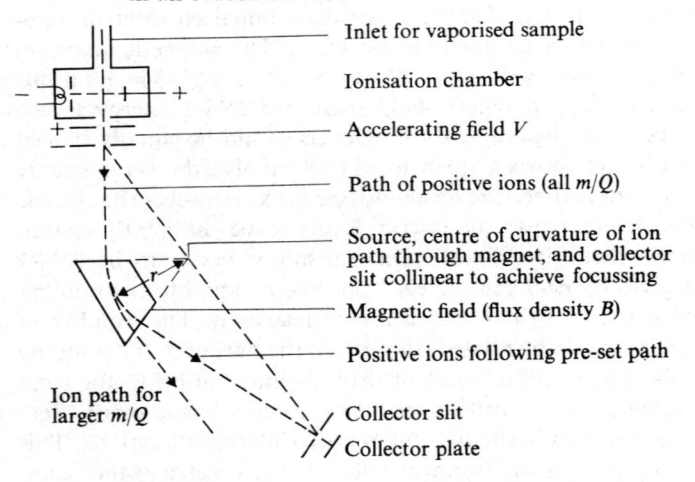

- Inlet for vaporised sample
- Ionisation chamber
- Accelerating field V
- Path of positive ions (all m/Q)
- Source, centre of curvature of ion path through magnet, and collector slit collinear to achieve focussing
- Magnetic field (flux density B)
- Positive ions following pre-set path
- Ion path for larger m/Q
- Collector slit
- Collector plate

singly charged but a small proportion become multiply charged. The particles then pass through slits in two successive plates between which a large electric potential difference V is applied. This accelerates the particles since the second slit is negative relative to the first. The beam of particles is then made to pass between the poles of an electromagnet. The magnetic field of flux density B deflects the particles and causes them to travel along the arc of a circle of radius r, related to the other quantities mentioned by

$$\frac{m}{Q} = \frac{B^2 r^2}{2V}.$$

For a given accelerating potential and magnetic field strength only ions of one particular value of m/Q will follow the path of preset radius r and pass through the collector slit as shown in Fig. 4.21. Here they fall on the collector plate and are recorded as an ion current whose strength depends on the number of ions per second of this particular m/Q which are being produced in the apparatus. The magnetic field produces direction focussing so that ion beams entering the region of the field in slightly different directions all pass through the collector slit together. By varying V or B beams of different values of m/Q follow the path of present radius r and pass through the collector slit to give an ion current. A recorder plots a graph of ion current versus m/Q and this is a mass spectrum. It shows the relative abundances of ions of different masses in the decomposition products produced by the electron beam. When m/Q is calculated in SI units from the equation quoted above it will have the units of kg C^{-1}. For practical purposes, however, m/Q is stated in units of m_u/e where m_u is the unified atomic mass unit, equal to 1.6605×10^{-27} kg and e is the protonic charge of 1.6022×10^{-19} C. In these units particles which have lost one electron (the usual situation) will give m/Q values numerically equal to the relative molecular mass of the particle.

This type of information is very valuable in helping to determine the molecular structure of an unknown organic substance. In most cases the empirical formula and the relative molecular mass (the molecular weight) are already known and the latter is confirmed by the recognition of the parent ion peak as one of the peaks of highest m/Q in the mass spectrum. This is the peak due to molecules which have not fragmented but simply lost an electron in the electron beam. Usually the parent ion peak is accompanied by a weaker peak one higher in relative molecular mass. This is due to the presence of the carbon isotope ^{13}C which is present to the extent of about 1.1 per cent of ^{12}C and is randomly distributed in the sample. Each fragment may give rise to a peak with m/Q one unit greater than the ^{12}C peak for the same reason but in such cases the peak may also be due to a fragment containing an extra hydrogen atom.

The main application of mass spectra in determining molecular structure arises from the recognition of the mode of fragmentation of the molecule in the electron beam. The parent ion is one electron short of a normal molecule whose bonds are formed by electron pairs. It is therefore a positive radical ion and will often decompose by splitting off a smaller radical to leave a positive electron-paired carbonium ion

$$[AB]^+\cdot \rightarrow A^+ + B\cdot \quad \text{or} \quad A\cdot + B^+$$

Only the carbonium ions A^+ or B^+ are accelerated through the mass spectrometer and are recorded as an ion current. A molecule may break at any point and the fragments may split further so a wide range of carbonium ions is detected. Those that are most stable are detected in greater abundance. Since the stabilities of carbonium ions are in the order tertiary > secondary > primary, ions resulting from a break at the point of branching of a molecule are often prominent in the mass spectrum. For example, if groups P,Q,R and S are all joined to a central carbon atom thus:

$$\begin{array}{c} Q \\ | \\ P-C-R \\ | \\ S \end{array}$$

fragments that are likely to be abundant in the mass spectrum are:

$$\begin{array}{c} Q \\ | \\ P-C^+ \\ | \\ S \end{array} \qquad \begin{array}{c} Q \\ | \\ P-C^+-R \\ | \\ S \end{array} \qquad \begin{array}{c} Q \\ | \\ {}^+C-R \\ | \\ S \end{array} \quad \text{and} \quad \begin{array}{c} Q \\ | \\ P-C-R \\ {}^+ \end{array}$$

because these are all tertiary carbonium ions. From the masses of these carbonium ions the masses and possibly the nature of the fragmenting groups P, Q, R and S can be deduced.

Most functional groups give rise to their own characteristic modes of cleavage. Thus a common mode of fragmentation of carbonyl compounds is to break between the carbonyl carbon atom and an adjacent carbon atom:

$$\begin{array}{c} R_1 \\ \diagdown \\ C=O^{\cdot+} \\ \diagup \\ R_2 \end{array} \rightarrow R_1-C\equiv O^+ + R_2^{\cdot} \quad \text{and} \quad R_2-C\equiv O^+ + R_1^{\cdot}$$

$$\begin{array}{c} R \\ \diagdown \\ C=O^{\cdot+} \\ \diagup \\ H \end{array} \rightarrow R-C\equiv O^+ + H^{\cdot} \quad \text{and} \quad H-C\equiv O^+ + R^{\cdot}$$

$$\begin{array}{c} R_1 \\ \diagdown \\ C=O^{\cdot+} \\ \diagup \\ R_2O \end{array} \rightarrow R_1-C\equiv O^+ + R_2O^{\cdot} \quad \text{and} \quad R_2O-C\equiv O^+ + R_1^{\cdot}$$

For alcohols the commonest form of cleavage is not one which expels a radical but an electron-paired molecule, water. This takes place so readily that the parent ion peak is not usually seen but one which is less by 18 units of m_u/e is observed. In addition the molecule breaks between the carbon atom attached to the OH group and an adjacent carbon atom.

These are just a few examples of common types of fragmentation. It is clear that experience is necessary for the successful derivation of a molecular structure from mass spectra. In practice this technique is used in conjunction with other techniques such as infra-red, ultra-violet, and nuclear magnetic resonance spectroscopy. Comparison of the fragmentation pattern (often called the cracking pattern) with those of closely related molecules provides the best confirmation of the molecular structure.

Two big advantages of mass spectrometry are that only a very small sample (microgram to nanogram quantities) is required and it can be in the form of a solid, liquid or gas. The smallness of the sample makes this technique a particularly suitable one for such studies as drug metabolism where only trace amounts of the material under investigation may be available. A very sensitive method for analysing mixtures is to inject a small quantity of the mixture into a gas-liquid chromatograph whose output is fed into the ionisation chamber of a mass spectrometer. As each component is eluted from the chromatographic column its mass spectrum is recorded and is subsequently used to identify the component.

All the deductions from mass spectra discussed so far are based on the relative abundances of ions of different m/Q values taken to the nearest integer in units of m_u/e and for such applications high resolution is not required. But some mass spectrometers are capable of determining m/Q values with an accuracy of about 1 in 10^5 and these open up further applications. The higher resolution required to achieve this accuracy is obtained by passing the ion beam through an electric field between two parallel plates curved to form the segments of two concentric circles before it enters the magnetic field. This has the effect of focussing ions of slightly different velocities and giving a much more narrowly defined beam for each m/Q value at the collector slit. A fragment whose accurate relative molecular mass is required is then compared with a fragment of known accurate relative molecular mass from a standard compound. The accurately measured ratio of the accelerating potentials V_1 and V_2 required to bring ion beams of these two fragments into the centre of the collecting slit gives the ratio of their m/Q values. From this ratio and the known relative molecular mass the required accurate relative molecular mass can be calculated. The advantage of knowing relative molecular masses accurately is that the empirical formula of any ion can be established unambiguously. This is because the presence of a nitrogen atom of relative mass 14.0031 can be distinguished from that of a CH_2 group of relative mass 14.0156, for example. Similarly, alternative molecular formulae giving relative masses that approximate to the same integer can always be distinguished. In this way ambiguities in the empirical formula of a compound due to inaccurate analyses can be resolved from a knowledge of the accurate m/Q value for the parent ion peak and the correct molecular formula for any fragment can always be selected from among alternatives of similar relative mass. This facility considerably increases the power of the mass spectrometric method for determining structural formulae.

The use of mass spectrometry for measuring relative atomic masses (atomic weights) of elements has already been mentioned (p. 58). This application depends on the ability to measure both the accurate relative mass and the relative abundance of each isotopic species of an element. Another type of application of mass spectrometry is in the determination of ionisation potentials of molecules and bond dissociation energies. These determinations require the measurement of the minimum energy of the bombarding electron beam that will produce ions of the required type in the mass spectrometer. The electrical potential accelerating the electron beam to a point where it will just produce a given ion is known as the appearance potential of that ion. If the process producing the ion is simply that of knocking out an electron from a molecule or radical the appearance potential is equal to the ionization potential in electronvolts (eV). Multiplication of this value by 96 487 gives the enthalpy of ionisation in J mol^{-1}. (See p. 271 for a discussion of enthalpies and p. 66 for a discussion of eV.) Measurement of a bond dissociation energy involves the comparison of the ionisation potential of a radical with the appearance potential of the same radical produced by breaking a bond in a larger molecule, thus:

$R_1 \cdot \rightarrow R_1^+ + e^-$ ΔH_1 (ionisation potential)
$R_1{-}R_2 \rightarrow R_1^+ + R_2 \cdot + e^-$ ΔH_2 (appearance potential of R_1^+ in $R_1{-}R_2$).

Subtracting the first equation from the second

$R_1{-}R_2 - R_1 \cdot \rightarrow R_2 \cdot$ $\Delta H_3 = \Delta H_2 - \Delta H_1$
or $R_1{-}R_2 \rightarrow R_1 \cdot + R_2 \cdot$ ΔH_3.

The enthalpy change ΔH_3 therefore corresponds to the bond dissociation energy of the $R_1{-}R_2$ bond. This method of measuring bond dissociation energies is not as accurate as, for example, the spectroscopic method but it can be applied to a wider range of molecules.

Biological Structure

By applying the techniques described in this chapter and the technique of X-ray diffraction of crystals described in Chapter 2 an enormous amount of information has been amassed about the structures of molecules. At the same time, but more slowly, information has been accumulated about the structures that occur in biological systems. This will be illustrated by a few important examples.

(a) *The structures of fibrous proteins*

From the X-ray diffraction patterns that they give, fibrous proteins can be classified into two main groups: (i) the keratin-myosin-epidermin-fibrinogen or k-m-e-f group and (ii) the collagen group.

Some members of the first group exist in what is called the α-conformation and some exist in the β-conformation. Sometimes transformation from α to β can be observed; for example, on stretching wool and hair under appropriate conditions of humidity and temperature. Pauling and Corey proposed structures for these α- and β-conformations mainly on the basis of theoretical considerations. These were that all the amino-acid residues should be equivalent apart from the differences in their side groups, that each amide group should be planar and have dimensions found in simple peptide structures, and that each NH

Fig. 4.22 The α-helix (from J. Paul, *Cell Biology*, Heinemann, 1967)

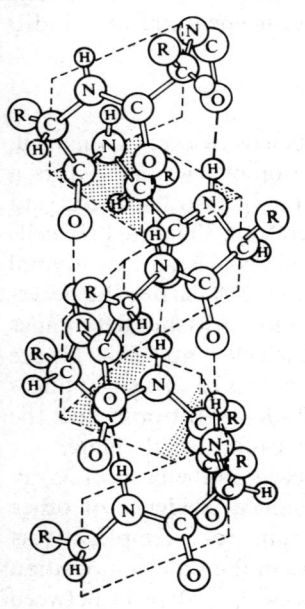

group should form a hydrogen bond to a C=O group. These criteria were best satisfied basically by two structures one being a helix in which each NH is hydrogen bonded within the helix to a C=O group of a residue three positions along (Fig. 4.22) and the other being an arrangement of partially extended chains in which the hydrogen bonds link adjacent chains. The first of these was found to have characteristic dimensions of spacings recorded on X-ray diffraction photographs of the α form and is therefore known as the α-helix. The second corresponds to the β X-ray diagram and is known as the β-pleated sheet because of the arrangement of adjacent folded chains into sheets. Each chain has the form shown in Fig. 4.23 and its N—H groups form hydrogen bonds to C=O groups in the next chain which may run in either the same or the opposite direction. In either case the planar segments are continuous from chain to chain and these two arrange-

Fig. 4.23 One chain of the β-pleated sheet (from J. Paul, *Cell Biology*, Heinemann, 1967)

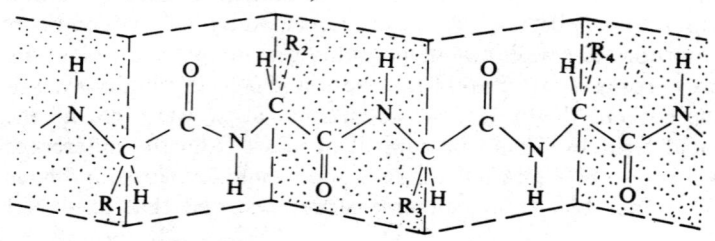

ments form the parallel and antiparallel pleated sheets respectively. Fragments of both the α-helix and the β arrangement are found also in many globular protein structures.

(b) *Deoxyribonucleic acid (DNA)*
DNA is an essential component of the nucleus of every living cell. It is found combined with protein in the chromosomes and before a cell divides the chromosomes and the DNA split into two parts and replicate themselves. After cell division, therefore, the daughter cells contain the same chromosomes and DNA molecules as the original parent cell. In this process it appears that the DNA carries the necessary genetic information from parent to daughter cells and determines which protein molecules are synthesised in the cells. Nucleic acids are polymers of nucleotides each of which consists of a base, a sugar and a phosphate group. In DNA the sugar is 2'-deoxy-D-ribose and the bases are almost entirely adenine, guanine, cytosine and thymine.

Although the structure of DNA was proposed on the basis of X-ray diffraction diagrams much physical and chemical evidence of other types provided a background for the proposal. For example, it was found that although there is a wide variation in the base composition of DNA from different sources there is always a 1 : 1 ratio between adenine and thymine and between guanine and cytosine. Watson and Crick found by model building that if adenine is always hydrogen-bonded to thymine and guanine is always hydrogen-bonded to cytosine as in Fig. 4.24 the sugar groups attached to these base pairs would be held a constant distance apart. The structure proposed therefore consists of two chains of alternating sugar and phosphate groups intertwined with each other as a double helix with the bases pointing inwards from the sugar groups towards the axis of helix. The bases link the two chains by the hydrogen bonding between base pairs across the axis. The planes of the bases being approximately at right-angles to the direction of the helix axis 0.34 nm apart and can be pictured as the rungs of a ladder held between the helical supports of the sugar and phosphate chains.

Fig. 4.24 Hydrogen bonding of bases in DNA

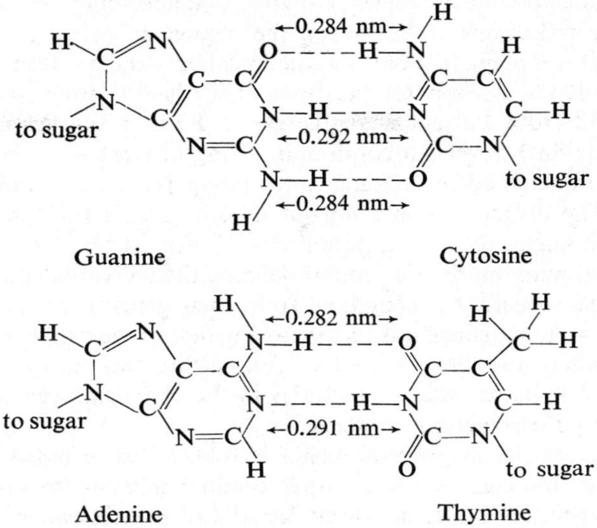

Guanine — Cytosine (0.284 nm, 0.292 nm, 0.284 nm)

Adenine — Thymine (0.282 nm, 0.291 nm)

Each chain contains all four types of base and the order and nature of each group of three bases determines which amino acid appears in the synthesis of proteins in the cell containing this particular DNA molecule. Once the arrangement of bases in one chain is fixed the arrangement in the chain intertwined with it is also determined because of the specificity of the base pairing. Before cell division, the two chains unwind from each other and each acts as a template for the synthesis of a new double helix. Because of the specificity of the base pairing the new complementary chain in each case must be an exact replica of the previous complementary chain. The two new complete DNA molecules must therefore be exact copies of the original molecule and will carry the same genetic information to the two daughter cells as was present in the original cell.

(c) *Crystalline proteins*
The primary structure of a polypeptide chain is the sequence of amino acids and this is often known with a fair degree of certainty from chemical studies. The secondary structure is the relative arrangement of adjacent amino acids, as in the α-helix for example, and some information on this can be derived from physico-chemical studies such as ORD and CD (p. 109). The tertiary structure is the detailed folding of the polypeptide chain as found for crystalline proteins by X-ray diffraction. Some proteins, such as haemoglobin, show quaternary structure in which sub-molecules unite together in a specific way to form the complete molecule.

Many globular proteins can be crystallised from aqueous solution and a number of them have had their crystal structures determined. This is made possible by a special method of determining the phases of the X-ray reflections by following the change of intensity brought about by introducing first one and then a different heavy atom into the molecule. Knowing only the positions of the heavy atoms in the unit cell (found from Patterson syntheses) a Fourier synthesis can be calculated which reveals a continuous string of electron density with periodic branches corresponding to the polypeptide chain and its side groups. The diffraction data do not usually extend to high enough diffraction angles to permit quite the resolution of separate atoms in the Fourier map but the side chains defining the individual amino acid residues can usually be recognised from their general shape. Usually the amino-acid sequence was known in advance of the crystal structure determination and this assisted in recognising the sequence in the crystal but in some cases ambiguities in the chemical sequence were resolved by the structural evidence.

In all cases the polypeptide chain is folded and refolded in what appears at first sight to be a rather random manner, to produce a roughly spherical molecule. Some lengths of α-helix can usually be discerned and occasionally adjacent lengths of chain form part of a β structure. Closer examination often shows that the chain is folded in such a way as to bring hydrophobic groups into the centre of the molecule and polar groups onto the outside. In addition there are sometimes clefts or channels which are associated with the specific activity of the protein. A good example of this is found in the structure of lysozyme, an enzyme which breaks up the walls of bacterial cells by hydrolysis of glycosidic links in polysaccharides formed from amino sugar molecules in the cell wall. The crystal structure determination of the enzyme revealed a deep cleft running down one side of the molecule. Structures of complexes which the enzyme forms with polysaccharides that competitively inhibit its activity confirm that this cleft is the binding site. By fitting scale models of larger polysaccharides (that are attacked) into the cleft in a model of the enzyme, it is found that the critical glycosidic link that is hydrolysed is adjacent to the hexose ring that must be distorted, and it is also near groups in the enzyme that can readily effect the hydrolysis. In ways such as this, knowledge of the structure of proteins is also throwing light on the way these molecules carry out their specific activities in life processes.

Problems

1 The bond lengths in HCl, HBr and HI are 0.127, 0.142 and 0.160 nm respectively. Given that the electronic charge is 1.602×10^{-19} C, calculate the dipole moments that these molecules would have, in

C m and in D, if they were in a completely ionic form (i.e. if there were a complete transfer of an electron from the hydrogen to the halogen atom in each case).

2 The dipole moment of the water molecule is 1.84 D (6.14×10^{-30} C m). Taking the moment for each H—O bond to be 1.51 D (5.04×10^{-30} C m) with hydrogen positive, find by scale drawing or by trigonometry the angle between the two H—O bonds which would be consistent with the observed total moment.

3 Methyl alcohol has a refractive index of 1.331 at 15° C, and its density at this temperature is 796 kg m^{-3}. Calculate the molar refraction, and compare it with the value calculated from the following bond refractions:

C—H = 1.69, C—O = 1.51, O—H = 1.88 cm^3 mol^{-1}.

4 For aqueous solutions of α-D-glucose and β-D-glucose, α_m (293.15 K, 589.3 nm) is +113.4 and +19.3° dm^2 kg^{-1}, respectively. A fully mutarotated solution, which is an equilibrium mixture of these two forms, has α_m(293.15 K, 589.3 nm) = +52.5° dm^2 kg^{-1}. Calculate the percentage of each form in a fully mutarotated aqueous solution of glucose at 293.15 K. Calculate the total mass concentration of glucose in a fully mutarotated solution for which α = +11.21° in a 4 dm cell.

5 Riboflavine in dilute sodium acetate solution shows a maximum in the absorption spectrum at 444 nm. Calculate the wavenumber and frequency of light at this peak. Calculate the energy of the corresponding electronic transition for one molecule and for one mole of riboflavine.

6 At the absorption maximum given in the previous problem E(1 per cent, 1 cm) is 323 for riboflavine and the absorbance is 0.127 for a solution of the compound in a 4 cm cell. Calculate the concentration of riboflavine in g per 100 cm^3 and in mol m^{-3} and the molar absorption coefficient for riboflavine. Also calculate the transmittance of the solution. The molecular weight (relative molecular mass) of riboflavine is 376.4.

7 The nucleus ^{19}F has a g_N value of 5.256. At what magnetic flux density would this nucleus resonate in a nuclear magnetic resonance experiment employing a radio frequency of 60 MHz?

8 Sketch the general appearances you would expect for the proton and ^{13}C nuclear magnetic resonance spectra of acetaldehyde, CH$_3$CHO, given that protons and carbon atoms in CHO groups have greater chemical shifts δ than protons and carbon atoms in CH$_3$ groups. Plot frequency increasing to the left (or field increasing to the right) on a horizontal axis and indicate the approximate relative intensities of the peaks by their vertical heights. Indicate

on each sketch the resonance position of the standard, tetramethylsilane.

9 A pure dense organic liquid gave peaks in the mass spectrum at the following m/Q values, in units of m_u/e, (approximate percentage relative abundances in brackets):

26 (2 per cent), 27 (18 per cent), 28 (2 per cent), 29 (36 per cent), 127 (14 per cent), 128 (5 per cent), 141 (1.5 per cent), 156 (100 per cent), 157 (2 per cent).

Identify the compound and give the most probable formulae for the ions giving rise to the peaks at $m/Q = 27, 29, 127, 128, 141, 156$ and 157.

5 Chemical Reaction

Introduction

So far in this book, attention has been focussed mainly on the development of our understanding of the structure of matter. Now we turn to one of the other main themes in physical chemistry, namely, the study of the way in which substances undergo chemical change. The experimental data for such a study are measurements of the rates at which chemical changes take place. Such data are interesting in themselves and are important, for instance, when designing chemical plant in which reactions are to take place or when estimating the shelf-life of a drug under various conditions. They are more important, however, in the information which they provide about the mechanisms of the reactions, and it is that aspect with which this chapter is mainly concerned. For example, in pharmacy it is necessary to know not only how rapidly a drug decomposes or reacts with the constituents of the organism, but also the nature and mechanisms of the processes involved.

It is only possible here to show how the simpler types of reaction are interpreted (though fortunately many chemical reactions, both in the gas phase and in solution, show a behaviour which at least approximates to these simpler forms). In biological systems the situation is complicated in many ways. The reactants are not normally used up and the products of the reaction do not normally accumulate as in an ordinary chemical reaction because diffusion of substances towards and away from the reaction site takes place continually. Also, it is more common to find sequences of reactions or reaction cycles than single reactions in biological media. Nevertheless, the fundamental principles now to be outlined can be extended to provide a theoretical understanding of these more complex systems.

Order of Reaction and Molecularity

The principle which forms the basis of the interpretation of reaction rates is that embodied in the Law of Mass Action enunciated by Guldberg and Waage. It states that the rate of a chemical reaction is proportional to the active masses of the reacting substances. The concentrations of the substances can be taken, in most cases, to represent the active masses and this usage will be adopted in this chapter. The task, in deriving reaction mechanisms from studies of reaction rates, is therefore to propose a reaction or series of reactions which would give an overall rate varying with the concentrations of the reactants in the manner observed.

If chemical reactions always took place in one stage, the rate in each case would be proportional to the concentrations of the various reactants raised to the same power as the stoichiometric number (i.e. number of moles involved) in the chemical equation. In practice, it is found that the rate always depends only on a very small number of concentration terms, and this number of concentration terms is called the 'order' of the reaction. Thus, if the rate is proportional to $[A][B][C]$, or to $[A][B]^2$, the reaction is of third order. If it is proportional to $[A][B]$, or to $[A]^2$, it is of second order; and if it is simply proportional to $[A]$, the reaction is of first order. The reason why the number of concentration terms which determine the rate of the reaction sometimes differs from the number to be expected from the stoichiometric equation is that many reactions occur in stages, of which only one is a slow stage. The concentrations of reactants in this slow stage will be the only ones affecting the rate of the reaction, because the reaction as a whole cannot go faster than its slowest stage. The order of a reaction is therefore the value determined experimentally from studies of reaction rate, and is not derived by inspection of the chemical equation. It either corresponds to the chemical equation for the slow stage of the reaction, or, in some cases, it indicates the reactants which, by a sequence of equilibrium processes, lead to the slow stage of the reaction. A further distinction is made between the 'order' and the 'molecularity' of a reaction. The molecularity of a reaction is the total number of molecules of all the reactants in the slowest stage of the reaction. For example, the saponification of a simple ester by alkali in aqueous solution, whose mechanism is given below, is a bimolecular reaction, since two molecules of reactant enter into the slowest stage of the reaction.

$$HO^- + \overset{CH_3}{\underset{\underset{O^{\delta-}}{\overset{\|}{C}}}{\delta+C}}-OC_2H_5 \underset{\rightleftharpoons}{\overset{slow}{\longrightarrow}} HO-\overset{CH_3}{\underset{\underset{O^-}{|}}{C}}-OC_2H_5 \quad \Big\downarrow \text{fast}$$

$$\overset{CH_3}{\underset{\underset{O}{\|}}{\overset{|}{\underset{}{-}}C}}-O^- + HOC_2H_5 \overset{\text{very fast}}{\longleftarrow} HO-\overset{CH_3}{\underset{\underset{O}{\|}}{C}} + {}^-OC_2H_5$$

Experimental Study of the Progress and Rate of a Reaction

The order of a reaction is, as mentioned above, the number of concentration terms affecting the rate (or to which the rate of reaction is proportional) under the conditions of the experiment. These conditions might be such that one of the reactants is in great excess (e.g. by being the solvent for the reaction), so that, although some of it is changed during the reaction, the amount used will not be sufficient to produce any significant change in its concentration. Hence under these conditions the effect of change of concentration of this reactant on the reaction rate cannot be observed. For example, the alkaline saponification of an ester, which is bimolecular as mentioned above, is of the first order in the presence of a great excess of either alkali or ester. When the concentrations of the alkali and ester are not too dissimilar, the order is equal to the molecularity, so that the reaction is of the second order. The reaction may therefore be described as a reaction of the first order with respect to ester, of the first order with respect to alkali and of the second order overall.

The order is less than the molecularity of a reaction, when a reactant which enters into the slowest stage is in great excess, as we have seen, or when it is not used up but is constantly regenerated during the reaction. In these instances the molecularity is found by repeating the reaction rate experiments at different total concentrations of the reactants which are in excess or of that which is regenerated. The molecularity is then equal to the total number of concentration terms to which the rate of the reaction is proportional.

Most reactions are bimolecular and therefore of the first or second order. Only a few unimolecular and termolecular reactions are known. Unimolecular reactions seem to be restricted to a few decompositions in the gas phase. Reactions between nitric oxide and either oxygen, hydrogen or a halogen in the gas phase are believed to be termolecular. Chain reactions, which are reactions involving free radicals, may have a fractional order owing to the complexity of their kinetics, but each stage in the chain is usually bimolecular (p. 175). Heterogeneous reactions, which are reactions which take place at the interfacial boundary between two phases, are usually bimolecular (p. 182). If the heterogeneous reaction has only one reactant and if the interface is saturated with the reactant, the rate of the reaction is a constant independent of the concentration of the reactant in the bulk phase; the reaction is then said to be one of zero-order (p. 161).

Experimental Study of the Progress and Rate of a Reaction

The progress of a chemical reaction is followed by estimating, at regular time intervals, the amount of reactants remaining, or the amount of products which have been formed. Since the numbers of moles of reactants and products taking part in the reaction are related by the chemical equation, it is usually sufficient to investigate the variation of concentration of only one of the reactants or products. Usually, the initial concentrations of the reactants must be known, but if only one reactant is involved, its concentration may be deduced instead from that of a product when the reaction has gone to completion. Each concentration is usually found by withdrawing an aliquot part of the reaction mixture, running it, if necessary, into a solution which immediately stops the reaction from proceeding in this portion, and then titrating it with some appropriate standard solution. Sometimes, however, the variation may be investigated of some physical property which depends upon the concentration of one or more of the substances taking part in the reaction.

The following are examples of this.

(a) The pressure of gases involved in biological processes and reactions, e.g. O_2 and CO_2, may be followed in a Warburg constant volume respirometer.

(b) The density of a solution may be followed in a dilatometer which measures the volume occupied by a fixed mass of the solution.

(c) The conductivity of an electrolyte (p. 343) may be followed.

(d) The optical rotation of a solution may be followed in a polarimeter (p. 105).

(e) The relative permittivity (dielectric constant) of a medium may be measured in a capacitor.

(f) The refractive index of a solution may be measured in a refractometer (p. 103).

(g) Absorption of ultra-violet, visible or infra-red light by a solution may be followed in a spectrophotometer (p. 120). Since many absorbing substances obey Beer's law (p. 116), the change in absorbance is often proportional to the concentration of reactant remaining or of product formed.

With suitable instruments any of the above physical properties can be measured continuously while the reaction is in progress without the need for sampling. Most rate processes, especially reaction rates, are strongly dependent on temperature which must therefore be kept constant within small limits.

Fast reactions may be followed and their rates determined by flow methods. For this purpose two reacting solutions are allowed to flow together at a constant speed through a tube and a physical property, such as light absorption, is measured at various distances along the tube from the point of mixing.

In the derivations which follow in this section x mole dm^{-3} is the concentration of the reactant which has decomposed (or of the product which has been formed) at a time t after the start of the reaction. It is obvious that x increases as the reaction proceeds (Fig. 5.1). The initial concentration of the reactant is a mole dm^{-3} and that of the product is zero. The concentration of the reactant remaining at time t is therefore $(a-x)$ mole dm^{-3}. As the reaction proceeds $(a-x)$ decreases (Fig. 5.1). The plot of x or $(a-x)$ against t is known as the progress

Fig. 5.1 Progress curve of a chemical reaction

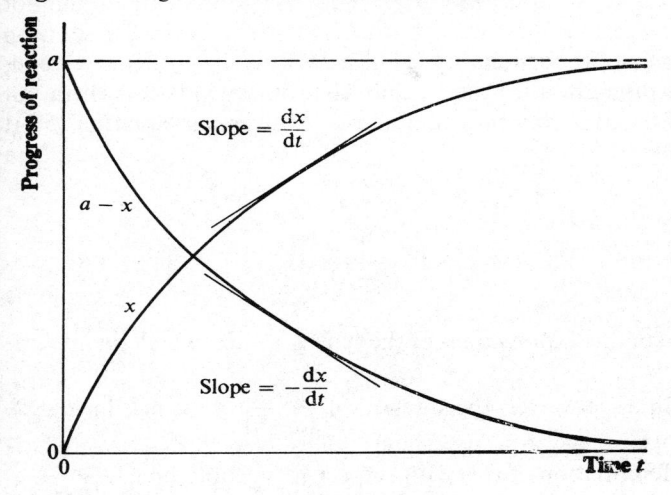

[Graph: Progress of reaction vs Time t, showing curve $a-x$ decreasing from a with Slope $= \frac{dx}{dt}$, and curve x increasing from 0 with Slope $= -\frac{dx}{dt}$]

a = initial concentration of reactant,
x = concentration of reactant decomposed at time, t,
$a-x$ = concentration of reactant present at time, t,
dx/dt = rate of reaction

curve of the reaction. For all types of reaction except those of zero order x approaches a asymptotically and $(a-x)$ approaches zero asymptotically (Fig. 5.1). The slope of the tangent to the progress curve of x against t at a given time is dx/dt, i.e. the rate of change of x with respect to t, and its dimension is concentration $\times$ time^{-1}. Similarly, the slope of the tangent to the progress curve of $(a-x)$ against t is $-dx/dt$. In the present treatment the rate of the reaction is taken to be dx/dt, but IUPAC now recommend a more complex definition. It can be seen from the progress curves that the rate of reaction, dx/dt, decreases as the reaction proceeds. This general rule applies to all reactions except those of zero order for which the rate of reaction is constant.

The law of mass action gives the rate of the reaction in terms of the concentration of the reactants and the corresponding equation is known as the rate equation. The kinetic equation or kinetic expression for a reaction indicates the concentration of reactant or product at any given time after the start of the reaction. It does not include rates or differentials and is derived by integrating the rate equation.

Chemical Reaction
First-order reactions

$$A \rightarrow \text{products.}$$

From the law of mass action the rate equation is:

$$\text{rate of reaction} \propto [A],$$

i.e.
$$\frac{dx}{dt} = k_1(a-x). \qquad [5.1]$$

The constant of proportionality, k_1, is called the first-order rate constant and its dimension is time^{-1}, since that of $(a-x)$ is concentration and that of dx/dt is concentration $\times$ time^{-1}. Inverting equation (5.1):

$$\frac{dt}{dx} = \frac{1}{k_1} \cdot \frac{1}{(a-x)}.$$

Hence,
$$t = \frac{1}{k_1} \int \frac{1}{(a-x)} \, dx.$$

The integral in this expression is of the type $\int \frac{1}{x} \, dx$, which on integration gives $\ln x$. However, the integral of $\frac{1}{(a-x)} \, dx$ is not $\ln(a-x)$, because differentiation of this with respect to x would give $-\frac{1}{(a-x)}$.[†]

Therefore a negative sign must be included in the integrated expression, which becomes:

$$t = \frac{1}{k_1}[-\ln(a-x)] + C,$$

where C is the integration constant which must be evaluated. We know that, at the beginning of the reaction, when $t = 0$, $x = 0$, because none of A is decomposed.

Hence:
$$0 = \frac{1}{k_1}[-\ln a] + C.$$

[†] To prove that, if $y = \ln(a-x)$, $\frac{dy}{dx} = -\frac{1}{(a-x)}$:

Put $\qquad (a-x) = u, \quad$ so that $\quad y = \ln u.$

Now, $\qquad \frac{dy}{dx} = \frac{dy}{du} \times \frac{du}{dx}.$

But, $\qquad \frac{dy}{du} = \frac{1}{u}, \quad$ and $\quad \frac{du}{dx} = -1,$

$\therefore \qquad \frac{dy}{dx} = -\frac{1}{u} = -\frac{1}{(a-x)}.$

By adding $\frac{1}{k_1}.\ln a$ to both sides, we have:

$$C = \frac{1}{k_1}.\ln a.$$

We may now substitute this value for C in the integrated expression obtained above. This gives:

$$t = \frac{1}{k_1}[-\ln(a-x)] + \frac{1}{k_1}.\ln a.$$

Multiplying both sides by k_1:

$$tk_1 = -\ln(a-x) + \ln a. \qquad [5.2]$$

And, putting the two ln terms together in the form of a quotient:

$$tk_1 = \ln\left(\frac{a}{a-x}\right), \qquad [5.3]$$

or, $$k_1 = \frac{1}{t}.\ln\left(\frac{a}{a-x}\right). \qquad [5.4]$$

This may be written in the alternative form, involving logarithms to the base 10, by including the factor 2.303, thus:

$$k_1 = \frac{2.303}{t}.\log\left(\frac{a}{a-x}\right). \qquad [5.5]$$

The last four equations are alternative kinetic expressions for a first-order reaction.

Equation [5.2] can be rearranged to give:

$$\ln(a-x) = \ln a - tk_1. \qquad [5.6]$$

From the form of this expression (cf. p. 2), it is seen that by plotting $\ln(a-x)$ against t, a straight line should be obtained, having a gradient $-k_1$. Or, putting the above expression in terms of logarithms to the base 10:

$$2.303 \log(a-x) = 2.303 \log a - tk_1, \qquad [5.7]$$

whence, dividing by 2.303, we obtain:

$$\log(a-x) = \log a - \frac{k_1}{2.303}.t. \qquad [5.8]$$

So that a graph of $\log(a-x)$ against t should be a straight line of gradient $-k_1/2.303$ and of intercept $\log a$ on the ordinate.

A rather surprising and most useful fact emerges from an examination of the form of the kinetic equation for a first-order reaction.

There are two concentration terms in equations [5.3–5.5] and these appear as the ratio $\left(\dfrac{a}{a-x}\right)$. This means that it is not necessary to express the concentrations in mole dm^{-3}. Any quantity or physical property which is proportional to the concentration may be used in the equation in place of the concentration. For example, if the concentrations are found by titration, the titres may be used directly in the equation. This simplification applies equally to the determination of k_1 from the gradient of the graph of log $(a-x)$ against time, since altering the units of concentration does not alter the gradient.

Experiment 5.1. Determination of k_1 for the hydrolysis of methyl acetate in the presence of an acid catalyst.

In order to obtain reliable results from this experiment, it is necessary to maintain a constant temperature by immersing the reaction flask in a thermostat. However, if this is not possible, the reaction may be carried out in a flask immersed in a bath of water at room temperature, or in a bath through which cold water from the tap is flowing.

50 cm³ of 2 mol dm^{-3} HCl are placed in a 250 cm³ flask in a thermostat, and 45 cm³ of fresh (CO_2-free) distilled water are added. The flask is stoppered and allowed to attain the temperature of the thermostat. Then, 5 cm³ of methyl acetate are pipetted into the flask, which is shaken to mix thoroughly, and the stop-clock is started. Samples of the reaction mixture are analysed at 15-minute intervals, the first sample being taken 5 minutes after the beginning of the reaction. To analyse the sample, 5 cm³ are withdrawn and run into about 50 cm³ of fresh distilled water and titrated against 0.2 mol dm^{-3} NaOH obtained from a burette fitted with a soda-lime tube. The reaction is slowed down considerably by running the sample into water, but it should nevertheless be titrated as soon as possible. Readings are continued for about two hours, and an 'infinity' reading is also made, preferably after about two days. The reaction is bimolecular, proceeding according to the equation:

$$CH_3COOCH_3 + H_2O \rightarrow CH_3COOH + CH_3OH,$$

but, since excess water is present, the reaction follows the first-order law. The final concentration of acetic acid is equal to the original concentration of methyl acetate (a), and the amount of methyl acetate decomposed (x) is equal to the amount of acetic acid formed at any given time t. Hence, the difference between the titres at 'infinity' and time t is proportional to $(a-x)$. The fact that the difference between two titres is taken allows automatically for that part of the titre required to neutralise the constant amount of HCl which is titrated at the same time as the acetic acid. A graph of the logarithm of the difference between these two titres plotted against t (in minutes) for each sample should give a straight line of gradient $-k_1/2.303$. From this gradient k_1 is calculated in units of min.$^{-1}$. If the reaction were repeated at different acid concentrations, it would be found that k_1 would be proportional to the acid concentration, thus demonstrating that the acid is acting as a catalyst for this reaction (cf. p. 387).

Another example of this principle of using a physical property in place of concentration is provided by the mutarotation of α-D-glucose (α-dextrose), a first-order reaction of measurable rate. Aqueous solutions of this substance undergo a fall in optical rotation as a result of a change in the configuration of one of the carbon atoms in the molecule. The final state, when the reaction proceeds no further, consists of an equilibrium mixture of the two forms of dextrose, α-dextrose and β-dextrose. This is shown by the fact that the optical rotation of the solution tends to a value between those expected for the two isomers. This reaction is an example of a 'reversible reaction' (see p. 315), since it proceeds at a finite rate in the reverse direction, with a rate constant k_{-1}, as well as in the forward direction, with a rate constant k_1. Other examples include *cis-trans* isomerisations. The mutarotation of glucose may be represented by the scheme:

$$\alpha\text{-D-glucose} \underset{k_{-1}}{\overset{k_1}{\rightleftharpoons}} \beta\text{-D-glucose}$$

Initial concentrations (at time $t = 0$)	a	0
Concentrations later (at time t)	$a-x$	x
Concentrations at equilibrium (at time $t = \infty$)	$a-e$	e

The rate of mutarotation according to the reaction scheme above, is the rate of formation of the β-isomer minus the rate of its removal, as follows:

$$\frac{dx}{dt} = k_1(a-x) - k_{-1}x. \qquad [5.9]$$

At equilibrium, the concentration of β-D-glucose, x, has the equilibrium value, e, and the rate of the overall reaction dx/dt, is zero or, in other words, the rates of the forward and reverse reactions are equal, thus:

$$0 = k_1(a-e) - k_{-1}e$$
$$\therefore \quad k_1(a-e) = k_{-1}e. \qquad [5.10]$$

The equilibrium constant, K, is the ratio of the equilibrium concentration of the β-isomer to that of the α-isomer which, according to equation [5.10], is equal to the ratio of the rate constant of the forward reaction to that of the reverse reaction, thus:

$$K = \frac{e}{a-e} = \frac{k_1}{k_{-1}} \quad [5.11]$$

$$\therefore k_{-1} = k_1\left(\frac{a-e}{e}\right).$$

Substituting into equation [5.9] we have:

$$\frac{dx}{dt} = k_1(a-x) - \frac{k_1 x(a-e)}{e}$$

$$= k_1 a - k_1 x - \frac{k_1 x a}{e} + k_1 x$$

$$\therefore \frac{dx}{dt} = \frac{k_1 a}{e}(e-x). \quad [5.12]$$

This rate equation may be integrated in the same way as equation [5.1] to give the following kinetic expression which is analogous to equation [5.3]:

$$\frac{k_1 a t}{e} = \ln\left(\frac{e}{e-x}\right). \quad [5.13]$$

Adding $k_1 e$ to both sides of equation [5.10] we obtain:

$$k_1 a = (k_{-1} + k_1)e.$$

Substitution of this expression for $k_1 a$ into equation [5.13] gives:

$$(k_1 + k_{-1})t = \ln\left(\frac{e}{e-x}\right). \quad [5.14]$$

Converting to logarithms to the base 10 and dividing throughout by $(k_1 + k_{-1})$, we obtain the following equation which parallels equation [5.5]:

$$t = \frac{2.303}{(k_1 + k_{-1})} \log\left(\frac{e}{e-x}\right). \quad [5.15]$$

Conversion to an equation analogous to equation [5.8] gives:

$$\log(e-x) = \log e - \frac{(k_1 + k_{-1})}{2.303} \cdot t \quad [5.16]$$

So a plot of $\log(e-x)$ against t should give a straight line of gradient $-(k_1 + k_{-1})/2.303$ and of intercept $\log e$ on the ordinate.

Experimental Study of the Progress and Rate of a Reaction

Let the equilibrium mixture, which contains a concentration e of β-dextrose, have an optical rotation α_∞. If the initial optical rotation is α_0 and that at a given time t is α_t, the decrease in optical rotation is $\alpha_0 - \alpha_t$, which is proportional to the concentration of α-dextrose which has mutarotated, namely, x, i.e.

$$(\alpha_0 - \alpha_t) = \text{constant} \times x. \tag{5.17}$$

When $t = \infty$, $\alpha_t = \alpha_\infty$ and $x = e$,

so that
$$(\alpha_0 - \alpha_\infty) = \text{constant} \times e \tag{5.18}$$

Subtraction of equation [5.17] from [5.18] gives:

$$(\alpha_0 - \alpha_\infty) - (\alpha_0 - \alpha_t) = (\text{constant} \times e) - (\text{constant} \times x)$$

$$\therefore (\alpha_t - \alpha_\infty) = \text{constant} \times (e - x). \tag{5.19}$$

Dividing equation [5.18] by equation [5.19] we obtain:

$$\frac{\alpha_0 - \alpha_\infty}{\alpha_t - \alpha_\infty} = \frac{e}{e - x}, \tag{5.20}$$

which may be substituted into equation [5.14] or [5.15] to enable $(k_1 + k_{-1})$ to be calculated. Alternatively, α_t can be measured at various times and $(k_1 + k_{-1})$ may be determined from the gradient of the graph of $\log(\alpha_t - \alpha_\infty)$ against time (cf. equation 5.16).

Knowledge of $(k_1 + k_{-1})$ and of the equilibrium constant, K, which according to equation [5.11] is equal to k_1/k_{-1}, enables both k_1 and k_{-1} to be evaluated. Alternatively, the rate constant k_1 of the forward reaction can be determined independently by measuring the initial rate of the forward reaction, v_1, when only the α-isomer is supplied at an initial concentration, a, thus:

$$v_1 = k_1 a. \tag{5.21}$$

Similarly, the rate constant, k_{-1}, of the reverse reaction can be found by measuring the initial rate of the reverse reaction, v_{-1}, when only the β-isomer is supplied at an initial concentration, b, thus:

$$v_{-1} = k_{-1} b. \tag{5.22}$$

The half-life (or time of half change), $t_{\frac{1}{2}}$, of a reaction is the time taken for the concentration of the reactant to fall to one half of its initial value. This quantity may be used to specify the rate of a reaction when there is only one reacting species or when two or more reactants have equal initial concentrations. The relationship between the half-life and the rate constant is determined by putting $t = t_{\frac{1}{2}}$ and $(a - x) =$

$\tfrac{1}{2}a$ into the kinetic equation. Equation [5.4] then gives:

$$k_1 = \frac{1}{t_{\frac{1}{2}}} \ln \frac{a}{\frac{1}{2}a}$$

$$= \frac{\ln 2}{t_{\frac{1}{2}}}$$

$$\therefore\ t_{\frac{1}{2}} = \frac{\ln 2}{k_1}.\qquad [5.23]$$

Since k_1 is a constant, the half-life of a first-order reaction is a constant independent of the initial concentration of reactant.

Second-order reactions

$$A + B \rightarrow \text{products.}$$

Let the initial concentration of A be a mole dm^{-3} and that of B be b mole dm^{-3}. If after a given time, t, x mole dm^{-3} of A has decomposed, x mole dm^{-3} of B must also have decomposed, provided that the stoichiometric equation for the overall reaction shows that equal numbers of moles of A and B react together. At this time, t, [A] $= (a-x)$ mole dm^{-3} and [B] $= (b-x)$ mole dm^{-3}. Application of the law of mass action gives the rate equation:

$$\text{rate of reaction} \propto [A][B],$$

i.e.
$$\frac{dx}{dt} = k_2(a-x)(b-x),\qquad [5.24]$$

where k_2 is the second order rate constant.

Inverting:
$$\frac{dt}{dx} = \frac{1}{k_2} \cdot \frac{1}{(a-x)(b-x)}$$

$$\therefore\ t = \frac{1}{k_2} \int \frac{1}{(a-x)(b-x)} \,dx.\qquad [5.25]$$

In order to integrate the function $\dfrac{1}{(a-x)(b-x)}$, it must first be converted into the partial fractions shown on the right hand side of the following identity:

$$\frac{1}{(a-x)(b-x)} \equiv \frac{Q}{a-x} + \frac{R}{b-x}.\qquad [5.26]$$

Multiplication of both sides of the identity by $(a-x)(b-x)$ enables Q and R to be evaluated, thus:

$$1 \equiv Q(b-x) + R(a-x)$$
$$\therefore\ 1 \equiv Qb - Qx + Ra - Rx.$$

Equating the terms in x:
$$0 = -Q - R \quad \therefore R = -Q.$$

Equating the constant terms:
$$1 = Qb + Ra.$$

Substituting $-Q$ for R:
$$1 = Qb - Qa = Q(b-a)$$
$$\therefore Q = \frac{1}{b-a}.$$

But
$$R = -Q.$$
$$\therefore R = \frac{-1}{b-a}.$$

Substituting for R and Q in the original identity [5.26] gives:
$$\frac{1}{(a-x)(b-x)} \equiv \frac{1}{b-a}\left[\frac{1}{a-x} - \frac{1}{b-x}\right].$$

This identity is valid for all values of a, b and x, except when $a = b$. If $a = b$, it can be seen that the right hand side of the identity becomes $\frac{1}{0} \times 0$, which has no meaning. Provided that $a \neq b$, the identity may be substituted into equation [5.25], thus:

$$t = \frac{1}{k_2} \cdot \frac{1}{b-a} \int \left(\frac{1}{a-x} - \frac{1}{b-x}\right) dx$$

$$\therefore t = \frac{1}{k_2} \cdot \frac{1}{b-a}\left[\int \frac{1}{a-x} \cdot dx - \int \frac{1}{b-x} \cdot dx\right]. \quad [5.27]$$

The functions $\frac{1}{a-x}$ and $\frac{1}{b-x}$ are integrated as previously described (p. 150) for first-order reactions, thus:

$$\int \frac{1}{a-x} \cdot dx = -\ln(a-x) + C$$

$$\int \frac{1}{b-x} \cdot dx = -\ln(b-x) + C'.$$

Substitution into equation [5.27] gives:

$$t = \frac{1}{k_2} \cdot \frac{1}{b-a}[-\ln(a-x) + \ln(b-x) + C_2]$$

$$\therefore t = \frac{1}{k_2} \cdot \frac{1}{b-a}\left[\ln\left(\frac{b-x}{a-x}\right) + C_2\right], \quad [5.28]$$

where $C_2 \ (= C+C')$ is an integration constant which is evaluated as before by putting $x = 0$ when $t = 0$, thus:

$$0 = \frac{1}{k_2} \cdot \frac{1}{b-a} \left[\ln \frac{b}{a} + C_2 \right].$$

Both sides of this equation may be multiplied by $k_2(b-a)$ which is not equal to zero because $a \neq b$.

$$\therefore \ 0 = \ln \frac{b}{a} + C_2$$

$$\therefore \ C_2 = -\ln \frac{b}{a}.$$

Introducing the value of C_2 into the integrated equation [5.28] gives the kinetic equation:

$$t = \frac{1}{k_2} \cdot \frac{1}{b-a} \left[\ln \left(\frac{b-x}{a-x} \right) - \ln \frac{b}{a} \right]. \qquad [5.29]$$

Putting this expression in terms of logarithms to the base 10:

$$t = \frac{2.303}{k_2} \cdot \frac{1}{b-a} \left[\log \left(\frac{b-x}{a-x} \right) - \log \frac{b}{a} \right], \qquad [5.30]$$

whence rearranging to give $\log \left(\frac{b-x}{a-x} \right)$ in terms of t we obtain:

$$\log \left(\frac{b-x}{a-x} \right) = \frac{k_2(b-a)}{2.303} \cdot t + \log \frac{b}{a}. \qquad [5.31]$$

If $\log \left(\frac{b-x}{a-x} \right)$ is plotted against t, a straight line graph is obtained, having a gradient $k_2(b-a)/2.303$ and intercepting the ordinate at $\log (b/a)$.

When the initial concentrations of the two reactants are equal, the above derivations are not valid and the following treatment must be applied. Let the initial concentration of each reactant **A** and **B** be a mole dm^{-3} (at $t = 0$). It is usually possible to arrange that this condition applies in practice. The rate equation [5.24] now becomes:

$$\frac{dx}{dt} = k_2(a-x)^2. \qquad [5.32]$$

Inverting:

$$\frac{dt}{dx} = \frac{1}{k_2} \cdot \frac{1}{(a-x)^2}.$$

$$\therefore \ t = \frac{1}{k_2} \int \frac{1}{(a-x)^2} \cdot dx.$$

Experimental Study of the Progress and Rate of a Reaction

The integral is of the form $\int \frac{1}{x^2} \, dx$, which becomes $-\frac{1}{x}$. However, the function to be integrated involves $(a-x)$ in place of x, and this involves the introduction of a further minus sign.† Hence, integration of the above expression for t gives:

$$t = \frac{1}{k_2} \cdot \frac{1}{(a-x)} + C_2.$$

C_2 is the integration constant which is found from the fact that $x = 0$ when $t = 0$.

$$\therefore \quad 0 = \frac{1}{k_2} \cdot \frac{1}{a} + C_2.$$

Subtracting $\frac{1}{k_2} \cdot \frac{1}{a}$ from both sides,

$$C_2 = -\frac{1}{k_2} \cdot \frac{1}{a}.$$

Substituting for C_2 in the integrated expression for t:

$$t = \frac{1}{k_2} \cdot \frac{1}{(a-x)} - \frac{1}{k_2} \cdot \frac{1}{a}.$$

Multiplying both sides by k_2,

$$tk_2 = \frac{1}{(a-x)} - \frac{1}{a}. \qquad [5.33]$$

Putting the right-hand side of the equation in a form having a common denominator:

$$tk_2 = \frac{a-(a-x)}{a(a-x)}$$

$$tk_2 = \frac{x}{a(a-x)}. \qquad [5.34]$$

Hence,
$$k_2 = \frac{1}{t} \cdot \frac{x}{a(a-x)}. \qquad [5.35]$$

† To prove that if $y = \frac{1}{(a-x)}$, $\frac{dy}{dx} = \frac{1}{(a-x)^2}$.

Put $(a-x) = u$, so that $y = \frac{1}{u}$.

$$\frac{dy}{dx} = \frac{dy}{du} \times \frac{du}{dx}.$$

Now $\frac{dy}{du} \equiv -\frac{1}{u^2}$, and $\frac{du}{dx} = -1$,

$$\therefore \quad \frac{dy}{dx} \equiv +\frac{1}{u^2} = \frac{1}{(a-x)^2}.$$

The last three numbered equations [5.33–5.35] are alternative kinetic equations for a second-order reaction with equal initial concentrations of reactants.

Equation [5.33] can be rearranged to give:

$$\frac{1}{a-x} = tk_2 + \frac{1}{a} \qquad [5.36]$$

This equation shows that a plot of $\frac{1}{a-x}$ against t gives a straight line of gradient k_2 intercepting the ordinate at $\frac{1}{a}$. The terms $(a-x)$ and a must be expressed in concentration units, i.e. mole dm^{-3}. Equation [5.35] shows that the dimensions of k_2 are concentration^{-1} time^{-1}, so convenient units are dm^3 mole^{-1} s^{-1} or dm^3 mole^{-1} minute^{-1}.

If both sides of equation [5.34] are multiplied by a, we have:

$$\frac{x}{a-x} = ak_2 t. \qquad [5.37]$$

When $\frac{x}{a-x}$ is plotted against t, we obtain a straight line with a gradient ak_2 and passing through the origin. The concentration terms x and $(a-x)$ in the ratio $\frac{x}{a-x}$ can be directly replaced by any quantity or physical property to which they are proportional, but a must be expressed in units of concentration, i.e. mole dm^{-3}. For example, the saponification of ethyl acetate in aqueous solution by an equal concentration of sodium hydroxide is a second-order reaction of measurable rate whose rate constant can readily be determined by conductivity measurements. Since sodium hydroxide and the product sodium acetate are completely ionised, the contribution of each to the conductivity of the solution is approximately proportional to its concentration (p. 346). Since ethyl acetate and the product ethanol are unionised they contribute nothing to the conductivity. Consequently, the conductivity of the solution at zero time, κ_0, is equal to the conductivity of the initial concentration of sodium hydroxide and the conductivity of the solution after an infinite time, κ_∞, is equal to the conductivity of the same concentration of sodium acetate. Since $\lambda(\text{OH}^-) > \lambda(\text{CH}_3\text{COO}^-)$ (p. 351), $\kappa_0 > \kappa_\infty$ and the conductivity falls during the reaction. If κ_t is the conductivity of the solution at a given time t, the fall in conductivity during this time is $(\kappa_0 - \kappa_t)$ which is proportional to concentration of substrate which has reacted, namely x, i.e.

$$(\kappa_0 - \kappa_t) = \text{constant} \times x. \qquad [5.38]$$

Experimental Study of the Progress and Rate of a Reaction 161

When $t = \infty$, $\kappa_t = \kappa_\infty$ and $x = a$

so that $(\kappa_0 - \kappa_\infty) = \text{constant} \times a.$ [5.39]

Subtracting equation [5.38] from [5.39], we obtain:

$$(\kappa_t - \kappa_\infty) = \text{constant} \times (a-x).$$ [5.40]

Division of equation [5.38] by [5.40] gives:

$$\frac{\kappa_0 - \kappa_t}{\kappa_1 - \kappa_\infty} = \frac{x}{a-x}.$$ [5.41]

Comparison with equation [5.37] indicates that, when $\dfrac{\kappa_0 - \kappa_t}{\kappa_t - \kappa_\infty}$ is plotted against t, a straight line of gradient ak_2 and passing through the origin is obtained. k_2 is therefore equal to the gradient divided by a, which must be expressed in concentration units.

Other examples of second-order reactions are the hydrolysis of esters when no reactant is in excess, the reaction betwen tertiary amines and alkyl iodides:

$$R_3N + R'I = R_3R'N^+ I^-$$

and the initial stages of the following reaction and of its reverse reaction:

$$2HI = H_2 + I_2.$$

The relationship between the half-life, $t_\frac{1}{2}$, and the second-order rate constant, k_2, for equal initial concentrations of reactant can be found by substituting $t = t_\frac{1}{2}$, $(a-x) = \frac{1}{2}a$ (or $x = \frac{1}{2}a$) into any of the kinetic equations, such as [5.33]:

$$t_\frac{1}{2} k_2 = \frac{1}{\frac{1}{2}a} - \frac{1}{a}$$

$$= \frac{2}{a} - \frac{1}{a} = \frac{1}{a}$$

$$\therefore \quad t_\frac{1}{2} = \frac{1}{ak_2}.$$ [5.42]

Since k_2 is a constant, the half-life of a second-order reaction with equal initial concentrations, a, of reactants is inversely proportional to a.

Zero-order reactions

$$\left. \begin{array}{c} \text{Constant quantity of} \\ \text{adsorbed reactant} \end{array} \right\} \rightarrow \text{products}$$

The rate of the reaction is a constant independent of the concentration of reactant $(a-x)$ in the bulk phase from which adsorption occurs (pp. 186, 187).

$$\therefore \frac{dx}{dt} = k_0(a-x)^0 = k_0, \quad [5.43]$$

where k_0 is the zero-order rate constant, whose dimension is that of dx/dt, namely mole dm^{-3} time^{-1}. Integration of this equation gives:

$$x = k_0 t + C_0$$

where C_0 is a constant of integration. At zero time no reactant has decomposed, i.e. when $t = 0$, $x = 0$ $\therefore C_0 = 0$.

$$\therefore x = k_0 t. \quad [5.44]$$

This is the kinetic equation for a zero-order reaction and implies that the concentration of reactant decomposed or of product formed is proportional to the time after the start of the reaction. The rate constant is the constant of proportionality and is equal to the gradient of the plot of x against t.

The half-life, $t_{\frac{1}{2}}$, is calculated as before by substituting $t = t_{\frac{1}{2}}$ and $(a-x) = \frac{1}{2}a$ (or $x = \frac{1}{2}a$) into the kinetic equation (i.e. 5.44), thus:

$$\frac{a}{2} = k_0 t_{\frac{1}{2}}$$

whence
$$t_{\frac{1}{2}} = \frac{a}{2k_0} \quad [5.45]$$

Since k_0 is a constant, the half-life of a zero-order reaction is proportional to the initial concentration of the reactant in the bulk phase.

Determination of Order of Reaction

One of the first steps in the investigation of the kinetics of a particular reaction is the determination of its order. One of the methods by which this may be found is the method of isolation. The reaction is carried out with excess of every reactant except one, and the effect is studied of having different initial concentrations of this one reactant. If the rate of reaction is proportional to the concentration of this reactant, the reaction must be of first order relative to this substance. Each reactant is studied in this way in turn, and the total order of reaction is the sum of the orders relative to the various reactants.

A similar method of determining the order, n_A, of a reaction with respect to a substance A is to determine the initial rate of the reaction at various initial concentrations, a, of that reactant, while keeping constant the initial concentrations of all the other reactants, e.g. B and C. The rate equation is:

$$\frac{dx}{dt} = k(a-x)^{n_A}(b-x)^{n_B}(c-x)^{n_C}, \quad [5.46]$$

Determination of Order of Reaction 163

where k is the rate constant and n_B and n_C are the orders relative to the reactants B and C whose initial concentrations are b and c respectively. The initial rate is the value of dx/dt when $t = 0$, and is therefore equal to the gradient of the tangent to the initial part of the plot of x against t or to the positive value of the gradient of the tangent to the initial part of the plot of $(a-x)$ against t (Fig. 5.1). The value of x is then zero, so equation [5.46] becomes:

$$\text{initial rate} = k a^{n_A} b^{n_B} c^{n_C}. \quad [5.47]$$

For constant values of b and c as stated above:

$$\text{initial rate} = k' a^{n_A}, \quad [5.48]$$

where k' is a constant. Taking logarithms we have:

$$\log (\text{initial rate}) = \log k' + n_A \log a. \quad [5.49]$$

A plot of log (initial rate) against log a gives a straight line of gradient n_A, the order of the reaction with respect to A. The order of the reaction with respect to the other reactants is similarly determined. The overall order, n, of the reaction is equal to $(n_A + n_B + n_C)$. This method is useful for finding the order of slow, complicated or enzymatic reactions, but requires that the early part of the progress curve be accurately determined. This is best carried out by measurement of a physical property, such as absorbance.

Another method of finding the order of reaction, which is particularly applicable when there is only one reacting species, is to determine how the half-life, $t_{\frac{1}{2}}$, varies with initial concentration, a. For a first-order reaction equation [5.23] shows that $t_{\frac{1}{2}}$ is independent of a. For a second-order reaction $t_{\frac{1}{2}}$ is inversely proportional to a (equation 5.42), and for a zero-order reaction $t_{\frac{1}{2}}$ is proportional to a (equation 5.45). In general, for a reaction in which all the reactants have the same initial concentration a, it can be shown that:

$$t_{\frac{1}{2}} = \text{constant} \times a^{(1-n)}, \quad [5.50]$$

where n is in order of the reaction. Taking logarithms of both sides gives:

$$\log t_{\frac{1}{2}} = \log \text{constant} + (1-n) \log a. \quad [5.51]$$

A plot of $\log t_{\frac{1}{2}}$ against $\log a$ gives a straight line of gradient $(1-n)$ from which the order of the reaction can be calculated. This method is particularly useful when the order is fractional.

A fourth method of finding the order is simply to employ calculation or graphical means to discover which kinetic equation best fits the experimental data. This method is essentially the same as the determination of the rate constant, k. If the wrong kinetic equation has been chosen, it will not be possible to obtain a constant value for k, or a straight line will not be obtained in the graphical method.

Temperature Dependence of Rate Constants

For most reactions the rate constant varies considerably with temperature. A rise in temperature produces an increase in reaction rate measured as an increase in k. That is why it is important, in rate experiments, to keep the temperature as uniform as possible by immersing the reaction vessel in a thermostat. It also explains why decomposition of labile drugs, metabolites, enzymes and organs is slower the lower the temperature and suggests that all the reactions occurring in the body are speeded up in fever. Such a variation with temperature suggests that reaction takes place more readily when the energy of the reacting molecules is raised.

Arrhenius found that the rate constant k of a given reaction varies with temperature according to the following equation:

$$k = Ae^{-E/RT}, \qquad [5.52]$$

where A is a constant known as the pre-exponential (or frequency) factor, which for practical purposes is independent of temperature,

$e^{-E/RT}$ is called the exponential factor,
E is the activation energy in energy units per mole,
R is the gas constant (8.314 J K^{-1} mol^{-1})
and T is the absolute temperature in K.

The activation energy may be expressed as ε ($= E/L$) energy units per molecule, where L is Avogadro's constant. The gas constant must then be replaced by Boltzmann's constant, k ($= R/L$), and the exponential factor is written as:

$$e^{-\varepsilon/kT}.$$

Writing equation [5.52] in the logarithmic form:

$$\ln k = \ln A - E/RT. \qquad [5.53]$$

Differentiating with respect to T:

$$\frac{d \ln k}{dT} = \frac{E}{RT^2}. \qquad [5.54]$$

This is of the same form as the van't Hoff isochore which relates the enthalpy of reaction to the variation of equilibrium constant with temperature (p. 320). In fact, just as the equilibrium constant is equal to the rate constant of the forward reaction divided by that of the reverse reaction, so the enthalpy of reaction is equal to the activation energy of the forward reaction minus that of the reverse reaction.

Since A is a constant, equation [5.53] suggests that E can be calculated from a linear plot. Dividing throughout by 2.303 and using the fact that $(\ln x)/2.303 \equiv \log x$, we have:

$$\log k = \log A - \frac{E}{2.303R} \cdot \frac{1}{T} \qquad [5.55]$$

If k is determined at a number of different temperatures, a graph of $\log k$ against $1/T$ should give a straight line of gradient $-E/2.303R$. A change in slope of the graph at a certain temperature indicates a change in activation energy and therefore a change in mechanism at that temperature. From the graph the rate constant at other temperatures may be readily determined by interpolation or sometimes by extrapolation. This method is used in pharmacy to predict the rates of decomposition of drugs at various ambient temperatures from the more rapid rates determined at two or more elevated temperatures. If only two values of k are found, k_1 and k_2, at temperatures T_1 and T_2 respectively, equation [5.55] can be applied to each pair of values k_2, T_2 and k_1, T_1, and the second result subtracted from the first to give:

$$\log k_2 - \log k_1 = \frac{-E}{2.303R}\left(\frac{1}{T_2} - \frac{1}{T_1}\right) \qquad [5.56]$$

or

$$\log \frac{k_2}{k_1} = \frac{E}{2.303R}\left(\frac{T_2 - T_1}{T_1 T_2}\right). \qquad [5.57]$$

So, either by the graphical method or by substituting into the equation, the activation energy can be found. If this and k at one temperature are known, k at any other temperature can be calculated. Since k enters into the plot as $\log k$ and into the equation as a ratio, any quantity to which k is proportional, such as the actual rate of reaction at a fixed concentration of reactants, can be used in place of k.

Collision Theory of Reaction Rates

This theory first states that molecules must come into contact by collision in order to react. Collision between two molecules is the general rule, which explains why bimolecular reactions are so common. Simultaneous collision between three molecules is most uncommon, which explains the rarity of termolecular reactions. The number of molecules reacting per unit volume per second is many orders of magnitude less than the number of molecules colliding per unit volume per second. Mere collision is therefore an insufficient condition for reaction. Further conditions must be fulfilled before reaction can take place.

The second condition of the collision theory is that molecules react only if their combined kinetic energy on collision is greater than, or equal to, a critical value known as the activation energy which constitutes a potential energy barrier to reaction (Fig. 5.2). After collision

Fig. 5.2 Activation energy as a barrier to reaction

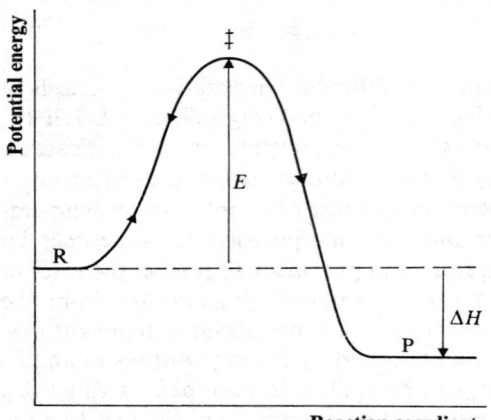

R = reactants, P = products, ‡ = transition state or activated complex, E = activation energy, ΔH = enthalpy of reaction (negative in this example)

the molecules slow down and their kinetic energy falls. The principle of conservation of energy demands that this energy is not lost but is converted into an equivalent quantity of potential energy. The colliding molecules, as it were, ascend to the appropriate 'height' up the potential energy 'hill'. Most colliding molecules have insufficient energy to reach the top of the barrier, so the potential energy is reconverted back into the original amount of kinetic energy. In other words, the molecules fly apart again with the same total momentum as before collision. The collision is then said to be an elastic collision.

Occasionally, a pair of colliding molecules have sufficient kinetic energy to overcome the barrier and reaction occurs. The Maxwell–Boltzmann distribution of molecular energies among a system of molecules continually exchanging energy by collision (p. 24) shows that the proportion of molecules having an energy greater than or equal to a given value ε J molecule^{-1} (or E J mol^{-1}) is $e^{-\varepsilon/kT}$ ($= e^{-E/RT}$). According to the collision theory this is also the ratio of the number N of molecules reacting in unit volume and in unit time to the number Z of bimolecular collisions in unit volume und in unit time, thus:

$$N = Ze^{-E/RT}. \qquad [5.58]$$

For the bimolecular reaction:

$$A + B \rightarrow \text{products},$$

Z should be the number of collisions in unit volume and in unit time between the different types of molecule and will vary as the concentration of both A and B and as the square root of the absolute temperature, T K, thus:

$$Z = \text{constant}' \times T^{\frac{1}{2}} [A] [B]. \qquad [5.59]$$

N is proportional to the rate of the reaction, i.e.:

$$N = \text{constant}'' \times \frac{dx}{dt}. \qquad [5.60]$$

Introducing these expressions for Z and N into equation [5.58]:

$$\text{constant}'' \times \frac{dx}{dt} = \text{constant}' \times T^{\frac{1}{2}} [A] [B] e^{-E/RT}$$

$$\therefore \frac{dx}{dt} = Y T^{\frac{1}{2}} [A] [B] e^{-E/RT}, \qquad [5.61]$$

where Y is constant ($Y = \text{constant}'/\text{constant}''$).
Comparison of this equation with the rate equation [5.24] shows that:

$$k_2 = Y T^{\frac{1}{2}} e^{-E/RT}. \qquad [5.62]$$

For the bimolecular reaction:

$$2A \rightarrow \text{products}$$

Z should be the number of collisions m^{-3} s^{-1} between the same type of molecule A and will be proportional to $T^{\frac{1}{2}}$ and $[A]^2$. Since the rate equation for the reaction shows that dx/dt is proportional to $[A]^2$, the equation relating k_2 and T has the same form as [5.62]. The constant Y in this equation can be calculated from the molecular characteristics by means of the kinetic theory of gases, and for gaseous bimolecular reactions is usually of the order 10^8 m^3 mol^{-1} s^{-1} K$^{-\frac{1}{2}}$, i.e. 10^{11} dm^3 mol^{-1} s^{-1} K$^{-\frac{1}{2}}$.

A very approximate empirical rule for many reactions is that an increase of temperature of 10° C roughly doubles k, i.e. increases k by about 100 per cent. If the temperature increases from 27° C ($T = 300$ K) to 37° C ($T = 310$ K), $T^{\frac{1}{2}}$ in equation [5.62] is increased by only 1.7 per cent, which is comparatively insignificant. The major factor which accounts for the very marked effect of temperature changes on the rate constants of chemical reactions is the exponential factor, $e^{-E/RT}$. The large variation of this factor with temperature overwhelms the relatively small variation in $T^{\frac{1}{2}}$. Consequently, $T^{\frac{1}{2}}$ is assumed to be constant and equation [5.62] can be written as:

$$k = Xe^{-E/RT}, \qquad [5.63]$$

where X is a constant approximately independent of temperature. This equation has the same form as the Arrhenius equation [5.52].

Using the experimental value of E and the calculated quantity Y, k may be calculated by means of equation [5.62] and compared with the value determined experimentally (p. 165). For some bimolecular reactions, e.g. $2\,HI = H_2 + I_2$ and its reverse, the calculated and experimentally determined values of k agree quite well. This agreement also applies to certain reactions in solution. At first sight this may seem surprising, since in solution solvent molecules predominate, with the result that the number of encounters between two reactant molecules is much less than in the gas phase. Having once come into contact, however, two reactant molecules stay near each other in solution much longer than in the gas phase, because the surrounding solvent molecules reduce the chances of their separation. The two reactant molecules repeatedly collide and the likelihood of their activation is increased. Thus, for many reactions in solution the former effect is compensated by the latter effect, so that the number of effective collisions and the rate constant are similar to those of the same reaction in the gas phase.

For many gaseous bimolecular reactions the experimental values of k are several orders of magnitude smaller than the values calculated from equation [5.62]. Therefore another, very variable condition must be fulfilled before reaction will take place. The third condition of the collision theory is that a collision of suitable energy only leads to reaction if the molecules are correctly oriented with respect to each other at the moment of collision. To allow for this, a constant known as the probability factor, P, is introduced into equation [5.63], thus:

$$k = PXe^{-E/RT}. \qquad [5.64]$$

This equation still has the same form as the Arrhenius equation [5.52] and it can be seen that $PX = A$. It is found that the greater the number and complexity of the spatial conditions or steric requirements which must be satisfied at the moment of reaction, the smaller is the probability of reaction and the smaller is P. For the formation of a quaternary ammonium halide in the reaction between an alkyl halide and a tertiary amine in solution, P is very small, presumably because the only effective collisions are those in which the alkyl group of the alkyl halide and the lone pair of the nitrogen atom of the amine come close together. For example $P = 10^{-9}$ for the following reaction in benzene:

$$Br\!-\!C_2H_5 + :N\!\begin{array}{l}\diagup C_2H_5 \\ \!-\!C_2H_5 \\ \diagdown C_2H_5\end{array} = Br^- \quad C_2H_5\!-\!\overset{+}{N}\!\begin{array}{l}\diagup C_2H_5 \\ \!-\!C_2H_5 \\ \diagdown C_2H_5\end{array}$$

For the gaseous reaction, $2\,HI = H_2 + I_2$ and its reverse, P is close to unity, which indicates that there are no steric requirements to be satisfied. The weakness of the collision theory is that it does not lend itself to the calculation of P and therefore of A in the Arrhenius equation [5.52].

Theory of Absolute Reaction Rates

This theory, due originally to Eyring, is also known as the transition state theory or activated complex theory and enables A to be calculated, if only approximately. The configuration of a reacting system of molecules at the point of maximum potential energy in Fig. 5.2 is known as the transition state (or activated complex) and is given the symbol $\ddagger$ and this can be regarded as being in equilibrium with normal reactant molecules. The increase in energy at constant pressure from the reactant to the transition state is the enthalpy of activation $\Delta H^{\ominus\ddagger}$ ($\approx E$), and the corresponding increase in entropy (or disorder) is the entropy of activation $\Delta S^{\ominus\ddagger}$. The Gibbs energy of activation $\Delta G^{\ominus\ddagger}$ is defined as follows (cf. Chapter 7),

$$\Delta G^{\ominus\ddagger} = \Delta H^{\ominus\ddagger} - T\Delta S^{\ominus\ddagger} \quad [5.65]$$

and is related to the equilibrium constant, $K^{\ddagger}$, for the formation of the transition state by the following relationship (cf. Chapter 7):

$$\Delta G^{\ominus\ddagger} = -RT \ln K^{\ddagger}, \quad [5.66]$$

whence $K^{\ddagger} = e^{-\Delta G^{\ominus\ddagger}/RT}$.
Substituting for $\Delta G^{\ominus\ddagger}$ using equation [5.65]:

$$K^{\ddagger} = e^{-\Delta H^{\ominus\ddagger}/RT} \cdot e^{\Delta S^{\ominus\ddagger}/R}. \quad [5.67]$$

A relationship between the rate constant and K can readily be derived by imagining that the reaction is bimolecular and by considering the equilibrium between reactants and transition state and the further conversion to products, as follows:

$$A + B \rightleftharpoons \text{transition state} \rightarrow \text{products}$$

so that:
$$K^{\ddagger} = \frac{[\text{transition state}]}{[A][B]}. \quad [5.68]$$

The rate of the reaction is given by equation [5.24], thus:

$$\frac{dx}{dt} = k_2 [A][B]$$

and by the frequency ν with which the molecules are carried through the transition state multiplied by the concentration of the molecules in this state, thus:

$$\frac{dx}{dt} = \nu \text{ [transition state]}. \quad [5.69]$$

Comparison of the last two equations shows that:

$$k_2[A][B] = \nu \text{ [transition state]}$$

$$\therefore k_2 = \nu \cdot \frac{\text{[transition state]}}{[A][B]} \quad [5.70]$$

Comparison of equations [5.68] and [5.70] indicates that:

$$k_2 = \nu K^{\ddagger} \quad [5.71]$$

The rise in the potential energy curve, from R to the transition state, $\ddagger$, at the maximum in the curve, corresponds to the bringing together of the reactant molecules against their forces of mutual repulsion and to the distortion of one (or more) bonds in the reactant molecules. The reactant molecule must then be carried through the transition state. This is achieved by a weak vibration of frequency ν in the stretched bond. In fact, the vibration is never completed because the bond breaks and the molecule descends the potential energy curve from $\ddagger$ to P. The fall in potential energy corresponds to the separation of the product molecules, which is assisted by their forces of mutual repulsion. The energy of the weak vibration is $h\nu$ according to the quantum theory (p. 80). This energy is in equilibrium with the thermal energy of the environment which is equal to kT, so that:

$$h\nu = kT, \quad [5.72]$$

where h is Planck's constant and k is Boltzmann's constant.

When ν is eliminated from equations [5.71] and [5.72], we have:

$$k_2 = \frac{kT}{h} \cdot K^{\ddagger}. \quad [5.73]$$

Introducing the expression for $K^{\ddagger}$ from equation [5.67], we obtain:

$$k_2 = \frac{kT}{h} e^{\Delta S^{\ominus \ddagger}/R} \cdot e^{-\Delta H^{\ominus \ddagger}/RT}. \quad [5.74]$$

If $\Delta S^{\ominus \ddagger}$ and $\Delta H^{\ominus \ddagger}$ can be calculated from the molecular characteristics and energy levels of the reacting molecules, equation [5.74] enables the rate constant of the reaction also to be calculated.

Comparison of equation [5.74] with the corresponding equation of the collision theory [5.64] and with the Arrhenius equation [5.52] shows that:

$$\frac{kT}{h} e^{\Delta S^{\ominus \ddagger}/R} \approx PX = A \quad [5.75]$$

and

$$\Delta H^{\ominus \ddagger} \approx E. \quad [5.76]$$

Thus, a reaction, which according to the collision theory has a very small value of P, has on the theory of absolute reaction rates a very small value of $e^{\Delta S^{\ominus\ddagger}/R}$ and therefore a large negative entropy of activation, $\Delta S^{\ominus\ddagger}$. According to the theory of absolute reaction rates, the transition state of such a reaction will be much more ordered than the molecules of reactant. This corresponds to the numerous and complex spatial conditions which, according to the collision theory, the collisions must fulfil if they are to be fruitful. Many association reactions belong to this category, including the reaction between an alkyl halide and a tertiary amine to form a quaternary ammonium salt, discussed above. The highly ordered structure of the transition state of this reaction probably corresponds to a highly ordered framework of solvent molecules which surround the region where the separation of charge occurs.

Enthalpy of activation, $\Delta H^{\ominus\ddagger}$, is found to be almost equivalent to the experimental activation energy, E (equation 5.76). $\Delta H^{\ominus\ddagger}$ and E are always positive, although for a reaction between free radicals or free atoms they may be close to zero. A reaction with a large enthalpy of activation is very temperature dependent and, if the entropy of activation is sufficiently negative, will only proceed at a measurable rate at high temperatures, e.g.

$$N_2 + 3H_2 = 2NH_3.$$

Some solution reactions with a large $\Delta H^{\ominus\ddagger}$ proceed at a measurable rate at moderate temperatures (0–100° C), provided that $\Delta S^{\ominus\ddagger}$ is also large and positive. According to equation [5.75] this corresponds to a large value of P on to the collision theory. Such reactions include many dissociations and the denaturation of macromolecules, such as proteins, nucleic acids and polysaccharides. In the native state the secondary, tertiary and quaternary structures of these biological macromolecules are maintained by a large number of weak hydrogen bonds (p. 139). Denaturation involves the loss of these higher orders of structure through breakage of the many hydrogen bonds, which requires much energy. Denaturation therefore has a large value of $\Delta H^{\ominus\ddagger}$ and its rate is very temperature dependent. The transition state, like the denatured product, is, however, more disordered than the native macromolecule, so that $\Delta S^{\ominus\ddagger}$ is positive and large as previously suggested. Cooking involves the denaturation of macromolecules. For example, to hard boil an egg takes about 8 minutes at 100° C and about 12 hours at 90° C. Application of the Arrhenius equation (in the form of 5.56) suggests that E ($\approx \Delta H^{\ominus\ddagger}$) is about 500 kJ mol^{-1}. The reason why the process is possible despite the extraordinarily high value of $\Delta H^{\ominus\ddagger}$ is that $\Delta S^{\ominus\ddagger}$ is also extremely high, namely about $+1000$ J K^{-1} mol^{-1}. Denaturation accounts for the very marked reduction in the rates of enzyme reactions and microbial growth with increasing temperature above the optimum value.

Unimolecular Reactions

Since molecules obtain their activation energy by collision between two molecules, it is at first sight difficult to see why the molecularity of a reaction can be less than two. Nevertheless, some decompositions in the gas phase have first order kinetics and appear to be unimolecular. Some examples are: the decomposition of cyclobutane, chloroethane, 1,1,1-trichloroethane, tertiary butyl bromide, tertiary butanol, ethyl chloroacetate and the isomerisation of cyclopropane to propane. The difficulty was first overcome by Lindemann, who suggested that there is a time lag between activation by bimolecular collision and unimolecular chemical change. In the time interval a collision between an activated molecule and a molecule poor in energy can result in deactivation. A mathematical treatment shows that the reaction is of the first order at normal pressures but is of the second order at very low pressures, as observed.

It was later suggested that during the time lag the acquired activation energy flows from one bond to another, and when the vibrational energy in the appropriate bond exceeds a certain value, the bond breaks. The temperature dependence of unimolecular reactions obeys the Arrhenius equation [5.52]. For most unimolecular reactions the value of A in that equation is approximately 10^{13} s^{-1}, which is the order of magnitude of molecular vibration frequencies, in agreement with this view.

Photochemical Reactions

Photochemical reactions are reactions which are initiated by light or by other forms of electromagnetic radiation. It is found that only radiations which are absorbed by the reacting system can produce a chemical change. Since photochemical reactions are brought about by visible and ultra-violet light, photochemistry is closely associated with visible and ultraviolet absorption spectroscopy. Each reaction is produced only by radiations of certain wavelengths, which, according to the quantum theory (p. 80), consist of quanta of definite energies. Each quantum of radiation absorbed causes one molecule to react in the primary stage of a photochemical reaction. This is a statement of the law of Stark and Einstein called the law of photochemical equivalence or of quantum activation. The absorbed quantum provides the activation energy for the primary stage of the reaction, probably by promoting an electron in a reactant molecule to a higher orbital, which is often an antibonding orbital. The course or mechanism of the reaction is bound up with the fate of the activated molecule. The quantum efficiency or quantum yield, ϕ, of a photochemical reaction is defined as follows:

$$\phi = \frac{\text{number of molecules reacting in a given time}}{\text{number of quanta absorbed in that time}} \quad [5.77]$$

According to the quantum theory (p. 80):
energy of one quantum = $h\nu = hc\bar{\nu}$,
so number of quanta absorbed = energy absorbed/$hc\bar{\nu}$.
If the time is now specified

$$\phi = \frac{\text{no. of moles reacted} \times L}{\text{energy absorbed}/hc\bar{\nu}}$$

$$\phi = \frac{\text{no. of moles reacted} \times Lhc\bar{\nu}}{\text{energy absorbed}} \quad [5.78]$$

where L is Avogadro's constant.
The quantity $Lhc\bar{\nu}$, for radiations of a certain wavenumber $\bar{\nu}$ (in m^{-1}), is the energy associated with L quanta and with the activation of 1 mole of reactant and is called the einstein. In SI units $L = 6.022\,2 \times 10^{23}$ mol^{-1}, $h = 6.626\,2 \times 10^{-34}$ J s and $c = 2.997\,9 \times 10^{8}$ m s^{-1}

$$\therefore Lhc = 0.119\,63 \text{ J m mol}^{-1}.$$

For absorption of radiation in the violet region for which $\bar{\nu}$ is about 25 000 cm^{-1} or 2.5×10^{6} m^{-1}, the einstein is about 300 kJ mol^{-1}. Thus, the magnitude of the einstein is of the same order as that of activation energy and bond energy when expressed in the same units. This is because each is associated directly or indirectly with the breakage of bonds.

The quantum efficiency of a reaction is found by means of equation [5.78]. The reaction mixture is contained in a suitable transparent cell which is irradiated with monochromatic light with one value of $\bar{\nu}$. The energy of the radiation before and after it has passed through the cell is determined using a thermopile or actinometer (see below) and the energy absorbed is calculated by difference. The number of moles reacted is determined analytically.

According to the Stark–Einstein law the quantum efficiency is unity for the primary stage of all photochemical processes. Consequently, ϕ is also unity for photochemical reactions consisting of only one stage, e.g.

$$h\nu + ClCH_2COOH + H_2O = HOCH_2COOH + H^+ + Cl^-.$$

Since this reaction and the photosensitised decomposition of oxalic acid, to be discussed below, have a known and reproducible quantum efficiency and lend themselves readily to chemical analysis, they may be used in actinometers to determine the incident and transmitted energy for other reactions.

The following reaction, which is brought about by ultraviolet light, also has a quantum yield of unity provided that uranyl ions (UO_2^{2+}) are present:

$$h\nu + H_2C_2O_4 = CO_2 + CO + H_2O_2.$$

In the absence of uranyl ions ϕ is much reduced and light of a lower wavelength is necessary to restore it to its previous value. For this reason the uranyl ion which is a light sensitive catalyst, is called a photosensitiser. A molecule of photosensitiser acts by absorbing a quantum of light, whereupon it becomes activated. The activated molecule passes on its energy to a molecule of reactant, which itself becomes activated and undergoes reaction.

The best known biological example of a photosensitiser is chlorophyll, the green pigment in plant leaves. Chlorophyll absorbs visible light, whereupon its electronic energy is increased. The electronic energy gained is transferred to other molecules and initiates a complex series of secondary biochemical reactions. The net result of these reactions is the conversion of carbon dioxide and water to carbohydrates and oxygen gas, known as photosynthesis. This is an example of a photochemical process which is associated with an overall increase in Gibbs energy provided by the absorbed radiation.

If the primary process of a photochemical reaction is followed by one or more secondary stages, the quantum efficiency of the overall reaction may differ from unity, although the Stark–Einstein law is still obeyed for the primary process. An example is provided by photochemically induced *cis-trans* isomerisation reactions. The primary stage is the breakage of the π bond linking the two carbon atoms. The excited molecule which results has a free electron associated with each carbon atom and can undergo more or less free rotation about the C—C σ bond. The secondary processes which follow are therefore conversion either to the geometrical isomer or to the original compound. The stages in the isomerisation reaction are as follows:

$$\begin{array}{c}
R^1\diagdown\diagup R^3 \\
C{=}C \\
R^2\diagup\diagdown R^4
\end{array}
\xrightleftharpoons{h\nu}
\begin{array}{c}
R^1\diagdown\diagup R^3 \\
\dot C{-}\dot C \\
R^2\diagup\diagdown R^4
\end{array}$$

$$\updownarrow$$

$$\begin{array}{c}
R^1\diagdown\diagup R^4 \\
C{=}C \\
R^2\diagup\diagdown R^3
\end{array}
\xrightleftharpoons{h\nu}
\begin{array}{c}
R^1\diagdown\diagup R^4 \\
\dot C{-}\dot C \\
R^2\diagup\diagdown R^3
\end{array}$$

Reactions of this type are involved in the synthesis of vitamin D in the skin under the influence of ultra-violet light. A particular case of this general type is the photochemically induced interconversion of maleic acid (the *cis*-isomer) and fumaric acid (the *trans*-isomer), for which $R^1 = R^3 = H$ and $R^2 = R^4 = COOH$. Using light of wavenumber 48 300 cm^{-1} (wavelength 207 nm), ϕ for the conversion of maleic to fumaric acid is 0.03 and for the reverse process is 0.11. Low quantum yields indicate that reconversion to the original compound predominates over other secondary processes.

In the photochemical decomposition of alkyl iodides, to give iodine, hydrocarbons and other compounds containing carbon, hydrogen and iodine, ϕ may be as low as 10^{-12}. The primary stage is the formation of an alkyl radical and an iodine atom. The minuteness of the quantum yield indicates that recombination is by far the most probable secondary process. Photochemical decomposition is often termed photolysis, and is reduced by storing organic compounds, particularly iodine compounds and drugs, in bottles made of brown glass or other opaque material.

Chain Reactions

There are some photochemical reactions for which the quantum efficiency is extremely high indeed. For example, the reaction between hydrogen and chlorine to produce hydrogen chloride has a quantum efficiency of the order of 100 000. If we assume that the Stark–Einstein law still applies to the primary process in the reaction, it follows that this process must start a sequence of reactions capable of bringing about the reaction of about 100 000 molecules without further activation. A reaction which proceeds by a sequence of steps of this type is called a chain reaction. The steps in this case are as follows:

(i) $Cl_2 + h\nu = 2Cl$ Primary process initiating chains.

(ii) $Cl + H_2 = HCl + H$ Chain propagating reaction. The Cl formed in (iii) reacts further by (ii), etc.

(iii) $H + Cl_2 = HCl + Cl$

(iv) $Cl + Cl + M = Cl_2 + M$ Chain termination after absorption of Cl atoms on the walls of the reaction vessel or by collision in the gas phase. M is a third body necessary to remove excess energy.

The quantum efficiency is determined by the number of times reactions (ii) and (iii) are repeated before the chain is broken by reaction (iv) between two chlorine atoms. Since (iv) is a relatively rare process, (ii) and (iii) may well repeat about 100 000 times before the chain is broken, thus explaining the observed quantum efficiency.

Chain reactions are always propagated by the removal and production of free atoms or free radicals, which are therefore known as *chain carriers*. Free radicals and atoms have one or more unpaired electrons and can consequently be detected, determined and studied by electron spin resonance spectroscopy (p. 132). This technique suggests that free radicals are involved in certain reactions in solution and in living cells.

Chain mechanisms are fairly common for reactions between gases. Such reactions need not be photochemical, since thermal activation is sometimes sufficient to produce the chain carriers, e.g. in the reaction between oxygen and hydrogen or hydrocarbons and in the thermal decomposition (or pyrolysis) of the vapours of many organic compounds. An example of the latter is provided by the pyrolysis of acetaldehyde vapour, for which methyl radicals ($CH_3\cdot$) are the chain carriers and the order is $\frac{3}{2}$. The overall reaction is:

$$CH_3CHO = CH_4 + CO$$

but small amounts of byproducts, such as ethane (C_2H_6), are produced. The mechanism appears to be:

(i) $CH_3CHO \xrightarrow{k_1} CH_3\cdot + \cdot CHO$ Chain initiation, by heat.

(ii) $CH_3\cdot + CH_3CHO \xrightarrow{k_2} CH_4 + CH_3CO\cdot$

(iii) $CH_3CO\cdot \xrightarrow{k_3} CH_3\cdot + CO$

⎫ Chain propagation

(iv) $CH_3\cdot + CH_3\cdot \xrightarrow{k_4} C_2H_6$ Chain termination, a comparatively rare reaction.

After the reaction has been in progress for a short time, the rate of production of a given type of free radical is equal to the rate of its removal. This is known as the *steady state approximation* and may be applied to $CH_3\cdot$ radicals, which are produced in equation (i) and removed in (iv) so that:

$$k_1[CH_3CHO] = k_4[CH_3\cdot]^2$$

whence
$$[CH_3\cdot] = \left(\frac{k_1}{k_4}\right)^{\frac{1}{2}} [CH_3CHO]^{\frac{1}{2}}. \qquad [5.79]$$

Since equations (ii) and (iii) together cause no overall gain or loss of methyl radicals, their rates do not enter into equation [5.79]. The rate of reaction may be taken to be the rate of formation of CH_4 by equation (ii), thus:

$$\text{rate of reaction} = k_2[CH_3\cdot][CH_3CHO]. \qquad 5.80]$$

Elimination of [$CH_3\cdot$] from equations [5.79] and [5.80] gives:

$$\text{rate of reaction} = k_2\left(\frac{k_1}{k_4}\right)^{\frac{1}{2}} [CH_3CHO]^{\frac{3}{2}} \qquad [5.81]$$

$$= \text{rate constant} \times [CH_3CHO]^{\frac{3}{2}}. \qquad [5.82]$$

The suggested mechanism corresponds to an order of reaction equal to $\frac{3}{2}$ as found experimentally. That methyl radicals are the chain carriers is shown by the fact that the reaction is accelerated by substances, such as tetramethyl lead, which produce them. In the absence of added radicals temperatures of 500° C and above are required for the pyrolysis but in their presence 300° C is sufficient.

Free alkyl radicals may be produced by passing vapour of a corresponding alkyl halide over heated sodium or by thermal decomposition of a metal alkyl, such as tetraethyl lead, the equations being:

$$RX + Na = NaX + R\cdot$$
$$Pb(C_2H_5)_4 = Pb + 4\, C_2H_5\cdot$$

Free alkyl radicals are very unstable. They have half-lives of the order milliseconds and rapidly react with each other to form hydrocarbons.

The reactions discussed above are called *stationary chain reactions*, because the number of chain carriers produced equals the number consumed. The steady state approximation can therefore be applied. Certain other reactions called *branching chain reactions* are known in which the chain carriers are produced in greater number than they are consumed. At certain temperatures and pressures these reactions continuously accelerate, thereby giving rise to explosions. The best examples of such reactions are those between O_2 and H_2 or hydrocarbons or other combustible gases or vapours. Branching chains arise here because the oxygen molecule has two unpaired electrons, and is therefore a diradical capable of reacting to cause an increase in the number of free radicals. The length of the branching chains in the internal combustion engine is regulated by additives such as tetraethyl lead, which decomposes to give free ethyl radicals.

Gaseous chain reactions of all types are inhibited by molecules, such as nitric oxide (NO, itself a radical), which react with the free radicals and atoms. It is sometimes necessary to inhibit undesirable chain reactions in pharmacy. Some unsaturated pharmaceuticals, such as oils and fats, in the liquid and solid phase are slowly oxidised by O_2 in the air through stationary chain reactions involving peroxides and free radicals. This phenomenon, known as autoxidation, is reduced by the addition of anti-oxidants such as amines and polyhydric phenols, which react readily with free radicals.

Mechanisms of Organic Reactions

The mechanism of a chemical reaction is commonly deduced from studies of the rate of reaction under various conditions. Illustrations of how this is done will serve to bring together diverse points made in this book.

Many reactions have one or more stages which involve the rupture or formation of a covalent bond. If the two fragments equally share or contribute to the two electrons of the bond, they must be free radicals or atoms, i.e.

$$A\!-\!B \underset{\text{colligation}}{\overset{\text{homolysis}}{\rightleftarrows}} A\cdot + B\cdot .$$

If one fragment takes (on splitting) or contributes (on uniting) both electrons of the bond, it is known as a Lewis base (p. 389), whereas the fragment which does not take or contribute any electrons is known as a Lewis acid, i.e.

$$A\!-\!B \underset{\text{co-ordination}}{\overset{\text{heterolysis}}{\rightleftarrows}} A^+ + :B^- .$$

If the bond is not covalent but dative, these processes become:

$$\overset{-}{A}\!-\!\overset{+}{B} \underset{\text{co-ordination}}{\overset{\text{heterolysis}}{\rightleftarrows}} A + :B.$$

In general, the rate and extent of ionisation, or charge separation, increase with increasing polarity of the solvent, being much smaller in hydrocarbon solvents than in water. The major contributing factors are relative permittivity (dielectric constant) of the solvent and solvation of the charged or polar products (p. 203). The neutralisation of opposite charges, be they ions or dipoles, is usually promoted by less polar media. Conversely, the rate and extent of homolysis generally increases with decreasing polarity and increasing polarisability of the solvent, being very great in carbon disulphide. Non-polar solvents which are highly polarisable, e.g. CS_2, probably solvate the free radicals, thereby stabilising them and encouraging their formation. The vapour state is also favourable to homolysis.

Many organic reactions are substitutions. If the attacking reagent is a free radical or atom, it will partake in a free radical substitution reaction, e.g.

$$CH_3\cdot + CH_3CHO = CH_4 + CH_3CO\cdot$$

If the attacking reagent has no unpaired electrons, it may attack a centre with a negative or a positive charge. Reagents which attack centres with an excess of electrons are termed *electrophilic reagents* and are generally cations or Lewis acids (p. 389), e.g. H^+, NO_2^+, Br^+, $AlCl_3$. Reagents which attack centres with a deficiency of electrons are termed *nucleophilic reagents* and are generally anions or Lewis bases, e.g. OH^-, $C_2H_5O^-$, I^-, H_2O, NR_3. Thus, depending on the attacking reagent, substitutions may be homolytic, S_H, electrophilic, S_E, or nucleophilic, S_N. The initial letter denotes the type of reaction, e.g. S for substitution, and the subscript denotes the type of reagent, e.g. N for nucleophilic. Addition and elimination reactions, denoted by the letters A and E respectively, are also known.

Ingold and Hughes found that the nucleophilic substitution reaction:

$$Y: + R—X = R—Y + X:$$

may be of the first order with respect to both Y: and R—X, i.e. of the second order overall and therefore bimolecular. This is symbolised by S_N2, the last figure denoting the molecularity. Alternatively, the reaction may be of the first order with respect to R—X only, i.e. of the first order overall and therefore unimolecular; this is symbolised by S_N1. These alternatives may be illustrated by the alkaline hydrolysis of alkyl halides. The S_N2 reaction probably takes place in one stage as follows:

$$HO:^- + H{\overset{CH_3}{\underset{H}{\searrow}}}\overset{\delta+}{C}{-}\overset{\delta-}{Br} \rightarrow \left[HO^{\frac{1}{2}-}{----}\underset{H\ \ H}{\overset{CH_3}{\overset{|}{C}}}{----}Br^{\frac{1}{2}-} \right] \rightarrow HO{-}\overset{\delta-}{\underset{H}{\overset{CH_3}{C}}}\overset{\delta+}{\underset{\diagdown}{{\diagup}H}} + :Br^-$$

transition state

The central carbon atom of the organic reactant and product molecule is sp³ hybridised (p. 97), whereas that of the transition state is sp² hybridised, the remaining p orbital being partially bonded to both the incoming and outgoing groups. The configuration of the central carbon atom, if asymmetric (p. 108), is always inverted in an S_N2 reaction. On the other hand, the S_N1 reaction probably proceeds through the intermediate formation of a carbonium ion, which is the unimolecular rate determining stage, as follows:

$$(CH_3)_3C{-}Br \rightleftharpoons \left[(CH_3)_3\overset{\delta+}{C}{----}\overset{\delta-}{Br} \right] \rightleftharpoons (CH_3)_3C^+ + Br^-$$

(slow step)   transition state

$$(CH_3)_3C^+ + H_2O \rightarrow \left[(CH_3)_3C{----}O{\diagdown\atop\diagup}{\overset{H}{H}} \right]^+ \rightarrow (CH_3)_3C{-}OH + H^+$$

(fast step)   transition state

The configuration of the central carbon atom is found to be either retained in all the molecules, inverted in all the molecules or retained in some molecules and inverted in others. The extent of inversion and racemisation is dependent on a number of factors.

The rate constant of a reaction is increased by conditions which stabilise the transition state rather than the reactants and is decreased by conditions which stabilise the reactants rather than the transition state. Thus, the rate constant of the S_N1 reaction is increased by factors which encourage charge separation, such as decreasing positive charge on the central carbon atom and increasing steric repulsion of the attached groups, as shown in Table 5.1. The rate constant of the S_N1 reaction is also increased by increasing polarity of the solvent. This is parallelled in addition reactions by a similar increase in the rate constant of the reaction between an alkyl halide and a tertiary amine (p. 168), leading to the formation of a quaternary ammonium halide, thus:

$$X\text{---}R + :NR_3 \rightarrow \left[\overset{\delta-}{X}\text{----}R\text{----}\overset{\delta+}{NR_3}\right] \rightarrow X^- + \overset{+}{N}R_4$$

transition state

A polar solvent will tend to encourage the formation of the highly polar transition state from the less polar reactants.

On the other hand, the factors which stabilise the transition state of the S_N1 reaction stabilise the reactants rather than the transition state of S_N2, and therefore reduce its rate constant as seen in Table 5.1. The rate constant of the S_N2 reaction is increased by factors favouring the spreading out of charge, by decreasing polarity of the solvent so that the charged reactant ions are not stabilised relative to the transition state, and by factors facilitating the attack by the nucleophilic reagent, such as increasing electron availability of the reagent and increasing positive charge on the central carbon atom that is being attacked.

Reasoning of the type just discussed is used in organic chemistry to explain reaction mechanisms and to predict the factors which facilitate or discourage a particular reaction.

Table 5.1. Rate constants of the alkaline hydrolysis of alkyl bromides in ethanol–water (80–20 per cent, v/v) at 55° C

Alkyl bromide	Second-order rate constant/ $dm^3\,mol^{-1}\,s^{-1} \times 10^{-5}$ (S_N2)	First-order rate constant/ $s^{-1} \times 10^{-5}$ (S_N1)
CH_3Br	2140	negligible
CH_3CH_2Br	170	negligible
$(CH_3)_2CHBr$	4.7	0.24
$(CH_3)_3CBr$	negligible	1010

Homogeneous Catalysis

Catalysis is the name of the phenomenon in which the rate of a chemical reaction is affected by an added substance without the substance being chemically changed in the reaction. If the added substance increases the rate constant, it may be called a positive catalyst or simply a catalyst, as in this book, but if it decreases the rate constant, it is called a negative catalyst or an inhibitor.

Since a catalyst is unchanged chemically in the reaction, a small amount of catalyst is sufficient to bring about a considerable amount of reaction. The catalyst may, however, undergo a physical change and may suffer losses in side reactions. For example, the surface properties of an inorganic solid may change.

A catalyst acts by directing the same overall reaction along an alternative mechanism in which each stage has a smaller Gibbs energy for the activation process, $\Delta G^{\ominus \ddagger}$, than has the slowest stage of the uncatalysed reaction (Fig. 5.3). The smaller $\Delta G^{\ominus \ddagger}$ results mainly from a smaller $\Delta H^{\ominus \ddagger}$ ($=E$) and partly from a larger $\Delta S^{\ominus \ddagger}$ (also A).

Fig. 5.3 Potential energy curve for an uncatalysed reaction (dotted line) and a catalysed reaction (full line)

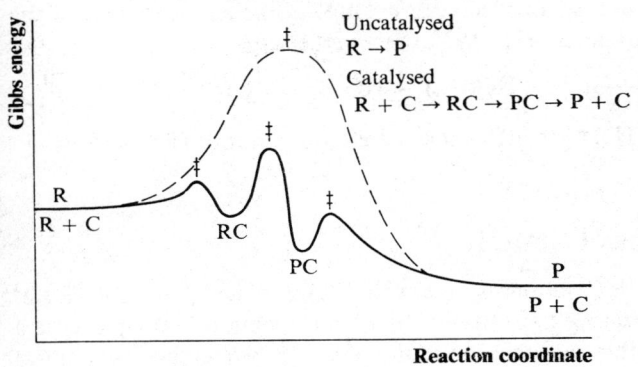

R = reactants, P = products, $\ddagger$ = transition state,
RC = reactant-catalyst complex, PC = product-catalyst complex. RC and PC may be intermediate compounds

A catalyst does not alter the equilibrium constant K of the reaction. This is because K is a state function, like ΔG and ΔU (Chapter 7), and therefore depends only on the nature and concentrations of the reactants and products and is independent of the mechanism or path of the reaction. An alternative point of view is that the catalyst increases the rate constants of the forward and the reverse reactions in the same proportion (cf. equation 5.11) so that their ratio, which is equal to K, remains unchanged.

A reaction in which all the reactants (and the catalyst, if present) are in the same phase is called a homogeneous reaction (and catalysis, if it occurs, is said to be homogeneous). Homogeneous catalysis often proceeds through the formation and subsequent reaction of intermediate compounds. Homogeneous reactions in the gas phase are frequently catalysed by substances which initiate chain reactions, i.e. by substances which produce free radicals. This sometimes applies to homogeneous reactions in solution, e.g. polymerisation reactions and certain oxidation-reduction reactions.

Owing to their variable oxidation states (valencies), the ions of transition metals are frequently powerful catalysts of oxidation-reduction reactions which involve the addition and removal of electrons. Such reactions are common in living cells in which the metal ion is in the active centre of an enzyme catalyst (p. 186).

Some of the most important homogeneous reactions in solution are catalysed by acids and bases, e.g. esterification and solvolysis, such as the acid hydrolysis of methyl acetate (p. 152) and the alkaline hydrolysis of ethyl acetate (p. 146). Acid-base catalysis will be discussed further in Chapter 9, but it may be mentioned here that the fundamental processes are the temporary donation of a proton, by an acid, to the reactant, followed by the reaction and the subsequent removal of the proton by a base (or vice versa). Thus, in the acid catalysis of the mutarotation of glucose (p. 153), there is the sequence:

$$H^+ + G(\alpha) \rightarrow HG^+(\alpha) \rightarrow HG^+(\beta) \rightarrow G(\beta) + H^+.$$

$HG^+(\alpha)$ and $HG^+(\beta)$ are intermediate compounds (RC and PC in Fig. 5.3).

Heterogeneous Catalysis

A heterogeneous reaction is one which takes place at the interfacial boundary separating two phases. Since the reaction takes place at a surface where the two phases are in contact in preference to a single phase, the surface is acting as a catalyst. Consequently, the terms, heterogeneous reaction, heterogeneous catalysis, surface reaction, surface catalysis and contact catalysis, are used synonymously. Many heterogeneous reactions take place on the surface of a solid, the catalyst, and the reactants are in the gas phase or in solution. The rate of a heterogeneous reaction is often proportional to the surface area of the catalyst, other factors being equal. This explains why heterogeneous catalysts are finely divided powders with a high specific surface (area per unit mass). The dependence of reaction rate on surface area also provides a convenient test for heterogeneous catalysis. Some reactions are catalysed by the walls of the reaction vessel, in which case the addition of finely divided wall material increases the reaction rate.

A heterogeneous catalyst acts by adsorbing the reactant molecules in a way suitable for reaction. Thus, hydrogenations (and dehydrogenations) are catalysed by substances, such as transition metals, particularly nickel and the platinum group, which adsorb hydrogen, e.g. the hydrogenation of unsaturated esters in the manufacture of margarine, and the dehydrogenation reaction:

$$C_2H_5OH \xrightarrow{Cu} CH_3CHO + H_2.$$

Dehydrations are catalysed by substances, such as alumina, which adsorb water, e.g. the conversion of an alcohol to an olefine, thus:

$$C_2H_5OH \xrightarrow{Al_2O_3} C_2H_4 + H_2O.$$

The making and breaking of carbon to carbon bonds are catalysed by strong Lewis acids (p. 389) which induce the formation of and adsorb carbonium ions. Carbonium ions are unstable and readily combine with other molecules, rearrange or decompose into smaller fragments. In the petroleum industry silica–alumina catalysts are used to induce such reactions among hydrocarbons. Since adsorption is very dependent on the nature of the solid and of the adsorbed material, heterogeneous catalysis is very specific. In many cases the course of the reaction is influenced by the catalyst (cf. the effect of copper and alumina on the decomposition of ethanol above).

A heterogeneous reaction only takes place at certain active centres on the surface of the catalyst where there are strong electrical forces or unsatisfied valencies (unpaired electrons or lone pairs) which influence or form a complex with the adsorbed reactant molecule. These active centres often occur at uneven surface features, such as certain faces, edges, peaks, cavities and cracks, or at the grain boundaries between crystals. Heterogeneous reactions are readily inhibited by substances which are strongly adsorbed at the active centres. The catalyst is then said to be poisoned. Inhibitors (or poisons) include compounds such as cyanides, arsenicals, mercury, mercurials, hydrogen sulphide and carbon monoxide, which are also active against the enzymes of living cells.

Heterogeneous catalysts may act as follows:
(a) The interaction between the catalyst and the reactant molecules which takes place on adsorption may weaken the appropriate chemical bonds, so that decomposition of the molecules is facilitated (i.e. the activation energy is lowered).
(b) Adsorption brings the reactant molecules closer together and with orientations more suitable for reaction than does random motion and occasional collision in one phase (i.e. the frequency factor is increased).
(c) Any energy evolved during a reaction will be stored in the catalyst, and is then available for activating further reactant molecules as they are adsorbed.

Adsorption Isotherms

Before reaction can take place on the surface of a heterogeneous catalyst, the reactant molecules must be adsorbed. A state of equilibrium exists between the number of molecules in the bulk of the medium and the number of molecules adsorbed on the catalyst surface. The equation which shows the relationship between the amount of substance adsorbed and the amount existing in the bulk phase at a given temperature is called an adsorption isotherm. We will consider two such equations which are relevant to the study of heterogeneous reactions. The first is an empirical equation known as the Freundlich adsorption isotherm. For the case of adsorption from the gas phase, where the pressure represents the concentration in the bulk phase, the equation takes the form:

$$\frac{x}{m} = kp^{\frac{1}{n}}, \qquad [5.83]$$

where x is the mass of gas adsorbed by a mass m of solid, and
p is the pressure of the gas.
k and n are constants for a given gas and a given solid.

The second equation, from which we can obtain rather more information, is one which can be derived theoretically, called the Langmuir adsorption isotherm. It is obtained as follows:

Suppose that, at a given pressure of gas p, the molecules are adsorbed on to the solid surface to an extent such that the fraction σ of the surface available for adsorption is covered with molecules. Then a fraction $(1-\sigma)$ remains available for further adsorption.

The rate at which further molecules are adsorbed on to the surface is proportional to the pressure of the gas (related to the total number of bombardments per second of the surface by the molecules–see p. 23), and to the amount of surface still available for adsorption (since the greater the amount of surface available, the greater is the chance that a molecule bombarding the solid will hit a spot where it can be adsorbed). Expressing this mathematically:

rate of adsorption $= k_1 p(1-\sigma)$, where k_1 is a proportionality constant. [5.84]

At the same time there will be a tendency for molecules already on the surface to leave it and return to the gas phase. The rate at which molecules leave the surface will be higher the larger the number of molecules on the surface, so it will be proportional to the fraction of surface covered with molecules. This may be expressed by:

rate of evaporation $= k_2 \sigma$, where k_2 is a second proportionality constant. [5.85]

When the solid surface and the gas have been together for some time, a state of dynamic equilibrium is set up such that the rate of adsorption is equal to the rate of evaporation.

i.e. $$k_1 p(1-\sigma) = k_2 \sigma.$$

Dividing both sides by k_2 and putting $k_1/k_2 = k$:

$$kp(1-\sigma) = \sigma$$

Collecting the σ terms on the right hand side:

$$kp = \sigma + kp\sigma = \sigma(1+kp).$$

Hence, dividing both sides by $(1+kp)$:

$$\frac{kp}{(1+kp)} = \sigma. \qquad [5.86]$$

This form of the adsorption isotherm gives the fraction of the surface covered with molecules at any given gas pressure p. In order to compare this theoretical expression with the empirical expression suggested by Freundlich, we observe that σ will be proportional to x/m of the Freundlich adsorption isotherm. If we put $x/m = k'\sigma$, the Langmuir adsorption isotherm becomes:

$$\frac{x}{m} = \frac{k'kp}{(1+kp)}. \qquad [5.87]$$

At low pressures, kp will be small compared with unity, so that $(1+kp) \approx 1$.
Hence, $$x/m \approx k'kp. \qquad [5.88]$$

At high pressures, 1 will be small compared with kp, so that $(1+kp) \approx kp$.
Hence, $$x/m \approx k'kp/kp = k'. \qquad [5.89]$$

So the relationship between x/m and p changes from one in which it is proportional to p (i.e. p^1) to one in which it is independent of p (i.e. proportional to p^0). At an intermediate pressure it is reasonable to suppose that x/m will be proportional to p raised to some power between 0 and 1. This is what is found in the Freundlich expression, where p is raised to the power $1/n$.

Although the adsorption isotherms have been considered only for the case of a gas, the same expressions apply to adsorption from solution, except that the pressure term is replaced by the concentration of the material to be adsorbed. Applying the information derived from a study of adsorption isotherms to the kinetics of heterogeneous reactions, it may be seen that as the pressure or concentration increases the order of reaction decreases until it becomes zero, when the catalyst is completely occupied by reacting molecules. A study of the kinetics at low pressures or concentrations will give the molecularity of the reaction, indicating, in the usual way, the mechanism of the reaction.

Enzyme Reactions

In the laboratory, it is common to use high temperatures and rather vigorous reagents to carry out reactions of synthesis and degradation. It seems remarkable, therefore, that the complex sequences of reactions which are necessary to maintain plant and animal life are all brought about through the agency of mild reagents and low temperatures. One reason why this is possible is that the majority of such reactions are catalysed by colloidal macromolecular substances, known as enzymes. Enzymes are proteins and, being separate hydrophilic molecules, they dissolve in water to give macromolecular solutions and therefore have some features in common with homogeneous catalysts. On the other hand, enzyme reactions proceed by adsorption of reactant, usually known as the substrate, at the active centre on the enzyme molecule so that enzymes have much in common with heterogeneous catalysts.

Experimentally, it is found that the rate of a reaction, catalysed by an enzyme, varies with both the concentration of substrate and the concentration of enzyme. If the concentration of enzyme is kept constant, and the rate of reaction is determined at different initial concentrations of substrate, a graph of reaction rate against substrate concentration has the form shown in Fig. 5.4(a). Increase of substrate

Fig. 5.4 Variation of reaction rate
   (a) with substrate concentration
   (b) with enzyme concentration

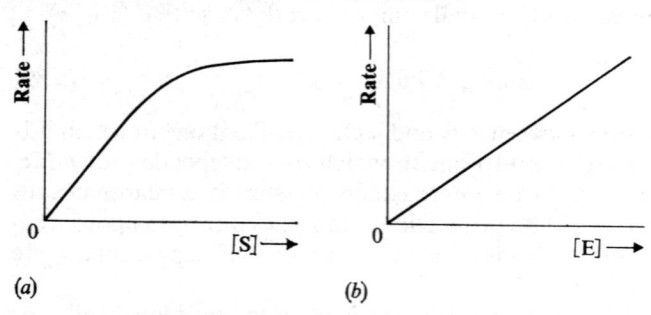

(a)   (b)

concentration at first increases the rate proportionally, but then to a smaller extent, until eventually the rate approaches an asymptotic value when there are sufficient reactant molecules to saturate the enzyme surface. The variation of reaction rate with enzyme concentration at constant substrate concentration is shown in Fig. 5.4(b).

Enzyme Reactions 187

It is seen that the rate is simply proportional to the enzyme catalyst concentration.

The form of these graphs can be deduced theoretically if it is assumed that the reaction is heterogeneous and that the adsorption of substrate molecules on to the surface of the colloidal particles of the enzyme follows the Langmuir adsorption isotherm.

If σ is the fraction of enzyme surface covered with adsorbed reactant molecules, we have (cf. p. 185):

$$\sigma = \frac{k[S]}{(1+k[S])}, \qquad [5.90]$$

where $[S]$ is the concentration of the substrate. Then, if $[E]$ is the concentration of enzyme, the total surface area, A, of enzyme covered with molecules is given by:

$$A = K'\sigma[E] \qquad [5.91]$$

where K' is a proportionality constant.

$$\therefore A = \frac{K'k[S][E]}{(1+k[S])}. \qquad [5.92]$$

The number of molecules reacting per second will be proportional to the number of molecules adsorbed, and hence to the total area covered. In other words, the rate of reaction, v, will be proportional to A, so that:

$$v = \frac{Kk[S][E]}{1+k[S]} = \frac{K[E]}{1+1/k[S]}, \qquad [5.93]$$

where K is another proportionality constant.

When $[E]$ is kept constant, the plot of v against $[S]$ has the form of Fig. 5.4(a). At low substrate concentrations $k[S]$ is small compared with unity, so that $1+k[S] \approx 1$, hence:

$$v = Kk[S][E] \qquad [5.94]$$

and the reaction is of the first order with respect to the substrate. At high substrate concentrations $k[S]$ will be large compared with unity, so that $1+k[S] \approx k[S]$, hence:

$$v = K[E] \qquad [5.95]$$

and the reaction is of zero order with respect to the substrate.

When $[S]$ is kept constant, $v \propto [E]$, so that reaction is of the first order with respect to the enzyme. This accounts for the straight-line relationship of Fig. 5.4(b).

An alternative theoretical treatment of enzyme reactions has been developed by Michaelis and Menten, in which the adsorption of the substrate on the enzyme is envisaged in terms of the formation of a definite enzyme–substrate complex. This treatment resembles that of homogeneous reactions, e.g. chain reactions.

$$\text{Enzyme} + \text{Substrate } (S) \underset{k_2}{\overset{k_1}{\rightleftharpoons}} \text{Complex } (X) \xrightarrow{k_3} \text{Products} + \text{Enzyme}.$$

If the initial concentration of enzyme were $[E]$, and after time t, a concentration $[X]$ of complex had been established, the concentration of free enzyme available for further reaction would be $[E]-[X]$. Also, if it is assumed that the substrate is present in excess, so that the amount used in forming the complex has a negligible effect on its concentration, the latter may be written as a constant, $[S]$, for a particular reaction. The rate of formation of complex is then proportional to the product of the two concentrations, i.e.:

$$\frac{d[X]}{dt} = k_1([E]-[X])[S].$$

The rate of disappearance of complex by the reverse of this first reaction would be:

$$-\frac{d[X]}{dt} = k_2[X].$$

The complex also disappears by the formation of products, at a rate given by:

$$-\frac{d[X]}{dt} = k_3[X].$$

When the reaction is proceeding at a steady rate, the complex will be decomposing, by both these processes, at the same rate at which it is being formed by the first process, hence:

$$k_1([E]-[X])[S] = k_2[X] + k_3[X].$$

This is the steady state treatment due to Briggs and Haldane. Dividing both sides by $k_1[X]$:

$$\frac{([E]-[X])[S]}{[X]} = \frac{k_2+k_3}{k_1} = K_m. \qquad [5.96]$$

K_m is called the Michaelis constant for the given substrate–enzyme system. It is a measure of the dissociation of the complex, and so its reciprocal indicates the affinity of the enzyme for the substrate. Equation [5.96] shows that K_m is expressed in concentration units. Multiplying both sides of equation (5.96) by $[X]$:

$$([E]-[X])[S] = K_m[X]. \qquad [5.97]$$

Collecting the terms in $[X]$ on the right-hand side

$$[E][S] = K_m[X] + [X][S] = [X](K_m+[S]).$$

Dividing both sides by $(K_m+[S])$:

$$[X] = \frac{[E][S]}{K_m+[S]} = \frac{[E]}{1+K_m/[S]}. \qquad [5.98]$$

The rate of the enzyme reaction is equal to the rate of formation of the products and is given by:

$$v = k_3[X], \qquad [5.99]$$

hence
$$v = \frac{k_3[E][S]}{[S]+K_m} = \frac{k_3[E]}{1+K_m/[S]}. \qquad [5.100]$$

This equation has the same form as the expression for the rate determined from the adsorption isotherm point of view. The value of the Michaelis constant can be found by comparing this expression for the rate with an expression for the maximum rate (V) at large substrate concentrations. This maximum rate occurs when virtually all the enzyme is in the form of complex, i.e. when $[X] = [E]$, the initial concentration of enzyme.

Thus, $\qquad V = k_3[E]. \qquad [5.101]$

Therefore, taking the ratio of the two rates:

$$\frac{V}{v} = \frac{[S]+K_m}{[S]} = 1+K_m/[S]. \qquad [5.102]$$

One method of finding K_m is to find what substrate concentration gives a rate of reaction equal to half the maximum rate, i.e. $v = \frac{1}{2}V$, or $V/v = 2$. At this point:

$$2 = 1+K_m/[S],$$
$$\therefore 1 = K_m/[S],$$

or $\qquad K_m = [S].$

This concentration of substrate is therefore equal to the Michaelis constant (Fig. 5.5(a)).

Fig. 5.5 Determination of the maximum rate, V, and the Michaelis constant, K_m, in enzyme kinetics

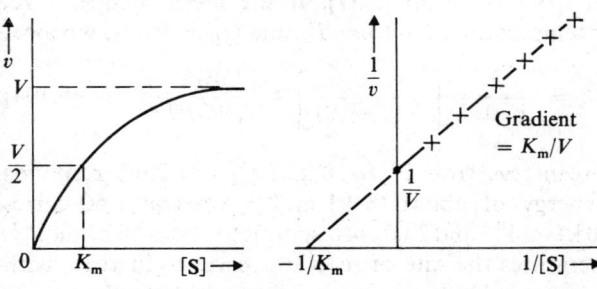

[S] = substrate concentration $\qquad v$ = rate of reaction

190 Chemical Reaction

Alternatively, taking the equation for the ratio of V/v, and dividing both sides by V:

$$\frac{1}{v} = \frac{1}{V} + \frac{K_m}{[S]} \cdot \frac{1}{V}. \qquad [5.103]$$

Thus, if the rate v is measured at a number of different substrate concentrations $[S]$, and a graph is plotted of $1/v$ against $1/[S]$, a straight line should be obtained of gradient K_m/V and intercept on the $1/v$ axis, where $1/[S] = 0$, equal to $1/V$ (Fig. 5.5(b)). From this plot, which is due to Lineweaver and Burk, $K_m =$ gradient/intercept. The hypothetical value of $[S]$ when $1/v = 0$ is found by substitution to be $-K_m$. Therefore, if the linear plot is extrapolated back, it intercepts the $1/[S]$ axis at $-1/K_m$. From either of these methods using the Lineweaver–Burk plot (Fig. 5.5(b)) a rather more reliable value of the Michaelis constant is obtained than from Fig. 5.5(a).

The validity of this treatment of enzyme reactions in terms of definite complex formation has been questioned, but, in some cases, independent evidence of complex formation has been obtained.

The presence of enzymes makes it possible for biological reactions to proceed at appreciable rates at ordinary temperatures. This is mainly due to reduction of the activation energy. For example, the decomposition of hydrogen peroxide without any catalyst requires an activation energy of 75 kJ mol^{-1}. With colloidal platinum as catalyst, the activation energy is reduced to 49 kJ mol^{-1}, but in the presence of the enzyme, liver catalase, it is further reduced to only 23 kJ mol^{-1}. The activation energies for enzyme reactions are determined in the usual way, by studying the variation of the rate constants with temperature.

Another way of expressing this variation is to quote the temperature coefficient Q_{10}. This is the ratio of the rate of reaction at the temperature $T+10$ K to the rate at T K. This coefficient usually has values between 1.4 and 2.0, showing that a 10 K rise in temperature approximately doubles the reaction rate. The higher the coefficient, the larger is the activation energy, E. The relationship between Q_{10} and E can be derived from equation [5.57]. If the lower temperature, $T = T_1$, the upper temperature, $T+10 = T_2$, and $Q_{10} = k_2/k_1$, whence:

$$\log Q_{10} = \frac{E}{2.303R} \left[\frac{10}{T(T+10)}\right] \approx \frac{10E}{2.303RT^2}. \qquad [5.104]$$

For the temperature interval from 20° to 30° C, a Q_{10} of 1.25 corresponds to an activation energy of about 16 kJ mol^{-1}, whereas 1.50 corresponds to about 30 kJ mol^{-1} and 2.00 corresponds to about 50 kJ mol^{-1}.

At higher temperatures the rate of reaction ceases to increase with temperature. The rate reaches a maximum value, at what is known as the optimum temperature, and then decreases rapidly with a further increase of temperature owing to denaturation of the enzyme (p. 171). For animal enzymes the optimum is often between 40° and 50° C, and for plant enzymes it is usually between 50° and 60° C. Besides having an optimum temperature, enzymes have an optimum pH for maximum activity, and care must be taken in studying enzyme reactions to keep the pH constant by means of a buffer solution.

Enzymes, like heterogeneous catalysts, increase the rate of reactions partly by increasing the frequency factor, A. Enzyme catalysis, however, is even more specific than heterogeneous catalysis. A given substrate can react in many ways depending on the enzyme which is present and a given enzyme can catalyse the reaction of one only or a small number of structurally related substrates.

Inhibition of Enzyme Reactions

The inhibition of enzymes is likewise even more specific than that of heterogeneous catalysts. Living cells control their own metabolism by producing organic compounds which selectively inhibit their own enzymes. Enzyme inhibitors are used in medicine to inhibit the growth of bacteria and of cancers and in biochemistry to study enzyme kinetics.

The kinetics of enzyme inhibition can be derived from the Michaelis–Menten treatment assuming that the enzyme undergoes the following reaction with the inhibitor:

Inhibitor (I) + Enzyme $\rightleftharpoons$ Inactive Complex (Y). The equilibrium constant for the reverse reaction, i.e. for the dissociation of the inactive complex, is known as the inhibitor constant and is given by:

$$K_i = \frac{[\text{Enzyme}][I]}{[Y]} \qquad [5.105]$$

whence $\qquad [\text{Enzyme}][I] = [Y]K_i. \qquad [5.106]$

The inhibitor can affect the kinetics of the enzyme reaction in various ways depending on how the inhibitor interferes with the binding between enzyme and substrate.

Competitive inhibition

The inhibitor reduces the reaction rate by occupying the substrate-binding site (active centre) and thus competes with the substrate for possession of the active centre. The concentration of free enzyme available for further reaction with the substrate is now $([E]-[X]-[Y])$. It can be seen that $[E]$ in the treatment of uninhibited reactions is here replaced by $([E]-[Y])$, so that the expression for K_m (equation 5.97) becomes:

$$([E]-[X]-[Y])[S] = K_m[X]. \qquad [5.107]$$

Equation [5.105] gives [Y]. Now [Enzyme] is the concentration of enzyme which is free of both substrate and inhibitor, i.e. ([E]−[X]−[Y]), because only then is the enzyme able to combine with the inhibitor. Thus equation [5.106] becomes:

$$([E]-[X]-[Y])[I] = K_i[Y]. \qquad [5.108]$$

The last two equations enable [X] to be expressed in terms of [S], K_m, [I] and K_i. Dividing both sides of equation [5.107] by [S] and adding [X] and [Y] to both sides gives:

$$[E] = [X]+[Y]+K_m[X]/[S]. \qquad [5.109]$$

Dividing equation [5.108] by equation [5.107], we have:

$$\frac{[I]}{[S]} = \frac{K_i[Y]}{K_m[X]},$$

so
$$[Y] = \frac{K_m[X][I]}{K_i[S]}. \qquad [5.110]$$

Substituting for [Y] in equation [5.109] gives:

$$[E] = [X] + \frac{K_m[X]}{[S]} \cdot \frac{[I]}{K_i} + \frac{K_m[X]}{[S]}$$

$$\therefore [E] = [X]\left[1 + \frac{K_m}{[S]}\left(\frac{[I]}{K_i}+1\right)\right] \qquad [5.111]$$

$$\therefore [X] = [E]\bigg/\left[1 + \frac{K_m}{[S]}\left(\frac{[I]}{K_i}+1\right)\right]. \qquad [5.112]$$

Since a given enzyme molecule cannot combine with both a substrate molecule and an inhibitor molecule, [X] always refers to the uninhibited enzyme–substrate complex and the rate of the inhibited reaction is given by the same equation [5.99] as for the uninhibited reaction. When the expression [5.112] for [X] is introduced, the rate of the inhibited reaction is given by:

$$v = \frac{k_3[E]}{1+\frac{K_m}{[S]}\left(\frac{[I]}{K_i}+1\right)}. \qquad [5.113]$$

This is the general equation for competitive inhibition.

On increasing the concentration of the inhibitor, equation [5.113] shows that the rate of the reaction is reduced and that this effect can in principle always be nullified by increasing the concentration of the substrate. This is what one would expect of a competitive phenomenon. With increasing concentration of substrate $K_m/[S]$ in equation [5.113] tends towards zero and v tends towards the maximum rate V,

Inhibition of Enzyme Reactions 193

which is independent of $[I]$ and is given by the same equation [5.101] as for the uninhibited reaction (Fig. 5.5).

Comparison of equation [5.113] with equation [5.100] for the uninhibited enzyme reaction shows that competitive inhibition causes $[S]$ to be divided by the factor $(1+[I]/K_i)$. Consequently, the gradient of the Lineweaver–Burk plot is $\dfrac{K_m}{V}\left(1+\dfrac{[I]}{K_i}\right)$ and increases with increasing $[I]$ (Fig. 5.6).

Fig. 5.6 Lineweaver–Burk plot for competitive inhibition

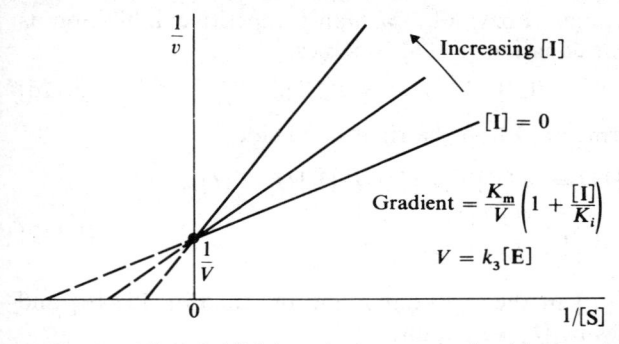

Competitive inhibition is common in biochemistry and usually occurs when the structure of the inhibitor strongly resembles that of the substrate, so that the substrate and inhibitor can compete for the same active site. For example, folic acid synthetase in bacteria is inhibited by sulphonamide drugs, which are structural analogues of the substrate p-aminobenzoic acid (pp. 383, 384).

Non-competitive inhibition

The inhibitor reduces the reaction rate by combining with the enzyme at a site different from the substrate binding site. It may be supposed that combination with the inhibitor distorts the enzyme molecule making it inactive as a catalyst while still enabling it to combine with the substrate. Thus, the enzyme is capable of combining with the substrate alone, with the inhibitor alone, or with both the substrate and inhibitor together, but a given enzyme molecule only acts as a catalyst if it is bound to the substrate alone.

As before, the total concentration of enzyme is $[E]$, the concentration of enzyme which is bound to the substrate is $[X]$ and that bound to the inhibitor is $[Y]$. The concentration of enzyme which is free of substrate is $([E]-[X])$, so that $[X]$ for non-competitive inhibition is given by the same equation [5.98] as for an uninhibited reaction. Since $[X]$ includes enzyme which is bound to both substrate and inhibitor, only a fraction of the enzyme–substrate complex, X, can give rise to the products and contribute to the rate of reaction. This fraction is the concentration of enzyme free of inhibitor, divided by the total concentration of enzyme, i.e. $([E]-[Y])/[E]$, so that equation [5.99] becomes

$$v = k_3[X]([E]-[Y])/[E]. \qquad [5.114]$$

Substituting for $[X]$, using equation [5.98], gives:

$$v = \frac{k_3([E]-[Y])}{1+K_m/[S]}. \qquad [5.115]$$

Equation [5.105] gives $[Y]$. The concentration of enzyme which is free of inhibitor, i.e. [Enzyme], for non-competitive inhibition is $([E]-[Y])$, so that equation [5.106] becomes:

$$([E]-[Y])[I] = K_i[Y]. \qquad [5.116]$$

Collecting the terms in $[Y]$ on the right-hand side:

$$[E][I] = K_i[Y]+[Y][I] = [Y](K_i+[I])$$

$$\therefore [Y] = \frac{[E][I]}{K_i+[I]}. \qquad [5.117]$$

Substituting for $[Y]$ in the right-hand side of equation [5.116] and dividing both sides by $[I]$ we obtain:

$$[E]-[Y] = \frac{[E]K_i}{K_i+[I]} = \frac{[E]}{1+[I]/K_i}. \qquad [5.118]$$

Substituting for $([E]-[Y])$ in equation [5.115] gives:

$$v = \frac{k_3[E]}{(1+K_m/[S])(1+[I]/K_i)}. \qquad [5.119]$$

This is the general equation for non-competitive inhibition.

With increasing concentration of substrate, $K_m/[S]$ in equation [5.119] tends towards zero and v tends towards the maximum rate which is here given by:

$$V = \frac{k_3[E]}{1+[I]/K_i}. \qquad [5.120]$$

Comparison of this equation with the corresponding equation [5.101] for the uninhibited reaction shows that non-competitive inhibition causes $[E]$ to be divided by $(1+[I]/K_i)$, which reflects the reduction of the effective concentration of the enzyme by the inhibitor. As $[I]$ increases, V decreases by this same factor. This reduction in V cannot be reversed by increasing $[S]$ and is what one would expect of a non-competitive phenomenon.

Dividing equation [5.119] by [5.120] for non-competitive inhibition results in the same equations [5.102] and [5.103] as for an uninhibited reaction. In the Lineweaver–Burk plot (Fig. 5.7) the hypothetical intercept on the abscissa is $-1/K_m$ and is unaltered by the inhibition, whereas the intercept on the ordinate and the gradient, which are still $1/V$ and K_m/V respectively, both increase as $[I]$ increases.

Fig. 5.7 Lineweaver–Burk plot for non-competitive inhibition

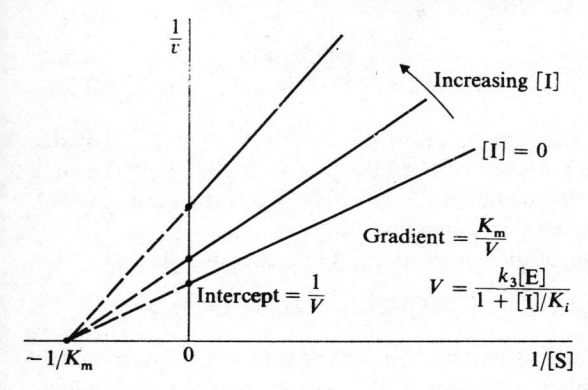

$$\text{Gradient} = \frac{K_m}{V}$$
$$\text{Intercept} = \frac{1}{V}$$
$$V = \frac{k_3[E]}{1 + [I]/K_i}$$

Non-competitive inhibition is extremely common in biochemistry and usually occurs when the structure of an organic inhibitor does not strongly resemble that of the substrate. In addition to competitive and non-competitive inhibition other types of inhibition are known, whose kinetics are more complicated.

Inhibition by covalent bonding

If any type of inhibitor forms a strong covalent bond with an enzyme molecule, the inactive enzyme–inhibitor complex can break down only to a minute extent. K_i is therefore very small and the reaction between inhibitor and enzyme is said to be 'irreversible' (p. 315). A relatively small concentration of inhibitor is sufficient to make the factor $(1+[I]/K_i)$ very large in equations [5.113] and [5.119] and so reduce the rate of reaction virtually to zero. In other words, very little uninhibited enzyme remains for catalysing the reaction. Examples include heavy metals which block the active centres, alkylating agents which bind to various groups, and arsenicals and mercurials which combine with thiol groups of enzymes.

Problems

1 The following are the concentrations $(a-x)$ of lactate at various times t during its enzymatic oxidation to pyruvate:

t/s	0	100	250	350	500	650
$(a-x)$/mmol dm^{-3}	32.00	27.24	21.38	18.21	14.29	11.22

Problems

Confirm that the reaction is of first order by plotting $\log(a-x)$ versus t and determine the rate constant k_1 from the gradient of the graph. Calculate the half-life of the reaction.

2 The acid-catalysed hydrolysis of sucrose into a mixture of glucose and fructose was followed by measuring the optical rotation, α, at invervals of time as follows:

time/minutes	0	20	60	100	160	200	∞
α/degrees	29.1	18.3	4.4	−3.3	−8.8	−10.5	−12.6

Show graphically that the reaction is of the first order, and hence determine the rate constant of the reaction. Calculate also the time at which half the sucrose has been hydrolysed and the time at which the solution is optically inactive.

3 The decomposition of diacetone alcohol in aqueous solution,

$$(CH_3)_2C(OH)CH_2COCH_3 = 2\ CH_3COCH_3,$$

is a first-order reaction catalysed by hydroxyl ions. The reaction is accompanied by a considerable increase in volume which is measured by following the movement of the meniscus of the solution in a capillary tube. The difference between the 'infinity' reading (r_∞) and the height of the meniscus (r) at any other time t is a measure of the extent of the reaction. Two sets of readings are given below, one for a solution in which the NaOH catalyst concentration is 0.1 mol dm^{-3} and one for which it is 0.02 mol dm^{-3}.

NaOH 0.1 mol dm^{-3}	r mm	0	61	98	120	133	141	146	154
	t min	0	10	20	30	40	50	60	∞
NaOH 0.02 mol dm^{-3}	r mm	0	15	28	40	51	61	70	154
	t min	0	10	20	30	40	50	60	∞

Plot graphs of $\log(r_\infty - r)$ versus t, and obtain k_1 from the gradient in each case. Describe and explain the relationship which exists between k_1 and the concentration of NaOH. Why is it possible to use $(r_\infty - r)$ in place of the concentration of diacetone alcohol remaining, in plotting the graphs?

4 From the results of problem 4 in Chapter 4 (p. 143), calculate the equilibrium constant for the reaction:

$$\alpha\text{-D-glucose} \rightleftharpoons \beta\text{-D-glucose}.$$

In a particular experiment in which the approach to equilibrium was studied by measurements of optical rotation, α, commencing with α-D-glucose, the gradient of the graph of $\log(\alpha_t - \alpha_\infty)$ versus time, t, was 0.015 min^{-1}. Calculate the rate constants for the forward and the reverse reactions.

Problems 197

5 The alkaline hydrolysis of ethyl acetate was studied after mixing equal concentrations of the ester and sodium hydroxide. The following are the concentrations $(a-x)$ of the alkali remaining at various times t:

t/min	0	5	12	20	26	32
$(a-x)$/mmol dm^{-3}		10.13	7.86	5.98	4.71	4.05	3.55

Show that the reaction is of the second order by plotting $1/(a-x)$ against t and determine the rate constant k_2 from the gradient of the graph. Calculate the half-life of the reaction.

6 The times of half change (half lives) in the thermal decomposition of nitrous oxide at a number of different initial pressures are given below:

Initial pressure/kPa	7.00	18.5	38.7
Time of half change/s	860	470	255

Plot a graph of the reciprocal of the time of half change versus the initial pressure, and confirm that the reaction is approximately of second order. From the gradient of the graph, calculate the rate constant in $Pa^{-1}s^{-1}$.

7 The rate constants at 323.2 K and 373.2 K for the decomposition of dibromosuccinic acid in solution are 1.80×10^{-6} and 2.08×10^{-4} min^{-1} respectively. Calculate the energy of activation for the reaction.

8 The antibiotic streptozotocin is slowly decomposed in aqueous solution. In phosphate buffer at pH 6.90 the first order rate constant k_1 for decomposition at various temperatures has the following values:

Celsius temperature	29.8	40.3	50.7	61.2
$10^5 k_1/s^{-1}$	4.27	19.1	77.6	291

Plot a graph of log k_1 against the reciprocal of the absolute temperature and confirm that the Arrhenius equation is obeyed. Hence calculate the activation energy and the frequency factor and predict the rate constant at 37.0° C and at 18.0° C.

9 After an aqueous solution of fumaric acid had absorbed 1.04 kJ of light energy of wavelength 207 nm, the solution was found to contain 0.198 mmol of maleic acid. Calculate the quantum efficiency of this photochemical reaction.

10 The hydrolysis of egg albumin is catalysed by the enzyme pepsin. The following table shows how the initial rate of reaction varies with the initial substrate (egg albumin) concentration at constant concentration of enzyme and constant temperature.

198 Problems

Per cent egg albumin   1.0   1.5   2.0   3.0   6.0
Initial rate (arbitrary units) 3.04 3.82   4.31   5.03   6.02

Plot a graph of the reciprocal of the initial rate versus the reciprocal of the concentration of egg albumin, and find the maximum rate of reaction V, and the Michaelis constant K_m. The latter is found in units of mol dm^{-3} by assuming that the molecular weight of egg albumin is 45 000.

11 The following results illustrate the inhibition of L-histidine ammonia-lyase by a tumour inhibitor:

Substrate concentration (mmol dm^{-3})	Rate of reaction/(nmol minute^{-1}) in the presence of the drug at:		
	0 mmol dm^{-3}	1.61 mmol dm^{-3}	5.08 mmol dm^{-3}
10.00	58.7	36.6	—
12.50	68.3	42.6	—
16.67	82.0	51.0	27.9
25.00	101.5	63.5	34.8
50.00	134.2	83.7	46.1

Plot the results on the same graph by the method of Lineweaver and Burk. Calculate the Michaelis constant, K_m, and the maximum rate, V, in the absence of drug and deduce whether the inhibition of the enzyme by the drug is competitive or non-competitive. Finally, calculate the inhibitor constant, K_i, for each concentration of the drug.

6 Properties of Mixtures and Solutions

Mixtures and Solutions

The word *mixture* is used to describe a gaseous, liquid or solid phase containing more than one substance, when the substances are all treated in the same way. The word *solution* is used to describe a liquid or solid phase containing more than one substance, when for convenience one of the substances, which is called the *solvent* (and may itself be a mixture), is treated differently from the other substances, which are called *solutes*. Whether the various substances are treated the same or differently refers in particular to the choice of their standard states in the definition of activities (see pp. 307 and 311).

A phase consisting of a solid or gas dissolved in a liquid is invariably treated as a solution with the liquid as the solvent and the gas or solid as the solute. A phase consisting of two or more liquids is often treated as a mixture. If such a liquid phase is treated as a solution, any component may be regarded as the solvent, though it is usual to reserve the term solvent for the component present in greatest quantity. A phase consisting of two or more solids may also be treated as a mixture or as a solution (see solid solutions p. 44). A phase consisting of two or more gases is always treated as a mixture (see Dalton's law of partial pressures p. 24).

The relative quantities of the substances in a mixture or solution may conveniently be expressed as mole fractions, mass fractions or volume fractions or as the corresponding percentages. The mole fraction of a given substance B, x_B, in a mixture or solution is the amount (moles) of B present, divided by the total amount (moles) of all the substances present in the mixture or solution. The mass fraction and volume fractions are similarly defined by replacing 'amount' by 'mass' or 'volume' respectively. Mole, mass and volume fractions are dimensionless quantities with values between zero (for an absent substance) and one (for a pure substance). It can be seen that mass or weight per cent is the same as per cent w/w and that volume per cent is the same as per cent v/v.

Concentration

Concentration is an important intensive property (p. 273) which measures in suitable units the relative proportions of solute and solvent in a solution. IUPAC recommends that the term concentration, c_B or [B], of solute substance B be restricted to the amount (in moles) of B divided by the volume of the solution. The unit mol dm^{-3} (= mol l^{-1}) is sometimes called *molar* and given the symbol M. It can be seen that 1 mol m^{-3} = 10^{-3} mol dm^{-3} = 10^{-3} mol l^{-1} = 1 millimolar = 1 mM.

IUPAC defines mass concentration, ϱ_B, of solute substance B as the mass of B divided by the volume of the solution. We note that 1 kg m^{-3} = 1 g dm^{-3} = 1 g l^{-1} = 0.1 per cent w/v (see below). The following units of mass (and volume) concentrations are also in common use in biology and pharmacy: per cent weight by volume (per cent w/v) which is the number of grams of solute dissolved in 100 cm³ of solution; per cent weight by weight (per cent w/w) which is the number of grams of solute dissolved in 100 g of solution; per cent volume by volume (per cent v/v) which is the number of cm³ of solute dissolved in 100 cm³ of solution.

A quantity known as molality, although strictly speaking a form of concentration, is distinguished by IUPAC from concentration. The molality, m_B, of a solute substance B is the amount (in moles) of B divided by the mass (in kg) of solvent. The unit mol kg^{-1} is often given the name *molal*, e.g. a solution having a molality equal to 0.1 mol kg^{-1} is called a 0.1 molal solution. Molality is much used in work with electrolytes.

Solubility of Solids in Liquids

If a solid and a liquid are shaken together for some time, some or all of the solid dissolves in the liquid. If there is excess solid present, a stage will be reached at which no more solid passes into solution at the temperature of the mixture. The solution is then said to be saturated and the concentration of the solute in the solution is the solubility of the solute in the solvent at that particular temperature. Solubility may be expressed in any of the above units of concentration of solute in a saturated solution at a given temperature. For sparingly soluble solutes the density of the solution may be assumed to be equal to that of the solvent so that the volume of the solution and of the solvent may be taken to be interchangeable; the solubility may then be given in milligrams per cent., i.e. milligrams of solute per 100 ml of solution or solvent at a given temperature. In the *British Pharmacopœia* the solubility is expressed as the number of parts (ml, i.e. cm³) of solvent which will form a saturated solution with one part of solute (1 ml of a liquid or 1 g of a solid) at 20° C.

It is possible, for most substances, to prepare, by careful cooling, a supersaturated solution, i.e. one which contains a higher concentration of solute than can normally exist in the presence of excess solid at that temperature. This is due to the difficulty of initiating crystallisation in the absence of solid nuclei. However, if such a supersaturated solution is caused to deposit solid by 'seeding' with a speck of solute, the solution which remains is a saturated solution.

Two methods are commonly used for the determination of solubility. In the first, excess solid is agitated with solvent at a temperature a few degrees higher than that at which the solubility is to be determined: this is to ensure that sufficient solid goes into solution. The solution and excess solid are then kept in a thermostat for about an hour, with occasional shaking. The presence of excess solid prevents the formation of a supersaturated solution, and ensures that equilibrium is established between solid and solution at the temperature of the thermostat. The solution is then separated from the solid by a special arrangement, allowing filtration through a plug of glass or cotton wool without removing the apparatus from the thermostat, or simply by drawing a portion of the solution into a pipette through a temporarily fitted filter. The solution is then analysed by any convenient method. The process may then be repeated at various temperatures.

The second method involves making up a set of tubes containing a range of different quantities of solute and solvent. The tubes are then slowly warmed in a water bath, and are shaken continuously. The temperature is noted at which the last crystal dissolves in each tube, and the composition in this tube gives the solubility at this temperature. From the results for all the tubes, a solubility curve may be drawn and the solubility at any temperature may be read off.

The solubility of most substances varies considerably with temperature, and this is usually represented by a solubility curve (solubility plotted against temperature), some examples of which are given in Fig. 6.1. The type of curve most commonly found is illustrated by the example of KNO_3: increasing the temperature from 10° C to 40° C brings about a three-fold increase in solubility. Sodium chloride is unusual in that there is only a slight increase in solubility with increasing temperature. In cases where the solubility curve is not smooth, but shows an abrupt change in slope, it indicates that the type of solid which would crystallise from the saturated solution has changed. An example of this behaviour is found in the curve for sodium sulphate. Crystallisation above 32.5° C would yield anhydrous Na_2SO_4, so the upper part of the curve represents the solubility of the anhydrous salt. Below 32.5° C, evaporation of the saturated solution would deposit the hydrated salt Na_2SO_4, 10 H_2O, so the lower part of the curve represents the solubility of the hydrated salt. The temperature 32.5° C is the transition point, above which the anhydrous salt is stable and below which the hydrated form is stable.

Fig. 6.1   Solubility curves

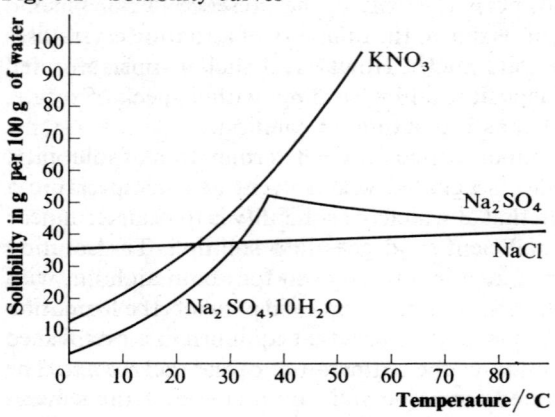

When a solute dissolves in a solvent, heat is usually absorbed or liberated. The differential enthalpy of solution (ΔH_s) is the heat absorbed under conditions of constant temperature and pressure when 1 mole of solute is dissolved in a very large (infinite) quantity of solvent. If heat is liberated, ΔH_s is a negative quantity. ΔH_s may be considered to be made up of two components, thus:

$$\Delta H_s = \text{lattice enthalpy} + \text{enthalpy of solvation} \qquad [6.1]$$

The lattice enthalpy is the heat absorbed at constant temperature and pressure, when the molecules or ions in 1 mole of a crystalline solid are completely separated (to infinity) against their forces of mutual attraction in the crystal lattice. Lattice enthalpy is always positive. The enthalpy of solvation is the heat absorbed at constant temperature and pressure when 1 mole of the completely separated molecules or ions become immersed in a very large (infinite) quantity of solvent. When a solute dissolves, its constituent molecules or ions will interact with or become bound to molecules of solvent, so that solvation is exothermic and the enthalpy of solvation is negative.

For most combinations of solute and solvent, including KNO_3, NaCl or Na_2SO_4, 10 H_2O in water (Fig. 6.1) and organic compounds in various solvents, the solubility increases with increasing temperature. This fact is exploited in the recrystallisation of substances. Application of Le Chatelier's principle (p. 316) to these cases shows that dissolution of solute is accompanied by the absorption of heat, i.e. ΔH_s is positive, because the positive lattice enthalpy outweighs the negative enthapy of solvation. For certain, less common cases the affinity of the solute for the solvent is so great that the negative enthalpy of solvation outweighs the positive lattice enthalpy, so that ΔH_s is negative, and solubility decreases with increasing temperature according to Le Chatelier's principle. Anhydrous Na_2SO_4 in water (Fig. 6.1) shows this behaviour. This salt and other substances with a strong affinity for water, e.g. anhydrous $CaCl_2$, $MgSO_4$ and concentrated sulphuric acid are used as drying agents and dissolve in water exothermically.

If a solute is to be appreciably soluble in a solvent, the intermolecular attractive forces between solute and solvent must approximately equal or exceed those between solute and solute and between solvent and solvent to enable the solute and solvent molecules to separate each other. Consequently, as a general rule, the enthalpy of solvation must be sufficiently exothermic in comparison with the endothermic lattice energy. Ionic substances, including salts, dissolve in polar solvents, such as water, partly because the forces of mutual attraction between the oppositely charged ions are weakened by the high relative permittivity (dielectric constant) of the solvent and partly because new forces of attraction are introduced between the ionic charges and the appropriate oppositely charged end of the permanent dipole of the solvent molecules (p. 30). The ions are then separated from each other and are surrounded by interacting molecules of solvent so that they are said to be solvated. If the cation is small, e.g. Mg^{2+}, and cations of the B subgroups or of the transition metals, the ion–dipole attraction gives way to a dative bond, in which a lone pair of electrons of the solvent molecule is donated to empty d, s or p orbitals, or dsp hybrid orbitals of the metal ion.

The ancient rule of solubility – like dissolves like – is still a valuable guide. Thus, polar substances tend to dissolve in each other partly as a result of dipole–dipole attractions. Substances containing the groups —OH, —NH$_2$, —COOH tend to be mutually soluble as a result of hydrogen bonding, which may sometimes be reinforced by dipole–dipole attractions at other places in the molecules. Furthermore, nonpolar substances such as hydrocarbons and substances whose molecules have a considerable hydrocarbon character, such as oils, fats and lipids, tend to be mutually soluble, because the forces between unlike molecules are similar to the forces between like molecules and therefore the two types of molecule can readily mix and are mutually soluble.

On the other hand, substances whose intermolecular forces are of a different nature or intensity tend to be mutually insoluble. For example, hydrocarbons, lipids (oils and fats), some halogenated compounds and inert substances, such as the noble gases, are only very sparingly soluble in polar solvents, particularly water, and tend to leave the polar phase and form a separate phase (i.e. layer). This has been attributed to a so-called 'hydrophobic interaction'. The molecular origin of this appears to be the virtual absence of polarity of the above solute molecules due to be similar electronegativities of the atoms in their bonds (e.g. C—H, C—Cl, C—Br and C—I). The thermodynamic explanation of the hydrophobic interaction is that, if more than a very small quantity of hydrocarbon material were to dissolve in water, the Gibbs energy (p. 297) of the system would increase so that the system would become unstable and would separate into two phases. The forbidden increase in Gibbs energy, if mixing occurred, is attributable mainly to a decrease in entropy (i.e. an increase in order, p. 286) of the system and partly to an increase in enthalpy (p. 271) of the system. The entropy would in this case decrease because the polar molecules, e.g. H_2O, would form an ordered clathrate type of structure around each inert solute molecule (see p. 450). The enthalpy would here increase because the inert solute molecules would separate some of the polar molecules against their relatively strong forces of mutual attraction. It is important to note that mixing and dissolution are always accompanied by an increase in entropy (i.e. a decrease in order) of the system (p. 292), except in the exceptional hypothetical case mentioned above. In general, both enthalpy and entropy play a part in deciding whether or not any chemical or physical process, including mixing and dissolution, is accompanied by a decrease in the Gibbs energy of the system to enable the process to take place spontaneously (see pp. 297–302).

Solubility of Liquids in Liquids

Although there are certain pairs of liquids, such as water and alcohol, alcohol and ether, which under normal conditions are miscible in all proportions, it is often found that a given pair of liquids is only micible within certain limits. Thus, if ether is added to water, the limit of miscibility is reached when the concentration of ether is 7 per cent, and, when water is added to ether, the limit occurs at 3 per cent of water. When the limit is exceeded, two liquid layers are formed, one being a saturated solution of ether in water, and the other a saturated solution of water in ether.

Liquid mixtures containing only two components, i.e. *binary liquid mixtures*, will first be considered. As in the case of solubility of solids, the solubility of a liquid in another liquid usually increases with increasing temperature, so the liquids become more miscible. The curve showing the variation of the limit of miscibility with temperature usually takes the form shown for the example of phenol and water in Fig. 6.2. (Although phenol is a solid at room temperature, a saturated solution of water in phenol is a liquid.) The curve is interpreted as follows. If successive amounts of phenol are added to water at 20° C, the phenol will all dissolve, until the composition represented by the point A is reached (approximately 8 per cent phenol and 92 per cent water). Further addition of phenol produces two liquid layers whose

Fig. 6.2 Miscibility curves for the phenol–water system

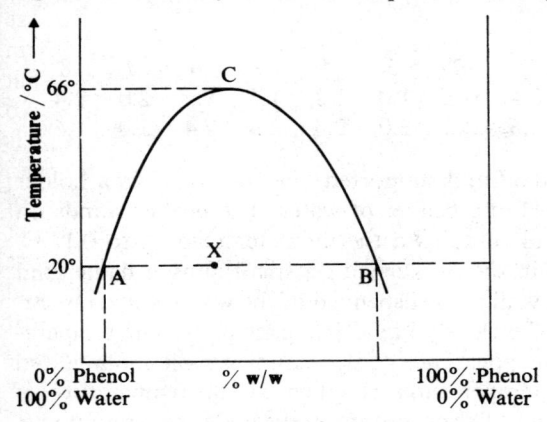

compositions are represented by the points A and B. The relative amounts of these two layers will alter as more phenol is added within the range A to B. Thus, when the overall composition is represented by the point X, we have:

$$\frac{\text{Amount of liquid A}}{\text{Amount of liquid B}} = \frac{\text{Distance XB}}{\text{Distance XA}}. \qquad [6.2]$$

The liquids A and B which exist in equilibrium with each other at a given temperature are called a conjugate pair. As the temperature increases, the solubilities increase, and the two limbs of the curve bend in towards each other, i.e. the range of immiscibility decreases. Eventually, at 66° C, the two limbs of the curve meet at the point C, so that above this temperature the two liquids are completely miscible in all proportions. The temperature corresponding to the point C is called the *critical solution temperature* for the system.

For some systems the two limbs of the curve also approach each other as the temperature is lowered as well as when it is raised. Eventually, the two limbs join again, so that there is a complete closed curve with both an upper and a lower critical solution temperature. An example of this type of behaviour is provided by the nicotine–water system, which has a lower critical solution temperature at 61° C as well as an upper critical solution temperature at 208° C. Cases where an increase in mutual solubility occurs on lowering the temperature are all found where complex formation between the components can take place, e.g. by hydrogen bonding (see p. 31). Formation of the complex is encouraged by lowering the temperature, because thermal motion of the molecules is reduced.

Experiment 6.1 To determine the critical solution temperature for the phenol–water system.

A set of ampoules is made up containing the following quantities of phenol and water:

Ampoule No.	1	2	3	4	5	6	7	8
Phenol (g)	0.4	0.8	1.0	1.2	1.4	1.6	2.0	2.4
Water (cm^3)	3.6	3.2	3.0	2.8	2.6	2.4	2.0	1.6

The ampoules are sealed off and supported, one at a time, in a holder made of wire, suspended in a beaker of water. The beaker stands on a tripod and gauze, and contains a thermometer reading to 0.1° C. By warming the water in the beaker with a small bunsen flame, and by shaking the ampoule while it is suspended in the water, the temperature is found at which the cloudy liquid just goes clear. The temperature is also found for reappearance of the cloudiness on cooling, and the mean of the two temperatures is taken as the temperature of miscibility for that mixture. The procedure is repeated for each ampoule, and a graph is plotted of temperature of miscibility against percentage composition for the various mixtures. This should give the curve shown in Fig. 6.2 and the maximum gives the critical solution temperature. The result may differ from 66° C, partly because of impurities in the phenol, and partly because a temperature of miscibility measured under pressure in a sealed tube differs slightly from the correct temperature measured at atmospheric pressure.

Addition of a third substance to a pair of liquids has a marked effect on their critical solution temperature which may therefore be used as a criterion of purity. Mixtures consisting of three components are called *ternary mixtures*. If the third substance is soluble only in one of the pair of liquids, it usually raises the upper critical solution temperature and lowers the lower critical solution temperature. If, for example, sufficient naphthalene is present in phenol to give a 0.1 mol dm^{-3} solution, the critical solution temperature of the phenol/water system is raised by about 20° C. If the third substance is soluble in both components, it usually lowers the upper critical solution temperature and raises the lower critical solution temperature. Use is made of this fact to increase the mutual solubility of two partially immiscible liquids by addition of a blending agent miscible with both liquids in all proportions. The blending agent, propylene glycol, may be used to blend a volatile oil with water. If a surface active agent is used to increase the solubility of a non-polar liquid in water or vice versa, solubilisation by micelle formation is responsible (p. 484), rather than simple blending.

The effects of changes in the relative amounts of the components of a ternary system at a given temperature and pressure may be shown most conveniently by plotting the mole, weight or volume fractions or the corresponding percentages on triangular graph paper ruled in triangular coordinates (Fig. 6.3). Each apex of the equilateral triangle corresponds to 100 per cent of the corresponding component, i.e. A, B or C, whereas the side of the triangle opposite the apex corresponds

Fig. 6.3 Triangular coordinates

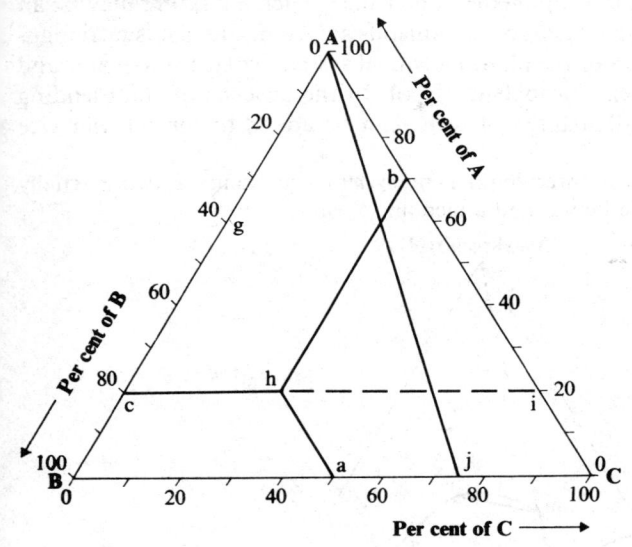

to the absence of that component and therefore to a binary liquid mixture, i.e. B+C, A+C or A+B respectively. In fact, the point g corresponds to 0 per cent C, 60 per cent A and 40 per cent B. A point, e.g. h, inside the triangle represents a ternary mixture. The length of a line drawn from the point to a given side of the triangle and parallel to either of the other sides is proportional to the percentage of the component represented by the opposite apex. The length of any side of the equilateral triangular diagram represents 100 per cent. In fact, the point h represents a ternary mixture containing A, B and C in the relative amounts ha, hb and hc respectively, which in this case are 20 per cent of A, 50 per cent of B and 30 per cent of C. The three percentages must, of course, add up to 100 per cent and the geometry of triangular diagrams conforms to this. Thus, for any point h, the sum of the three distances, ha ($=$ Ci), hb ($=$ ib), hc ($=$ bA), is equal to the length of one side of the triangular diagram (e.g. CA) and therefore to 100 per cent. When plotting a point on the diagram, it is only necessary to consider the percentage of two of the components; the percentage of the third component may be used to check the plotting. All points on any straight line drawn from an apex to the opposite side of the triangle represent the compositions of mixtures in which the ratio of the quantities of two components remain constant whilst the proportion of the third component may have any value from 0 per cent at the base to 100 per cent at the apex. Thus jA represents a constant ratio of B/C ($= \frac{25}{75} = \frac{1}{3}$), whilst A can vary from 0 per cent at j to 100 per cent at A.

A mixture of substances whose proportions are constant may be treated as a single component. In pharmacy such a mixture may be an oil, a surface active agent or an aqueous solution. Fig. 6.4 is a triangular phase diagram of the pharmaceutical system, volatile oil, water and the blending agent, propylene glycol. In the absence of the blending agent only a small quantity of water can be added to the oil, and vice

Fig. 6.4 A typical three-component system consisting of two partially miscible liquids and a blending agent

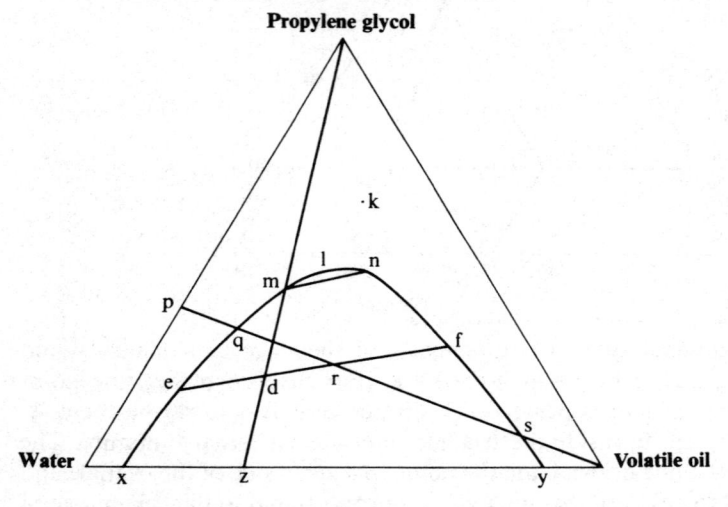

versa, before the partially miscible mixture separates into two liquid layers. When the overall composition of the mixture is z or any other value between x and y, one of the immiscible layers or conjugate binary solutions consists of a saturated solution of oil in water, and has the composition x, and the other consists of a saturated solution of water in oil and has the composition y. The addition of increasing quantities of propylene glycol to the binary mixture of overall composition z causes the overall composition of the whole mixture to move along the line from z, through points d and m, towards the propylene glycol apex. At d the mixture consists of two immiscible layers or conjugate ternary solutions, e and f, each containing oil, water and propylene glycol such that

$$\frac{\text{Amount of liquid e}}{\text{Amount of liquid f}} = \frac{\text{Distance df}}{\text{Distance ed}} \qquad [6.3]$$

The line ef is called a tie line. A tie line is parallel to the base of the triangle only on those rare occasions when the blending agent distributes itself equally between the oil and water. At m the two immiscible layers or conjugate ternary solutions have compositions m and n respectively and when more propylene glycol is added, layer n disappears and a homogeneous solution is left. At all points outside the curve, e.g. k, there is only one liquid phase, whereas at all points under the curve, e.g. d, r and z, two separate liquid phases coexist in equilibrium with each other. At point 1, which is often known as the plait point, the tie lines disappear and the two liquid phases have identical compositions and merge into one. It is an instructive exercise to describe the changes which take place when the volatile oil is gradually added to a mixture of water and propylene glycol represented by the point p in Fig. 6.4.

Experiment 6.2 To investigate the limits of miscibility of the system toluene–water–ethanol.

To a 100 cm^3 stoppered flask are added from burettes 5 cm^3 of toluene and 0.5 cm^3 of water. Ethanol is then run into the flask from a burette with shaking after each addition and the minimum volume of ethanol required to give a clear solution is carefully determined. A further 0.5 cm^3 of water is then added to the solution and more ethanol is run in as before until the mixture is again just clear. This is repeated adding different quantities of toluene and water. Suggested volumes for the commencement of the titrations are given in the first part of Table 6.1. For mixtures whose compositions are close to the sides of the triangle, it is extremely difficult to determine the end-point when ethanol is run in. Accordingly, known volumes of ethanol and toluene are mixed and water is run in from the burette until the mixture is just cloudy, as suggested in the second part of Table 6.1. Alternatively, toluene could be added to mixtures of ethanol and water.

The compositions of just miscible solutions expressed as per cent v/v for each component are plotted on triangular graph paper. Since the results depend on the room temperature and atmospheric pressure, these quantities are recorded on the completed ternary phase diagram. This experiment does not determine the tie lines which can only be obtained by analysis of the conjugate ternary solutions.

By reference to the ternary phase diagram it is instructive to describe the changes which take place when water is gradually added to a mixture consisting of 60 per cent (v/v) of ethanol and 40 per cent (v/v) of toluene. It is also instructive to determine from the diagram the minimum volume of ethanol which would have to be added to a mixture consisting of 70 cm^3 of water and 30 cm^3 of toluene in order to make the mixture homogeneous.

Table 6.1 Suggested volumes for the commencement of titrations in the determination of the limits of miscibility of the system toluene–water–ethanol

cm^3 toluene	cm^3 water	cm^3 ethanol
5	0.5	—
5	1	—
5	1.5	—
5	2	—
5	3	—
5	4	—
5	6	—
5	8	—
2	5	—
1	5	—
1	—	8
1	—	11
1	—	15
1	—	20
0.5	—	20
0.2	—	20
20	—	0
20	—	2
20	—	4
20	—	6
20	—	8
20	—	10
20	—	12
—	20	0

Solubility of Gases in Liquids

The solubility of a gas in a liquid is the number of volumes of gas required to saturate one volume of the liquid under given conditions of temperature of the system and partial pressure of the gas. If the solubility is measured under such conditions that the gas is exerting a partial pressure of 1 atm (101 325 Pa), and the number of volumes of gas required to saturate one volume of liquid under these conditions is reduced to NTP, the resulting quantity is called the *Bunsen absorption coefficient* or simply the absorption coefficient of that gas–liquid system. Bunsen absorption coefficient and solubility of a gas, as defined above, are dimensionless quantities.

Solubility of Gases in Liquids

The *mass* of any gas absorbed by a given volume of liquid varies according to the partial pressure of the gas above the liquid and the temperature of the system. This is expressed by *Henry's Law* which states that the mass w of gas dissolved by a given volume V of liquid at a given temperature is proportional to the partial pressure p of the gas above the liquid, thus:

$$\frac{w}{V} = hp, \qquad [6.4]$$

where h is a constant which depends on the gas, the liquid, the temperature and the units of w, V and p.

The *volume* v of a single gas dissolved by a given volume V of liquid is independent of the pressure p of the gas above the liquid at a given temperature, T K, provided that Henry's law applies to the system and the ideal gas laws apply to the gas. This may be proved as follows. V m³ of liquid dissolves w kg of gas which is the same as w/M mol or v m³, where M kg mol^{-1} is the molar mass of the gas. For w/M mol of gas before its loss from the gas phase, we can write:

$$pv = \frac{w}{M}.RT. \qquad [6.5]$$

Substituting for w, using Henry's law (equation 6,4) in the form $w = Vhp$:

$$pv = VhpRT/M.$$

Dividing both sides by p, we have:

$$v = VhRT/M. \qquad [6.6]$$

The final expression for the volume of gas absorbed is independent of the pressure.

Henry's Law is an 'ideal' law in the sense that the behaviour of gases in practice shows deviations from the law. However, at low pressures, and therefore at low solubilities, many gases obey the law fairly well. Large deviations occur when the conditions of the gas are close to those for liquefaction, if the gas associates or dissociates in solution or if the gas reacts chemically with the solvent, e.g. formaldehyde, hydrogen chloride and ammonia react readily with water, thus:

$$HCHO + H_2O = CH_2(OH)_2$$
$$HCl + H_2O = H_3O^+Cl^-$$
$$NH_3 + H_2O = NH_4OH$$

If there is a mixture of gases above a liquid, the solubility of each component of the gas is governed separately by Henry's Law, with the appropriate value of the constant h for each gas. In most cases the partial pressure and the absorption of one gas will not affect the partial pressure and the absorption of another gas. To illustrate the principles involved in this type of problem, we shall consider the solubilities of the three main constituents of air in blood plasma, assuming that we can apply Henry's law in terms of the partial pressures of the constituents in the lung.

Let the pressure of air in the lung be 96.0 kPa and let the temperature be that of the body, i.e. 37.5° C. The vapour pressure of water in the lung is 6.4 kPa so the total pressure is to be taken as 102.4 kPa. The absorption coefficients at 37.5° C for N_2, O_2 and CO_2 are 0.012, 0.024 and 0.51 respectively and the partial pressures, from the calculation of p. 26, are 76.8, 13.4 and 5.8 kPa respectively. Consider nitrogen first. An absorption coefficient of 0.012 means that if N_2 were exerting a partial pressure of 101 325 Pa (1 atm), each dm^3 of blood plasma would dissolve 0.012 dm^3 = 12 cm^3 of nitrogen, if this were measured at 0° C (273.15 K) and 101 325 Pa pressure. Suppose that the mass of this quantity were x[†]. Then, for a partial pressure of 76.8 kPa the mass of N_2 dissolved would be $x \times \dfrac{76.8}{101.3}$. Now if a mass x of N_2 occupies 12 cm^3 at 0° C and 101 325 Pa, a mass $\dfrac{x \times 76.8}{101.3}$ of N_2 occupy $\dfrac{12}{x} \times \dfrac{x \times 76.8}{101.3}$ cm^3 = 9.1 cm^3. Or, measuring under the conditions of the experiment, namely 37.5° C (310.65 K) and 102.4 kPa total pressure, the volume of N_2 dissolved per dm^3 of plasma would be $9.1 \times \dfrac{101.3}{102.4} \times \dfrac{310.65}{273.15}$ = 10.2 cm^3. In a similar manner we calculate that the corresponding volumes of O_2 and CO_2 which would dissolve in 1 dm^3 of blood plasma would be 3.6 cm^3 and 32.9 cm^3 respectively. The fact that blood actually contains considerably more O_2 and CO_2 than these figures suggest indicates that these gases must be held in the blood mainly by chemical combination rather than by physical solution.

Vapour Pressure of Liquid Mixtures

Since all the components A, B, etc., of a liquid mixture may be volatile, the total vapour pressure P is the sum of the partial vapour pressures of each of the components p_A, p_B, etc. (Dalton's law of partial pressures,

---

[†] It is not necessary to introduce the mass of N_2 in this calculation, since the volume of gas at constant temperature and pressure is proportional to the mass. It has been done to emphasise again the fact that Henry's law applies to solubility expressed as mass of gas dissolved.

p. 25), thus:

$$P = p_A + p_B +, \text{etc.} \quad [6.7]$$

The way in which the total vapour pressure varies with the composition of the liquid mixture depends upon the interactions between the molecules of the components.

(a) *Miscible liquids*

If a pair of liquids, A and B, have similar intermolecular forces, the forces between the various molecules in the mixture will be much the same as the forces between the molecules in the pure liquids A and B. Pairs of liquids of this type are exemplified by: ethylene dibromide and propylene dibromide; benzene and toluene; two paraffins. On adding one liquid to the other, the liquids will be miscible, so the disorder (entropy) will increase, but there will be practically no liberation or absorption of heat and no change in volume. Addition of B to A will reduce in strict proportion the fraction, x_A, of molecules of A in the bulk and at the surface and hence the escaping tendency or partial vapour pressure of A, p_A. This is expressed as follows by Raoult's law for component A:

$$p_A = x_A p_A^\ominus. \quad [6.8]$$

When x_A is unity, only pure liquid A is present whose vapour pressure is $p_A^\ominus$. Exactly analogous reasoning can be applied to the addition of A to B, leading to Raoult's law for component B:

$$p_B = x_B p_B^\ominus. \quad [6.9]$$

If one component obeys Raoult's law, the other component will also do so. The mole fraction of A, x_A, and of B, x_B, are related to the number of moles of A, n_A, and of B, n_B, as follows:

$$x_A = \frac{n_A}{n_A + n_B}; \quad x_B = \frac{n_B}{n_A + n_B}. \quad [6.10]$$

It can be seen that:

$$x_A + x_B = 1.$$

The partial vapour pressures of A and B above the mixture, p_A and p_B respectively, are related to the respective mole fractions, x_A and x_B, by the Raoult's law relationships given above and illustrated in Fig. 6.5. The total vapour pressure, $P = p_A + p_B$, varies linearly from $p_A^\ominus$ to $p_B^\ominus$, where $p_A^\ominus$ and $p_B^\ominus$ are the saturated vapour pressures of the pure liquids A and B respectively. Liquid mixtures which obey Raoult's law are called ideal mixtures (or ideal solutions).

The boiling-point diagram can be derived from the vapour-pressure diagram if the temperature variations of the vapour pressures of the two components are known. Since A has the lower vapour pressure at

Fig. 6.5 Distillation of ideal miscible liquids
(a) vapour pressure—composition diagram
(b) boiling-point—composition diagram

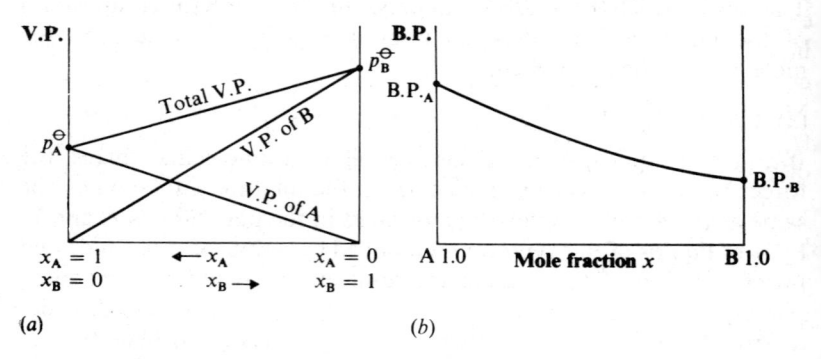

(a) V.P. diagram with Total V.P., V.P. of B, V.P. of A lines between $p_A^\ominus$ ($x_A=1, x_B=0$) and $p_B^\ominus$ ($x_A=0, x_B=1$).

(b) B.P. diagram showing B.P.$_A$ and B.P.$_B$ curves vs mole fraction x from A 1.0 to B 1.0.

any given temperature, it has the higher boiling-point. The boiling-points of the mixtures then fall regularly but not linearly, from that of A to that of B, as the composition changes from 100 per cent A to 100 per cent B.

Liquid mixtures which do not obey Raoult's law are known as non-ideal mixtures (or solutions). We define a quantity called the activity (or thermodynamic activity), a, which, when substituted for the mole fraction in the equations for Raoult's law, makes the equations valid under all circumstances (both ideal and non-ideal), thus:

$$p_A = a_A p_A^\ominus \quad \text{and} \quad p_B = a_B p_B^\ominus, \qquad [6.11]$$

where a_A and a_B are the activities of A and B respectively in the mixture. The activity of a substance is always unity, when the substance is in its standard state. The above definition of activity requires that the standard state of a given liquid component is its pure liquid state. Thus, for pure liquid A (i.e. $x_A = 1$, $x_B = 0$) $p_A = p_A^\ominus$, so that $a_A = 1$. Similarly, for pure liquid B (i.e. $x_B = 1$, $x_A = 0$) $p_B = p_B^\ominus$, so that $a_B = 1$.

For ideal mixtures $a_A = x_A$ and $a_B = x_B$. For non-ideal mixtures $a_A \neq x_A$ and $a_B \neq x_B$. If the solutions are dilute, i.e. consist of a great excess of one component, $a_A \approx x_A$ and $a_B \approx x_B$, and Raoult's law is approximately obeyed. On mixing two substances which give non-ideal mixtures, the disorder (entropy) increases as for ideal mixtures but there is also an exchange of heat with the surroundings and a volume change.

If the liquids A and B have intermolecular forces of a different nature and have little affinity for each other, the forces between A and B molecules in the mixture will be weaker than the forces between the molecules in the pure liquids A and B. Molecules of A and B will therefore tend to leave the condensed phase, the liquid state, and enter the vapour phase. Consequently, p_A, p_B and P will exceed the values expected from Raoult's law (Fig. 6.6.a) and the system is said to exhibit positive deviations from the law. The activities of A and B will always exceed their respective mole fractions (i.e. $a_A > x_A$, $a_B > x_B$) and the liquids may be partially immiscible. Mixing, when it occurs, is accompanied by an increase in volume and by the absorption of heat. Most miscible liquid mixtures show relatively small positive deviations from Raoult's law, e.g. an alcohol and a hydrocarbon. Water and diethyl ether give larger positive deviations which result in partial

Fig. 6.6 Vapour-pressure curves showing deviations from Raoult's law (*a*) positive deviation (*b*) negative deviation

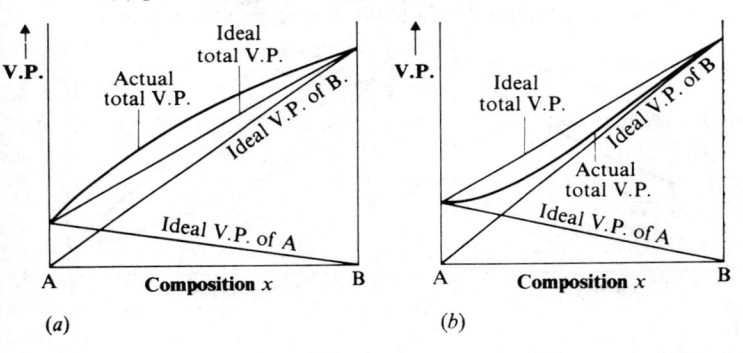

miscibility over a certain range of mole fractions. The systems water–benzene and water–chlorbenzene give such large positive deviations that these pairs of liquids are immiscible over virtually the entire concentration range (discussed later in this section).

If the molecules of A and B have a strong affinity for each other, so that they tend to form a complex, an ion pair or a compound, the forces between A and B molecules in the mixture will be stronger than the forces between the molecules in the pure liquids A and B. Acid-base systems, e.g. acetic acid/pyridine, and acetone/chloroform are examples. The molecules of A and B will resist leaving the liquid phase. As a result p_A, p_B and P will be less than those expected from Raoult's law (Fig. 6.6b) and the system is said to exhibit negative deviations from the law. The activities of A and B will always be less than their respective mole fractions (i.e. $a_A < x_A$, $a_B < x_B$). Mixing is accompanied by the liberation of heat and by a decrease in volume.

Except in the special case of an azeotropic mixture (p. 219), the composition of the vapour distilling from the liquid mixture is not the same as the composition of the liquid. As might be expected, the vapour is richer in the more volatile component. It is common practice to represent, on a vapour-pressure or boiling-point diagram, both the composition of the liquid and that of the vapour. In Fig. 6.7, both the liquidus and the vapour composition lines are plotted, on a vapour-pressure graph and on a boiling-point graph, for mixtures giving a small positive deviation from Raoult's law. In each diagram, a horizontal line has been drawn showing, in (a), the compositions of the liquid and vapour in equilibrium with one another giving one particular vapour pressure, and, in (b), the compositions of liquid and vapour in equilibrium with one another at one particular boiling-point. This is a tie line (cf. p. 45). It should be noted that, for each diagram, the tie line indicates that the vapour contains more of the more volatile component B than the liquid in equilibrium with it.

Fig. 6.7 Liquidus and vapour composition curves for small positive deviation
(a) vapour-pressure diagram (b) boiling-point diagram

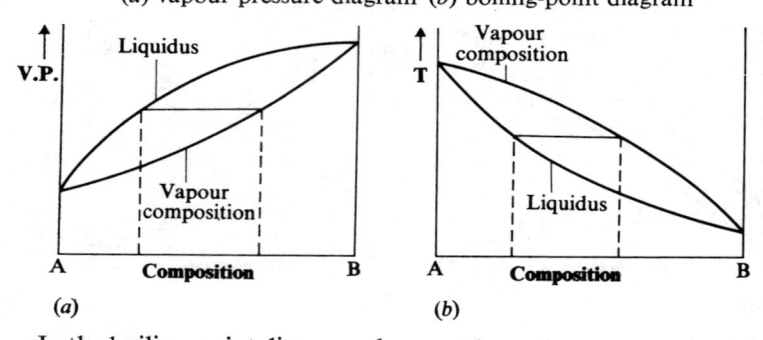

In the boiling-point diagram, the area above the vapour composition line indicates conditions of temperature and composition where only vapour can exist. The area below the liquidus line represents conditions where only liquid can exist. The area between the two lines is a heterogeneous region where both liquid and vapour exist in equilibrium with each other. The tie line drawn through any point in this heterogeneous region indicates the composition of the liquid and vapour into which a mixture of the specified composition would separate at that particular temperature. Bearing this in mind, the process of fractional distillation may be followed with the aid of such a boiling-point diagram.

Suppose a liquid mixture of composition X is heated to its boiling-point t_1 (Fig. 6.8), the vapour distilling off will have a composition represented by the point v_1, obtained by drawing the tie line at t_1. This now enters the fractionating column and cools to a temperature t_2. The tie line at this temperature shows that the heterogeneous mixture, under these conditions, will separate out into a liquid of composition l_2 (which will run back into the flask), and a vapour of composition v_2. This vapour will pass up the fractionating column, and will be cooled again to a temperature t_3. Here it will separate out into a liquid l_3 and vapour v_3. Again, the liquid will run down the column, and the vapour will move up to be cooled to t_4, and the process continues. Eventually, if the column is efficient enough to effect complete separation, the vapour will have the composition of pure B and will pass out of the top of the column to be condensed. As B continually leaves the system in this way, the composition of the liquid remaining in the distillation flask will alter from X, and will approach A. Finally, when all the component B has been distilled off the pure A remaining in the flask can be distilled.

In practice, in a laboratory distillation column, the temperature decreases fairly uniformly up the column and the steps in the process outlined above become infinitely small. Also, at every point there is

Fig. 6.8 Boiling-point diagram illustrating fractional distillation

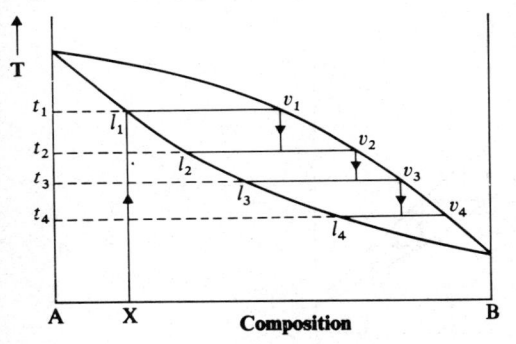

interchange of heat and molecules between rising vapour and descending liquid. The composition of the vapour therefore varies gradually up the column from that of the point v_1 to that of the final distillate. The composition of the liquid flowing down the column varies from that in equilibrium with the final distillate at the top to X at the bottom. However, at any given point in the column the compositions of the vapour and liquid still approximate to those represented by the ends of a tie line drawn at the appropriate temperature on the boiling-point diagram.

Industrial fractionation columns are of the plate and bubble-cap type and these often approach the opposite extreme in which separation takes place on each plate into liquid and vapour in equilibrium with each other. The vapour then passes up to the next plate, where it is condensed to liquid of the same composition which proceeds to give off its own equilibrium vapour, and so on. In such a process, the boiling-point diagram is traversed by alternate horizontal and vertical lines between the liquidus and the vapour composition line, as in Fig. 6.9.

This ideal is not actually attained but it is used as a measure of the degree of separation of the components during fractionation and as a measure of the efficiency of a fractionation column. Each step on the boiling-point diagram (or, strictly, each horizontal line) is counted as a 'theoretical plate' and the number of steps from the composition of the original liquid to that of the final distillate gives the number of theoretical plates for this degree of separation. In Fig. 6.9 there are 5 theoretical plates between the original liquid of composition X and the distillate of composition Y. Of these, one separation takes place at the surface of the boiling liquid and four are due to the fractionating column. The degree of fractionation attainable by any type of column working under specified conditions is quoted in terms of the number of

Fig. 6.9 Separation produced by 5 theoretical plates

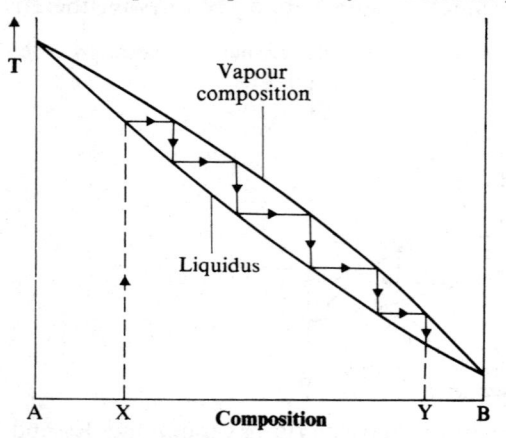

theoretical plates, meaning the number of equilibrium separations of liquid and vapour which would be equivalent to the total effect of the column. Although laboratory fractionation columns often have efficiencies of only a few theoretical plates, efficiencies of 100 or more theoretical plates can be attained.

Some liquid mixtures have large positive or negative deviations from Raoult's law, so that their boiling-point diagrams have definite minima or maxima. The curves then are of the types illustrated in Fig. 6.10. The mixtures in (a), with a large positive deviation, pass through a minimum boiling-point, whereas the mixtures in (b), having a large negative deviation, pass through a maximum boiling-point. In both these cases it is impossible to effect complete separation of the components by fractional distillation, however efficient. This can be shown by following the process as before. Each of the two arms of the positive deviation or negative deviation diagram now resemble the complete diagram of Fig. 6.8. This means that separation along one arm of each diagram is possible, but the process cannot be taken further. Thus, in Fig. 6.10(a) mixture P could be separated into the minimum boiling mixture and pure A; or a mixture Q could be separated into the minimum boiling mixture and pure B. In Fig. 6.10(b) mixture R could be separated into the maximum boiling mixture and pure A; and S could be separated into the maximum boiling mixture and pure B. The maximum or minimum boiling mixtures are known generally as constant boiling mixtures, or as azeotropic mixtures. Common examples are: ethyl alcohol and water, giving a minimum boiling mixture containing 96 per cent alcohol; and hydrochloric acid and water, giving a maximum boiling mixture containing 20 per cent HCl. Such mixtures can only be separated into their components by chemical or by other physical means. For example, lime is used to remove water from the azeotropic mixture in the preparation of

Fig. 6.10 Boiling-point curves showing large deviations from Raoult's law (a) positive deviation (b) negative deviation

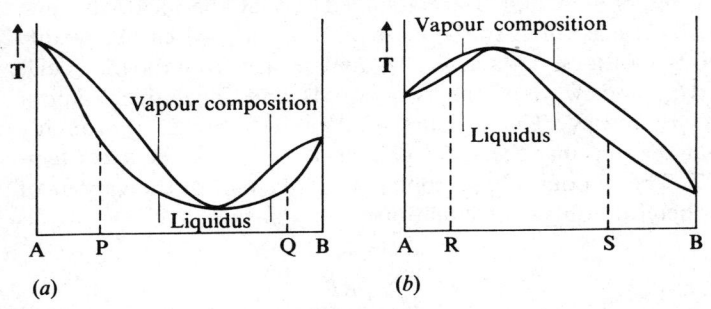

absolute alcohol. An alternative method of preparing absolute alcohol is to add benzene to the azeotropic mixture, and then redistil. A new azeotropic mixture is formed, of lower boiling-point, involving all three substances. The first fraction which distils off will therefore be this ternary azeotropic mixture, and this will remove all the water. The second fraction is an azeotropic mixture of alcohol and benzene, and this removes all the excess benzene. The final distillate is then absolute alcohol. Other physical methods of separating azeotropic mixtures are to carry out alternate fractional crystallisations and fractional distillations, or to subject the mixture to fractional extraction with an immiscible solvent.

It is interesting to note that the principle of azeotropic evaporation is made use of in the control of body temperature by perspiration. The perspiration is a mixture of about 99 per cent water and 1 per cent fatty acid, and it has a boiling-point much lower than either of its components and is therefore more readily evaporated.

(b) *Immiscible liquids*

If two liquids which are completely immiscible are placed together, neither will influence the vapour pressure of the other, because there is no solution of one in the other. Each liquid therefore exerts the vapour pressure which it would normally have at that temperature, irrespective of the presence of the other. The total vapour pressure at any composition of the mixture is then simply the sum of the two vapour pressures.

If such a mixture is heated, it will boil at the temperature at which the sum of vapour pressures is equal to atmospheric pressure. Since the total vapour pressure must be higher than that of either component, the boiling-point will be lower than that of either component. This principle is applied in steam distillation, enabling organic compounds which normally have a high boiling-point to be distilled at a temperature somewhat less than that of boiling water, providing they are immiscible with water. By distilling at this lower temperature, decomposition of the organic material is prevented.

If the vapours of A and B are assumed to obey the ideal gas laws, the ratio of the mass of A to the mass of B, i.e. w_A/w_B, can be readily calculated from the molar masses, M_A and M_B (or from the molecular weights, M_{rA} and M_{rB}) of the two liquids and from their vapour pressures, p_A and p_B. The vapours of A and B are so thoroughly mixed that they occupy the same volume V and have the same temperature T. The vapour phase contains w_A/M_A mol of the vapour of A which therefore obeys the equation:

$$p_A V = \frac{w_A}{M_A} . RT.$$

Similarly the vapour phase contains w_B/M_B mol of B vapour which therefore obeys the equation:

$$p_B V = \frac{w_B}{M_B} . RT.$$

Dividing the first equation by the second:

$$\frac{p_A}{p_B} = \frac{w_A}{M_A} \cdot \frac{M_B}{w_B}$$

$$\therefore \frac{w_A}{w_B} = \frac{M_A p_A}{M_B p_B} = \frac{M_{rA} p_A}{M_{rB} p_B}. \qquad [6.12]$$

If an organic compound A is being steam distilled, B is water. The last equation shows that the process will be more efficient (i.e. will give the larger ratio w_A/w_B), the larger is M_{rA}. Since M_{rB} is only 18, whereas M_{rA} may be many times greater, the ratio of molecular weights is generally quite favourable. However, this is usually partly offset by the fact that the high molecular weight organic substance A has rather a low vapour pressure relative to water.

Equation [6.12] may be used to obtain an approximate molecular weight of A, from an investigation of the composition of the distillate and a knowledge of the appropriate vapour pressures. This is illustrated in the following experiment.

Experiment 6.3 To determine the molecular weight of chlorobenzene by steam distillation.

A normal steam distillation apparatus is fitted up and is so arranged that the distillate is delivered into a measuring cylinder. About 150 cm³ of chlorobenzene and 50 cm³ of water are placed in the round-bottomed distilling flask, and the mixture is heated both directly with a bunsen and gauze, and also by passing steam, until the temperature is about 80 °C. The heating is then continued by passing steam only. When distillation starts, it should be maintained at about one drop per second, by adjusting the rate of passing steam. Fractions of distillate are collected in a 50 cm³ measuring cylinder previously cleaned with chromic acid. The volumes of chlorobenzene and water in each fraction are read off in the cylinder, and the temperature of distillation is noted. From the density of chlorobenzene (1.11 g cm⁻³) and of water (1.00 g cm⁻³), the mass of each component in each fraction may be calculated. From tables, the vapour pressure of water at the temperature of distillation is found. The total vapour pressure is equal to atmospheric pressure, and this is read on a barometer. The difference between the total vapour pressure and that of water is the vapour pressure of chlorobenzene, at the temperature of distillation. Hence, the molecular weight of chlorobenzene (A) is calculated from the expression given above.

During the distillation, the composition of the distillate should remain constant (i.e. there should be constant relative volumes in each fraction), even though the composition of the liquid mixture in the distilling flask is continually altering. The temperature of distillation should also remain constant.

If a vapour-pressure–composition graph were plotted for distillation of a pair of immiscible liquids, it would have the form shown in Fig. 6.11(a). The full line represents the compositions of the liquids having the corresponding vapour pressures. The point marked on the line represents the composition of vapour in equilibrium with these liquids. The graph shows that, immediately the composition differs from 100 per cent A, the vapour pressure rises from the value for pure A to that for A+B, and remains at this value until it drops suddenly, at 100 per cent B, to the vapour pressure for pure B. All compositions of liquid mixture are therefore capable of being in equilibrium with the fixed composition of vapour, at the temperature of distillation. This is merely another way of saying that, whenever a mixture of A and B is distilled, the vapour always has the composition

Fig. 6.11 Distillation of immiscible liquids
 (a) vapour pressure—composition diagram
 (b) boiling-point—composition diagram

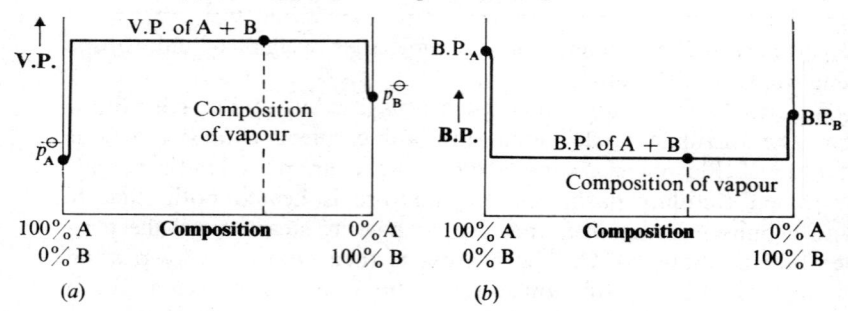

(figure 6.11a: vapour pressure–composition diagram; figure 6.11b: boiling-point–composition diagram)

given by $\dfrac{M_A}{M_B} \times \dfrac{p_A}{p_B}$, whatever the relative amounts of A and B in the liquid phase.

A boiling-point–composition diagram drawn for the same system has the form shown in Fig. 6.11(b). Pure A, with lower vapour pressure than pure B, has the higher boiling-point, and the boiling-point of the mixture is less than either A or B. Again, the full line represents the relationship between boiling-point and composition of the liquid mixture. The point marked on the line represents the composition of vapour in equilibrium with all the liquid mixtures.

Relationships between Activities, Solubilities and Drug Action

When two or more solutions (or phases) are in equilibrium with each other, the partial vapour pressure p_X, of a given component X must be the same in each solution (or phase), since otherwise X would distil from one phase to another in order to equalise the partial pressures and restore equilibrium. To a given component X in any solution, ideal or non-ideal, equation [6.11] may be applied, thus:

$$a_X = p_X/p_X^{\ominus}, \qquad [6.13]$$

where $p_X^{\ominus}$ is the saturated vapour pressure of X, which is constant at a given temperature. Consequently, when two or more solutions (or phases) are in equilibrium, the activity a_X of a given substance X must be the same in each phase. This is proved thermodynamically on p. 310. Similar reasoning and conclusions apply to other substances A, B etc., but, in general, there is no reason why the partial pressures or activities of different substances should be equal.

If, however, excess liquid or solid X is in equilibrium with its solution, the solution will be saturated with respect to X and the

Relationships between Activities, Solubilities and Drug Action

partial vapour pressure, p_X, of X above the solution will equal the saturated vapour pressure, $p_X^{\ominus}$, of pure liquid or solid X, so that a_X will equal unity. Similar reasoning and conclusions apply to other substances A, B, etc. Therefore, according to the above definition of the standard state of unit activity, all liquid and solid solutes in the pure state, in saturated solutions and in their saturated vapours have unit activity, i.e.

$$1 = a_X = a_A = a_B = \ldots \text{etc.}$$

To a first approximation the activity, a_X, of X may be taken to be proportional to its mole fraction x_X and hence to its concentration c_X in appropriate units, thus:

$$a_X = k'_X c_X, \qquad [6.14]$$

where k'_X is a constant. For a saturated solution of X, however, $a_X = 1$ and c_X is equal to the solubility, s_X, of X, so that:

$$1 = k'_X s_X. \qquad [6.15]$$

Dividing equation [6.11] by [6.15] gives:

$$a_x = c_X/s_X. \qquad [6.16]$$

Similar reasoning and conclusions apply to the other constituents A, B, etc.

Equation [6.13], in principle, enables activities to be determined accurately from vapour pressure measurements, provided that the constituents of the vapour phase behave almost ideally, whereas equation [6.16] provides an approximate method of determining activities from concentrations and solubilities.

Biological media of known water activity, a_w, can be prepared by equilibrating the solution with a relatively large volume of a standard aqueous solution of the required a_w in an evacuated vessel for several days. The concentrations of the constituents of the standard solution, usually salts or sulphuric acid, required to give a certain a_w can be calculated or read from tables. During equilibration water distils from the solution of higher vapour pressure to that of lower vapour pressure until the partial vapour pressure of water, and hence a_w, of both solutions are equal. The solutions are then said to be isopiestic (and iso-osmotic, p. 244). A relatively large volume of the standard solution ensures that its a_w does not change appreciably.

Increasing concentrations of polar or ionised solutes decreases a_w partly by binding water molecules, thereby reducing the concentration of free water, and partly by impeding the mobility of water molecules. Most bacteria prefer a water activity close to unity and are inhibited by low a_w values which make it difficult for them to maintain the correct a_w values inside the cells. For this reason high concentrations of common salt or sugar have been used for centuries to preserve foods from bacterial spoilage.

A knowledge of the thermodynamic activities, solubilities or vapour pressures of substances is valuable in correlating results of investigations of the degree of biological action of non-specific drugs, such as anaesthetics, depressants, certain insecticides and phenolic antibacterials. Non-specific drugs are believed not to function by interacting with specific receptors as do specific drugs, but by generally impeding the mobility of other molecules in various cellular processes and by generally distorting the structure of macromolecules.

Ferguson derived an important principle by assuming that the biological effect B of each of a series of nonspecific drugs on a given organism is proportional to the thermodynamic activity a of the drug at the site of action, thus:

$$B = ka, \qquad [6.17]$$

where k is a constant for the whole series of non-specific drugs and for the organism. Since these drugs can diffuse from the external medium into the cell to the site of action, each drug will be in equilibrium in all these phases or regions and in the vapour phase; the thermodynamic activity of the drug will therefore have the same value, a, in all these regions. The activities of all non-specific drugs in their respective saturated solutions have a constant maximum value, equal to unity according to the above definition of the standard state. Consequently, when the cells are in contact with each of these solutions, B has a constant maximum value, M. Thus, when $a = 1$, $B = M$ and equation [6.17] becomes $M = k$. Hence:

$$B = Ma, \qquad [6.18]$$

where M is a constant for the organism and for the whole series of non-specific drugs and a is the thermodynamic activity of the drug in the biological medium. This is a generalised form of Ferguson's principle which states that a series of non-specific drugs have the same degree of biological effect on a given organism, if their amounts or concentrations are adjusted so that their thermodynamic activites are equal.

For substances in solution the thermodynamic activity is given approximately by equation [6.16]. Substituting for a in equation [6.18], gives:

$$B = Mc/s \quad \text{or} \quad \frac{B}{c} = \frac{M}{s}. \qquad [6.19]$$

This is a form of Ferguson's principle useful for drugs in solution. It implies that non-specific drugs which are present at the same proportional saturation, c/s, in a given medium have the same biological effect on a given organism. Since the medium is usually aqueous, an alternative statement might be that the concentrations of a series of non-specific drugs required to produce the same biological effect in a given organism is a constant fraction of their respective water solubilities. Furthermore, for such drugs, the potency, if measured by B/c, is inversely proportional to the solubility in water. There are, however, limits to how far one can increase the potency of a drug by modifying its structure to reduce its solubility. Drugs must have a certain solubility in water in order to penetrate the polar protein component of the cell membrane and diffuse through the aqueous parts of the cell.

For substances in the vapour phase and in solution the thermodynamic activity is given by equation [6.13]. Substituting for a in equation [6.18] gives:

$$B = Mp/p^\ominus. \qquad [6.20]$$

This is a form of Ferguson's principle useful for drugs administered in the vapour phase, e.g. anaesthetics. It implies that non-specific drugs which are present at partial vapour pressures that are the same fraction, $p/p^\ominus$, of their saturated vapour pressures give the same biological effect on a given organism.

Ferguson's principle is obeyed quite well by series of non-specific drugs but only very roughly by drugs in general. Any drug which is more effective than is predicted by Ferguson's principle is almost certain to be specific in its action; this fact makes the principle useful in drug research.

Distribution (or Partition) between Immiscible Solvents

Solubility, and the effects of the solute on the properties of the solution, have so far been discussed in terms of the simplest system, of one solvent and one solute. It often happens that a solute finds itself in the presence of more than one solvent. If the solvents are miscible, the solubility may be taken, to a first approximation, to be the mean of the solubilities in the separate solvents – weighted according to the proportions of the respective solvents in the whole solution. If the solvents are partially miscible, the situation is rather complex, because the presence of the solute will alter the miscibilities of the solvents (e.g. the effect of solute on critical solution temperature, p. 206).

When the solvents are immiscible, or nearly so, the situation is greatly simplified. The solute distributes itself between a pair of solvents according to the partition or distribution law, which states that the ratio of the concentrations in the two solvents at equilibrium is a constant, known as the partition (or distribution) coefficient K, thus:

$$\frac{c_A}{c_B} = K, \quad [6.21]$$

where c_A and c_B are the concentrations of the solute in solvent A and B, respectively, in any units. The law does not apply if the temperature is not constant throughout the system or if, in either solvent, the solute associates, dissociates, ionises or reaches its solubility limit.

The experimental law can be justified on theoretical grounds, if it is assumed that the activity of the solute in each solution are given by equation [6.16], which is now written in the forms:

$$a_A = c_A/s_A \quad \text{and} \quad a_B = c_B/s_B,$$

where the subscripts now refer to the solvents and not, as before, to the solutes. Since the solute in one solution is in equilibrium with the solute in the other solution, $a_A = a_B$ (p. 222), so that:

$$\frac{c_A}{s_A} = \frac{c_B}{s_B}$$

whence
$$\frac{c_A}{c_B} = \frac{s_A}{s_B} = \text{constant} (= K). \quad [6.22]$$

Since the solubility of the solute in each solvent is a constant, c_A/c_B is a constant, i.e. the partition coefficient K. K is therefore approximately equal to the ratio of the solubilities and may roughly be estimated from the rules of solubility (p. 203).

Overton and Meyer found that the greater the oil/water partition coefficient of an anaesthetic or depressant drug, the stronger the action. Since the brain and nervous tissues have a high lipid content, they preferentially take up anaesthetics. Cell membranes contain a lipid component and a protein component and a high oil/water partition coefficient is necessary for penetration through the lipid part. The Overton and Meyer rule is closely linked with Ferguson's principle, because a high oil/water partition coefficient implies a low water solubility which, according to Ferguson's principle, would give a high potency.

The partition law as stated only applies if the solute has the same molecular form in the two solvents. If, for example, the solute X is associated to form dimers X_2 in solvent A (as do many carboxylic acids in non-hydroxylic solvents, due to hydrogen bonding, p. 31) but does not associate in B (e.g. carboxylic acids in water), the partition law only applies to the monomeric form X, thus:

$$\frac{[X] \text{ in A}}{[X] \text{ in B}} = K. \quad [6.23]$$

Distribution (or Partition) between Immiscible Solvents

In each solvent there will be an equilibrium between the monomer and the dimer. The law of mass action enables the concentration of the monomer to be calculated.

$$2X \rightleftharpoons X_2$$

$$\frac{[X_2]}{[X]^2} = K_c, \qquad [6.24]$$

where K_c is the equilibrium constant for dimerisation. In solvent A the solute is present almost entirely as the dimer, so that K_c is large. Equation [6.24] gives the concentration of the monomer thus:

$$[X]^2 \text{ in } A = [X_2] \text{ in } A/K_c.$$

In solvent B the solute is already present almost entirely as the monomer. Introducing these results into equation [6.23] gives:

$$\frac{\sqrt{[X_2] \text{ in } A}}{\sqrt{K_c}[X] \text{ in } B} = K.$$

Thus, we have:

$$\frac{\sqrt{\text{concentration of solute in A}}}{\text{concentration of solute in B}} = \text{constant.} \qquad [6.25]$$

On the other hand, if each solute molecule dissociates completely into two parts in solvent A, but does not dissociate in B, the result is

$$\frac{(\text{concentration of solute in A})^2}{\text{concentration of solute in B}} = \text{constant}'. \qquad [6.26]$$

Experiment 6.4 To determine the partition coefficient of iodine between 1,1,1-trichloroethane (TCE) and water.

A saturated solution of iodine in approximately 50 cm³ TCE is first prepared as a stock solution. About 20 cm³ of this solution are withdrawn by means of a pipette, using a pipette filler, taking care not to draw any excess solid iodine into the pipette. The portion is run out into a stoppered bottle, and 200 cm³ of water are added. After shaking for about 5 minutes (and if possible allowing to stand in a thermostat at about 25° C for about 20 minutes), the liquids are transferred to a separating funnel. The TCE layer is run off, and a 5 cm³ portion is withdrawn, by means of a pipette, for analysis. Approximately 5 cm³ of 10 per cent KI solution are added, and the titration of the iodine is carried out with 0.025 mol dm⁻³ standard sodium thiosulphate solution. The iodine colour is discharged from both the aqueous and the TCE layers in the flask at the end point. After repeating the titration of the TCE layer, the aqueous layer in the separating funnel is analysed. This is done by withdrawing 50 cm³ portions and titrating with 0.0025 mol dm^{-3} standard sodium thiosulphate solution. A little starch solution may be added, if desired, just before the end point. The ratio of the concentration in the TCE layer to that in the aqueous layer gives the partition coefficient.

The experiment is repeated using the following quantities:

Saturated solution of iodine in TCE (cm³)	TCE (cm³)	Water (cm³)	Volumes for titration	
			TCE layer (cm³)	Water layer (cm³)
15	5	200	5	50
10	10	200	10	100
5	15	200	10	100

The ratio of the concentrations in the two layers should give the same partition coefficient as before.

Distribution of a solute between two solvents is often used as a means of extracting material from impurities and unwanted solvent after a chemical preparation. Most organic compounds are much more soluble in ether than in water, whereas inorganic byproducts of the reaction are usually ionic, and therefore insoluble in ether. It is the practice in many preparations, therefore, to extract the product with ether, and then to distil off the ether.

A general formula will now be derived for the removal of a solute from solvent B, e.g. water, by repeated extraction with solvent A, e.g. an organic solvent, such as ether. The mathematics are simplified if the amount left in the original solvent B is calculated. Let w_0 be the original mass of solute in volume V_B of solvent B. Let w_n be the mass of solute remaining in solvent B after n extractions, each with volume V_A of solvent A.

For the first extraction $n = 1$ and the mass of solute left in solvent B is w_1, therefore the mass concentration in solvent B is w_1/V_B. The mass of solute extracted by V_A is $(w_0 - w_1)$, therefore the mass concentration of solute in solvent A is $(w_0 - w_1)/V_A$. Application of the partition law (equation 6.21) to the above mass concentrations gives:

$$K = \frac{(w_0 - w_1)/V_A}{w_1/V_B}.$$

Multiplying both sides by V_A/V_B, we have:

$$K \cdot \frac{V_A}{V_B} = \frac{w_0 - w_1}{w_1} = \frac{w_0}{w_1} - 1.$$

Distribution (or Partition) between Immiscible Solvents

Adding 1 to both sides and inverting:

$$\frac{w_1}{w_0} = \frac{1}{1+KV_A/V_B} = \frac{V_B}{V_B+KV_A}.$$

For the second extraction, w_0 is replaced by w_1 and w_1 by w_2, thus:

$$\frac{w_2}{w_1} = \frac{V_B}{V_B+KV_A}.$$

Multiplying the last two equations, we have:

$$\frac{w_2}{w_0} = \left[\frac{V_B}{V_B+KV_A}\right]^2.$$

After the nth extraction 2 is replaced by n in this equation, giving:

$$\frac{w_n}{w_0} = \left[\frac{V_B}{V_B+KV_A}\right]^n, \qquad [6.27]$$

which is the general formula required.

The following example shows that it is better to extract using several small portions of solvent A than with one large portion. Consider the extraction of an organic substance X from 100 cm³ of aqueous solution, using a total of 100 cm³ of ether, the partition coefficient K being given by:

$$\frac{\text{Concentration of X in ether}}{\text{Concentration of X in water}} = 5.0.$$

Thus $V_B = 100$ and $K = 5$.

If the ether is added in *one* amount $V_A = 100$ and $n = 1$. Equation [6.27] gives:

$$\frac{w_1}{w_0} = \left[\frac{100}{100+(5\times 100)}\right]^1 = \frac{1}{6} = 16.7 \text{ per cent.}$$

Thus 16.7 per cent of X remains unextracted.

If extraction is carried out *twice*, using equal amounts (50 cm³) of ether each time, $V_A = 50$ and $n = 2$. From equation [6.27] we have:

$$\frac{w_2}{w_0} = \left[\frac{100}{100+(5\times 50)}\right]^2 = \left[\frac{100}{350}\right]^2 = 0.0816 = 8.16 \text{ per cent.}$$

Now rather less, 8.2 per cent, of X remains unextracted.

If the extraction is carried out *four* times, using equal amounts (25 cm³) of ether each time, $V_A = 25$ and $n = 4$. Equation [6.27] now gives:

$$\frac{w_4}{w_0} = \left[\frac{100}{100+(5\times 25)}\right]^4 = \left[\frac{100}{225}\right]^4 = 0.0390 = 3.90 \text{ per cent.}$$

Now only 3.9 per cent of X remains unextracted.

Thus, the efficiency of extraction by a given volume of extracting solvent increases with the number of extractions, but the practical inconvenience also increases, so a compromise must be made.

A useful development in extraction methods is the technique of multiple fractional extraction. This is carried out in an apparatus consisting of several identical tubes in each of which portions of the two immiscible solvents can be shaken together. The tubes are also connected together in such a way that, after equilibration, the upper phase from each tube is transferred first to a reservoir and then to the next tube by appropriate tipping movements. The lower phase in each tube is not transferred. A new portion of solvent for the upper phase is introduced into the first tube (either manually or automatically) at each time of transfer of the upper phases. In this way the upper phases progress along the apparatus and can be collected, if desired, in a reservoir at the end of the series of tubes.

If the partition coefficient for the distribution of a particular solute between the two solvents is K and the volumes of the upper and lower layers in each tube are V_U and V_L respectively, then the masses w_U and w_L in the two layers after equilibration are given by:

$$K = \frac{w_U/V_U}{w_L/V_L}.$$

Therefore $\quad w_U/w_L = KV_U/V_L.$

The mass fraction of solute p found in the upper layer after each equilibration is therefore given by:

$$p = \frac{w_U}{w_U+w_L} = \frac{KV_U}{KV_U+V_L}. \qquad [6.28]$$

The mass fraction q in the lower layer is, of course, $(1-p)$. Whatever the total mass of the solute in a given tube during the process of shaking there will be a fraction p of this total mass in the upper phase and a fraction q of this total mass in the lower phase after equilibration.

Suppose that an initial mass x of the solute is present in the first tube at the beginning of the process. After the first equilibration a mass px is in the upper phase and a mass qx is in the lower phase. The upper phase is now transferred to the next tube and is shaken with fresh lower phase. At the end of this equilibration a fraction p of the mass px is in the upper phase of this second tube (i.e. a mass p^2x) and a fraction q of px (i.e. a mass pqx) is in the lower phase. At the same time, fresh upper phase is shaken with the lower phase remaining in the first tube and containing a mass qx of the solute. At the end of this equilibration, therefore, a fraction p of this mass qx (i.e. a mass pqx) is in the upper phase and a fraction q of qx (i.e. a mass q^2x) is in the lower phase. When the upper phases are now all moved on one step the tubes

Distribution (or Partition) between Immiscible Solvents

contain the following masses of solute:

tube 1, lower phase q^2x, upper phase 0;
tube 2, lower phase pqx, upper phase pqx;
tube 3, lower phase 0, upper phase p^2x.

During the next period of equilibration the same principles apply as before, but care must be taken in counting the total mass of material in the second tube. Some of the solute is initially in the upper phase (brought over from tube 1) and some is in the lower phase (remaining from the previous equilibration). There is therefore a total mass of $2pqx$ of solute to be distributed such that a fraction p is in the upper phase and a fraction q is in the lower phase after equilibration. The masses in the lower (L) and upper (U) phases of the first few tubes after various stages in the process are shown in Table 6.2. This Table could be extended indefinitely to give the mass of solute in any phase at any stage in the process.

So far, only one solute has been considered, but the technique is actually applied to the separation of two or more components of a mixture. Suppose there are two substances whose partition coefficients are such that p for the first solute is 0.7 (so that q is 0.3) and p for the second solute is 0.3 (so that q is 0.7). The masses of the two solutes after the fourth transfer are shown in the last two lines of Table 6.2. It can be seen that, after only four equilibrations, a fair degree of separation has been attained, since nearly a half of the total mass of solute 1 is to be found in tube 4 whereas the second solute has only progressed to the extent of nearly a half of it being present in tube 2. Already hardly any of solute 1 remains in tube 1 though a quarter of solute 2 is still there. In practice, at least 12 tubes in series are normally used and effective separations of chemically similar substances can be achieved. The technique is very useful in the separation of labile substances extracted from natural sources.

If the complete removal of organic compounds from a solution is desired rather than their separation from each other, a continuous extractor will reduce the tedium of repeated extraction. The design of the apparatus depends on whether the extracting liquid (the extractant) is less or more dense than the original solution (the raffinate). The extractant is boiled in a flask and the vapour is condensed. The liquid produced is allowed to percolate through the raffinate. The extractant now contains a small amount of the compounds to be extracted and is returned to the heated flask through a tube. The extractant is automatically recycled through the raffinate and the compounds to be extracted are gradually concentrated in the extractant in the flask. In practice, several days of continuous extraction are necessary for complete removal of the solute(s) and the compounds being extracted are always at the temperature of the boiling liquid, so

Table 6.2. Distribution of Solute during Multiple Fractional Extraction

Tube Number	1		2		3		4		5	
Phase	L	U	L	U	L	U	L	U	L	U
Initial Mass	x	—	—	—	—	—	—	—	—	—
1st Equilibration	qx	px	—	—	—	—	—	—	—	—
1st Transfer	qx	—	—	px	—	—	—	—	—	—
2nd Equilibration	q^2x	pqx	pqx	p^2x	—	—	—	—	—	—
2nd Transfer	q^2x	—	pqx	pqx	—	p^2x	—	—	—	—
3rd Equilibration	q^3x	pq^2x	$2pq^2x$	$2p^2qx$	p^2qx	p^3x	—	—	—	—
3rd Transfer	q^3x	—	$2pq^2x$	pq^2x	p^2qx	$2p^2qx$	—	p^3x	—	—
4th Equilibration	q^4x	pq^3x	$3pq^3x$	$3p^2q^2x$	$3p^2q^2x$	$3p^3qx$	p^3qx	p^4x	—	—
4th Transfer	q^4x	—	$3pq^3x$	pq^3x	$3p^2q^2x$	$3p^2q^2x$	p^3qx	$3p^3qx$	—	p^4x
Total Mass in Tube	q^4x		$4pq^3x$		$6p^2q^2x$		$4p^3qx$		p^4x	
e.g. (see text):										
Solute 1	$0.008x_1$		$0.075x_1$		$0.265x_1$		$0.412x_1$		$0.240x_1$	
Solute 2	$0.240x_2$		$0.412x_2$		$0.265x_2$		$0.075x_2$		$0.008x_2$	

this method may be unsuitable for certain labile compounds. If the extractant is ether which boils at 34.5° C, decomposition will in most cases be negligible.

Partition Chromatography

An important application of the principles of distribution between two solvents is to be found in the separation of substances by partition chromatography. One of the two immiscible solvents is held stationary on a column of some inert material, such as cellulose or kieselguhr. At the top of this column is placed the mixture of solutes to be separated. The second immiscible solvent is allowed to flow slowly down the column, and, as it moves over the mixture of solutes, they distribute themselves between the two solvents. If the stationary solvent is firmly held by the supporting material, it does not matter if this solvent is miscible with the moving solvent. The amount of any solute entering the moving solvent will depend upon its partition coefficient between the two solvents. The continuation of the distribution, as the solvent moves forward, results in the solutes with high partition coefficient in favour of this solvent being carried forward further than those with low partition coefficient. The technique is, in fact, similar in many respects to that of multiple fractional extraction, described above, and the distribution of solutes follows a similar pattern to that described in detail there. The main difference is that there is a continuous variation in concentrations along a chromatographic column in place of the step-wise variation from one extraction tube to the next.

As an alternative experimental arrangement, the separation may be carried out on a strip of filter paper, instead of on a column. In this case, the paper is exposed to the vapour of the stationary solvent, until it has become saturated, before the beginning of the experiment. Spots of a solution of the mixture of solutes, in any volatile solvent, are placed on a pencil line drawn at one end of the paper. The moving solvent can either be arranged to flow down the paper from a trough at the top, or it may rise by capillary action from a trough at the bottom. The strip and trough are kept in an enclosed space saturated with the vapours of both solvents, to prevent evaporation.

A similar but more rapid, versatile and reproducible method is thin layer chromatography (TLC) developed by Stahl. A slurry of a fine powder, such as silica gel, alumina or kieselguhr, suspended in the stationary solvent, such as water, is spread on a number of glass plates to give a uniform layer about 0.1 mm to 1 mm thick. After drying in air, the plates may be activated by heating in an oven to give a lower content of the stationary solvent. Spots of the solution of the mixture of solutes are placed near one end of the plate. This end is then placed in a trough of the second solvent in an airtight container. This solvent moves upwards by capillary action. Using thicker layers, substances can be separated in cg quantities for preparative work. Using thin layers and paper chromatography the methods can be made very sensitive and substances in μg quantities can be separated. The rate of movement of a given substance for partition between a particular pair of solvents is expressed numerically by its R_F value. This is the ratio of the distance travelled by the spot of solute, to the distance travelled by the solvent front. The higher the partition coefficient, the greater is the R_F value. In order to identify the components of a mixture of solutes, their R_F values are measured and compared with values for known substances. It is also necessary to check the identification by carrying out a run with adjacent spots of the unknown mixture and a similar mixture of known substances. At the end of the experiment the spots of solute may be identified by their fluorescence in ultraviolet light, by their quenching of the fluorescence of a fluorescent substance originally added to the stationary phase or by spraying with an appropriate reagent to make the spots visible.

Experiment 6.5 To illustrate the relationship between partition coefficient and R_F value in thin-layer partition chromatography using aromatic acids.

The partition coefficients are determined as follows. 200 mg of 4-hydroxybenzoic acid are dissolved in 50 cm³ of diethyl ether saturated with water in a separating funnel, taking care to avoid losses by evaporation. 50 cm³ of water saturated with ether is added and the mixture is shaken for about 5 minutes. 25 cm³ of the upper ethereal layer is transferred to a weighed 50 cm³ beaker and the ether carefully evaporated and dried on a hot water bath. The weight of the residue is obtained. The mass concentration of the acid in the ethereal and aqueous layers is determined and from this the partition coefficient is calculated. This procedure is repeated for benzoic acid and 3,4,5-trihydroxybenzoic acid (gallic acid). The latter is dissolved in the aqueous instead of the ethereal solution.

The R_F values are determined as follows. Ethanolic solutions (1 per cent, w/v) of each of the three acids and a mixture of them are prepared. By means of capillary tubes about 1 mm³ of each solution (containing about 10 μg of each acid) are separately spotted in a row on a scratch about 2 cm from the end of two labelled 10 × 20 mm thin-layer plates of silica gel 0.25 mm thick. The end of each plate is stood in a 1 cm layer of diethyl ether in a tank which is immediately made air-tight by covering it with a glass plate. When the ether has moved 15 cm up the plates, they are removed, and the solvent fronts marked. When the ether has evaporated, one plate is sprayed very lightly and evenly with 5 per cent (w/v) aqueous ferric chloride solution and the other with bromocresol green solution (B.P.). The plates are then dried in a current of warm air, leaving the acids as either yellow or blue-black spots. To test whether the indicators are satisfactory and whether sufficient of each acid is being spotted onto the plates, an initial trial should be carried out in which two plates are sprayed immediately after spotting. From the developed plates the R_F values are determined by measurements from the base line to the centre of the spots and the components of the mixture are identified.

The order of the R_F values is compared with the order of the partition coefficients. In the series, benzoic acid, 4-hydroxybenzoic acid, 3,4,5-trihydroxybenzoic acid, the number of hydroxyl groups and the polarity increases, so that the solubility in water increases and the solubility in ether decreases; consequently the partition coefficient ether/water decreases (p. 226), and therefore the R_F value decreases.

Another sensitive technique for analysis, separation and preparation is that of vapour phase or gas–liquid chromotography (GLC). It is a type of partition chromatography in which the various substances to be studied undergo interchange between a moving stream of a carrier gas and an involatile liquid adsorbed on a suitable packing material in a long, often coiled, column. The liquid must be chosen so as to have a low vapour pressure at the temperature of the column and to have some affinity, but not too great an affinity, for the substances to be separated. These substances must be sufficiently volatile to remain gaseous in the carrier gas and this is often aided by maintaining the column at an elevated temperature. As the substances are eluted in turn from the column, they are detected by changes in some physical property, such as thermal conductivity or the ability to become ionised in a flame or in an electron beam. Each substance is characterised by its retention time, which is the time interval between injection of the original material into the incoming carrier gas stream and elution of the substance from the column. The physical property mentioned above is converted into an electric current which governs the position of a pen on the ordinate of a pen recorder. The abscissa is the time. The plot of the physical property against time has the form of a series of peaks, ideally one for each substance in the original mixture. The horizontal position of the peak gives an accurate measure of the retention time and the area under the peak is usually proportional to the amount of substance in the injected material. Organic substances are usually injected in solution in a volatile solvent. Micrograms of material are sufficient for many instruments, but if mass spectrometry (p. 134) is used to detect and characterise the constituent substances, nanograms of material are sufficient. The more important instrumental variables are the nature and concentration of the liquid stationary phase, the column temperature, which must be accurately controlled, and the rate of flow of carrier gas. At high electronic sensitivity settings any slight volatility of the stationary phase causes a considerable response on the pen recorder. This effect is balanced out by the use of a second column of type and conditions identical to those of the first column into which the mixture is injected. Each column is followed by a detector and the two detectors are connected in opposition.

Other forms of chromatography are described on pp. 455 and 457. Chromatographic techniques are particularly suitable for separating and determining small quantities of chemically similar compounds in mixtures.

Effect of Non-volatile Solutes on Vapour Pressure and Related Properties

The physical properties of solutions vary with the activity or concentration of the solvent and therefore with the activities or concentrations of dissolved solutes. There is an important group of properties, often known as colligative properties, which are all approximately proportional to the concentration of solute particles in a dilute solution and are largely independent of their nature. The colligative properties are lowering of vapour pressure, elevation of boiling point and depression of freezing point of the solvent, and osmotic pressure.

In the following discussion, the system will consist of one solute, indicated by the subscript 2, dissolved in one solvent, indicated by the subscript 1. The mole fractions x_1 and x_2 therefore add up to unity. If the solution is ideal or non-ideal but dilute, Raoult's law can be applied to the solvent thus:

$$p_1/p_1^\ominus = x_1 = 1 - x_2. \qquad [6.29]$$

Subtracting both sides from unity, Raoult's law becomes:

$$\frac{p_1^\ominus - p_1}{p_1^\ominus} = x_2 = \frac{n_2}{n_1 + n_2} = \frac{w_2/M_2}{w_1/M_1 + w_2/M_2}, \qquad [6.30]$$

where

$p_1^\ominus$ is the saturated vapour pressure of the pure solvent,
p_1 is the partial vapour pressure of the solvent above the solution,
n_1 and n_2 are the number of moles of solvent and solute,
w_1 and w_2 are the masses of solvent and solute,

and M_1 and M_2 are the molar masses of solvent and solute.
If the solution is dilute, n_2 is small, so that $n_1 + n_2 \approx n_1$ and:

$$p_1^\ominus - p_1 \approx p_1^\ominus \cdot \frac{n_2}{n_1} = p_1^\ominus \cdot \frac{w_2 M_1}{w_1 M_2} = p_1^\ominus \cdot \frac{w_2 M_{r1}}{w_1 M_{r2}}, \qquad [6.31]$$

where M_{r1} and M_{r2} are the molecular weights of the solvent and solute, respectively. Thus, for a series of solutions consisting of the same solvent and solute the lowering of the vapour pressure is proportional to the concentration of the solute. These equations may be used to determine the molecular weight of the solute M_{r2}.

In the isopiestic method of determining the molecular weight, M_{rX}, of a non-volatile solute X, a vessel containing a solution of X is equilibrated in a closed, preferably evacuated space with a solution of a non-volatile substance Y of known molecular weight, M_{rY}. The solvent must be the same for both solutions so that M_{r1} and $p_1^{\ominus}$ have the same value. The solvent will distil from one solution to the other until they have the same vapour pressure p_1 (i.e. are isopiestic, p. 223). Then, according to equation [6.31]:

$$\frac{M_{r2}w_1}{w_2} = \frac{p_1^{\ominus}M_{r1}}{(p_1^{\ominus}-p_1)} = \text{constant for solutions of X and Y} \quad [6.32]$$

whence
$$\frac{M_{rX}w_{1X}}{w_X} = \frac{M_{rY}w_{1Y}}{w_Y}$$

or
$$M_{rX} = \frac{M_{rY}w_{1Y}w_X}{w_{1X}w_Y} \quad [6.33]$$

where w_X and w_Y are the predetermined masses of the respective solutes in the two solutions and w_{1X} and w_{1Y} are the masses of solvent in the respective solutions, which are measured at equilibrium. Substitution into equation [6.33] enables the unknown molecular weight M_{rX} to be calculated. If the solutions are dilute, it may be assumed that their densities are identical, so that:

$$\frac{w_{1Y}}{w_{1X}} = \frac{\text{volume of solution Y}}{\text{volume of solution X}}.$$

It is necessary to know when the system has reached equilibrium. For this purpose the volumes of the solutions may be measured from time to time in calibrated glassware without opening the apparatus.

The relationships between the pressure and the temperature of a pure substance are shown by the phase diagram (Fig. 6.12). The addition of a non-volatile solute, which is soluble in the liquid phase but which does not form a solid solution, does not modify the solid–vapour equilibrium curve because the solute does not enter these phases. The solid–liquid and liquid–vapour equilibrium curves are, however, displaced, as shown in Fig. 6.12, because the solute enters the liquid phase. At a given temperature these displacements obey Raoult's law (equations 6.29 to 6.31), if the solution is dilute.

The boiling point is the temperature to which the solvent must be raised before its vapour pressure is equal to atmosphere pressure. Therefore, if the vapour pressure is lowered by the presence of the

Fig. 6.12 Phase diagram showing the relationship between lowering of vapour pressure, elevation of boiling point and depression of freezing point of water by a non-volatile solute

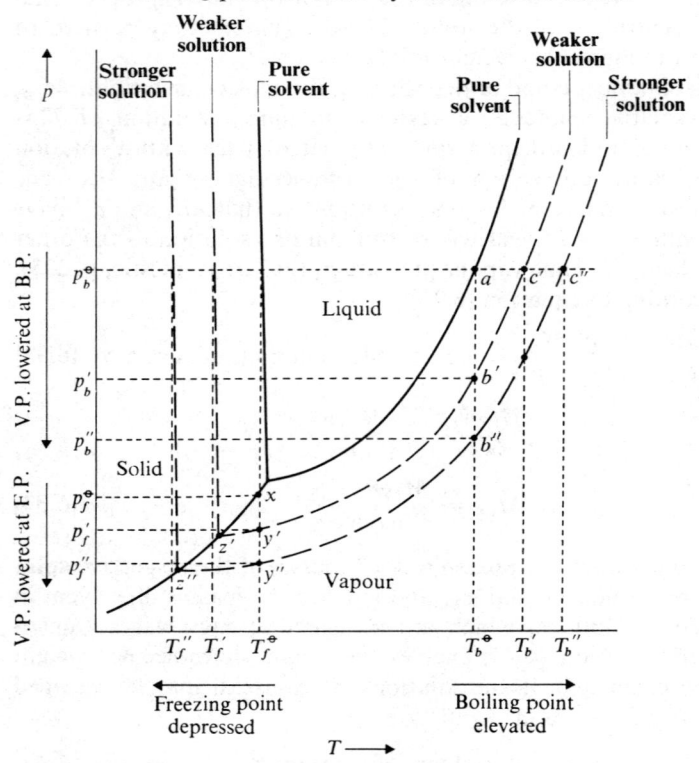

solute, the solvent must be raised to a higher temperature before it will boil, as shown in Fig. 6.12. Since the solutions are dilute, $b'c'$ and $b''c''$ are virtually straight lines and triangles $ab'c'$ and $ab''c''$ are similar, so that:

$$\frac{ac''}{ac'} = \frac{ab''}{ab'} \quad \text{or} \quad \frac{T_b'' - T_b^\ominus}{T_b' - T_b^\ominus} = \frac{p_b^\ominus - p_b''}{p_b^\ominus - p_b'},$$

where T_b is the boiling point and p_b is the vapour pressure of the solvent at the boiling point and the superscripts are as follows: $^\ominus$ for the pure solvent, $'$ for the weaker solution and $''$ for the stronger solution. Consequently, for dilute solutions, the elevation of boiling point $(T_b - T_b^\ominus = \Delta T)$ is proportional to the lowering of the vapour pressure $(p_b^\ominus - p_b)$ and hence to the concentration of the solute (from equation 6.31), thus:

$$\Delta T_b = k(p_b^\ominus - p_b) \qquad [6.34]$$

$$= k p_b^\ominus \cdot \frac{w_2 M_{r1}}{w_1 M_{r2}}. \qquad [6.35]$$

Effect of Non-volatile Solutes on Vapour Pressure

For any given solvent k, $p_b^{\ominus}$ and M_{r1} are constants and it is usual to replace $kp_b^{\ominus} M_{r1}$ by $1000K_b$, therefore;

$$\Delta T_b = \frac{1000 K_b w_2}{w_1 M_{r2}}.\qquad [6.36]$$

The number of moles of solute is given by:

$$n_2 = \frac{w_2}{M_2} = \frac{1000 w_2}{M_{r2}}\qquad [6.37]$$

where w_2 is the mass of solute in kg and M_2 is its molar mass in kg mol^{-1}.

So the molality (p. 200) of the solute is given by:

$$m_2 = \frac{n_2}{w_1} = \frac{1000 w_2}{w_1 M_{r2}}\qquad [6.38]$$

and equation [6.36] can be written in the form:

$$\Delta T_b = K_b m_2.\qquad [6.39]$$

Equations (6.36 and 6.39) provide means of determining either the molecular weight M_{r2} of the solute or its molality m_2. The constant K_b is called the molal elevation constant of the solvent and is equal to the elevation of boiling point which would be produced by a molal solution (1 mole of any solute dissolved in 1 kg of the solvent). K_b for the particular solvent can either be determined by experiment using a solute of known molecular weight or can be calculated from the formula:

$$K_b = \frac{R(T_b^{\ominus})^2 M_{r1}}{1000 \Delta H_l^v}\qquad [6.40]$$

where

R is the gas constant, 8.314 J mol^{-1} K^{-1},
$T_b^{\ominus}$ is the boiling point of the solvent in K,
M_{r1} is the molecular weight of the solvent
and ΔH_l^v is the molar enthalpy of vaporisation of the solvent in J mol^{-1}.

This equation arises in a more complete derivation involving the Clapeyron–Clausius equation in the form [7.80] as well as Raoult's law. For water $K_b = 0.512$ K kg mol^{-1}.

Since the vapour pressure of the solvent is lowered by the presence of the solute, the pressure at the triple point, x in Fig. 6.12, will be lowered and the temperature of the triple point will also be depressed. Because of the steep slope of the solid-liquid equilibrium line, the triple point x can be taken to be virtually the same as the point marked on the solid-vapour equilibrium curve at the pressure $p_f^{\ominus}$, the vapour pressure of the solid at the freezing point. Since the solutions are dilute xz', xz'', $y'z'$ and $y''z''$ are virtually straight lines and the triangles $xy'z'$ and $xy''z''$ are similar, so that

$$\frac{p_f^\ominus - p_f''}{p_f^\ominus - p_f'} = \frac{xy''}{xy'} = \frac{y''z''}{y'z'},$$

where $p_f^\ominus$ is the partial vapour pressure at the freezing point of the pure solvent and the superscripts have the same meaning as above. The freezing point of the solvent, $T_f^\ominus$, or of each solution, T_f' or T_f'', as normally measured, is the temperature at which the appropriate liquid phase is in equilibrium with the solid at a pressure of 1 atm. Since the liquid–solid equilibrium curves are almost parallel straight lines, the freezing point is depressed by a given concentration of the solute to the same extent as the triple point, so that:

$$\frac{y''z''}{y'z'} = \frac{T_f^\ominus - T_f''}{T_f^\ominus - T_f'}.$$

Substitution in the above equation gives:

$$\frac{T_f^\ominus - T_f''}{T_f^\ominus - T_f'} = \frac{p_f^\ominus - p_f''}{p_f^\ominus - p_f'}.$$

Consequently, for dilute solution the depression of freezing point, ΔT_f, is proportional to the lowering of vapour pressure. The following equations for the depression of freezing point are of exactly the same form as those above for the elevation of boiling point and are derived in exactly the same way.

$$\Delta T_f = \frac{1000 K_f w_2}{w_1 M_{r2}} = K_f m_2. \qquad [6.41]$$

The equations provide a most convenient method of determining either the molecular weight or the concentration of the solute. K_f is the molal depression constant for the solvent, e.g. 1.858 K kg mol^{-1} for water, and can be determined experimentally or calculated from the formula:

$$K_f = \frac{R(T_f^\ominus)^2 M_{r1}}{1000 \Delta H_s^l} \qquad [6.42]$$

where R is the gas constant, 8.314 J mol^{-1} K^{-1},
  $T_f^\ominus$ is the freezing point of the solvent in K,
  M_{r1} is the molecular weight of the solvent,
  ΔH_s^l is the molar enthalpy of fusion of the solvent in J mol^{-1}.

Because of the approximations made in its derivation, and its dependence on Raoult's law, equation [6.41] applies only to dilute solutions. For most solvents, dilute solutions will have freezing-points differing only slightly from the freezing-point of the pure solvent, so a sensitive thermometer is necessary in order to obtain an accurate measure of this difference. A sensitive thermometer is one in which a large movement of mercury in the capillary corresponds to a small difference in temperature, and in order to fulfil this condition, it is not possible to make the thermometer cover a large temperature range without having an inconveniently long scale. This difficulty is overcome in the Beckmann thermometer by having a mercury reservoir above the top of the scale, as well as the bulb at the bottom. For use over a higher temperature range, some of the mercury is transferred

Fig. 6.13 Beckmann method for the determination of freezing point

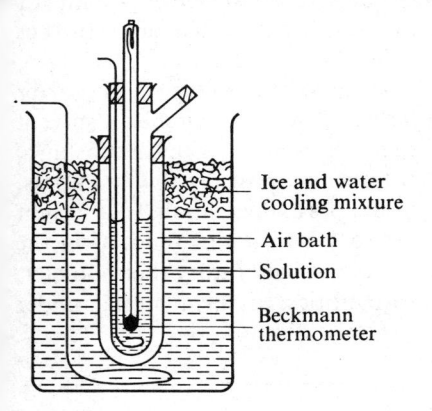

Ice and water cooling mixture
Air bath
Solution
Beckmann thermometer

from the bulb to the reservoir to bring the required range of temperature on to the scale. This is done by heating the bulb to a temperature about 4° C above the temperature to be measured, and then tapping the stem to break the mercury thread where the capillary widens to join the reservoir. To reset for a lower temperature range, the bulb is first warmed, until the mercury thread reaches the end of the capillary at the entrance to the reservoir, and it is caused to join the mercury in the reservoir by temporarily inverting the thermometer and tapping gently. The bulb is then cooled again to a temperature about 4° C above that to be measured, and the mercury thread broken as before. In using the Beckmann thermometer for measuring a depression in freezing-point, it must be remembered that, since the bulb is larger than that of an ordinary thermometer, it does not take up the temperature of its surroundings as quickly, so the solution must not be cooled too rapidly through its freezing-point. In order to prevent this, the apparatus shown in Fig. 6.13 is usually employed.

A known amount of solvent is placed in the inner tube, and a rough freezing-point is first determined by placing this tube directly into the cooling mixture. A more accurate determination of the freezing-point is then carried out by allowing the tube and its contents to warm up to a temperature just above the freezing-point, and then inserting it in the air bath which has previously been cooled in the melting ice. The solvent cools much more slowly this time, and in spite of continuous stirring, supercooling to a temperature below the freezing-point usually occurs. However, as soon as crystallisation begins, the temperature rises to the true freezing-point, and remains constant until all the liquid has become solid. After again allowing the solvent to warm up to just above its freezing-point, a weighed pellet of solute is introduced through the side arm of the inner tube. The freezing-point of the solution is then found in the same way by slow cooling in the air bath. The difference between the two freezing-points indicated on the Beckmann thermometer is the depression of freezing-point for this solution in °C. The experiment may be repeated after further addition of solute.

It is possible to avoid the use of a sensitive Beckmann thermometer in measuring depressions of freezing-point by employing as a solvent a substance with a high molal depression constant K_f. Such a substance is camphor, and it is also a good solvent for most organic materials. It is common, therefore, in carrying out approximate molecular weight determinations of organic substances to use Rast's method, as in the following experiment, employing an ordinary laboratory thermometer, camphor as a solvent, and small quantities of the material under examination.

Experiment 6.6 To determine the molecular weight of p-toluidine by Rast's method.

The freezing-points are determined by using the apparatus normally employed for the determination of the melting-point of an organic substance – a capillary tube attached to a thermometer suspended in a paraffin bath. For a pure substance the freezing-point and melting-point are the same, but for a mixture of miscible substances there is usually an appreciable range of temperature over which solid and liquid can exist together in equilibrium. The melting-point is then taken as the temperature at which liquid first appears on warming the solid, and the freezing-point is the temperature at which solid first appears on cooling the liquid or, more appropriately for this experiment, that at which the last trace of solid becomes liquid on heating.

A little pure camphor is first placed in a capillary tube and its freezing-point determined. Then a solution of p-toluidine in camphor is prepared by weighing out about 0.02 g of solute into an ignition tube and, after adding about 20 times as much camphor, weighing again. The tube is corked and shaken for a few seconds in an oil bath at about 180° C until a homogeneous solution is obtained.

When the liquid has solidified, the solid is powdered, and its freezing-point determined by heating a little in a capillary tube until the last trace of solid disappears. The difference between this temperature and the freezing-point of pure camphor is the depression of freezing-point for that particular solution. In order to calculate the molecular weight of p-toluidine, it is necessary to know the molal depression constant K_f for camphor. It has been found that it is not reliable to take a quoted value for this because it varies with the concentration of the solution. The constant must be determined by preparing a solution of similar concentration of a substance of known molecular weight, preferably having a molecular weight of the same order as that to be determined. Hence the procedure described above is repeated with about 0.02 g of naphthalene as solute. From the depression of freezing-point brought about by naphthalene and its known molecular weight of 128, the molal depression constant K_f is calculated for camphor. This value of K_f is then used to calculate the molecular weight of p-toluidine from the freezing-point depression brought about with this substance as solute. The method can only be regarded as being accurate to within about 10 per cent.

Osmotic Pressure

Another colligative property which is related to the variation of vapour pressure with concentration of solute is osmotic pressure. Osmosis is the spontaneous diffusion of a solvent from a more dilute solution (or pure solvent) to a more concentrated solution through a semipermeable membrane which separates them. A perfect semipermeable membrane allows the solvent to pass through but not dissolved solutes; in practice it is often difficult to obtain one which is completely impermeable to solutes. A good semipermeable membrane, if the solute particles are small, e.g. a sugar or a salt, is cupric ferrocyanide carefully prepared by chemical reaction in the pores of a porous pot. If the solute particles are macromolecules, e.g. a protein, various cellulose derivatives are often used, but such membranes are permeable to small solute particles. Osmosis takes place in such a direction as to attempt to equalise the activity of the solvent and hence the total concentration of the solute particles on both sides of the membrane. Osmosis can be prevented by exerting an external pressure on the solution containing the smaller quantity of the solvent. The external pressure increases the activity of the solvent by forcing the molecules closer together. The external pressure which must be applied to the solution to prevent osmosis, when the pure solvent is on the other side of the semipermeable membrane, is the osmotic pressure of the solution. When this has been done, the activity of the solvent in the solution equals the activity of the pure solvent.

Fig. 6.14 Simple osmometer for macromolecules

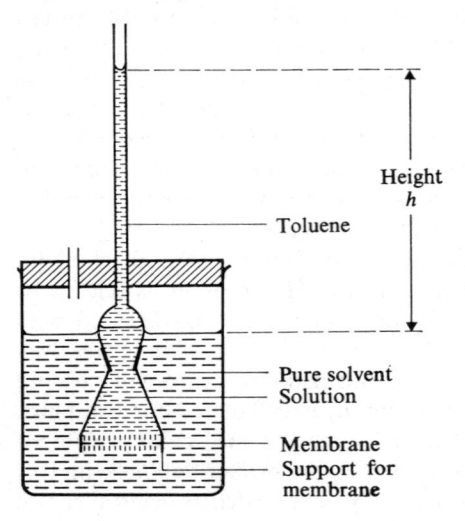

[Diagram showing a simple osmometer with labels: Toluene, Height h, Pure solvent, Solution, Membrane, Support for membrane]

A simple osmometer for measuring the osmotic pressure of solutions of macromolecules is shown in Fig. 6.14. The membrane must be firmly supported and the temperature must be accurately controlled. At equilibrium the osmotic pressure is equal to the hydrostatic pressure, $h\varrho g$, of the liquid in the vertical tube, where h is the height of the liquid column, ϱ is the density of the liquid and g is the acceleration due to gravity. This liquid should have a low surface tension to reduce the capillary rise correction (p. 459) and sticking of the meniscus; toluene or petroleum ether are suitable. The meniscus may be set slightly above the expected equilibrium level and the height plotted against time. This may be repeated with the liquid level set slightly below the anticipated position. A plot of the mean height against time enables the equilibrium value to be determined quite quickly.

If, under the same conditions of temperature and applied pressure, the activity of the solvent is the same in two solutions containing the same solvent, the two solutions will have the same colligative properties, viz. osmotic pressure, vapour pressure, boiling point and freezing-point, and are said to be iso-osmotic or isopiestic (p. 223). In dilute iso-osmotic solutions the total concentration of the solute particles is the same. There is no net transfer of solvent between two iso-osmotic solutions, when they are separated by a semi-permeable membrane or when they are placed in separate vessels in contact with the same vapour phase in an enclosed space.

The law of osmosis was discovered experimentally by Pfeffer using aqueous solutions of sucrose and may be stated as follows:

$$\pi V = nRT, \qquad [6.43]$$

Ionisation of Solutes and Colligative Properties

where π is the osmotic pressure (in Nm^{-2}, i.e. Pa),
V is the volume (in m^3) of solution occupied by n mol of solute,
R is the gas constant (8.314 J $mol^{-1} K^{-1}$) and
T is the absolute temperature (in K).

This equation is seen to be identical in form with the ideal gas equation for n moles, except that π and V refer to a pressure and volume for the liquid state and not for the gaseous state. It indicates that particles of solute in a solution behave rather like the particles of an ideal gas. Also, just as the ideal gas equation only applies to ordinary gases at very low pressures, so the osmotic pressure equation only applies to solutions at very low concentrations, because only under these conditions are the solute or gas particles sufficiently far apart so as not significantly to influence each other. A rigid thermodynamic treatment leads to an equation which approximates to the above equation [6.43] at low concentrations and seems to indicate, however, that the above analogy between the liquid and gaseous states is merely formal.

A measurement of osmotic pressure may be used to determine the molecular weight of a solute in the following way. If w kg of solute are placed in V m^3 of solution and the unknown molecular weight is M_r, we have:

$$n = \frac{1000w}{M_r} \qquad [6.44]$$

$$\therefore \qquad \pi V = \frac{1000w}{M_r} \cdot RT \qquad [6.45]$$

Thus, from the observed value of the osmotic pressure π at absolute temperature T, M_r may be calculated. A more accurate value for M_r is obtained by repeating the determination at a number of different low concentrations and extrapolating to zero concentration. This is done by plotting $\pi V/w$ against w/V to obtain an almost linear graph, which may easily be extrapolated to $w/V = 0$. This extrapolated value of $\pi V/w$ is then inserted into the equation above to calculate M_r. The osmotic-pressure method is suitable for the determination of molecular weights of colloidal substances. Further difficulties are encountered in such cases, but they are discussed in Chapter 11 (see p. 489).

Ionisation of Solutes and Colligative Properties

Ionisation of solutes, discussed in Chapter 8, affects all the colligative properties of a solution, since these properties are approximately proportional to the concentration of all the particles of solute, ions as well as molecules, in dilute solutions. The colligative property equations in their simplest form [6.31–6.39; 6.41; 6.43–6.45] only apply to unionised solutes (or non-electrolytes). In order to adapt them to ionised solutes (or electrolytes), van't Hoff introduced the correction factor i defined by:

$$i = \frac{\text{the observed colligative property}}{\text{the colligative property expected if no ionisation occurred}} \quad [6.46]$$

The same factor i may be placed in all the colligative property equations on the side opposite to the actual colligative property. Thus, the vapour pressure equations [6.31] become:

$$p_1^\ominus - p_1 \approx p_1^\ominus \cdot \frac{n_2 i}{n_1} = p_1^\ominus \cdot \frac{w_2 M_1 i}{w_1 M_2} = p_1^\ominus \cdot \frac{w_2 M_{r1} i}{w_1 M_{r2}}, \quad [6.47]$$

the freezing point equations [6.41] become:

$$\Delta T_f = \frac{1000 K_f w_2 i}{w_1 M_{r2}} = K_f m_2 i \quad [6.48]$$

and the osmotic pressure equations [6.43] and [6.45] become:

$$\pi V = ni RT = \frac{1000 w i}{M_r} \cdot RT \quad [6.49]$$

Strong electrolytes, i.e. salts, strong acids and strong alkalis, are completely ionised in aqueous solutions (p. 326) and i is then simply the number of ions produced by one (formula) molecule of solute, e.g. NaCl, $i = 2$; Na_2CO_3, $i = 3$; $MgSO_4$, $i = 2$; H_2SO_4, $i = 3$; for $NaHCO_3 \rightarrow Na^+ + HCO_3^-$, $i = 2$. For non-electrolytes, e.g. dextrose, urea, i is obviously unity. Most weak electrolytes (p. 341) are usually so slightly ionised in aqueous solution that $i \approx 1$.

In biochemistry and medicine the quantity ni ($= 1000wi/M_r$) in the osmotic pressure equation [6.49] and the quantity $n_2 i$ ($= 1000 w_2 i/M_{r2}$) in the other equations [6.47] and [6.48] are sometimes called the number of osmoles of solute. By analogy with the mole, an osmole is Avogadro's number (6.022×10^{23}) of osmotically active solute particles (molecules or ions). Thus, 1 mole of sodium carbonate contains 2 osmoles of sodium ions and 1 osmole of carbonate ions, i.e. 3 osmoles of osmotically active solute particles altogether. By analogy with molality the osmolality of a solution is the number of osmoles of solute particles divided by the mass (in kg) of solvent, and should therefore be equal to $m_2 i$ ($= 1000 w_2 i/w_1 M_{r2}$), e.g. in equation [6.48] which can be written as:

$$\Delta T_f / K_f = m_2 i. \quad [6.50]$$

Since, however, this and all other colligative property equations are not obeyed exactly, $m_2 i$ may be defined as the ideal osmolality, while for freezing point depression $\Delta T_f / K_f$ is equal to the real osmolality. To correct the colligative property equations for non-ideal behaviour a

Ionisation of Solutes and Colligative Properties

correction factor, known as the practical osmotic coefficient, ϕ is introduced and is defined by:

$$\phi = \frac{\text{the observed colligative property}}{\text{the calculated colligative property allowing for ionisation}} \qquad [6.51]$$

ϕ is close to unity for many dilute solutions encountered in practice but deviates from unity with increasing concentration of all solutes and with increasing valency of dissolved ions. The fact that:

$$\phi = \frac{\text{real osmolality}}{\text{ideal osmolality}} \qquad [6.52]$$

can be readily deduced by application of the above principles to freezing point depression, thus:

$$\phi = \frac{\Delta T_f}{K_f mi} = \frac{\Delta T_f/K_f}{mi}$$

$$\therefore \quad \Delta T_f/K_f = mi\phi. \qquad [6.53]$$

For a mixture of osmotically active particles, a, b, c, etc., from one or more solutes, each type of particle will have its own value of m, i and ϕ hence:

$$\Delta T_f/K_f = m_a i_a \phi_a + m_b i_b \phi_b + \ldots = \Sigma mi\phi. \qquad [6.54]$$

The quantity $\Sigma mi\phi$ is equal to the colligative property divided or multiplied by an appropriate constant factor. $\Sigma mi\phi$ may be called the total real osmolality or simply the osmolality of the solution and is a practical measure of the total activity of all the osmotically active particles of solute, molecules and ions. Biological solutions consist of a mixture of many solutes dissolved in water, so osmolality is used in biological research and in medical diagnosis. For these purposes osmolality is determined by measurements of any of the colligative properties in an osmometer, e.g. in a freezing-point, vapour pressure or membrane osmometer. These instruments usually require only 0.1 to 0.5 ml of sample and are sometimes calibrated directly in osmolality. In a freezing-point osmometer the sample is refrigerated and the freezing-point is accurately determined by means of a calibrated electrical resistance thermometer of the thermistor type. By analogy with molality, the SI unit of osmolality is mol kg^{-1}. Osmolality is, however, usually expressed in the practical units of osmoles per kg of solvent (osm/kg) or milliosmoles per kg of solvent (mosm/kg). For example, application of equation [6.54] shows that, if a sample of human blood serum freezes at $-0.54°$ C, its osmolality is $\frac{0.54}{1.858}$.osm/kg = 290 mosm/kg. Concentration of osmotically active particles may also be expressed in other units, e.g. osmoles per litre of solution, which for dilute solutions approximates to osmolality.

Biological Importance of Colligative Properties

Osmotic pressure is of considerable importance in biological systems, since it is one of the main influences involved in the movement of water and solutes through cell membranes. The cell contains substances, such as salts, sugars, amino acids and other metabolites, in solution. These are retained by the cell membrane which is supported and surrounded by the cell wall. A cell stripped of its wall is referred to as a protoplast. Cell walls are freely permeable to most solutes. The walls of plant cells form a comparatively rigid protective network of cellulose which surrounds the protoplasts. The membranes of animal and bacterial cells are surrounded by a less rigid, more elastic cell wall. In the normal healthy cell the membrane is kept expanded against the cell wall by osmotic pressure developed by diffusion of water through the membrane into the cell. Such a cell is said to be in a turgid state.

An aqueous solution is said to be *isotonic* with the fluids in a cell, if the cell neither gains nor loses water, i.e. if the volume of the cell or protoplast does not change, when placed in the solution. An alternative statement is that the external solution and the fluids in the cell have the *same tonicity*. Solutions for injection and eyedrops should be isotonic with the cells in the blood, tissues and eyes. Solutions which are not isotonic with the relevant cells are referred to as *paratonic* solutions and cause pain and cell damage. When a cell is placed in a *hypertonic* solution, i.e. a solution of higher tonicity than the internal cellular fluids, water diffuses out of the cell and the volume of the cell or protoplast decreases. The membrane of plant cells collapses inwards away from the cell wall and the cell is said to be *plasmolysed*, whereas red blood cells shrink and become crenated in appearance. When a cell is immersed in a *hypotonic* solution, i.e. one of lower tonicity than the internal fluids of the cell, water diffuses into the cell and the volume of the cell or protoplast increases. Plant cells become turgid as mentioned above, whereas protoplasts and red blood cells may burst, so that their contents escape into the surrounding hypotonic solution. Red blood cells are then said to be *haemolysed* and the medium becomes uniformly reddened as the haemoglobin escapes. The concentration of salt at which rupture of the membrane *(haemolysis)* first occurs is a measure of the strength of the membrane.

As mentioned above, two solutions are said to be iso-osmotic if they have the same osmotic pressure. Solutions which are iso-osmotic with blood serum or tissue fluids are sometimes paratonic (hypotonic or hypertonic) with respect to the blood cells or tissue cells. This happens when the cell membrane is permeable to one or more solutes as well as to the solvent. Cell membranes are not perfectly semi-permeable, because otherwise no nutrients or waste products would diffuse through them and the cell would die.

Biological Importance of Colligative Properties 249

The distinction between osmotic pressure and tonicity is illustrated by the examples shown in Fig. 6.15, in which a hypothetical cell membrane is assumed to be permeable to the solvent, water and to urea but impermeable to glycogen and to a neutral protein. The solutes are assumed to be present at equal and low concentration (10 m mol dm^{-3}) so that their solutions are iso-osmotic. The solutions inside and outside the cell are therefore initially iso-osmotic in examples

Fig. 6.15 Illustration of the difference between osmotic pressure and tonicity
(a) iso-osmotic and isotonic
(c) initially iso-osmotic and hypertonic
(b) initially iso-osmotic and hypotonic
(d) initially neither iso-osmotic nor isotonic. At equilibrium iso-osmotic and isotonic

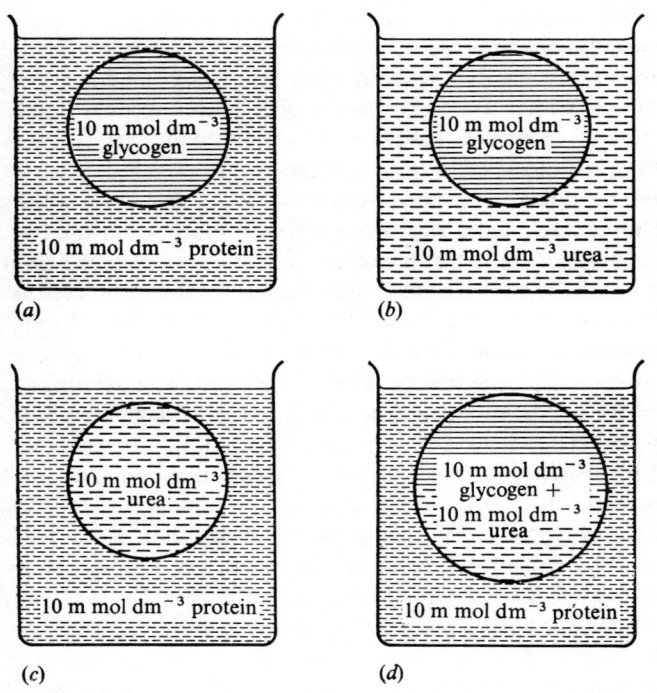

(a) Inside: 10 m mol dm^{-3} glycogen; Outside: 10 m mol dm^{-3} protein

(b) Inside: 10 m mol dm^{-3} glycogen; Outside: 10 m mol dm^{-3} urea

(c) Inside: 10 m mol dm^{-3} urea; Outside: 10 m mol dm^{-3} protein

(d) Inside: 10 m mol dm^{-3} glycogen + 10 m mol dm^{-3} urea; Outside: 10 m mol dm^{-3} protein

(a), (b) and (c). In (a) the solutes cannot pass through the cell membrane, therefore the external solution is always isotonic as well as iso-osmotic with the internal solution. In case (b) urea diffuses into the cell; water also diffuses into the cell in an attempt to equalise the total concentration of solute molecules on both sides of the membrane. The external solution is therefore hypotonic with respect to the internal solution. Case (c) is the reverse of (b). In (c) both urea and water diffuse out of the cell so the external solution is hypertonic relative to the internal solution. In both (b) and (c) the external and internal solutions will cease to be iso-osmotic when some urea has diffused through the cell membrane. In example (d) the total concentration of solute molecules inside the cell is initially twice that outside, so the solutions are not initially iso-osmotic. Urea diffuses out of the cell and water diffuses in, therefore the solutions are not initially isotonic. Eventually urea and water distribute themselves so that their concentrations are equalised on both sides of the membrane. Consequently, at equilibrium the external and internal solutions will be both iso-osmotic and isotonic. In summary, the osmotic pressure of a solution is determined by the total concentration of solute particles, whereas the tonicity of a solution is governed by the concentration of those solute molecules which cannot cross the membrane.

Example (b) in Fig. 6.15 is particularly relevant to pharmacy. Many pharmaceutical formulations contain active substances which pass through cell membranes, because otherwise they would not be able to reach their sites of action within the cell. Therefore solutions for injection and eye-drops which are iso-osmotic with blood serum, tissue fluids and lachrymal secretions (tears) may be hypotonic with respect to the contents of the blood cells or cells of the eye. For example, boric acid, hexamine, saponins, urea and urethane in solutions iso-osmotic with blood serum cause haemolysis because they rapidly penetrate the membrane of red blood cells. Nevertheless, extreme examples like this are rare and it is usually sufficient to ensure that solutions for injection and eye-drops are iso-osmotic with, and therefore have the same freezing-point depression as, tissue fluid, blood serum and tears by adding an adjusting substance, if necessary. Various values for the freezing-point of blood serum have been quoted; the British Pharmaceutical Codex gives $-0.52°$ C and has a table of the freezing-point depressions produced by 1 per cent (w/v) concentrations of many medicaments and adjusting substances.

As an example of osmotic adjustment, we may calculate what concentration of sodium chloride is required to cause a 3 per cent solution of quinine hydrochloride to be iso-osmotic with blood serum. A 1 per cent solution of quinine hydrochloride has a freezing-point depression, relative to pure water, of $0.076°$ C. The depression is proportional to concentration for dilute solutions, so a 3 per cent solution will have a depression of $0.228°$ C. Now, blood serum freezes at $-0.52°$ C, so the further depression required to make the quinine hydrochloride solution iso-osmotic is $0.29°$ C. A 1 per cent solution of sodium chloride has a freezing-point depression of $0.576°$ C, so to produce the extra depression of $0.29°$ C the sodium chloride must be present to an extent of $\frac{0.29}{0.576} \times 1$ per cent, i.e. 0.50 per cent.

The Supplement to the International Pharmacopœia contains, in

Appendix 7, graphs with the concentration of the drug as the abscissa and the concentration of the adjusting substance as the ordinate. In the above ways solutions for injection and eye-drops may be readily adjusted so as to be iso-osmotic with tissue, blood and lachrymal fluid.

Besides being of importance in the transfer of substances into and out of cells, osmotic pressure is also one of the main factors influencing the movement of body fluids on a larger scale. Thus, if the blood becomes diluted by drinking too much water, its osmotic pressure is lowered and water begins to pass into the tissues. But, at the same time, water begins to pass through the kidneys to be excreted until the normal blood concentration is restored. If the body loses water from the blood and tissues, perhaps as a result of fever, the osmotic pressure of the blood may become so high that the kidneys can no longer excrete water and the permeable solutes which are normally excreted with it, and the patient suffers from anuria. Filtration through the kidneys ceases if the osmotic pressure of the blood becomes higher than the hydrostatic blood pressure in the kidneys. Thus, anuria may be caused either by a high osmotic pressure in the blood, or by low blood pressure.

Since most membranes are not completely semipermeable, the transfer of substances across the membranes is usually a combination of osmosis and diffusion. When diffusion of a solute causes a difference in osmotic pressure between the two sides of a membrane, solvent will also diffuse to attempt to equalise the osmotic pressures. For example, suppose a solution of NaCl is separated from a solution of $MgSO_4$, of the same osmotic pressure, by a membrane permeable to both solutes and to water. The NaCl diffuses into the $MgSO_4$ solution more rapidly than the $MgSO_4$ diffuses into the NaCl solution, so the osmotic pressure on the side of the $MgSO_4$ solution increases. Water will tend to diffuse into this solution to attempt to equalise the osmotic pressures. Now, if the system is not allowed to attain equilibrium but the two solutions are constantly renewed, water will continue to diffuse with the NaCl into the $MgSO_4$ solution. It is this type of mechanism which is of importance in controlling the flow of solutes and water into and out of the alimentary canal. The laxative action of Epsom salts may also be explained by the same sort of mechanism. A large concentration of $MgSO_4$ is set up in the alimentary canal. Water diffuses into the canal to attempt to reduce the osmotic pressure set up, and this assists the process of excretion.

Diffusion

Diffusion is the tendency of a gas, liquid or dissolved solute to distribute itself uniformly at constant temperature over the space available to it and is therefore an important property of solutions.

There is evidence that even solids can diffuse into each other although at an extremely low rate. Diffusion is a spontaneous process (Chapter 7) which continues until the activity of the diffusing substance is uniform throughout the volume available to it. For normal purposes the activity of a substance can be taken to be equal to or proportional to its concentration or partial pressure. A gas tends to diffuse from a region in which its partial pressure is higher to one in which its partial pressure is lower. Similarly, a dissolved solute tends to diffuse from a region in which its concentration is higher to one in which its concentration is lower. Diffusion is a phenomenon of great importance in many biological and physical processes and in chemical and pharmaceutical engineering.

Matter flow (diffusion and viscosity), heat flow (thermal conductivity) and electricity flow (electrical conductivity) are all examples of transport (or flow) phenomena and are described by equations of the same general form:

$$\text{flux} = -\text{constant} \times \text{grad potential}, \qquad [6.55]$$

where the *flux* is the rate of flow across unit area with respect to time and *grad potential* is the potential gradient or rate of change of the potential, which causes or drives the flow, with respect to distance. The *minus* sign indicates that the flow proceeds in the direction of *decreasing* potential.

For matter flow equation [6.55] is known as Fick's first law of diffusion which may be written as:

$$J_n = -D \cdot \text{grad } c = -D \cdot \frac{dc}{dx}, \qquad [6.56]$$

where c is the concentration (in mol m^{-3}), grad c is the concentration gradient dc/dx (in mol m^{-4}) for flow in one direction, x being the distance (in m) in that direction, and J_n is the flux (in mol m^{-2} s^{-1}) of the amount of substance in moles. From the definition of flux we have:

$$J_n = \frac{dn}{dt} \cdot \frac{1}{A}, \qquad [6.57]$$

whence Fick's first law may be written as:

$$\frac{dn}{dt} \cdot \frac{1}{A} = -D \cdot \frac{dc}{dx}, \qquad [6.58]$$

where dn/dt is the rate of flow of amount of substance n (in mol) with respect to time t (in s) and A is the area (in m^2) of the plane across which flow occurs. D is a constant for a given solute and solvent at a given temperature and is called the coefficient of diffusion of the solute (in $m^2 \ s^{-1}$). The quantities stated above have been expressed in SI units. D in c.g.s. units ($cm^2 \ s^{-1}$) is numerically 10^4 times the SI value. Multiplying the numerator and denominator of equation [6.57] by dx gives:

$$J_n = \frac{dn}{dt} \cdot \frac{1}{A} \cdot \frac{dx}{dx}. \qquad [6.59]$$

Rearranging the denominator we have:

$$J_n = \frac{dn}{A.dx} \cdot \frac{dx}{dt}. \qquad [6.60]$$

Since $A.dx$ is a small element of volume dV (in m^3) occupied by dn mol of diffusing substance,

$$\frac{dn}{A.dx} = \frac{dn}{dV} = c, \qquad [6.61]$$

where c is the concentration of the substance (in mol m^{-3}),

$$\therefore \qquad J_n = c \cdot \frac{dx}{dt}. \qquad [6.62]$$

Since the velocity u (in $m \ s^{-1}$) of a given particle of diffusing substance in the direction of flow is dx/dt,

$$J_n = cu. \qquad [6.63]$$

If the quantity of matter is measured in mass units m (in kg), the concentration becomes the mass concentration ϱ (in kg m^{-3} i.e. g dm^{-3}) and Fick's first law is written as:

$$J_m = -D.\text{grad } \varrho \qquad [6.64]$$

where
$$J_m = \frac{dm}{dt} \cdot \frac{1}{A}, \qquad [6.65]$$

$$\text{grad } \varrho = \frac{d\varrho}{dx}, \qquad [6.66]$$

and
$$J_m = \varrho u. \qquad [6.67]$$

The quantities t, A, D, x and u have the same values and SI units as before (equations 6.56 to 6.63).

Like all rate processes diffusion is strongly dependent on temperature. Diffusion coefficients increase with increasing temperature and may be determined by means of the free boundary method or by the porous disc method.

Diffusion by the Free Boundary Method

In this method the two solutions or liquids are brought into direct contact in a suitable cell with a sharp boundary between them (Fig. 6.16). The free boundary method is the most accurate technique for studying diffusion but requires considerable expertise. Vibration and temperature differences must be eliminated in order to ensure that broadening of the boundary is due only to diffusion. Diffusion may be followed *either* by measuring concentrations c at various times t at a

Fig. 6.16 Diffusion studies by the free boundary method
        (*a*) boundary and flux in a shear cell
        (*b*) concentration at various times
        (*c*) concentration gradient at various times

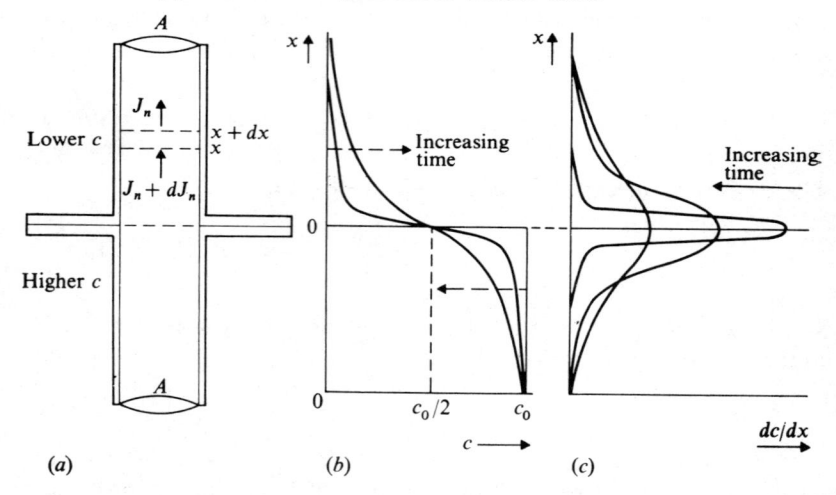

fixed distance x from the boundary (Fig. 6.16*b*) using spectrophotometry or conductivity measurements, *or* by determining the concentration gradient dc/dx in the diffusing column at various times (Fig. 6.16*c*) from measurements of refractive index gradients. In Fig. 6.16 the lower liquid has at zero time a uniform concentration c_0 of solute, while the upper liquid is initially pure solvent, in which the concentration of solute is zero. As time goes on, solute diffuses from the lower to the upper solution, and the concentration differences in the cell tend to become smaller (Fig. 6.16*b* and *c*). The cell must be sufficiently long for the concentrations at its extremities to remain at their initial values. Equations relating the measurable quantities c, x and t or dc/dx, x and t are now required.

The difference in flux dJ_n between the lower and upper limits of a slice of column of thickness dx, of area A and of distance x from the boundary (Fig. 6.16*a*), may be calculated in two ways. One method is to differentiate Fick's first law (equation [6.56]) with respect to x, thus:

$$\frac{dJ_n}{dx} = +D \cdot \frac{d^2c}{dx^2}. \quad [6.68]$$

The minus sign disappears because the flux J_n and the concentration gradient dc/dx *both decrease* as x increases (cf. Fig. 6.16a and c). The quantity dJ_n/dx may also be calculated by differentiating J_n as defined by equation [6.57], thus:

$$\frac{dJ_n}{dx} = \frac{1}{A} \cdot \frac{d}{dx}\left(\frac{dn}{dt}\right). \quad [6.69]$$

But $A \cdot dx = dV$, the volume of the slice, therefore

$$\frac{dJ_n}{dx} = \frac{d}{dV}\left(\frac{dn}{dt}\right). \quad [6.70]$$

When a function is differentiated more than once, the result is independent of the order in which the differentiations are carried out, therefore:

$$\frac{dJ_n}{dx} = \frac{d}{dt}\left(\frac{dn}{dV}\right). \quad [6.71]$$

But $dn/dV = c$ (equation [6.61]), therefore:

$$\frac{dJ_n}{dx} = \frac{dc}{dt}. \quad [6.72]$$

Substituting for dJ_n/dx using equation [6.68] gives:

$$\frac{dc}{dt} = D \cdot \frac{d^2c}{dx^2}. \quad [6.73]$$

This equation is known as Fick's second law of diffusion. When it is integrated between the boundary conditions appropriate to the problem (e.g. in Fig. 6.16, when $t = 0$, $c = c_0$ in the lower column and $c = 0$ in the upper column), a complicated equation is obtained which relates c to x and t. If this equation is differentiated, the following expression for the concentration gradient is obtained:

$$\frac{dc}{dx} = \frac{-c_0}{\sqrt{4\pi Dt}} e^{-x^2/4Dt}. \quad [6.74]$$

Substitution of the experimental results for a given solute into the complicated integrated equation or into [6.74] enables the diffusion coefficient of the solute to be determined. For example, dc/dx has a maximum value, $(dc/dx)_{max}$, at the boundary, where $x = 0$ (Fig. 6.16c),

so that:
$$\left(\frac{dc}{dx}\right)_{max} = \frac{-c_0}{\sqrt{4\pi Dt}} \times 1 \qquad [6.75]$$

whence
$$D = \frac{c_0^2}{4\pi t (dc/dx)_{max}^2}. \qquad [6.76]$$

Diffusion by the Porous Disc Method

A simpler but less accurate method of studying diffusion is the porous disc method in which the two liquids are separated by a sintered glass disc, (Fig. 6.17a). The diffusion column is held rigidly in the disc

Fig. 6.17 Diffusion studies using the porous disc method (a) apparatus (b) concentration at various times

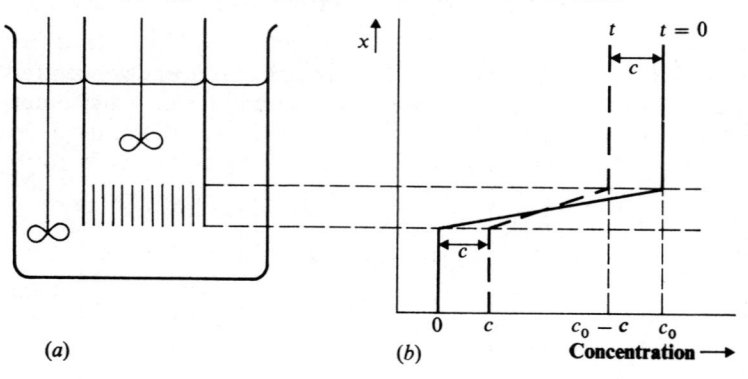

and is almost independent of small external changes. Diffusion causes movement of solute through the disc. Both solutions are stirred so that at any given time the concentration of each is constant throughout and can be determined by any convenient method. If the porous disc is uniform the concentration gradient can be made constant, i.e.

$$\frac{dc}{dx} = \text{constant} = \frac{\Delta c}{\Delta x}. \qquad [6.77]$$

Substituting for dc/dx in equation [6.58] and multiplying throughout by A gives:

$$\frac{dn}{dt} = -DA\frac{\Delta c}{\Delta x}. \qquad [6.78]$$

Concentration is defined by $c = n/V$ and since the volume V of the porous disc is constant:

$$\frac{dc}{dn} = \frac{1}{V}. \qquad [6.79]$$

Since differential coefficients may be treated as ratios:

$$\frac{dc}{dt} = \frac{dc}{dn} \cdot \frac{dn}{dt}. \qquad [6.80]$$

Introducing equations [6.78] and [6.79] into this equation, we have:

$$\frac{dc}{dt} = -\frac{DA}{V} \cdot \frac{\Delta c}{\Delta x}. \qquad [6.81]$$

Since the area A, the volume V and the thickness Δx of the porous disc are constant, $A/V\Delta x$ is also constant for a given disc and is known as the *cell constant* K (SI unit m^{-2}).

$$\therefore \qquad \frac{dc}{dt} = -DK\Delta c. \qquad [6.82]$$

Initially, at $t = 0$, the concentration in the inner vessel is c_0 and the concentration in the outer vessel is 0. At a later time t, the concentration in the inner vessel is c_0-c and the concentration in the outer vessel is c (Fig. 6.17b). At time t the difference in concentration Δc across the porous disc in the direction of flow is negative and is given by:

$$\Delta c = c-(c_0-c) = 2c-c_0$$

or by

$$-\Delta c = c_0-2c. \qquad [6.83]$$

Substituting for $-\Delta c$ in equation [6.82] gives:

$$\frac{dc}{dt} = DK(c_0-2c). \qquad [6.84]$$

Inverting:
$$\frac{dt}{dc} = \frac{1}{DK} \cdot \frac{1}{c_0-2c}.$$

Integrating:
$$t = \frac{1}{DK} \int \frac{1}{c_0-2c} \cdot dc,$$

where D, K and c_0 are constants.

The integral is of the type $\frac{1}{x}.dx$, which on integration gives $\ln x$. However, the integral of $\frac{1}{c_0-2c}.dc$ is not $\ln(c_0-2c)$, because differentiation of this with respect to c would give $-2/(c_0-2c)$ (cf. p. 150). Therefore, the factor $1/-2$ must be included in the integrated expression, which becomes:

$$t = \frac{1}{DK}[-\tfrac{1}{2}\ln(c_0-2c)+\text{constant}].$$

The constant of integration is evaluated from the initial conditions, by putting $c = 0$ at $t = 0$, thus:

$$0 = \frac{1}{DK}[-\tfrac{1}{2}\ln c_0 + \text{constant}]$$

$\therefore$ $\quad\quad\quad\quad\text{constant} = \tfrac{1}{2}\ln c_0\quad\text{and}$

$$t = \frac{1}{DK}[-\tfrac{1}{2}\ln(c_0-2c) + \tfrac{1}{2}\ln c_0].$$

Multiplying both sides of this equation by DK and converting ln to log we have:

$$DKt = \frac{2.303}{2}[\log c_0 - \log(c_0-2c)] \quad\quad\quad [6.85]$$

whence

$$D = \frac{2.303}{2Kt}\log\left(\frac{c_0}{c_0-2c}\right). \quad\quad\quad [6.86]$$

This equation enables the diffusion coefficient D of a solute to be determined from an analytical determination of c at a suitable time t, provided that c_0 and K are also known. The cell constant K is best determined by introducing the values of c, t, c_0 and D for an experiment with a solute of known diffusion coefficient as in the experiment to be described. Since the concentration terms appear in a ratio of concentrations, they may be expressed in any units or in any quantity to which they are proportional, without affecting the values or the units of K and D (cf. p. 152).

Experiment 6.7 Determination of the diffusion coefficient of hydrochloric acid (0.4 mol dm^{-3}) at ambient temperature by means of the porous disc method.

The inner vessel, called the diffusion cell, consists of a sintered glass disc of pore diameter 5–15 μm and of overall diameter 2–3 cm sealed on to a funnel of volume about 60 ml and fitted with a tap of wide bore. In this simple experiment the solutions are not stirred during the diffusion process and the ambient temperature is assumed to be constant. Distilled water, 0.1 mol dm^{-3} potassium chloride and 0.4 mol dm^{-3} hydrochloric acid are freed of dissolved gases by gentle boiling under reduced pressure. Gas-free water is sucked into the porous disc to remove gas from its pores. The inside of the cell is rinsed with the gas-free potassium chloride solution and filled with 50 cm^3 of this solution. The cell is immersed in 50 cm^3 of gas-free water in a 100 cm^3 beaker and is left undisturbed for $\tfrac{1}{2}$ to 1 hour so that a diffusion column of constant concentration gradient is set up within its sinter. The cell is then withdrawn from the beaker and any adhering water is removed with a small glass rod. The cell is immersed in 50 cm³ of fresh gas-free water in the outer vessel which consists of a 100 cm³ beaker and the time is noted. The position of the inner vessel is adjusted so that the liquid level in the outer vessel is the same as that in the inner vessel. After a sufficient time has elapsed to allow accurate analysis of the external solution (3–5 hours), the cell is carefully withdrawn from the beaker and the time is again noted. The time interval is t seconds. The solution in the beaker is titrated in 20 cm³ portions with 0.05 mol dm⁻³ silver nitrate and the concentration c of solute is calculated. The solution in the cell is slowly blown out through the disc by applying positive air pressure through the tap. The first 10 cm³ is discarded. The remainder of the solution is blown into a beaker and titrated in 10 cm³ portions with the silver nitrate solution and the concentration $(c_0 - c)$ of solute is calculated. Taking a mean value of c, the cell constant K (in m⁻²) is calculated by inserting the following data into equation [6.86]: $c_0, c, t, D = 1.448 \times 10^{-9}$ m² s⁻¹ at 25° C.

The experiment is repeated with 50 cm³ of gas-free hydrochloric acid ($c_0 = 0.4$ mol dm⁻³) in the cell. The concentrations c and $(c_0 - c)$ are determined by titration with 0.2 mol dm⁻³ sodium hydroxide. The diffusion coefficient D (in m² s⁻¹) is calculated by inserting the experimental data and K into equation [6.86].

Dissolution Rates of Solids

Before a drug being administered in a solid form can exert its action, it must first dissolve in the body fluids and then be absorbed from them. A tablet or capsule may also disintegrate to give granules or aggregates which may undergo deaggregation to give fine particles. Dissolution may take place from the undisintegrated solid, from the aggregates or from the fine particles. When the rate of dissolution is much less than the rate of the other processes (< 5 per cent), it is the rate determining step. Since this is often the case, the rate at which solids dissolve is very important in pharmacy. Agitation of the liquid medium invariably increases the dissolution rate and presumably occurs in the body.

In 1897 Noyes and Whitney reported that the kinetics of dissolution of cylindrical sticks of benzoic acid and lead chloride rotated in water were of the first order if the surface area of the solid was assumed constant and if the solute was allowed to accummulate in the liquid medium. The latter condition is referred to as the non-sink condition. The Noyes–Whitney law has the same form as the rate equation [5.1] for a first order reaction (p. 150) and may be written:

$$\frac{d\varrho}{dt} = k_1(\varrho_s - \varrho), \qquad [6.87]$$

where ϱ is the mass concentration of solute in the solution at time t, ϱ_s is the solubility in units of mass per unit volume, k_1 is the first order rate constant and $d\varrho/dt$ gives a measure of the dissolution rate. Now

$$m = \varrho V, \qquad [6.88]$$

where m is the mass of solute dissolved in volume, V, of liquid medium at time, t. Since V is constant, this equation on differentiation gives:

$$\frac{dm}{dt} = V \cdot \frac{d\varrho}{dt}. \qquad [6.89]$$

The dissolution rate of the solid is usually taken to be dm/dt. The mass flux, J_m, from the solid phase to the liquid medium is the rate of dissolution per unit area of solid (see equation [6.65]) and is usually known as the 'intrinsic dissolution rate' of the solid. Combining equations [6.65] and [6.89] gives

$$J_m = \frac{V}{A} \cdot \frac{d\varrho}{dt}, \qquad [6.90]$$

so that the Noyes–Whitney equation [6.87] may be expressed as:

$$J_m = k_1 \cdot \frac{V}{A}(\varrho_s - \varrho). \qquad [6.91]$$

We shall discuss three useful models or theories of dissolution and compare the Noyes–Whitney relationship with the equation derived from each. The diffusion layer model of Nernst and Brunner (1904) assumes that the rate of dissolution is controlled by diffusion. A solid which is dissolving is surrounded by an infinitesimally thin film of saturated solution whose mass concentration is the solubility, ϱ_s. According to the theory of Nernst and Brunner this film is surrounded by a stationary diffusion layer of finite but minute thickness h across which the mass concentration decreases linearly from ϱ_s to ϱ, where ϱ is the uniform mass concentration in the bulk of the solution (Fig. 6.18a). The more the solution is agitated, the smaller is h. The mass concentration gradient across the diffusion layer in the direction of flow is negative and is given by:

$$\text{grad } \varrho = \frac{d\varrho}{dx} = \frac{\varrho - \varrho_s}{h}. \qquad [6.92]$$

Applying Fick's first law (equations [6.64] and [6.65]) to the diffusion layer we have:

$$\frac{dm}{dt} \cdot \frac{1}{A} = J_m = \frac{D}{h}(\varrho_s - \varrho), \qquad [6.93]$$

Fig. 6.18 Models for the dissolution of solids by liquids
(a) diffusion layer model
(b) interfacial barrier model
(c) Danckwerts' model

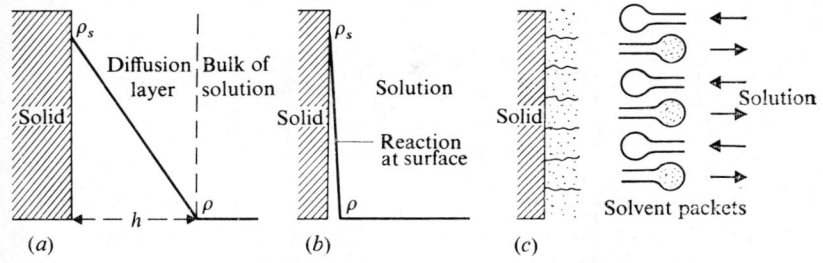

where D is the diffusion coefficient of the solute from the saturated solution to the bulk of the solution. Since the solute does not accumulate in the diffusion layer, dm/dt is the rate at which solute enters and leaves the diffusion layer and enters the bulk of the solution, and J_m is the corresponding mass flux. Thus, as stated above, dm/dt is the dissolution rate and J_m is the intrinsic dissolution rate of the solid. A is the surface area of each side of the diffusion layer and therefore of the solid itself.

According to the interfacial barrier model the dissolution rate is not controlled by diffusion but by a dissolution reaction at the solid surface (Fig. 6.18b). This reaction consists of breaking up of the crystal lattice to give free molecules or ions in solution and solvation of these molecules or ions. The dissolution rate equation according to this theory is simply

$$\frac{dm}{dt} \cdot \frac{1}{A} = J_m = k_i(\varrho_s - \varrho), \qquad [6.94]$$

where k_i is known as the 'effective interfacial transport rate constant' which has the dimension length time^{-1}.

Danckwerts developed a theory for the dissolution of gases by liquids, which may also apply to the dissolution of solids by liquids, particularly in the absence of agitation. In this model (Fig. 6.18c) packets of solvent move towards the surface of the solid, etch the surface and then move away from it carrying dissolved solute. The dissolution rate equation in this case is:

$$\frac{dm}{dt} \cdot \frac{1}{A} = J_m = S^{\frac{1}{2}} D^{\frac{1}{2}} (\varrho_s - \varrho), \qquad [6.95]$$

where S is a constant with the dimension time^{-1} and known as the 'mean rate of production of fresh surface'. The other symbols have the same significance as above.

The dissolution rate equation for each model has the same form as the Noyes–Whitney equation [6.91] and all may be written simply as

$$\frac{dm}{dt} \cdot \frac{1}{A} = J_m = K(\varrho_s - \varrho) \qquad [6.96]$$

where $K \equiv k_1 V/A$ (Noyes–Whitney equation 6.91),
$K \equiv D/h$ (diffusion-layer equation 6.93),
$K \equiv k_i$ (interfacial barrier equation 6.94),
$K \equiv S^{\frac{1}{2}} D^{\frac{1}{2}}$ (Danckwerts' equation 6.95).

No dissolution model is universally valid or invalid. The dissolution system and its conditions determine which model is operating. It may even be possible for two or more models to operate together.

If the mass concentration, ϱ, of the dissolved solute is very small compared with the solubility, ϱ_s, then $(\varrho_s - \varrho) \approx \varrho_s$ and the system is said to operate under sink conditions. In practice $(\varrho_s - \varrho)$ in the equations is replaced by ϱ_s whenever ϱ is less than about 5 per cent of ϱ_s. Sink conditions apply *in vivo* whenever the dissolution of the solid drug in body fluids is appreciably slower than absorption of the drug from the fluids by the tissues. The experimental conditions *in vitro* are often arranged so that sink conditions apply and may be carefully controlled so that ϱ_s and the constants in the dissolution rate equations ([6.91] and [6.93] to [6.96]) remain invariable. The intrinsic dissolution rate, J_m, is then constant, i.e.

$$\frac{dm}{dt} \cdot \frac{1}{A} = J_m = \text{constant}. \qquad [6.97]$$

Since, under sink conditions in which A is also constant, J_m and A are independent of m, this equation is a rate equation of zero order (see p. 162). Integrating this equation with respect to t gives:

$$\frac{m}{A} = J_m t + C, \qquad [6.98]$$

where C is a constant of integration. Since the mass m of dissolved material is zero when the solid is first brought into contact with the liquid medium at zero time, i.e. $m = 0$ when $t = 0$, it follows that $C = 0$. Hence:

$$m = A J_m t. \qquad [6.99]$$

The intrinsic dissolution rate, J_m, is found by determining, as described below, the mass m of solute dissolved in the liquid medium at various times t. A plot of m/A against t is linear and its gradient gives J_m directly. J_m has the dimension mass length^{-2} time^{-1}.

Studies of the intrinsic dissolution rate of pure substances must be carried out on a solid form which does not disintegrate, such as a single crystal or a highly compressed disc or tablet. This may be mounted in a holder so that only one face of constant area, A, is exposed to the liquid medium, or all the faces may be exposed and the change in surface area of the solid may be allowed for in the calculation. The liquid medium, which may be a solvent, a buffer solution or a biological fluid, is made to flow past the solid at a constant velocity. For this purpose the liquid is commonly stirred at a constant rate. The mass concentration, ϱ, of dissolved solid is determined at various times by analysis of samples taken by a reproducible technique from the liquid medium. The mass dissolved, m, may be calculated using equation [6.88] or can be determined directly by measuring the decrease in mass of the sample. Many different types of apparatus have been designed for dissolution rate studies and each may give different results, e.g. different plots of m/A against t. This is because J_m depends in a complex manner on the surface geometry of the sample and on the flow field of the solvent past the liquid. A most important feature of any dissolution rate apparatus is reproducibility, particularly in the degree of agitation. Column, or flow, methods have recently been developed in which the dissolution medium flows at a constant rate dV/dt, through a column containing a tablet or disc of the solid held in a fixed orientation relative to the direction of flow. The mass concentration ϱ, of the dissolved solute is continuously monitored by analysing the effluent. Then $dm/dt = \varrho \, dV/dt$. These methods have a number of advantages over the stirred solvent method described above. In all work on dissolution rates the detailed experimental conditions should be stated.

The theoretical equations [6.93], [6.94] and [6.95] suggest that the dissolution rate dm/dt is affected by several factors. Increasing the temperature increases the rate quantities D, k_i and S and usually the solubility, ϱ_s, and consequently increases dm/dt. Since D is usually inversely proportional to the viscosity, dm/dt decreases as the viscosity of the liquid medium increases. The greater the degree of agitation of the solution, the smaller is h and the greater is dm/dt. Reduction in the mean particle size increases A and to a lesser extent increases ϱ_s and therefore increases dm/dt. Factors such as pH (p. 361), ionic strength (salt concentration, p. 340) and interfacial tension at the solid-liquid interface (p. 458) will affect ϱ_s and other quantities and will consequently affect dm/dt. Reaction between the solid and the liquid medium will increase k_i and ϱ_s and will therefore increase dm/dt, unless the reaction yields an insoluble solid at the interface. Reactions which increase dm/dt include those resulting from a difference between the pK_a of the solid and the pH of the medium (Chapter 9) and micellar solubilisation of the solid by a surface active agent dissolved in the liquid medium (p. 484). Factors having a complex influence on the dissolution rate are the presence of other substances in the solid state and the state of compression, the crystalline form, including polymorphism, and the extent of solvation of the dissolving solid. The less stable crystalline form, polymorph or solvate will have a higher ϱ_s and a higher dm/dt than the more stable polymorph, other factors being equal.

Problems

1. Two liquids A and B, which form mixtures which behave ideally, have vapour pressures at 330 K of 50.0 and 30.0 kPa, respectively. Calculate, for a temperature of 330 K, (a) the pressure at which a mixture containing mole fractions 0.4 of A and 0.6 of B will boil, (b) the mole fraction of the more volatile component in the vapour evolved from this boiling mixture, assuming that the vapours obey Dalton's law of partial pressures, and (c) if this vapour were condensed, the pressure at which it would boil. Sketch the graph of vapour pressure versus composition for mixtures of A and B from pure A to pure B indicating both the vapour composition and the liquidus lines.

2. Calculate the mass percentage composition of the distillate when nitrobenzene ($C_6H_5NO_2$) is steam distilled. When the atmospheric pressure is 101.3 kPa the mixture boils at 372 K and the vapour pressure of water at this temperature is 97.7 kPa.

3. The effect of different partial pressures of various narcotic gases and vapours was studied. In the following table ϕ is the minimum volume fraction of drug in the gaseous phase capable of causing narcosis and $p^\ominus$ is the saturated vapour pressure of the drug.

Gas or Vapour	ϕ	$p^\ominus$/kPa
nitrous oxide	1.00	7910
ethylene oxide	0.058	253.4
chloroethane	0.050	237.4
diethyl ether	0.034	110.8
1,2-dichloroethane	0.0095	60.1
chloroform	0.005	43.2

The total pressure of the gaseous phase was always 101 325 Pa, and the temperature was constant at 37° C. For each drug calculate the partial pressure and the thermodynamic activity just capable of causing narcosis and comment on the consistency of the results with reference to Ferguson's principle.

4. Calculate for the following gases and vapours the minimum volume fraction of each drug in cell lipid capable of causing narcosis, assuming the values of ϕ tabulated in the previous problem and the following values of the solubility s (v/v) in cell lipid at 37° C.

Gas or Vapour	s
nitrous oxide	1.4
ethylene oxide	31.0
chloroethane	40.5
1,2-dichloroethane	130
chloroform	265

(The solubilities were measured in olive oil, which is here assumed to have solvent properties similar to those of cell lipid.) From the volume fractions in cell lipid calculate for each drug the minimum narcotic concentration in cell lipid, assuming Avogadro's hypothesis and the gas laws. Comment on the consistency of your results with reference to the Overton–Meyer theory.

5. The effect of different concentrations of various narcotic drugs on tadpoles was investigated. In the following table c is the minimum drug concentration capable of causing paralysis in the organisms and P is the partition coefficient of the drug between oleyl alcohol (numerator) and water (denominator):

Drug	c/mol dm^{-3}	P
ethanol	0.33	0.10
n-butanol	0.03	0.65
phenazone	0.07	0.30
barbitone	0.03	1.38
phenobarbitone	0.008	5.9
thymol	4.7×10^{-5}	950

Assuming that oleyl alcohol represents satisfactorily the solvent properties of cell lipid, calculate the minimum narcotic concentration of each drug in the lipid. Comment on the consistency of the results with reference to the Overton–Meyer theory.

6. The following are the concentrations in mol dm^{-3} for the distribution of phenol between chloroform and water:

Aqueous solution	0.094	0.126	0.205	0.297	0.383
Chloroform solution	0.254	0.455	1.19	2.54	4.17

Plot a graph of the square root of the concentration in the chloroform solution versus the concentration in the aqueous solution, and confirm that the phenol is associated to form double molecules in the chloroform solution.

7. The vapour pressure of pure water at 293 K is 2.34 kPa. Calculate the lowering of the vapour pressure for a solution of 50 g of glucose ($C_6H_{12}O_6$) in 950 g of water, assuming Raoult's law to be obeyed.

8. The elevation of the boiling-point of acetone produced by dissolving 0.4538 g of camphor in 20 g of acetone is 0.250 K. Given that the

molal elevation constant for acetone is 1.72 K kg mol^{-1}, calculate the molecular weight of camphor.
9. A solution of 1.049 g of naphthalene in 100 g of benzene freezes at $4.98°$ C. The molecular weight of naphthalene is 128, and the freezing-point of pure benzene is $5.40°$ C. Calculate the molal depression constant for benzene.
10. Calculate the osmotic pressure (in Pa and atmospheres) at 311 K of a solution which contains 12 g of urea (CON_2H_4) per dm^3.
11. A 5 per cent (w/v) solution of a protein (free from ionic material) gives an osmotic pressure of 2.482 kPa at 293 K. Calculate the molecular weight of the protein. Given than the molal depression constant of water is 1.85 K kg mol^{-1}, calculate the depression of the freezing point of water which would be caused by the presence of this concentration of protein.
12. 1 per cent (w/w) aqueous solutions of emetine hydrochloride and sodium chloride have freezing point depressions of $0.062°$ C and $0.576°$ C respectively. Calculate the weight percent of sodium chloride required to make a 6 per cent (w/w) aqueous solution of emetine hydrochloride iso-osmotic with blood serum (freezing-point $-0.52°$ C).
13. A 2.0 mmol dm^{-3} solution of a simple potassium salt has an osmotic pressure of 14.6 kPa at $20°$ C. Assuming that the practical osmotic coefficient is unity, calculate the van't Hoff factor, i, of the salt and hence deduce the valency of its anion.
14. An aqueous solution of salt has an ideal osmolality of 333 mmol kg^{-1} and freezes at $-0.56°$ C. Assuming that the molal depression constant for water is 1.86 K kg mol^{-1}, calculate the real osmolality and the practical osmotic coefficient of the solution.
15. In a study of the diffusion of sodium chloride in water at $25°$ C by the moving boundary method, pure water was layered on to a 0.984 mol dm^{-3} aqueous sodium chloride solution. 3.0×10^4 seconds later the concentration gradient at the boundary was 0.417 mol dm^{-3} cm^{-1}. Calculate the diffusion coefficient of sodium chloride.
16. Calculate the diffusion coefficient of lysozyme in water from the following data obtained at $20°$ C in the porous disc method. An aqueous lysozyme solution, initially of concentration 60 mg dm^{-3}, was allowed to diffuse through a porous disc into the same volume of distilled water, which after 26 minutes was found to contain 17 mg dm^{-3} lysozyme. When an aqueous potassium chloride solution, initially of concentration 75 mg dm^{-3}, was allowed to diffuse through the same disc for 448 minutes, the distilled water was found to contain 20 mg dm^{-3} potassium chloride. The diffusion coefficient of potassium chloride is 1.76×10^{-9} m^2 s^{-1} at $20°$ C.

17. The dissolution rate of potassium chloride from a compressed circular disc of diameter 13 mm was found to be 1.41 mg s^{-1} at a given solvent flow rate at 37° C. Calculate the intrinsic dissolution rate. Assuming that sink conditions apply and that the solubility and diffusion coefficient are 334 g dm^{-3} and 3.2×10^{-5} cm^2 s^{-1} respectively at 37° C, calculate the constants h, k_i and S in the various model theories of dissolution.

7 Energy and Equilibrium

Heat and Work

The interconversion of various forms of energy, including heat, work and 'chemical energy', is of fundamental importance to physical chemistry and is dealt with in a subject known as thermodynamics. We may define a *system* in thermodynamics at that part of the universe chosen for consideration. For our purposes a system is a physical or biophysical entity of limited size or is a chemical or biochemical reaction in a suitable container, which is often open to the atmosphere.

The term *heat* is applied to the exchange of that form of energy associated with the disordered or chaotic movement of atoms, molecules and ions, including translational and rotational motion and intermolecular and intramolecular vibrations, in all forms of matter. Quantity of heat, q, is positive, if heat is absorbed by the system and negative if heat is evolved by the system.

The term *work*, w, is the change in energy associated with the ordered or co-ordinated movement of particles in one or more definite directions outside the system under consideration. The following are examples of work: the movement of a piston by an expanding or contracting gas; the movement of a muscle fibre or of the flagellum of a bacterial cell; the flow of electrons along a wire; the motion of an engine or an electric motor; the raising or lowering of a weight; the expansion or compression of a spring; the movement of a body; the performance of an action. Chemical reactions can sometimes be made to do work under suitable conditions. Thus, mechanical work is performed by the hydrolysis of adenosine triphosphate (ATP) to adenosine diphosphate (ADP) and inorganic phosphate ion in the complex between actin and myosin in muscle and by the oxidation of hydrocarbons by air to carbon dioxide and water in an internal combustion engine. Electrical work is done by the chemical reactions taking place in electrochemical cells (Chapter 10) and in the living cells in the electric organs of certain fish.

Mechanical work is proportional to the applied force and the distance moved by the point of application in the direction of the force. The SI unit of work, heat and all other forms of energy is the joule. One joule (J) is the mechanical work done when the applied force is 1 newton (N) and the distance moved in the direction of the force is 1 metre (m), thus: 1 J = 1 N m. One newton (N) is that force which when applied to a mass of 1 kilogram (kg) gives it an acceleration of 1 metre per second squared (m s^{-2}) in the direction of the force, thus: 1 N = 1 kg m s^{-2}, therefore 1 J = 1 m^2 kg s^{-2}. The c.g.s. unit of heat, work and energy is the erg (1 erg = 10^{-7} J) and the c.g.s. unit of force is the dyne (1 dyne = 10^{-5} N). Work, w, is

Fig. 7.1 Expansion of a gas against a constant external pressure

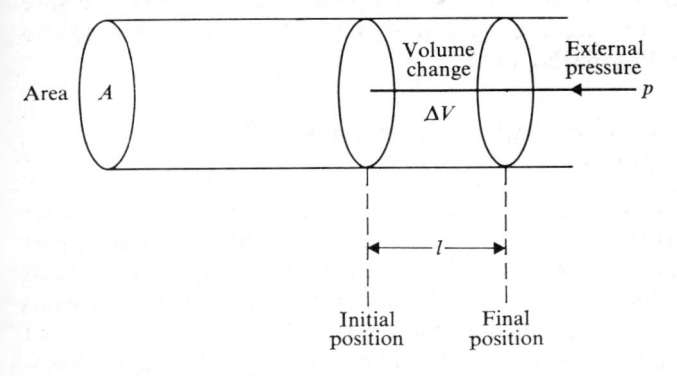

positive, if work is done *on* the system and is negative, if work is done *by* the system. The mechanical work ($-w$) done by a gas or a liquid expanding by a volume ΔV against a constant external pressure p may be calculated by reference to Fig. 7.1 as follows. If the piston has an area A and moves through a distance l, the volume change, $\Delta V = Al$. Since force = pressure × area, then, by definition of mechanical work,

$$-w = \text{force} \times \text{distance}$$
$$-w = p \times A \times l$$
$$-w = \text{pressure} \times \text{volume change}$$
$$\therefore \quad -w = p\,\Delta V \qquad [7.1]$$

$\Delta V = V_2 - V_1$, where V_1 is the initial volume and V_2 is the final volume. Equation [7.1] also applies to the expansion of a solid. If the system expands, $V_2 > V_1$ and ΔV is positive, so that w is negative and work is done by the system on the surroundings. If the system contracts, $V_2 < V_1$ and ΔV is negative, so that w is positive and the surroundings do work on the system. Under certain conditions, particularly for appreciable changes of volume in a gaseous system, the external pressure may not be constant. The overall change in volume can, however, be divided into a large number of infinitesimally small volume changes. If p is the applied pressure for a given volume element, dV, the work done during the infinitesimal volume increment is given by

$$-\mathrm{d}w = p.\mathrm{d}V. \qquad [7.2]$$

The total work done during an appreciable volume change at a variable applied pressure, p, is given by

$$-w = \int_1^2 -\mathrm{d}w = \int_{V_1}^{V_2} p.\mathrm{d}V. \qquad [7.3]$$

In SI units w is expressed in J ($=$ N m) and V in m^3. Since pressure is force divided by area, the SI unit of pressure is N m^{-2}, which is also known as the Pascal (Pa), i.e. 1 Pa $=$ 1 N m^{-2}.

Conservation of Energy, the First Law of Thermodynamics

The state of a system is defined by the temperature, pressure, volume and composition of its phases. For each phase these quantities are connected by an appropriate equation of state. A system in a fixed state has a definite and constant amount of internal energy (or thermodynamic energy), U, which is the sum of the kinetic energy and potential energy of the system. The kinetic energy is the energy associated with the motion of the system and its parts, e.g. the molecules, ions and atoms, whereas the potential energy is the energy associated with the position and arrangement of the system and its parts. Since the value of the potential energy depends on the reference point chosen, so does the absolute value of the internal energy, which is therefore arbitrary. A change in internal energy, ΔU is, however, quite definite and ascertainable, because the reference point is the same for both the initial state, 1, and the final state, 2, of the system, thus:

$$\Delta U = U_2 - U_1, \qquad [7.4]$$

where U_1 and U_2 are the internal energies of the initial and final states respectively.

The First Law of Thermodynamics is the law of conservation of energy, which states that energy can neither be created nor destroyed, although it may be changed from one form into another. The First Law is a law of experience; if it were not obeyed, impossible phenomena, such as perpetual motion, could take place. In chemistry, biology and pharmacy we are mainly concerned with heat, work and internal energy, so that the First Law becomes:

$$\Delta U = q + w. \qquad [7.5]$$

The sign convention for energy changes is to give each acquisition by the system a positive sign, and each loss from the system a negative sign. Thus heat absorbed by the system, work done *on* the system and an increase in U of the system are positive, whereas heat evolved by the system, work done *by* the system and a decrease in U of the system are negative.

In certain studies it may be necessary to introduce other forms of energy into the First Law equation [7.5]. For example, certain photochemical reactions (p. 174) are not only initiated by light but must continually absorb it in order for continual photochemical change to take place, e.g. the photochemical decomposition of alkyl iodides, and photosynthesis in the chloroplasts of green plants. A term of the type $+Nh\nu$ must be introduced on the right-hand side of equation [7.5], where N is the number of quanta absorbed and $h\nu$ is the energy of each quantum (p. 80). Certain chemical and biochemical reactions take place with the emission of light energy, e.g. the oxidation of luminol (5-aminophthalazin-1,4-diol) in alkaline solution by ferricyanide, the hydrolysis of adenosine triphosphate catalysed by the enzyme system luciferin–luciferase in the lantern of the fire-fly. A term of the type $-Nh\nu$ must be introduced on the right-hand side of the First Law equation [7.5], where N is now the number of quanta emitted. Furthermore, nuclear physics makes possible the interconversion of matter and energy and therefore regards mass, m, as a form of energy equal to mc^2, where c is the speed of light *in vacuo*.

Pressure, p, temperature, T, and volume, V, depend only on the state of the system whilst changes in these quantities, Δp, ΔT and ΔV, depend only on the initial and final states of the system and are independent of the path between them. For this reason p, T and V are called state functions. An important corollary of the First Law is that U is a state function and ΔU depends only on the initial and final states of the system. If this were not so, i.e. if ΔU were dependent on the path between states 1 and 2, it would be possible in proceeding from 1 to 2 and then back to 1 by a suitable choice of paths to create or destroy energy, which is contrary to the First Law. Although ΔU depends only on the initial and final states of the system (equation [7.4]) q and w (equation [7.5]) depend also on the path between these states and are not state functions. One path may be such that more work is done and less heat is exchanged, whereas another path may be one in which less work is done but more heat is exchanged. For both paths, however, ΔU will be the same.

Enthalpy

If a physical process or chemical reaction is carried out at constant volume, e.g. in a bomb calorimeter, so that no work is done, i.e. $w = 0$, q becomes q_v which is the heat exchanged at constant volume. Equation [7.5] then becomes:

$$\Delta U = q_v. \qquad [7.6]$$

Thus, the heat exchanged at constant volume is simply the change in internal energy during the process or reaction.

If the process or reaction is carried out at constant pressure, e.g. in a calorimeter open to the atmosphere, so that the work done is mechanical due only to a change in volume of the system at the constant pressure p (equation [7.1]), q becomes q_p which is the heat exchanged at constant pressure and equation [7.5] becomes:

$$\Delta U = q_p - p\,\Delta V$$
$$\therefore \quad \Delta U + p\,\Delta V = q_p. \qquad [7.7]$$

A quantity known as enthalpy, H, is defined by the equation

$$H = U + pV. \qquad [7.8]$$

For a process taking place at constant pressure

$$H_2 = U_2 + pV_2 \quad \text{in the final sate}$$

and

$$H_1 = U_1 + pV_1 \quad \text{in the initial state.}$$

Subtracting, $\quad (H_2 - H_1) = (U_2 - U_1) + p(V_2 - V_1).$

Using the Δ notation:

$$H_2 - H_1 = \Delta H; \quad U_2 - U_1 = \Delta U; \quad V_2 - V_1 = \Delta V.$$
$$\therefore \quad \Delta H = \Delta U + p\,\Delta V. \qquad [7.9]$$

Comparison of equations [7.7] and [7.9] shows that

$$\Delta H = q_p. \qquad [7.10]$$

Thus the heat exchanged at constant pressure is equal to the change in enthalpy during the process or reaction. Since U, p and V are state functions, H is by definition (equation [7.8]) a state function. ΔH, like ΔU, therefore depends only on the initial and final states of the system and is independent of the path between them.

Extensive and Intensive Quantities

The following quantities depend on the amount or mass of the substance or system: V, U, H (and also S and G considered later in this chapter and the heat capacities C, C_p and C_v discussed in the next section). These quantities are called *extensive* quantities or capacity factors. An extensive quantity of a substance divided by the amount of the substance, n, is called the *molar* quantity. A molar quantity for an extensive property may be regarded as the value of the property when the amount of substance considered is 1 mole. When it is necessary to emphasise that a molar quantity is under consideration, the subscript m is used, e.g. the molar volume, molar internal energy and molar enthalpy are given by:

$$V_m = V/n; \quad U_m = U/n; \quad H_m = H/n. \quad [7.11]$$

More often, however, the subscript $_m$ is omitted and the quantity is assumed to refer to 1 mole unless otherwise stated. The substance to which the molar quantity refers may then be indicated by a subscript or by a suffix in parenthesis. For example, the molar enthalpy of substance B may be written as H_B and the molar volume for carbon dioxide may be written as V_m (CO_2).

An extensive quantity of a substance divided by the mass, m, of the substance is called the *specific* quantity and is denoted by the corresponding lower case letter. For example, the specific heat capacity and specific volume are given by:

$$c = C/m; \quad v = V/m = 1/\varrho. \quad [7.12]$$

It can be seen that the specific volume of a substance is the reciprocal of its density. A specific quantity, when m is in kg, is equal to the corresponding molar quantity divided by the molar mass, M, in kg mol^{-1} ($= M_r/1000$), e.g.

$$v = V_m/M = 1000 V_m/M_r. \quad [7.13]$$

Quantities which do not depend on the amount or mass of the substance or system, e.g. T and p, are called *intensive* quantities or intensity factors.

Heat Capacity

The heat capacity, C, of a body or system is the heat it must absorb to increase its temperature by 1 K. Since C varies slightly with temperature, T K, it is usually defined by a differential notation. Since C depends on the conditions under which the heat, q, is absorbed, one condition is usually kept constant and this condition is indicated by a subscript for C or q. Thus the heat capacity at constant volume is given by

$$C_v = \frac{dq_v}{dT} = \left(\frac{\partial U}{\partial T}\right)_v \quad [7.14]$$

and the heat capacity at constant pressure is given by

$$C_p = \frac{dq_p}{dT} = \left(\frac{\partial H}{\partial T}\right)_p. \quad [7.15]$$

The internal energy, U, of a system depends on the reference point chosen. For convenience U will be set at zero at 0 K in the following discussion.

For ideal monatomic gases, e.g. the noble gases and mercury vapour, U is equal to the kinetic energy of translation resolved along 3 axes and corresponding to 3 degrees of freedom. U_m for the gas (p. 270) is given by

$$U_m = \tfrac{3}{2}RT \quad [7.16]$$

$$\therefore \quad C_{v,\,m} = \left(\frac{\partial U_m}{\partial T}\right)_v = \frac{3}{2}R = 12.5 \text{ J K}^{-1}\text{ mol}^{-1}. \quad [7.17]$$

We note that each translational degree of freedom contributes $\tfrac{1}{2}RT$ to U_m and $\tfrac{1}{2}R$ to $C_{v,\,m}$ and that C_v is independent of temperature.

For a monatomic solid, e.g. a solid metallic element, U is equal to the kinetic energy of vibration plus the potential energy of vibration of the atoms about their mean positions each resolved along 3 axes. There are therefore 3 vibrational degrees of freedom per atom, each consisting of 2 'square terms' of energy (kinetic and potential), and each square term contributes $\tfrac{1}{2}RT$ to the total energy at normal and high temperatures, so that:

$$U_m = 6 \times \tfrac{1}{2}RT = 3RT \quad [7.18]$$

$$C_{v,\,m} = \left(\frac{\partial U_m}{\partial T}\right)_v = 3R = 25.0 \text{ J K}^{-1}\text{ mol}^{-1}. \quad [7.19]$$

For diatomic or polyatomic gases $C_{v,\,m}$ is not given by equation [7.17], because the molecules can rotate and their atoms can undergo intramolecular vibrations, so that the molecules have rotational and vibrational degrees of freedom in addition to those of translation. As the number of atoms in the molecule increases, the number of vibrational degrees of freedom increases, and U and C_v increase. For most diatomic and all polyatomic molecules C_v increases with increasing temperature, because the rotational and vibrational degrees of freedom can take up more and more energy from the environment. At very high temperatures a bond in a molecule may vibrate so strongly that it breaks. The number of vibrational degrees of freedom lost by the original molecule on decomposition is always exceeded by the number of translational and rotational degrees of freedom gained by the molecular fragments. Therefore, when the molecules decompose, C_v increases still further.

Gases can often be assumed to behave ideally, so that $pV_m = RT$, and equation [7.8] applied to 1 mole of gas becomes:

$$H_m = U_m + RT$$

$$\therefore \quad \frac{dH_m}{dT} = \frac{dU_m}{dT} + R. \quad [7.20]$$

Using equations [7.14] and [7.15] we have

$$C_{p,\,m} = C_{v,\,m} + R. \quad [7.21]$$

R represents the excess energy required for the gas to expand against a constant pressure for a rise in temperature of 1 K. For monatomic gases

$$C_{v,\,m} = \tfrac{3}{2}R\,(=12.5\text{ J K}^{-1}\text{ mol}^{-1}),$$

so $\qquad C_{p,\,m} = \tfrac{5}{2}R\,(=20.8\text{ J K}^{-1}\text{ mol}^{-1})$

and $\qquad C_p/C_v = \tfrac{5}{3} = 1.667.$

For polyatomic gases $C_{v,\,m} > \tfrac{3}{2}R$, so that C_p/C_v is nearer unity, and approaches it more closely with increasing molecular complexity and temperature.

The specific heat capacity, c_p, of water is an important quantity because it relates the joule to the calorie (cal), which is an earlier and much used unit of heat, work and energy. The calorie (cal) is most conveniently defined as that amount of heat which, when absorbed by 1 g of water at 15° C, increases its temperature to 16° C (i.e. by 1 K). Thus, c_p for water is 1 cal g^{-1} K^{-1} from the definition of the calorie and is 4186 J kg^{-1} K^{-1}, i.e. 4.186 J g^{-1} K^{-1}, from calorimetric measurements, hence 1 cal = 4.186 J. Other definitions of the calorie are possible and these have given rise to two further values expressed in joules. The international calorie (cal$_{\text{IT}}$) is equal to 4.1868 J and the thermochemical calorie (cal$_{\text{th}}$) is 4.1840 J. The latter is in common use in chemical thermodynamics and is the basis for the values given in joules in the next section.

Thermochemistry

Thermochemistry is the study of heat changes which accompany chemical reactions and which are therefore often known as heats of reaction. Most authors consider the heat of reaction to be positive when heat is absorbed, i.e. when the reaction is endothermic, and negative when heat is liberated, i.e. when the reaction is exothermic. This agrees with the sign convention for q. In earlier works the opposite sign convention may be found. If the heat of reaction is denoted by the appropriate quantity, ΔU or ΔH, no ambiguity arises and the sign convention is that of q, i.e. positive for endothermic reactions and negative for exothermic reactions. Since most reactions in biology and pharmacy are carried out in vessels open to the atmosphere, and therefore at constant atmosphere pressure, the heat exchanged is usually ΔH and is called the 'enthalpy of reaction', see p. 277.

The general form of chemical equation may be written

$$a\text{A} + b\text{B} + \ldots = q\text{Q} + r\text{R} + \ldots, \qquad [7.22]$$

where the capital letters refer to individual substances and each small letter refers to the stoichiometric number of the corresponding substance. Changes in extensive properties, such as ΔV, ΔU, ΔH, ΔC_V, ΔC_p, ΔS and ΔG, during a chemical reaction nearly always refer to that amount of chemical change when the stoichiometric numbers of moles of all the reactants are completely converted into the stoichiometric number of moles of all the products. If H_A, H_Q, etc., are the molar enthalpies of the substances involved in the reaction,

$$H(\text{reactants}) = aH_A + bH_B + \ldots \qquad [7.23]$$
$$H(\text{products}) = qH_Q + rH_R + \ldots \qquad [7.24]$$

If the reaction goes to completion, the initial state corresponds to the reactants and the final state corresponds to the products. The enthalpy of reaction is given by:

$$\Delta H = H(\text{products}) - H(\text{reactants}) \qquad [7.25]$$
$$\therefore \quad \Delta H = qH_Q + rH_R + \ldots - aH_A - bH_B - \ldots \qquad [7.26]$$

Partial differentiation with respect to temperature at constant pressure gives:

$$\left(\frac{\partial(\Delta H)}{\partial T}\right)_p = q\left(\frac{\partial H_Q}{\partial T}\right)_p + r\left(\frac{\partial H_R}{\partial T}\right)_p + \ldots$$
$$- a\left(\frac{\partial H_A}{\partial T}\right)_p - b\left(\frac{\partial H_B}{\partial T}\right)_p - \ldots \qquad [7.27]$$

Introducing the corresponding values of C_p (equation [7.15]), we have:

$$\left(\frac{\partial(\Delta H)}{\partial T}\right)_p = qC_{p,Q} + rC_{p,R} + \ldots$$
$$- aC_{p,A} - bC_{p,B} - \ldots \qquad [7.28]$$

H in equations [7.23] to [7.27] may be replaced throughout by any other extensive property to give a set of analogous and equally valid equations. Replacement of H in equation [7.26] by C_p gives:

$$\Delta C_p = qC_{p,Q} + rC_{p,R} + \ldots - aC_{p,A} - bC_{p,B} - \ldots \qquad [7.29]$$

Comparing this equation with [7.28], we have:

$$\left(\frac{\partial(\Delta H)}{\partial T}\right)_p = \Delta C_p. \qquad [7.30]$$

Integration of this equation at constant pressure between the limits ΔH_1 at temperature T_1 to ΔH_2 at temperature T_2, we have:

$$\Delta H_2 - \Delta H_1 = \int_{T_1}^{T_2} \Delta C_p . \mathrm{d}T. \qquad [7.31]$$

If ΔC_p for the reaction is assumed to be independent of temperature:

$$\Delta H_2 - \Delta H_1 = \Delta C_p(T_2 - T_1). \qquad [7.32]$$

This is known as the Kirchhoff equation. The last two equations enable ΔH_2 at any temperature, T_2, to be calculated, when ΔH_1 at any other temperature, T_1, and ΔC_p are known. Equation [7.31] is applied when ΔC_p is known as a function of temperature. The effect of small changes of temperature on C_p, on ΔC_p and even on ΔH can often be ignored.

The *enthalpy of reaction* is the increase in enthalpy, ΔH, i.e. the quantity of heat absorbed at constant pressure, when the number of moles of reactants shown in the chemical equation for the reaction have reacted completely to give the products.

In thermochemical equations the value of ΔH is written after the chemical equation to which it refers and its SI unit is J. The temperature in K may be written after ΔH, e.g. ΔH_{298} or, better, $\Delta H(298.15 \text{ K})$ at 25° C. All changes in extensive quantities depend on the physical states of the materials involved. The states of each reactant and product may be indicated by an appropriate lower case letter abbrevation placed after the formula of the substance in parenthesis, thus: g = gas; l = liquid; s = solid; c = crystalline; aq = dissolved in water, sometimes in infinitely dilute solution; e.g.

$$C(\text{graphite}) + O_2(g) = CO_2(g); \qquad \Delta H(298 \text{ K}) = -393.5 \text{ kJ}.$$

Energy changes are extensive properties and are proportional to the stoichiometric numbers in the balanced chemical equation for the reaction. Therefore, if the stoichiometric numbers of a chemical equation are multiplied by a factor, ΔH must be multiplied by the same factor, e.g.

$$2C(\text{graphite}) + 2O_2(g) = 2CO_2(g);$$
$$\Delta H(298 \text{ K}) = 2 \times -393.5 \text{ kJ} = -787.0 \text{ kJ}.$$

The same, of course, applies to changes in other extensive properties during a reaction, e.g. ΔV, ΔU, ΔC_v, ΔC_p, ΔS and ΔG.

Changes in extensive quantities, e.g. ΔH, during different reactions are best compared under standard conditions. The standard state of a pure element or compound is the most stable form of the pure substance at a pressure of 1 atm (101 325 Pa) and usually at a temperature of 25° C (298.15 K). When the reactants and products are in their standards states, the word *standard* is placed before the term to which it applies, e.g. enthalpy of reaction, and the superscript $^\ominus$ or $^\circ$ is placed after the symbol for the property, e.g. $\Delta H^\ominus$. The following are examples of standard states of various substances: gas at 1 atm for H_2, O_2, CO_2 and CH_4; liquid for H_2O, C_2H_5OH, Br_2 and Hg; the crystalline or solid state for other metals and for sodium chloride, dextrose, urea and acetylsalicylic acid. For substances capable of existing in different polymorphic forms, e.g. carbon, the standard state is the most stable polymorph, e.g. graphite not diamond.

Certain enthalpy changes have special names. The (standard) *enthalpy of combustion*, ΔH_c (or $\Delta H_c^\ominus$), of an element or compound is the increase in enthalpy when 1 mole of the substance is completely burned in oxygen gas (in their standard states to give the products in their standard states), e.g. from the above reaction, ΔH_c (C, graphite, 298 K) = -353.5 kJ mol^{-1}. The (standard) *enthalpy of formation*, ΔH_f (or $\Delta H_f^\ominus$), of a compound is the increase in enthalpy when 1 mole of the compound (in its standard state) is formed from its elements (in their standard states), e.g. from the above reaction, $\Delta H_f^\ominus(CO_2$, g, 298 K) = -353.5 kJ mol^{-1}.

ΔH_f and $\Delta H_f^\ominus$ are not normally directly observable quantities but can be calculated by an indirect route using the fact that ΔH is a change in a state function. This fact gives rise to Hess's law, which states that the total enthalpy change during a chemical process is independent of the intermediate stages through which it may proceed, and depends only on the chemical nature, and the physical state, of the initial reactants and the final products. The significance of this law is that, if we can produce a series of changes which, when added together, will give the chemical equation for the process whose ΔH we cannot measure directly, the sum of the ΔH values for that series of changes will be equal to the ΔH we wish to find. This is best understood by means of an example:

Let us calculate $\Delta H_f^\ominus(CH_4$, g, 298 K) given that $\Delta H_c^\ominus(CH_4$, g, 298 K) = -890.35 kJ mol^{-1}, $\Delta H_c^\ominus$(C, graphite, 298 K) = -393.5 kJ mol^{-1} and $\Delta H_c^\ominus(H_2$, g, 298 K) = -285.85 kJ mol^{-1}.

These data correspond to the following thermochemical equations at 298 K:

$$CH_4(g) + 2O_2(g) = 2 H_2O(l) + CO_2(g); \quad \Delta H = -890.35 \text{ kJ} \quad \text{(i)}$$
$$C(graphite) + O_2(g) = CO_2(g); \quad \Delta H = -393.5 \text{ kJ} \quad \text{(ii)}$$
$$H_2(g) + \tfrac{1}{2}O_2(g) = H_2O(l); \quad \Delta H = -285.85 \text{ kJ} \quad \text{(iii)}$$

We wish to find ΔH for the reaction

$$C(graphite) + 2H_2(g) = CH_4(g); \quad \Delta H = ? \quad \text{(iv)}$$

Now, the chemical equation in [iv] may be obtained by adding the chemical equation in [ii] and twice that in [iii] and subtracting that in [i], thus:

$$C(graphite) + O_2(g) + 2H_2(g) + O_2(g) - CH_4(g) - 2O_2(g)$$
$$= CO_2(g) + 2H_2O(l) - 2H_2O(l) - CO_2(g).$$

This reduces to:

$$C(graphite) + 2H_2(g) - CH_4(g) = 0$$

which is another form of the chemical equation in [iv]. We must therefore add and subtract the ΔH values in the same way to obtain the required ΔH:

$$\Delta H = -393.5 - (2 \times 285.85) + 890.35 = -74.85 \text{ kJ}.$$

Whence, $\Delta H_f^{\ominus}(CH_4, g, 298 \text{ K}) = -74.85 \text{ kJ mol}^{-1}$.

The more negative is $\Delta H_f^{\ominus}$ for a compound, the more heat must be absorbed to convert it into its elements in the reverse reaction and therefore the more stable is the compound. However, compounds with positive values of $\Delta H_f^{\ominus}$, called endothermic compounds, do exist, e.g. acetylene, C_2H_2, for which $\Delta H_f^{\ominus}(298 \text{ K}) = +226.75 \text{ kJ mol}^{-1}$. Such compounds tend readily to react with other substances in order to lose their excess energy but do not usually break down spontaneously because energy must be absorbed to break bonds in a molecule. When all the bonds are broken, the molecule is converted into its component atoms and such a reaction is invariably endothermic.

Bond Energies

The value of ΔH for the reaction in which 1 mole of a compound is split completely into its component atoms in the free gaseous state may be referred to as the *enthalpy of atomisation*, ΔH_a, of the compound. Thus, $\Delta H_a(CH_4, g)$ is ΔH for the reaction:

$$CH_4(g) = C(g) + 4H(g). \quad \text{(v)}$$

Calculation of this quantity requires the application of Hess's law to the following thermochemical equations:

$$C(\text{graphite}) + 2H_2(g) = CH_4(g); \quad \Delta H(298 \text{ K}) = -74.85 \text{ kJ} \quad \text{(vi)}$$
$$(= \Delta H_f^{\ominus}(CH_4, g, 298 \text{ K}); \text{ see above})$$
$$H_2(g) = 2H(g); \quad \Delta H(298 \text{ K}) = +436.0 \text{ kJ} \quad \text{(vii)}$$
$$(= \text{molar enthalpy of dissociation of hydrogen gas})$$

$$C(\text{graphite}) = C(g); \quad \Delta H(298 \text{ K}) = +716.7 \text{ kJ} \quad \text{(viii)}$$
$$(= \text{molar enthalpy of sublimation of graphite}).$$

The chemical equation in (v) is equal to $2(\text{vii}) + (\text{viii}) - (\text{vi})$, thus:

$$2H_2(g) + C(\text{graphite}) - C(\text{graphite}) - 2H_2(g) =$$
$$4H(g) + C(g) - CH_4(g)$$

which reduces to

$$0 = 4H(g) + C(g) - CH_4(g)$$

which is another form of (v). Treating the ΔH values in the same way, we obtain the required quantity, thus:

$$\Delta H(v) = 2\Delta H(vii) + \Delta H(viii) - \Delta H(vi)$$
$$= (2 \times 436.0) + 716.7 + 74.85 = 1663.55 \text{ kJ}$$

whence, $\Delta H_a(CH_4, g, 298 \text{ K}) = 1663.6 \text{ kJ mol}^{-1}$.

When methane is split into its component atoms, all four of its C—H bonds are broken. $\Delta H_a(CH_4, g)$ may be taken to be the energy per mole (at constant pressure) which must be absorbed to break four C—H bonds. One quarter of this quantity, i.e. 415.9 kJ, may be considered to be the average energy per mole required to break one C—H bond and may be termed the mean bond energy of the C—H bond. (A slightly lower value is quoted in Table 7.1 but that is an average for several compounds.)

The *mean bond energy*, $E(A—B)$, of the bond A—B may be defined as the energy required to break the bond A—B in a molecule when all the other bonds are broken at the same time. In this way the molecule is split completely into its component atoms in their free gaseous state. Bond energies are always positive, because heat must be absorbed to break a bond. If the structure of a compound can be represented by conventional valency bonds, ΔH_a is equal to the sum of the mean bond energies:

$$\Delta H_a = \sum E(A-B). \qquad [7.33]$$

The justification for this statement is that, provided that the structure of the compound is not modified by resonance, a self-consistent table (e.g. Table 7.1) of mean bond energies can be computed from a series of equations of the type:

$$\Delta H_a(CH_4, g) = 4E(C—H);$$
$$\Delta H_a(C_2H_6, g) = 6E(C—H) + E(C—C);$$
$$\Delta H_a(C_2H_4, g) = 4E(C—H) + E(C=C);$$
$$\Delta H_a(C_2H_2, g) = 2E(C—H) + E(C\equiv C); \text{ etc.}$$

Table 7.1. Mean bond energies at 298 K $(E(A-B)/\text{kJ mol}^{-1})$

C—C	347.7	C—N	291.6	C—O	351.5
C=C	615.0	C=N	615.0	C=O (R.CHO)	715.5
C≡C	811.7	C≡N	891.2	C=O (RR'CO)	728.0
C—H	413.4	N—H	390.8	O—H	462.8

The *bond dissociation energy*, $D(A—B)$, of the bond A—B in a given molecule, radical or ion may be defined as the energy required to break the bond A—B and thereby give rise to two fragments A and B, which may be free radicals, atoms or sometimes ions. Bond dissociation energies are determined using gases by thermodynamic and kinetic methods, by electron impact studies in a mass spectrometer or from molecular spectra in the ultraviolet and visible regions. Bond dissociation energies are very sensitive to differences in electronic and atomic configuration, because such differences greatly affect the forces between the atoms. Thus, for bonds between a given pair of atoms (e.g. C and H) in different molecules or free radicals, bond dissociation energies may differ appreciably as shown by the following experimental data:

$CH_4 = \cdot CH_3 + \cdot H$; $\Delta H(298\ K) = 431$ kJ mol^{-1} = $D(CH_3$—$H)$
at 298 K

$\cdot CH_3 = :CH_2 + \cdot H$; $\Delta H(298\ K) = 356$ kJ mol^{-1} = $D(\cdot CH_2$—$H)$
at 298 K

$:CH_2 = \mathbf{:}CH + \cdot H$; $\Delta H(298\ K) = 536$ kJ mol^{-1} = $D(:CH$—$H)$
at 298 K

$\mathbf{:}CH = \mathbf{:}C + \cdot H$; $\Delta H(298\ K) = 339$ kJ mol^{-1} = $D(\mathbf{:}C$—$H)$
at 298 K

The sum and the arithmetic mean of these bond dissociation energies are 1662 kJ mol^{-1} and 415.5 kJ mol^{-1} respectively, which compare favourably with 1663.6 kJ mol^{-1} for ΔH_a (CH_4, g, 298 K) and 413.4 kJ mol^{-1} for the mean bond energy of the C—H bond, as expected from Hess's law.

The term *phosphate bond energy* (p. 419) is used in discussions of energy changes in biological systems, but differs in three ways from the two types of bond energy discussed above. Firstly, phosphate bond energy refers not to a bond breaking reaction but to the hydrolysis of phosphate esters to inorganic phosphate. Secondly, it is a Gibbs energy change (p. 297) not an enthalpy change. Thirdly, the energy is not absorbed by the molecules but is liberated and may be passed on to synthetic reactions which require Gibbs energy to drive them.

Resonance Energy

The stabilisation of a molecule by resonance can be seen by the fact that ΔH_a of the substance is greater than the sum of the mean bond energies for any canonical structure. The difference between ΔH_a and the sum of the mean bond energies for the most stable canonical structure is called the resonance energy, E_r, of the molecule and is a measure of its stabilisation by resonance:

$$E_r = \Delta H_a - \sum E(A\text{—}B). \qquad [7.34]$$

For example, in the case of benzene the sum of the bond energies of each Kekulé structure (Fig. 4.1) is equal to

$6E(C{-}H) + 3E(C{-}C) + 3E(C{=}C)$
$= (6 \times 413.4) + (3 \times 347.7) + (3 \times 615.0) = 5368.5$ kJ mol^{-1}.

From experimental thermochemistry and Hess's law $\Delta H_a(C_6H_6$, g, 298 K) = 5531.9 kJ mol^{-1}. The difference is 163.4 kJ mol^{-1}, the resonance energy for benzene.

A resonance energy calculated as above appears as a relatively small difference between two large quantities and is therefore particularly subject to error. This difficulty is largely overcome if enthalpies of hydrogenation are considered instead of enthalpies of atomisation. The *enthalpy of hydrogenation*, ΔH_h, of an unsaturated hydrocarbon may be defined as the increase in enthalpy when 1 mole of the compound is completely converted into the corresponding saturated hydrocarbon by reaction with hydrogen gas. This process can be carried out fairly readily by catalytic means (p. 183), and so ΔH_h can be measured experimentally. ΔH_h of cyclohexene

is found to be -119.62 kJ mol^{-1}. This molecule possesses only one double bond so resonance is assumed to be negligible. Consequently, if there were no resonance in benzene, which contains three double bonds, we would expect ΔH_h of benzene to be three times that of cyclohexene, i.e. $3 \times (-119.62) = -358.86$ kJ mol^{-1}. Experimentally, ΔH_h is found to be -208.36 kJ·mol^{-1}. The difference, $-208.36 - (-358.86) = 150.5$ kJ mol^{-1}, is the resonance energy of benzene and is in fair agreement with, but more accurate than, the value calculated previously.

Reversible and Irreversible Processes

Equilibrium is a state of balance. The system shown in Fig. 7.2 will be in equilibrium at constant temperature if the pressure, p, exerted at the solution side of the membrane is equal to the osmotic pressure, π, of the solution and if the solution is homogeneous with respect to concentration and temperature (p. 244). Let the solution contain 1 mole of solute and occupy a volume, V. Initially, let V equal V_1, p equal p_1 and

π equal π_1 and let the solution be in equilibrium so that $p_1 = \pi_1$. When the pressure $(p - h\rho g)$ applied to the solution, is reduced to zero, osmosis will occur and the solution will be diluted and will rise until π equals the hydrostatic pressure, $h\rho g$. If the solution is maintained at constant temperature, the process is said to be isothermal and heat may pass into the solution from the thermostat. After a certain amount

Fig. 7.2 System to illustrate reversible and irreversible dilution of a solution in osmosis

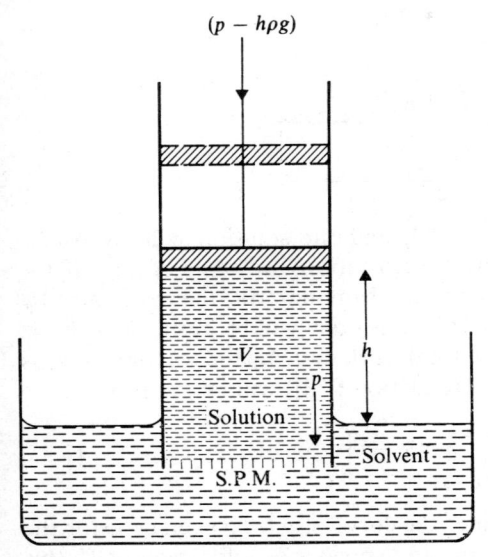

of dilution let V equal V_2, p equal p_2 and π equal π_2 and let the solution be in equilibrium, so that $p_2 = \pi_2$. The system may pass from the initial state, 1, to the final state, 2, by a variety of isothermal paths.

Let us consider the isothermal path in which p is reduced infinitely slowly, so that p is infinitesimally lower than π at all times. The system, viz. the solution between the membrane and the piston, will always be in equilibrium with its surroundings, viz. the solvent and the thermostat. Furthermore, an infinitesimal increase in p will cause an infinitesimal amount of solvent to flow in a direction opposite to osmosis and an infinitesimal quantity of heat to pass from the solution into the thermostat. There is no reason why the dilution process cannot be exactly reversed so that the system proceeds from state 2 to state 1 by an infinitely slow increase in p from p_2 to p_1. At all times p will then be infinitesimally greater than π. Such a process proceeds from p_1 to p_2 (or from p_2 to p_1) by a series of equilibrium states and in called a *reversible process*. Thermodynamic reversibility therefore implies that only an infinitesimal change in the environment is sufficient to cause the process to go in reverse. At all times p will be virtually

Fig. 7.3 Reversible and irreversible changes in concentration of a solute in solution during osmosis

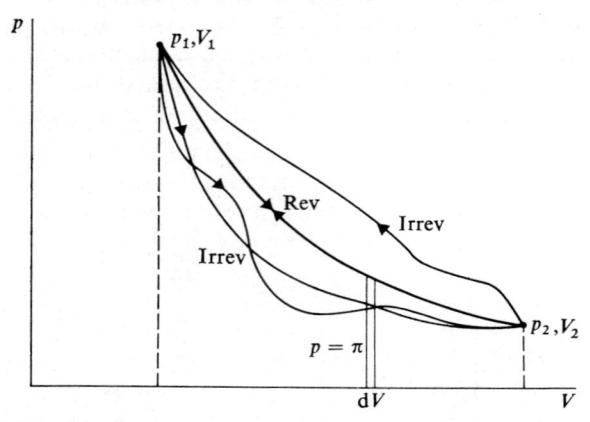

indistinguishable from π (Fig. 7.3) and the solution will always be homogeneous with respect to concentration and temperature. If the volume is increased by an infinitesimal amount, dV, the work done by the solution in pushing up the piston, $-dw$, will be given by $-dw = p \cdot dV$ (equation [7.2]). The total work done by the solution, $-w$, as it passes from state 1 to state 2 (equation [7.3]) will be given by

$$-w = \int_{V_1}^{V_2} p \cdot dV = \int_{V_1}^{V_2} \pi \cdot dV, \qquad [7.35]$$

which is equal to the area under the curve of $p (= \pi)$ versus V for the reversible path (Fig. 7.3). If we can express π as a function of V, this integral can be evaluated. Let the solution be so dilute that it obeys the limiting laws of osmosis given by equation [6.43], which, for $n = 1$ mol of solute, may be written as $\pi = RT/V$. Introducing this function of π into equation [7.35] we have:

$$-w = \int_{V_1}^{V_2} \frac{RT}{V} \cdot dV. \qquad [7.36]$$

Since T is constant, RT may be placed outside the integral

$$\therefore \; -w = RT \ln V_2 - RT \ln V_1 = RT \ln \frac{V_2}{V_1}. \qquad [7.37]$$

Since $\quad RT/V_1 = \pi_1 = p_1 \;$ and $\; RT/V_2 = \pi_2 = p_2$

$$-w = RT \ln \frac{V_2}{V_1} = RT \ln \frac{\pi_1}{\pi_2} = RT \ln \frac{p_1}{p_2}. \qquad [7.38]$$

An exactly analogous set of equations may be derived for the isothermal reversible expansion and compression of 1 mole of an ideal gas, where π is replaced by the internal pressure of the gas. Other examples of reversible processes are phase changes occurring at constant pressure, e.g. melting, freezing, vaporisation, condensation, sublimation and polymorphic transitions. An analogy of a reversible process is the slow movement of two almost equally matched teams in a tug-of-war.

If the pressure on the membrane is reduced from p_1 to p_2 at a finite rate by reducing the external pressure at a finite rate, the solvent which passes through the membrane will not have sufficient time to distribute itself equally through the solution by diffusion. There will then be a layer of more dilute solution in contact with the upper side of the membrane and p will be equal to the osmotic pressure of this more dilute layer and will be less than π of a homogeneous solution occupying the same volume V. Furthermore, heat may be adsorbed so rapidly by the solution from the thermostat that temperature gradients will be set up in the system. These inhomogeneities and the fact that $p < \pi$ indicate that the solution is neither in internal equilibrium nor in equilibrium with its surroundings. The various stages in the process cannot be exactly reversed except by a virtually impossible coincidence. Such a process is therefore termed an *irreversible process*.

All spontaneous or natural processes are irreversible and experience shows that their direction is always such that the system could do work. An irreversible process can only be completely defined and therefore properly understood if the various quantities describing the internal changes within the system are known. These quantities are known as internal parameters and each helps to describe a process of flow, such as diffusion, thermal conductivity or mechanical movement, within the system. Non-equilibrium or irreversible thermodynamics attempts to describe in detail spontaneous (or natural) processes, e.g. diffusion and flow through living cells and their membranes, muscle and nerve action, electrokinetic phenomena and fluctuations within systems.

To return to the osmotic system illustrated in Fig. 7.2, there will be a variety of irreversible paths from state 1 to 2, depending on how the applied pressure, p, at the membrane, is reduced, but there is only one reversible path. Since p for any irreversible process is always less than p for the reversible process, the work done by the solution, $-w$, (i.e. the area under the appropriate curve in Fig. 7.3) is a maximum for the reversible process, $-w_{rev}$, and is always less than this for any irreversible process, $-w_{irr}$. Mathematically stated:

$$-w_{irr} < -w_{rev} = \text{a maximum.} \qquad [7.39]$$

If the system is made to return to state 1 from state 2 by increasing p from p_2 to p_1, osmosis will be opposed and solvent will pass out of the solution. If this process is carried out at a finite rate, the solution will again not be homogeneous with respect to concentration or temperature and the process will again be irreversible. As the solvent passes out of the solution through the membrane, a more concentrated layer of higher osmotic pressure than the rest of the solution will be formed on the upper side of the membrane. The pressure p on the membrane will equal the osmotic pressure of this layer and will be greater than π of a homogeneous solution occupying the same volume. The direction of change is now opposite to the spontaneous or natural change and work must now be done on the system to make this possible. Since p for any irreversible path from state 2 to state 1 is greater than p for the reversible path, the work done on the system, $+w$, (i.e. the area under the appropriate curve in Fig. 7.3) is a minimum for the reversible process, $+w_{rev}$, and is always more than this for any irreversible process, $+w_{irr}$. Mathematically stated:

$$+w_{irr} > w_{rev} = \text{a minimum.} \qquad [7.40]$$

This formula may be derived from [7.39] by multiplying throughout by -1. We note that a change in the plus or minus sign requires a reversal of the inequality sign and a change from maximum to minimum or vice versa. A negative quantity is always smaller than a positive quantity, and the more negative is the quantity the smaller it is algebraically.

In the discussion of the First law, $\Delta U = q + w$ (equation [7.5]), we have seen that w and q depend on the path taken between the two states in question. In a reversible process, however, the path is defined, so that w_{rev} depends only on the states of the system and is therefore a state function. Since ΔU is also a state function, q for a reversible process is a state function too, known as q_{rev}. For given initial and final states and for a given value of ΔU of the system, the fact that $-w_{rev}$ is a maximum (and $+w_{rev}$ is a minimum) means that $-q_{rev}$ is a minimum (and $+q_{rev}$ is a maximum). The heat evolved in an irreversible process, $-q_{irr}$, must be more than $-q_{rev}$, in order for $-w_{irr}$ to be less than $-w_{rev}$ (equation [7.39]). Similarly, the heat absorbed in an irreversible process, $+q_{irr}$, must be less than $+q_{rev}$, in order for $+w_{irr}$ to be more than $+w_{rev}$. Stated mathematically:

$$-q_{irr} > -q_{rev} = \text{a minimum} \qquad [7.41]$$

$$+q_{irr} < +q_{rev} = \text{a maximum.} \qquad [7.42]$$

Entropy and the Second Law of Thermodynamics

We note that mechanical work is a product of force (an intensive quantity) and distance moved (an extensive quantity) or of pressure (an intensive quantity) and volume change (an extensive quantity). Similarly, electrical work (p. 397) is a product of potential difference or electromotive force (an intensive quantity) and charge transferred (an extensive quantity). It is convenient to regard heat, another form of energy, as a product of an intensive quantity and an extensive quantity. For heat the intensive quantity is the absolute temperature, T, and the extensive quantity is defined as a change in a state function which is given the name *entropy* and the symbol S. In order for entropy to be a state function, it is defined in terms of the state functions, q_{rev} and T, as follows:

$$q_{rev} = T \Delta S = T(S_2 - S_1), \qquad [7.43]$$

where q_{rev} and T refer to the isothermal reversible process from the initial state 1 of entropy S_1 to the final state 2 of entropy S_2. ΔS is the change in entropy. For an infinitesimal reversible process, equation [7.43] becomes

$$dq_{rev} = T\, dS. \qquad [7.44]$$

Entropy is a very valuable concept because it helps in the understanding of spontaneous change. We shall calculate the entropy changes for a system (subscript Y) exchanging heat with its surroundings (subscript R). The heat gained (or lost) by the system, q_Y, must equal the heat lost (or gained) by the surroundings, q_R, thus:

$$q_Y = -q_R \quad \text{or} \quad q_Y + q_R = 0. \qquad [7.45]$$

From equation [7.43] for an isothermal reversible process:

$$q_{Y, rev} = T \Delta S_Y \quad \text{and} \quad q_{R, rev} = T \Delta S_R$$
$$\therefore\ T \Delta S_Y = -T \Delta S_R \quad \text{or} \quad T \Delta S_Y + T \Delta S_R = 0.$$

Dividing throughout by T which is not equal to zero:

$$\Delta S_Y = -\Delta S_R \quad \text{or} \quad \Delta S_Y + \Delta S_R = \Delta S_{total} = 0. \qquad [7.46]$$

Therefore S_{total} does not change.

For an isothermal irreversible process equations [7.42] and [7.43] may be combined and applied to the system and surroundings to give:

$$q_{Y, irr} < T \Delta S_Y \quad \text{and} \quad q_{R, irr} < T \Delta S_R.$$

Since $q_{Y, irr} + q_{R, irr} = 0$ by equation [7.45]:

$$T \Delta S_Y + T \Delta S_R > 0$$
$$\therefore\ \Delta S_Y + \Delta S_R = \Delta S_{total} > 0. \qquad [7.47]$$

Therefore ΔS_{total} is positive and S_{total} increases.

These derivations may be shown to apply to processes in which the temperature is not constant. Such processes can be divided into a large number of infinitesimal isothermal processes (equation [7.44]) whose heat and entropy changes are dq and dS instead of q and ΔS. For reversible processes this leads to

$$dS_Y + dS_R = dS_{total} = 0, \qquad [7.48]$$

which on integration gives equation [7.46]. For irreversible processes this leads to

$$dS_Y + dS_R = dS_{total} > 0, \qquad [7.49]$$

which on integration gives equation [7.47]. Equations [7.46] and [7.47] mean that the total entropy of the system and its surroundings, i.e. of the whole universe, remains constant for a reversible process but increases for an irreversible, spontaneous or natural process.

We note that all irreversible, spontaneous or natural processes are accompanied by an overall increase in disorder, and entropy gives a measure of the amount of disorder. In thermal conduction heat passes spontaneously from a hotter region to a colder region until both are at the same intermediate temperature, because the molecules become distributed among the various energy levels in a more disordered manner, or, in other words, because the thermal entropy increases. A dissolved solute diffuses spontaneously from a region of higher concentration to a region of lower concentration and a gas diffuses spontaneously so as to fill the space available to it, because the molecules become distributed among the various positions available to them in a more disordered manner, or in other words, because the configurational entropy increases. Similarly, in osmosis a solvent diffuses spontaneously from a region in which its activity is higher to one in which its activity is lower, because the molecules of solvent become more randomly distributed among the various environments available to them; i.e. because the configurational entropy increases. The spontaneous or natural tendency for disorder or entropy to increase is quite general provided that the whole of the region within which any energy changes are taking place is considered (so that the region as a whole represents a system of constant energy). Such a system always tends to undergo spontaneous and hence irreversible changes which are accompanied by increases in entropy. In other words, a system of constant energy always tends towards a condition of maximum entropy.

It is sometimes possible to bring about an increase in orderliness in a system, for instance when a liquid crystallises, but this is always accompanied by a greater disordering effect on the surroundings (in the example of a liquid crystallising the effect on the surroundings is brought about by the enthalpy of fusion which is liberated) so that when all the changes are taken into account there is a net increase in entropy. When a state of equilibrium has been attained by a system of constant energy there is no further change of entropy. Living organisms consist of highly ordered structures, e.g. macromolecules, such as proteins, polysaccharides and nucleic acids, in a highly ordered arrangement. To maintain a steady state of low entropy far removed from a state in which the organism is in equilibrium with its environment, the living organism causes an increase in entropy of its surroundings, e.g. by breaking down molecules into smaller molecules in the environment and by producing heat. The rate of increase in entropy of the surroundings always exceeds the rate of decrease in entropy of the organism so that the total entropy steadily increases. After death the steady state is no longer stable and the organism decays thereby coming into thermodynamic equilibrium with its environment. This process involves a further increase in the total entropy, until at equilibrium a constant maximum value is attained.

It has been seen earlier that the First Law of Thermodynamics is the law of conservation of energy. The Second Law of Thermodynamics is also a law of experience and states the restrictions which Nature imposes on the conversion of energy from one form into another. The Second Law was stated implicitly in the above discussions and the following are some alternative statements of it. The direction of spontaneous change is such that work can be obtained under suitable circumstances. A spontaneous process in a system can only be reversed by supplying work from the surroundings. Heat cannot pass from a colder to a hotter body without the expenditure of work. In an isothermal reversible process, $q_{rev} = T \Delta S$ (equation [7.43]). All irreversible or spontaneous processes are accompanied by an overall increase in entropy of the system and its surroundings.

Statistical Thermodynamics and the Third Law

Figure 7.4 shows the distribution of the number, n, of indistinguishable molecules of a pure gas among the various quantized energy levels, ε. Such a distribution is known as the Maxwell–Boltzmann distribution (see p. 24). A similar distribution of molecular energies would apply to the liquid and solid states of a pure substance. Exchange of energy between the molecules constantly takes place by collision in a gas or liquid and by coupled vibration of the molecules in the solid state. An individual molecule may jump from one energy level to another, but the overall statistical distribution of a large number of molecules among the various energy levels is virtually constant at equilibrium and at constant temperature, pressure and volume.

Statistical thermodynamics (or statistical mechanics) enables the thermodynamic functions (extensive and intensive quantities) and other properties of a system consisting of one or more substances to be calculated from the properties of the individual molecules, such as their energy levels and their freedom to move around in space. In general, the calculated values agree well with those determined ex-

Fig. 7.4 Maxwell–Boltzmann distribution of molecular energies

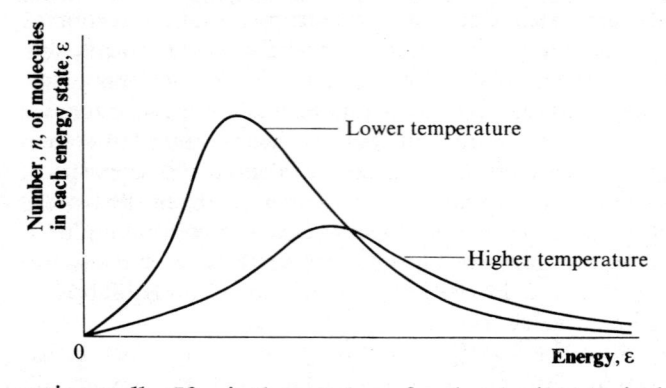

perimentally. If n_i is the number of molecules in a typical energy level of energy ε_i, the total number of molecules, N, is given by:

$$N = \sum_{\varepsilon_i=0}^{\varepsilon_i=\infty} n_i \qquad [7.50]$$

If 1 mole of substance is considered, N is equal to Avogadro's constant, L, and all the other extensive quantities become molar quantities.

The internal energy of the system is given by:

$$U = \sum_{\varepsilon_i=0}^{\varepsilon_i=\infty} n_i \varepsilon_i \qquad [7.51]$$

Both ε_i and U have a kinetic energy and an arbitrary potential energy component (p. 270) which for convenience is set at zero in Fig. 7.4. As the temperature increases, the kinetic energy tends to increase and the higher energy levels become increasingly occupied at the expense of the lower energy levels. Consequently, the overall distribution becomes broader but less tall and, of course, U increases. The fraction of the total number of molecules which have a kinetic energy greater than or equal to a given value ε per molecule or E per mole is $e^{-\varepsilon/kT}$ or $e^{-E/RT}$, where k is the gas constant per molecule, known as the Boltzmann constant. We note that

$$k = \frac{R}{L} = \frac{8.314 \text{ J K}^{-1} \text{ mol}^{-1}}{6.022 \times 10^{23} \text{ mol}^{-1}} = 1.381 \times 10^{23} \text{ J K}^{-1}$$

and $\varepsilon = E/L$.

Entropy measures the degree of disorder. Since an irreversible process is associated with an increase in disorder, a disordered state is more probable than one of complete order. The entropy, S, and the thermodynamic probability, Ω, of a system must therefore be related mathematically, as suggested by Boltzmann. Let us consider two systems, A and B, having entropies S_A and S_B and probabilities Ω_A

and Ω_B respectively. Since entropy is proportional to the amount of substance, entropies are additive and the entropy, S of the combined system is (S_A+S_B). The thermodynamic probability, Ω, of a system may be defined as the number of distinguishable arrangements of the molecules of the system among the various energy levels and positions in space available to them. Each of the Ω_A arrangements of system A can be chosen in combination with any of the Ω_B arrangements of system B. The total number of distinguishable arrangements of the combined system is $(\Omega_A \times \Omega_B)$. Probabilities are therefore multiplicative. The simplest form of equation which relates the additive nature of entropy and the multiplicative nature of probability is a logarithmic relationship as proposed by Boltzmann:

$$S = k \ln \Omega + \text{constant}, \quad [7.52]$$

where k is constant (the Boltzmann constant, see above).

A perfectly crystalline solid of a pure substance at 0 K is the most highly ordered structure imaginable, since all the molecules have the same environment and since each molecule has a virtually fixed position in space. At 0 K all the molecules in the crystal will be in the lowest or *zero point* energy level. All the ways of arranging the molecules are then indistinguishable, therefore only one arrangement can be discerned. So $\Omega = 1$, $\ln \Omega = 0$ and from equation [7.52] $S = $ constant. Since it is found experimentally that all entropy changes involving solids approach zero as the temperature approaches 0 K, all pure perfectly crystalline substances at 0 K have the same entropy. That is to say that the constant in equation [7.52] is the same for all substances and entropy is therefore seen to depend only on Ω and not on the nature of the substances concerned.

Since only changes in entropy can be measured (equation [7.43]), it is impossible to ascertain absolute values of entropy. We must therefore make an arbitrary choice of the universally constant value of entropy at 0 K. No entropy can be lower than this, since pure crystalline solids at 0 K have minimum disorder. The universal minimum value of entropy is put equal to zero. This choice is incorporated in the Third Law of Thermodynamics, which states that the entropy of a perfect crystal of a pure substance of 0 K is zero. Equation [7.52] now becomes

$$S = k \ln \Omega. \quad [7.53]$$

S cannot be negative, since its minimum value is zero (or since Ω cannot be less than unity).

Entropy may conveniently be separated into thermal entropy and configurational entropy components. The thermal entropy is greater the wider is the spread or distribution or molecules among the energy levels, i.e. the higher the temperature (Fig. 7.4). The configurational entropy is greater the larger is the number of positions in space available to the molecules, i.e. the greater the volume. The number of spatial positions of the molecules is never infinite because the translational energy is quantised as are all other forms of energy. For a gas or a liquid the number of spatial positions is usually very large, because the translational energy levels are very closely packed. In all cases, the greater the number of distinguishable arrangements, Ω, the greater the disorder of the system, measured by S. For one mole of substance at normal temperatures and volumes, Ω is astronomically large, whereas standard molar entropies, $S_m^\ominus$, are of the order 100 J K^{-1} mol^{-1}, on account of the logarithmic relationship (equation [7.53]). Molecules containing different isotopes, with the possible exception of those containing the isotopes of hydrogen, are chemically indistinguishable and are therefore usually treated as being thermodynamically indistinguishable.

Entropy of Mixing

Ideal gases have negligible intermolecular forces. When two ideal gases A and B are at the same temperature and pressure, their molecules have the same distribution of molecular energies and an indiscriminate interchange of their molecules A and B does not change the internal energy of either gaseous system. In liquid or solid ideal mixtures of substances A, B, etc., at the same temperature and applied pressure, the forces between two neighbouring like or unlike molecules is the same. This applies to ideal solutions with the same solvent but with different solutes A and B or with the same solute at different concentrations A and B. In any of these ideal systems an interchange of molecules between A and B leaves the internal energy and volume of the system strictly unchanged. In other words, on mixing two ideal gases or two substances which give ideal mixtures or ideal solutions, $\Delta U_{mix} = 0$ and $\Delta V_{mix} = 0$. According to equation [7.9], the enthalpy change on mixing, known as the enthalpy of mixing, ΔH_{mix}, is zero. Since the various molecular sites are equally available to molecules of A and B, removal of a barrier separating pure A and pure B (Fig. 7.5) will result in an interchange of molecules between the equally probable molecular positions and random mixing will occur. The configurational entropy of the system will increase by an amount known as the entropy of mixing, ΔS_{mix}. Since there is no change in U, V or H, there is no exchange of heat or work with the surroundings and so the thermal entropy of the system and the entropy of the surroundings will not change.

Let N_A molecules of A and N_B molecules of B undergo mixing when the barrier separating them is removed. The total number of molecular sites to be filled is $(N_A + N_B)$ and the *total* number of ways in which

Fig. 7.5 Schematic representation of the mixing of molecules of two ideal gases or ideal solutions A and B

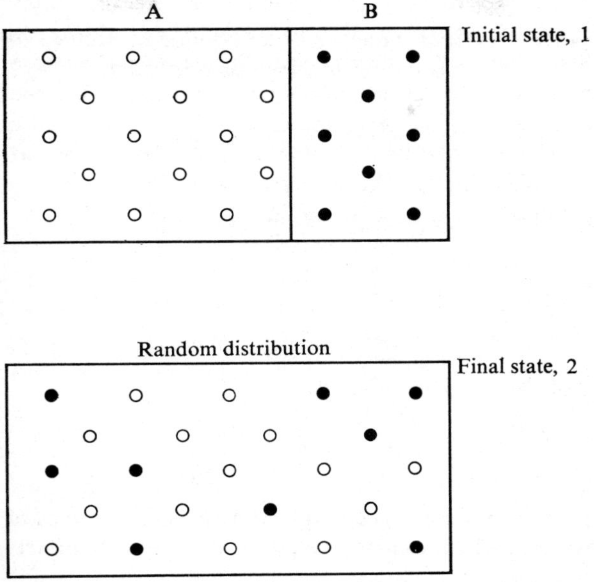

```
        A           B
   ┌─────────────┬─────────┐
   │ ○  ○  ○    │ ●  ●    │   Initial state, 1
   │   ○  ○  ○  │ ●  ●    │
   │ ○  ○  ○    │ ●  ●    │
   │   ○  ○  ○  │   ●     │
   │ ○  ○  ○    │ ●  ●    │
   └─────────────┴─────────┘
```

```
       Random distribution
   ┌─────────────────────────┐
   │ ●  ○  ○    ●  ●         │   Final state, 2
   │   ○  ○  ○    ●          │
   │ ●  ●    ○  ○    ○       │
   │   ○  ○    ●    ○        │
   │ ○  ●    ○  ○    ●       │
   └─────────────────────────┘
```

this can be done is equal to the total number of permutations of (N_A+N_B) objects taken all together, which is given by:

$$(N_A+N_B) \times (N_A+N_B-1) \times (N_A+N_B-2) \times \ldots \times 3 \times 2 \times 1$$
$$= (N_A+N_B)!$$

To obtain the number of *distinguishable* ways in which the (N_A+N_B) sites can be filled we must divide $(N_A+N_B)!$ by one factor to allow for the fact that all the molecules of A are indistinguishable from each other and by another factor to allow for the fact that all the B molecules are indistinguishable from each other. Since these factors are $N_A!$ and $N_B!$ respectively, the number of distinguishable ways in which (N_A+N_B) sites could be filled will be

$$\frac{(N_A+N_B)!}{N_A! N_B!}.$$

This is the factor by which Ω in equation [7.53] is increased on mixing. Before mixing, $S_1 = k \ln \Omega_1$; after mixing $S_2 = k \ln \Omega_2$.

$$\therefore \Delta S_{mix} = S_2 - S_1 = k \ln \Omega_2 - k \ln \Omega_1 = k \ln \frac{\Omega_2}{\Omega_1}$$

$$\therefore \Delta S_{mix} = k \ln \frac{(N_A+N_B)!}{N_A! N!_B}$$

$$= k[\ln (N_A+N_B)! - \ln N_A! - \ln N_B!].$$

Stirling's approximation states that

$$\ln y! = y \ln y - y$$

when y is a large positive integer. Since the numbers of molecules of A and B are very large, this approximation may be applied to $\ln N_A!$, $\ln N_B!$ and $\ln (N_A + N_B)!$

$$\begin{aligned}\Delta S_{mix} &= k[(N_A+N_B)\ln(N_A+N_B)-(N_A+N_B)\\ &\quad -(N_A \ln N_A - N_A) - (N_B \ln N_B - N_B)]\\ &= k[N_A \ln(N_A+N_B) + N_B \ln(N_A+N_B)\\ &\quad - N_A \ln N_A - N_B \ln N_B]\end{aligned}$$

or, $\quad \Delta S_{mix} = k\left[N_A \ln \dfrac{N_A+N_B}{N_A} + N_B \ln \dfrac{N_A+N_B}{N_B}\right].$

The mole fractions of A and B are given by

$$x_A = \frac{N_A}{N_A+N_B} \quad \text{and} \quad x_B = \frac{N_B}{N_A+N_B}.$$

The number of molecules of A, N_A, is equal to the number of moles of A, n_A, multiplied by Avogadro's constant, L, thus $N_A = n_A L$. Similarly for B, $N_B = n_B L$.

$$\therefore \quad \Delta S_{mix} = kL\left(n_A \ln \frac{1}{x_A} + n_B \ln \frac{1}{x_B}\right)$$

But $kL = R$ and $\ln \dfrac{1}{x} = -\ln x$

$$\therefore \quad \Delta S_{mix} = -R(n_A \ln x_A + n_B \ln x_B). \qquad [7.54]$$

Since x_A and x_B are both less than unity, the logarithmic terms are negative and ΔS_{mix} is positive. Thus, as expected, the entropy or disorder increases on mixing.

Entropy Changes from Calorimetry

Entropy changes and absolute entropies of substances may be determined from calorimetric measurements and their values agree well with those determined statistically. The equations to be derived in this section apply to entropy changes at constant pressure. Analogous equations at constant volume can be derived by replacing q_p by q_V, C_p by C_v and ΔH by ΔU.

We consider a substance absorbing an infinitesimal amount of heat, dq_{rev}, in a reversible process, thereby increasing its entropy by an infinitesimal amount, dS. According to equation [7.44] $dq_{rev} = T\,dS$. As a result of the absorption of heat, the temperature, T, of the substance increases by an infinitesimal amount, dT, determined by the heat capacity. According to equation [7.15]:

$$dq_p = C_p.dT \quad \text{at constant pressure.}$$

Equating the heat changes, we have:

$$T\,dS = C_p.dT$$

$$\therefore\ dS = C_p.\frac{1}{T}.dT \quad \text{at constant pressure.}$$

The overall increase in entropy, ΔS, which results from an increase in temperature from T_1 to T_2, is given by integrating between these temperature limits thus:

$$\Delta S = S_2 - S_1 = \int_{S_1}^{S_2} dS = \int_{T_1}^{T_2} C_p.\frac{1}{T}.dT \qquad [7.55]$$

$$\Delta S = \int_{T_1}^{T_2} C_p.d(\ln T) = 2.303 \int_{T_1}^{T_2} C_p.d(\log T). \qquad [7.56]$$

C_p/T is plotted against T (equation [7.55]) or C_p is plotted against $\log T$ (equation [7.56]) and ΔS at constant pressure is calculated from the area under the curve between the temperature limits.

A change of phase from α to β is usually associated with changes in thermodynamic functions (e.g. S, H, V, G, etc.). The change in a given thermodynamic function, say X, may be conveniently represented by the symbol ΔX_α^β, which is equal to the value of X for phase β (X_β) minus the value of X for phase α (X_α), thus $\Delta X_\alpha^\beta = X_\beta - X_\alpha$. Returning now to the previous paragraph, at every change of phase (from α to β) between T_1 and T_2 an entropy change ($\Delta S_\alpha^\beta = S_\beta - S_\alpha$) in addition to that calculated from equation [7.55] or [7.56] must be taken into account and added in order to obtain the total entropy change between these temperatures. This is because the disorder of the molecules, i.e. the entropy of the substance, increases as the substance changes at constant temperature, T, from solid to liquid, from liquid to gas (or from solid to gas directly if sublimation occurs), and during a transition from a polymorph more stable at a lower temperature to one more stable at a higher temperature. To bring about the phase change, i.e. the increase in disorder, heat must be absorbed, formerly known as 'latent heat', which at constant pressure is equal to the increase in enthalpy ($\Delta H_\alpha^\beta = H_\beta - H_\alpha$). Since changes of phase are reversible at the constant pressure and temperature, T, of the phase change, equation [7.43] may be applied, where $q_{\text{rev}} = \Delta H_\alpha^\beta$, so that:

$$\Delta S_\alpha^\beta = \frac{\Delta H_\alpha^\beta}{T}. \qquad [7.57]$$

Melting or fusion (s to 1), evaporation, vaporisation or boiling (1 to g) and sublimation (s to g) are accompanied by the absorption of heat and by an increase in disorder so that ΔH_s^l, ΔH_1^g, ΔH_s^g, ΔS_s^l, ΔS_1^g and ΔS_s^g are positive. Freezing (1 to s) and condensation (g to 1) are accompanied by the liberation of heat and by a decrease in disorder so that ΔH_1^s, ΔH_g^l, ΔS_1^s, ΔS_g^l are negative.

Fig. 7.6 Molar entropy of nitrogen, at one atmosphere pressure (101 325 Pa), as a function of temperature

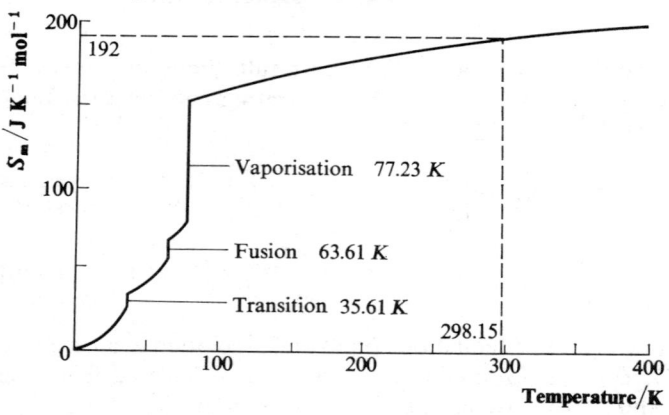

Assuming the Third Law, i.e. that the entropy of the solid substance at 0 K is zero, and using equations [7.56] and [7.57] the entropy, S, or the molar entropy, S_m, can be determined at any temperature and plotted against temperature (Fig. 7.6). The standard molar entropy, $S_m^\ominus$, of a substance is the absolute entropy of 1 mole of the substance in its standard state (1 atm or 101 325 Pa pressure p. 277), e.g. $S_m^\ominus$ (N_2, 298.15 K) = 192 J K^{-1} mol^{-1}.

Entropy and Enthalpy as Criteria for Spontaneity

When discussing the Second Law we saw that one criterion for spontaneity of a process or reaction is that the total entropy of the system and its surroundings must increase. Criteria based upon observations of the system only are also required.

For an infinitesimal process in a system the First Law (equation [7.5]) may be written

$$dU = dw + dq. \qquad [7.58]$$

If the only external work done is due to a change in volume, $dw = -p.dV$ (equation [7.2]). According to equations [7.42] and [7.44] $dq \leq T.dS$. We note that $dq = T.dS$, if the process is reversible, i.e. if the system is in internal equilibrium, and $dq < T.dS$, if the process is irreversible, i.e. if the system is undergoing a spontaneous change.

Equation [7.58] may now be written

$$dU \leqslant -p\,dV + T\,dS. \qquad [7.59]$$

Differentiating $H = U + pV$ (equation [7.8]) gives

$$dH = dU + p\,dV + V\,dp.$$

Eliminating dU we have

$$dH \leqslant V\,dp + T\,dS. \qquad [7.60]$$

For a system at constant pressure (e.g. atmospheric) and at constant enthalpy (the energy term used at constant pressure), $dp = 0$ and $dH = 0$, whence $0 \leqslant T\,dS$ ∴ dS (or ΔS) $\geqslant 0$, i.e. dS or ΔS is zero for a reversible process or at equilibrium but is positive for an irreversible or spontaneous change. In other words, the entropy (or disorder) of a system at constant p and constant H is unchanged for a reversible process, i.e. in an equilibrium state, but increases for an irreversible or spontaneous change until it reaches a maximum at equilibrium and thereafter does not change. Mixing of ideal gases or of the components of an ideal mixture takes place under these conditions as we have seen. These criteria in fact apply to any system of molecules at constant pressure and enthalpy, e.g. conduction of heat along a bar. The criteria for equilibrium and spontaneity are here the same as those in the case of a system and its surroundings taken together, and indeed for the whole universe.

For a system at constant pressure (e.g. atmospheric) and constant entropy, $dp = 0$ and $dS = 0$, and equation [7.60] gives dH (or ΔH) $\leqslant 0$, i.e. dH or ΔH is zero for a reversible process or at equilibrium but is negative for an irreversible or spontaneous change. In other words, the enthalpy of a system at constant p and constant S is unchanged for a reversible process, i.e. in an equilibrium state, but decreases for an irreversible or spontaneous change until it reaches a minimum at equilibrium and thereafter does not change. Mechanical systems of weights, pulleys, wheels and levers usually operate under these conditions, the major decrease in H being in potential energy and in heat losses due to friction.

Gibbs Free Energy

It has been seen that, for a mechanical system, the approach to equilibrium is characterised by a decrease in enthalpy (or energy). Also, for a system of particles such as molecules maintained at constant enthalpy (or energy), the approach to equilibrium is characterised by an increase in entropy. When chemical and physical changes take place at constant pressure in the laboratory or in a living system it is usual for both the enthalpy and the entropy of the system to change simultaneously. The approach to equilibrium in such cases must be characterised by some other property which takes account of both these changes. This may be seen by considering again the simple case of equilibrium between a solid and a liquid at the melting-point. If the system is taken to be the solid and the liquid and the changes in the surroundings are ignored the system is no longer one of constant energy. If the liquid freezes, energy is released (the latent heat of fusion), so the system is falling to a condition of lower energy (lower enthalpy). At the same time, however, the disordered arrangement of molecules in the liquid is becoming an ordered one in the crystal. The distribution of molecular energies which is rather random in the liquid is also becoming more ordered in the crystal lattice so the entropy must be decreasing. The direction in which the system changes to reach its equilibrium state depends on temperature, because above the melting-point solid changes to liquid (energy and entropy increase) whereas below the melting-point liquid changes to solid (energy and entropy decrease). If either the energy or the entropy changes were operating separately as a criterion for an approach to equilibrium the change could take place in only one direction. The energy criterion would suggest that all liquids should solidify and the entropy criterion would suggest that all solids should liquefy. In practice, it is as though these two criteria were operating against each other, and the factor determining which has the greater effect must be the temperature. Any function which is to be a true criterion of the approach to equilibrium under normal conditions must express in some way this opposition of the energy and entropy terms and must also take account of the variation of their relative magnitudes with temperature. The function expressing these facts which is most appropriate for the normal laboratory and biological conditions of constant pressure is the Gibbs free energy (sometimes called the Gibbs energy or the free energy). It is given the symbol G and is defined by:

$$G = H - TS \qquad [7.61]$$

where H is the enthalpy
T is the absolute temperature
and S is the entropy.

By analogy with the reasoning used in the derivation of equation [7.9] from [7.8], we can show that, for a change in G, H and S at constant temperature:

$$\Delta G = \Delta H - T \Delta S. \qquad [7.62]$$

From the form of this equation it can be seen qualitatively that ΔG fulfils the requirements mentioned above. The energy term ΔH and the entropy term $T \Delta S$ are opposed (i.e. related by a minus sign) and it is the temperature which determines which term is the larger. It will now be shown formally that ΔG fulfils the requirements because the decrease in free energy is found to represent that part of the total energy change of the system which can actually be made available for doing work or for being converted into other forms of energy.

A process or reaction may often be carried out at constant pressure, p, in such a way that only some of the work done is mechanical due to a change in volume of the system, $p\,\Delta V$ according to equation [7.1]. The rest of the work then appears in another form, w', called *net work* or *useful work* (e.g. electrical work). The total work done by the system is given by:

$$-w = p\,\Delta V - w'. \qquad [7.63]$$

Substituting for w in the First law equation [7.5], we have:

$$\Delta U = q + w' - p\,\Delta V$$
$$\therefore \ \Delta U + p\,\Delta V = q + w'.$$

Comparison with equation [7.9] shows that at constant pressure:

$$\Delta H = q + w'. \qquad [7.64]$$

For a reversible process at constant temperature and pressure, $q = q_{rev} = T\,\Delta S$ (equation 7.43) and w' has a minimum value (see p. 286) symbolised by w'_{rev}, so that

$$\Delta H = T\,\Delta S + w'_{rev}$$

whence $\qquad w'_{rev} = \Delta H - T\,\Delta S.$

Comparison with equation [7.62] indicates that:

$$w'_{rev} = \Delta G \quad \text{or} \quad -w'_{rev} = -\Delta G. \qquad [7.65]$$

$(-\Delta G)$ is therefore equal to the net work done by the system in a reversible process at constant temperature and pressure, or in other words $(-\Delta G)$ is the maximum work which can be done by the system other than that associated with a change in volume of the system at constant pressure.

Since H and S are state functions, G is a state function by definition (equation [7.61]). ΔG therefore depends only on the initial and final states of the system and is independent of the path between them. Hence, for a given system with given initial and final states, ΔG will always be the same, although the work done by the system may have any value from zero to a maximum. $(-\Delta G)$ is sometimes known as the 'available energy' because it represents the energy which is made available by the system for any purpose; in a reversible process it is the net work done against the constraint which causes the process to go in reverse and has a maximum value; when measuring enthalpy of reaction it all appears as a contribution to the heat exchanged;

in many irreversible processes, part is used as work and part appears as heat.

The meaning of a change in Gibbs energy cannot be appreciated fully until it is realised how the amount of available energy differs from the total change in energy. If we write equation [7.62] in the form:

$$-\Delta G = -\Delta H + T\Delta S,$$

it can be seen that the amount of energy made available during a change ($-\Delta G$) will be less than the total decrease in energy of the system ($-\Delta H$) if the system is forced to undergo a decrease in entropy (i.e. if $T\Delta S$ is negative). This will be the case if the system is changing in such a way as to satisfy the spontaneous tendency for energy to decrease at the expense of a decrease in entropy, as in the example of a liquid solidifying. The quantity $-T\Delta S$ is sometimes spoken of as the 'unavailable energy' and it represents the amount of energy that has to be expended within the system as it becomes more ordered. If at a higher temperature the term $T\Delta S$ becomes larger than $-\Delta H$, the spontaneous tendency to increase entropy can drive the system to a state of higher total energy but again the overall effect is to decrease the Gibbs energy. This is found in the example of a solid liquefying. A system which is not being driven by some external supply of energy always undergoes changes in such a way as to cause a decrease in Gibbs energy, i.e. so as to cause the available energy to be liberated. The changes continue until no more energy can be made available by the system, i.e. until $\Delta G = 0$, at which point equilibrium is reached and the enthalpy and entropy terms balance exactly. This will now be proved formally and some important equations will be derived.

Since $G = H - TS$ (equation [7.61]) and $H = U + pV$ (equation [7.8]) we have

$$G = U + pV - TS. \qquad [7.66]$$

This expression may be differentiated as follows:

$$dG = dU + p\,dV + V\,dp - T\,dS - S\,dT.$$

Since $dU \leqslant -p\,dV + T\,dS$ (equation [7.59]) we have:

$$dG \leqslant V\,dp - S\,dT. \qquad [7.67]$$

For a system at constant pressure (e.g. atmospheric) and at constant temperature, $dp = 0$ and $dT = 0$, therefore dG (or ΔG) $\leqslant 0$, i.e. dG or ΔG is zero for a reversible process or at equilibrium but is negative for an irreversible or spontaneous change. In other words, the Gibbs energy of a system is unchanged in a reversible process or in an equilibrium state, but decreases for an irreversible or spontaneous change until it reaches a minimum at equilibrium and thereafter does not change. Biochemical, biophysical and many chemical and physical changes often operate under these conditions. If ΔG is positive, the process or reaction to which it refers proceeds spontaneously in the reverse direction, i.e. from state 2 to state 1.

In a process of isothermal mixing the Gibbs energy of mixing, ΔG_{mix}, is given by equation [7.62] in the form

$$\Delta G_{mix} = \Delta H_{mix} - T \Delta S_{mix},$$

where ΔH_{mix} is the enthalpy of mixing and ΔS_{mix} is the entropy of mixing. For ideal gases and ideal mixtures or solutions $\Delta H_{mix} = 0$ and $\Delta S_{mix} = -R (n_A \ln x_A + n_B \ln x_B)$ according to equation [7.54],

$$\therefore \quad \Delta G_{mix} = RT (n_A \ln x_A + n_B \ln x_B). \qquad [7.68]$$

Since the mole fractions, x_A and x_B, of the two components A and B are each less than unity, ΔG_{mix} is negative, as expected from the fact that mixing is a spontaneous process.

Most mixtures exhibit positive deviations from Raoult's law, in which case mixing is endothermic, i.e. ΔH_{mix} is positive. In accordance with this, the differential enthalpy of solution, ΔH_s, (p. 202) is positive for solutions of many organic compounds in organic solvents and many salts in water. Mixing is spontaneous only when ΔG_{mix} is negative, i.e. when $T \Delta S_{mix} > \Delta H_{mix}$; for this to happen the temperature must be sufficiently high. At lower temperatures $T \Delta S_{mix} < \Delta H_{mix}$ and ΔG_{mix} is positive, in which case the spontaneous process is the reverse of mixing, and the mixture separates into two phases, which may be a pair of conjugate saturated solutions or a crystalline solid and its saturated solution. As the mixture separates or as precipitation or crystallisation proceeds, G falls and reaches a minimum at equilibrium in which state no further change occurs, so that $\Delta G_{mix} = 0$ and $T \Delta S_{mix} = \Delta H_{mix}$ at equilibrium.

A minority of mixtures show negative deviations from Raoult's law, in which case mixing is exothermic. The negative sign of ΔH_{mix} and the positive sign of ΔS_{mix} ensure that ΔG_{mix} is negative at all temperatures. For such solutions mixing is therefore always spontaneous.

If the system is in internal equilibrium, i.e. at a uniform temperature and pressure, an infinitesimal change will obey equation [7.67] in the form:

$$dG = V\,dp - S\,dT. \qquad [7.67a]$$

If the system undergoes an infinitesimal change in temperature at constant pressure, $dp = 0$, so $dG = -S\,dT$, i.e.

$$\left(\frac{\partial G}{\partial T}\right)_p = -S. \qquad [7.69]$$

In the initial state, 1, $\left(\dfrac{\partial G_1}{\partial T}\right)_p = -S_1.$

In the final state, $2, \left(\dfrac{\partial G_2}{\partial T}\right)_p = -S_2$.

Subtracting, $\left(\dfrac{\partial G_2}{\partial T} - \dfrac{\partial G_1}{\partial T}\right)_p = \left(\dfrac{\partial (G_2 - G_1)}{\partial T}\right)_p = -(S_2 - S_1)$

whence $\left(\dfrac{\partial (\Delta G)}{\partial T}\right)_p = -\Delta S,$ [7.70]

for any change, process or reaction at constant pressure. This equation enables ΔS to be calculated from a knowledge of ΔG at various temperatures.

For any change or process in a system at constant temperature $\Delta G = \Delta H - T\,\Delta S$ according to equation [7.62]. Introducing the above function for $-\Delta S$, we obtain

$$\Delta G = \Delta H + T\left(\dfrac{\partial (\Delta G)}{\partial T}\right)_p. \qquad [7.71]$$

This is one form of the Gibbs–Helmholtz equation and is valid for any change or process in a system at constant temperature and pressure. It is used to determine ΔH from measurements on electrochemical cells or to calculate the temperature coefficient of ΔG for a process or reaction at a given temperature and pressure.

If the system undergoes an infinitesimal change in pressure at constant temperature, $\mathrm{d}T = 0$ and equation [7.67a] shows that $\mathrm{d}G = V\,\mathrm{d}p$, i.e.

$$\left(\dfrac{\partial G}{\partial p}\right)_T = V \qquad [7.72]$$

for the system. Using the method for deriving equation [7.70] from [7.69] we obtain

$$\left(\dfrac{\partial (\Delta G)}{\partial p}\right)_T = \Delta V \qquad [7.73]$$

for any change, process or reaction at constant temperature.

Effects of External Conditions on Phase Changes

It is possible for two phases, α and β, of a pure substance to be in equilibrium with each other at a definite temperature, T, and at a definite pressure, p, e.g. liquid water and steam at $100°\,\mathrm{C}$ (373 K) and at 1 atm (101 325 Pa), see pp. 39 and 40. The phases may be gaseous, liquid, solid or two solid polymorphs. By slowly supplying or removing heat (formerly known as latent heat) one phase may be changed reversibly into the other and the system will always be in internal equilibrium. Since the only work done is that due to a possible

Effects of External Conditions on Phase Changes 303

change in volume, w' and therefore ΔG are zero, no matter how much of the substance undergoes the phase change. The free energy lost by one phase is therefore equal to that gained by the other phase. Furthermore, the amount of substance lost by one phase is equal to that gained by the other. Consequently, equal amounts of a given substance have the same Gibbs energy in phases which are in equilibrium at constant temperature and pressure, i.e.

$$G_\beta = G_\alpha.$$

If one mole of substance is considered, G_α and G_β are the molar Gibbs energies of the substance in the respective phases, α and β, which are in equilibrium at constant temperature and pressure. If the temperature is changed by an infinitesimal amount, dT, equation [7.67a] shows that the pressure undergoes a corresponding change, dp, in order for equilibrium to be maintained. The corresponding change in the molar Gibbs energy of phase α is given by

$$dG_\alpha = V_\alpha\, dp - S_\alpha\, dT,$$

whereas that of phase β is given by

$$dG_\beta = V_\beta\, dp - S_\beta\, dT,$$

where V_α and V_β are the molar volumes and S_α and S_β are the molar entropies of the respective phases. Since the two phases are still in equilibrium after the small change in temperature and pressure, their molar Gibbs energies are still equal, i.e.

$$G_\beta + dG_\beta = G_\alpha + dG_\alpha.$$

But since $\quad G_\beta = G_\alpha, \quad dG_\beta = dG_\alpha.$

$$\therefore\ V_\beta\, dp - S_\beta\, dT = V_\alpha\, dp - S_\alpha\, dT$$

$$\therefore\ (V_\beta - V_\alpha)\, dp = (S_\beta - S_\alpha)\, dT$$

$$\therefore\ \frac{dp}{dT} = \frac{S_\beta - S_\alpha}{V_\beta - V_\alpha} = \frac{\Delta S_\alpha^\beta}{\Delta V_\alpha^\beta}. \qquad [7.74]$$

ΔS_α^β and ΔV_α^β are the changes in entropy and volume respectively when a given amount (usually one mole) of the substance is converted from phase α to phase β (see p. 295). According to equation [7.57]

$$\Delta S_\alpha^\beta = \frac{\Delta H_\alpha^\beta}{T}$$

where ΔH_α^β is the molar enthalpy for the phase change (e.g. molar latent heat of evaporation, sublimation, melting or transition at constant pressure). Substitution for ΔS_α^β in equation [7.74] gives

$$\frac{dp}{dT} = \frac{\Delta H_\alpha^\beta}{T\,\Delta V_\alpha^\beta} \quad \text{or} \quad \frac{dT}{dp} = \frac{T\,\Delta V_\alpha^\beta}{\Delta H_\alpha^\beta}, \qquad [7.75]$$

where dT/dp is the rate of change of the temperature of the phase change, e.g. boiling, sublimation, melting or transition point, with respect to the applied pressure. This equation is applicable to any phase change of a pure substance and is known as the Clapeyron–Clausius equation in its exact form. Since $\Delta H_\alpha^\beta / \Delta V_\alpha^\beta$ or its reciprocal is a ratio of two extensive quantities, it is independent of the amount of substance undergoing the change of phase. This amount may be 1 mole as stated above or may be 1 kg, in which case:

$$\frac{dp}{dT} = \frac{\Delta h_\alpha^\beta}{T \Delta v_\alpha^\beta} \quad \text{or} \quad \frac{dT}{dp} = \frac{T \Delta v_\alpha^\beta}{\Delta h_\alpha^\beta}, \qquad [7.76]$$

where Δv_α^β and Δh_α^β are the changes in specific volume and in specific enthalpy respectively associated with the phase change.

For the evaporation of a liquid or the sublimation of a solid, α represents the liquid or solid phase and β represents the gas phase. The molar or specific volume of the liquid or solid phase is negligible compared with that of the gas phase, except in the case of a liquid near the critical point, i.e. $V_\alpha \ll V_\beta$ or $\Delta V_\alpha^\beta \approx V_\beta$. If the vapour produced is assumed to behave ideally

$$\frac{RT}{p} = V_\beta \approx \Delta V_\alpha^\beta.$$

Substitution for ΔV_α^β in equation [7.75] gives

$$\frac{dp}{dT} = \frac{p \, \Delta H_\alpha^\beta}{RT^2}$$

$$\therefore \frac{dp}{dT} \cdot \frac{1}{p} = \frac{\Delta H_\alpha^\beta}{RT^2}. \qquad [7.77]$$

Since
$$\frac{d \ln p}{dT} = \frac{1}{p} \cdot \frac{dp}{dT}$$

$$\frac{d \ln p}{dT} = \frac{\Delta H_\alpha^\beta}{RT^2}, \qquad [7.78]$$

where p is the saturated vapour pressure of the liquid or solid at the temperature T K, or T is absolute boiling or sublimation point of the liquid or solid at the applied pressure p, and ΔH_α^β is the molar enthalpy of evaporation or sublimation. This equation is sometimes known as the Clausius equation; it is an approximate form of the Clapeyron–Clausius equation [7.75] and is applicable only to evaporation, sublimation or to the reverse condensation processes. It enables enthalpies of evaporation or sublimation to be calculated directly from the variation of p with T. For this purpose equation [7.78] is integrated with respect to T, assuming that ΔH_α^β does not vary significantly over the temperaure range studied, thus:

$$\log p = -\frac{\Delta H_\alpha^\beta}{2.303R} \cdot \frac{1}{T} + \text{constant}. \qquad [7.79]$$

The derivation of this equation is analogous to that of the integrated form of the van't Hoff isochore (equation [7.113] p. 320). By plotting $\log p$ against $1/T$, ΔH_α^β may be calculated from the gradient or p may be obtained at any temperature within or slightly beyond the range studied. The following alternative equation:

$$\log p_2 - \log p_1 = \log \frac{p_2}{p_1} = \frac{\Delta H_\alpha^\beta}{2.303R}\left(\frac{T_2-T_1}{T_1 T_2}\right) \qquad [7.80]$$

(cf. equations [5.46], [5.47] and [7.114]) enables ΔH_α^β to be calculated from measurements of p at two temperatures or p to be calculated at any temperature if ΔH_α^β and p at another temperature are known. The Clausius equation accords with Le Chatelier's principle (p. 316). Since ΔH_α^β are positive for evaporation or sublimation, the saturated vapour pressures of liquids and solids increase with increasing temperature. That is to say, conversion to the gas phase is favoured by increasing temperature, as observed.

Partial Molar Quantities of Substances in Mixtures

In a gaseous mixture, the total pressure, P, is the sum of the partial pressures, p_A, p_B, etc., of its constituent substances A, B, etc., (p. 24) thus:

$$P = p_A + p_B + \text{etc.}$$

Pressure is an intensive quantity and is therefore independent of the amount of each substance. For any mixture the total value of an extensive quantity, on the other hand, does depend on the amount of each substance in the mixture, e.g. n_A mol of A, n_B mol of B, etc. For mixtures we therefore introduce partial molar quantities for the extensive state functions, e.g. V, H, S and G. The partial molar quantity, X_A, (sometimes denoted $\bar{X}_A$) for substance A in a given mixture may be thought of as the value of the extensive state function X associated with each mol of A *in the mixture*. Similarly X_B is the partial molar quantity for substance B in the mixture. Thus the total value of the extensive state function, X_2, for the mixture at constant temperature and pressure is given by

$$X_2 = n_A X_A + n_B X_B + \text{etc.} \qquad [7.81]$$

The subscript, 2, indicates that the mixture is the final state obtained after mixing A, B, etc.

The strict definition of the partial molar quantity, X_A, of substance A is the rate of change of the corresponding extensive state function of the mixture with respect to the amount of substance A, n_A mol, when the pressure, temperature and the amounts of all the other substances in the mixture are kept constant. X_B is similarly defined. Thus:

$$X_A = \left(\frac{\partial X}{\partial n_A}\right)_{p,T,n_B, \text{etc.}} \qquad X_B = \left(\frac{\partial X}{\partial n_B}\right)_{p,T,n_A, \text{etc.}}, \qquad [7.82]$$

or, in words, X_A is equal to the slope at the point of interest on the curve obtained by plotting X against n_A at constant values of p, T, n_B, etc. Indeed, partial molar quantities are usually determined from such a graph. Integration of equations [7.82] followed by their addition gives the total value of X for the mixture as stated in equation [7.81].

Before mixing, the initial state, 1, refers to n_A mol of pure substance A with n_B mol of pure B, etc., for which the molar quantities are X_A^*, X_B^*, etc., respectively. The asterisk indicates that the substance concerned is in the pure state for which ordinary molar quantities (p. 272) are still used. The absence of the asterisk indicates a partial molar quantity in the mixture. The total extensive state function for the initial state is given by:

$$X_1 = n_A X_A^* + n_B X_B^* + \text{etc.} \qquad [7.83]$$

For the process of mixing at constant temperature and pressure, the change in the extensive state function is given by:

$$\Delta X_{\text{mix}} = X_2 - X_1$$
$$\therefore \Delta X_{\text{mix}} = n_A(X_A - X_A^*) + n_B(X_B - X_B^*) + \text{etc.} \qquad [7.84]$$

Chemical Potential and Activities

Partial molar Gibbs energy is defined by equation [7.82] in which X is replaced by G. The Gibbs energy of mixing is given by [7.84] in the form:

$$\therefore \Delta G_{\text{mix}} = n_A(G_A - G_A^*) + n_B(G_B - G_B^*) + \text{etc.} \qquad [7.85]$$

Partial molar Gibbs energy is so important that it is given a special name, chemical potential, and a special symbol, μ. Thus, the respective chemical potentials of A and B in the mixture are given by $\mu_A = G_A$ and $\mu_B = G_B$ and in the pure state by $\mu_A^* = G_A^*$ and $\mu_B^* = G_B^*$.

$$\therefore \Delta G_{\text{mix}} = n_A(\mu_A - \mu_A^*) + n_B(\mu_B - \mu_B^*) + \text{etc.}, \qquad [7.85a]$$

under conditions of constant temperature and pressure. For ideal mixtures under these conditions the Gibbs energy is also given by equation [7.68]:

$$\Delta G_{\text{mix}} = RTn_A \ln x_A + RTn_B \ln x_B.$$

Since each substance makes its own contribution to the Gibbs energy of the mixture, the terms referring to any given substance in the last two equations [7.85a] and [7.68] may be equated, thus:

$$n_A(\mu_A - \mu_A^*) = RT n_A \ln x_A.$$

Dividing throughout by n_A which is not equal to zero, rearranging and repeating the procedure for B, we obtain:

$$\mu_A = \mu_A^* + RT \ln x_A; \quad \mu_B = \mu^* + RT \ln x_B. \quad [7.86]$$

These and analogous equations hold for each among a number of substances in an ideal gaseous, liquid or solid mixture at equilibrium at constant temperature and pressure. Since x_A and x_B are less than unity, the logarithmic terms are negative and so $\mu_A < \mu_A^*$ and $\mu_B < \mu_B^*$.

If the mixture does not behave ideally, the mole fraction, x_B, of component B must be replaced by the activity, a_B, of B, and similarly for A. The activity, a_B, of substance B is always defined in terms of an appropriate standard state of B at which $a_B = 1$. Appropriate standard states of unit activity are defined for all the other components of the mixture. If the standard states chosen for A and B are the respective pure substances, then when $a_A = 1$, $x_A = 1$ and $\mu_A = \mu_A^*$ and when $a_B = 1$, $x_B = 1$ and $\mu_B = \mu_B^*$. In this case equations [7.86] may be written:

$$\mu_A = \mu_A^* + RT \ln a_A; \quad \mu_B = \mu_B^* + RT \ln a_B. \quad [7.86a]$$

This standard state is adopted for the components of mixture in which there is no distinction between solute and solvent, and also for the solvent of a solution (see Chapter 6). For solutes in solution an alternative standard state will be defined later in this section. The standard state for the solute B is not pure B, so that when $a_B = 1$, $x_B \neq 1$ and $\mu_B \neq \mu_B^*$ and equation [7.86a] is not valid for B. An equation valid for any standard state is obtained by replacing μ_B^* by a constant, which is written as $\mu_B^\ominus$, and μ_A^* by another constant, $\mu_A^\ominus$, etc., thus:

$$\mu_A = \mu_A^\ominus + RT \ln a_A; \quad \mu_B = \mu_B^\ominus + RT \ln a_B. \quad [7.87]$$

$\mu_A^\ominus$ and $\mu_B^\ominus$ are known as the standard chemical potentials of A and B respectively and are constants for given standard states at a given temperature and pressure. Equations [7.86a] are special cases of [7.87].

Equations [7.87] have been derived from the entropy and enthalpy of mixing. They may also be derived from equation [7.72] written in the following form (in the case of substance B) using partial molar quantities:

$$\left(\frac{\partial \mu_B}{\partial p}\right)_T = \left(\frac{\partial G_B}{\partial p}\right)_T = V_B. \quad [7.88]$$

308 Energy and Equilibrium

Each component, e.g. B, present in the mixture or solution in question, which may or may not be ideal, is assumed to be in equilibrium with a vapour phase. The vapour of B is for simplicity assumed to behave ideally, so the partial vapour pressure, p_B, replaces p in equation [7.88] and the partial molar volume, V_B, is equal to the molar volume which, according to the gas law, is given by RT/p_B.

$$\therefore \left(\frac{\partial \mu_B}{\partial p_B}\right)_\tau = \frac{RT}{p_B}.$$

Integrating, we have, at constant temperature:

$$\mu_B = \mu'_B + RT \ln p_B,$$

where μ'_B is a constant of integration. Since the vapour of B is in equilibrium with the mixture or solution, $p_B = p_B^\ominus a_B$ (equation [6.11]), whence:

$$\mu_B = \mu'_B + RT \ln p_B^\ominus + RT \ln a_B,$$

where a_B is the activity of B in the mixture or solution. Since μ'_B and $RT \ln p_B^\ominus$ are constants for a given substance B at constant temperature and pressure, they may be combined to give a new constant $\mu_B^\ominus$. This constant is equal to the chemical potential of B in the standard state of unit activity as before. The resulting equation is therefore identical with equation [7.87] for B. The equation for A can be derived in a similar way.

Differentiation of equations [7.87], at constant temperature and pressure, gives:

$$d\mu_A = RT\, d \ln a_A \quad \text{and} \quad d\mu_B = RT\, d \ln a_B. \qquad [7.89]$$

Since $\mu_A^\ominus$ and $\mu_B^\ominus$ are constants under these conditions, they disappear on differentiation. This fact and the logarithmic function mean that changes in chemical potential are independent of the standard state chosen. Moreover, any quantity proportional to activity may be substituted for activity in equation [7.89].

A quantity known as 'absolute activity', which is given the symbol λ, is often found in modern texts. The absolute activity λ_B of any substance B is defined by:

$$\mu_B = RT \ln \lambda_B \quad \text{or} \quad \lambda_B = e^{\mu_B/RT}. \qquad [7.90]$$

Since chemical potentials (except $\mu^\ominus$ values) are independent of a standard state, so are absolute activities. When using absolute activities it is good practice to give the name 'relative activities' to the common 'activities' (symbol a) which do depend on a standard state. Application of equation [7.90] to the standard state of B gives:

$$\mu_B^\ominus = RT \ln \lambda_B^\ominus,$$

where $\lambda_B^\ominus$ is the value of λ_B when B is in its standard state and is known as the 'standard absolute activity' of B. Like $\mu_B^\ominus$, $\lambda_B^\ominus$ is a constant at constant temperature and pressure. Subtracting the last equation from equation [7.90] we have:

$$\mu_B - \mu_B^\ominus = RT \ln \lambda_B - RT \ln \lambda_B^\ominus$$

$$\therefore \mu_B = \mu_B^\ominus + RT \ln \left(\frac{\lambda_B}{\lambda_B^\ominus}\right).$$

Comparison of this equation with equation [7.87] shows that for any substance B:

$$a_B = \lambda_B / \lambda_B^\ominus. \qquad [7.91]$$

Thus at constant temperature and pressure the relative activity of a substance in a series of mixtures or solutions is proportional to its absolute activity. Throughout this book relative activity is given the common name 'activity' (symbol a).

Fig. 7.7 A system of phases, α, β, γ, and δ, in equilibrium with each other

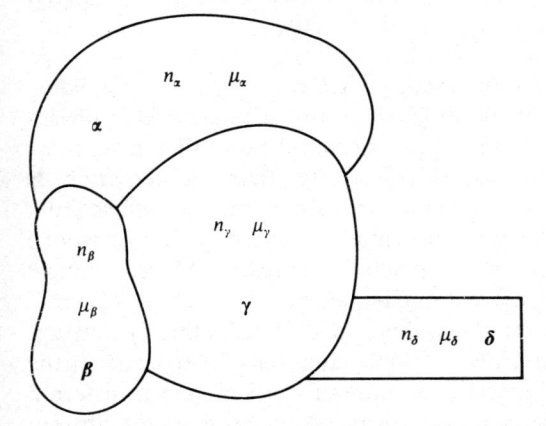

Chemical potential is a valuable concept when considering the transfer of matter from one phase to another at constant temperature and pressure. Figure 7.7 shows a group of phases which may be gaseous, liquid or solid phases, various liquid layers or particles, various organs in a whole organism, various compartments or organelles in a living cell, etc. We shall consider the consequences which result when a given component B can permeate through the phase boundaries and distribute itself between the phases. A system or phase which can gain or lose matter is known as an open system. Each phase in Fig. 7.7 is an open system. Let there be n_α moles of B at chemical potential μ_α in phase α, n_β moles of B at chemical potential μ_β in phase β, etc. If the total Gibbs energy of phase α is G_α, and that of phase β is

G_β, etc., then according to equation [7.81] we have:

$$G_\alpha = n_\alpha \mu_\alpha + \text{constant for all other substances in phase } \alpha$$
$$G_\beta = n_\beta \mu_\beta + \text{constant for all other substances in phase } \beta.$$

Let an infinitesimal amount of B, dn moles, be transferred from phase α to phase β. Since the amount transferred is small, the chemical potentials will not change appreciably, so the above expressions for G_α and G_β can be differentiated at constant temperature, pressure and amounts of the other components, taking μ_α and μ_β to be constant, thus:

$$dG_\alpha = \mu_\alpha \, dn_\alpha \quad \text{and} \quad dG_\beta = \mu_\beta \, dn_\beta.$$

These equations are special cases of equation [7.82]. Since phase α loses dn mol of B, $dn_\alpha = -dn$. Since phase β gains dn mol of B, $dn_\beta = +dn$. The total change in Gibbs energy, dG, which results from the transfer of matter is given by

$$dG = dG_\alpha + dG_\beta$$
$$\therefore \quad dG = -\mu_\alpha \, dn + \mu_\beta \, dn. \qquad [7.92]$$

The transfer of B will be spontaneous when dG is negative, i.e. when $\mu_\alpha > \mu_\beta$. Application of equation [7.87] to the substance B in phases α and β, assuming that $\mu^\ominus$ and T are constant, shows that, if $\mu_\alpha > \alpha_\beta$, the activity of B in phase α, a_α, is also greater than the activity of B in phase β, a_β, i.e. $a_\alpha > a_\beta$. A substance therefore tends to diffuse from a region in which its chemical potential or activity is higher to one in which its chemical potential or activity is lower. More practical aspects of this law are discussed in Chapter 6.

If the phases α and β are in equilibrium with each other at constant temperature and pressure, $dG = 0$, and equation [7.92] leads to the result, $\mu_\alpha = \mu_\beta$. If an infinitesimal amount of B is similarly transferred between any of the phases which are in contact and in equilibrium with each other, it can similarly be proved that

$$\mu_\alpha = \mu_\beta = \mu_\gamma = \mu_\delta.$$

Application of equation [7.87] to the substance B in each phase, assuming that $\mu^\ominus$ and T are constant, shows that, if the chemical potential of B in each phase is the same, the activity of B in each phase must be the same, i.e.

$$a_\alpha = a_\beta = a_\gamma = a_\delta.$$

It should be stressed that this is only true if the same standard state is used for the substance in question in each phase.

Activities and Activity Coefficients

In order to describe correctly the properties and behaviour of a non-ideal mixture or solution the 'strength', i.e. the mole fraction, molality, concentration or partial pressure, of each substance must be replaced by the activity of the substance in the appropriate formulae. The correction factor which is used to correct the 'strength' so as to give the activity is known as the activity coefficient, i.e.

$$\text{activity} = \text{activity coefficient} \times \text{'strength'}.$$

The quantities and units by which the 'strength' of a given mixture or solution is specified, e.g. mole fraction, molality, concentration or partial pressure, will affect the activity coefficient but not the activity. The activity of a substance in mixtures or solutions depends on the amount of the substance present and on the standard state chosen to correspond to unit activity.

For solvents of solutions and for components of liquid, solid or gaseous mixtures in which no distinction is made between solvent and solute, the standard state is taken to be the pure substance for which the mole fraction is also unity (see p. 223). In this case:

$$\text{activity} = \text{activity coefficient} \times \text{mole fraction}$$

which for substance B may be stated symbolically

$$a_B = f_B \times x_B. \qquad [7.93]$$

As the solution becomes richer and richer in component, B, which may be the solvent such as water, $x_B \to 1.0$ (as stated), $a_B \to 1.0$ (by definition), therefore $f_B \to 1.0$. For ideal solutions at all concentrations, $a_B = x_B$, therefore, $f_B = 1.0$. For solutions showing positive deviations from Raoult's law, $a_B > x_B$, therefore $f_B > 1$. For solutions exhibiting negative deviations, $a_B < x_B$, therefore $f_B < 1$, provided that $x_B \neq 1$ (see p. 215).

In the case of solutes in solution, such as drugs and electrolytes dissolved in water, discussed in Chapters 8, 9 and 10, the activity coefficient is taken to be unity for an infinitely dilute solution. Then:

activity = 'molal' activity coefficient × molality (mol kg^{-1})

$$a_B = \gamma_B \times m_B \qquad [7.94]$$

$$= \text{'molar' activity coefficient} \times \text{concentration (mol m}^{-3}$$
$$\text{or mol dm}^{-3}, \text{ etc.)}$$

$$= y_B \times c_B. \qquad [7.95]$$

For a given solution, the activity, a_B, is the same using either equation, but the activity coefficients are generally different (except for dilute aqueous solutions when concentration expressed in mol dm^{-3} approximates to mol kg^{-1}). The standard state of the solute in a solution is such that it has unit activity (e.g. for aqueous solutions of HCl this is approximately 1.2 mol kg^{-1}). Because the deviation from ideal behaviour varies with m (or c), γ (or y) vary similarly. Since very dilute solutions behave almost ideally, a approaches m (or c) and therefore γ (or y) approaches unity for such solutions. As the concentration increases from zero γ deviates more and more from unity. For solutions of electrolytes, γ first decreases but then increases again at fairly high concentrations. For non-electrolytes, γ may either increase or decrease according to whether the vapour pressure of the solution is greater or less than the ideal predicted by Raoult's law. The type of variation found in practice for an electrolyte and for a non-electrolyte is exemplified by the data in Table 7.2.

Table 7.2. Variation of Activity Coefficient with Molality in Water

m/mol kg^{-1}	γ for NaCl	γ for glycerol
0	1.000	1.000
0.005	0.930	—
0.1	0.778	1.006
0.2	0.732	1.012
0.5	0.679	1.032
1.0	0.656	1.068
2.0	0.670	1.132
3.0	0.719	—
5.0	—	1.348

Lest it should be thought that activity coefficients are merely empirical correction factors, it should be mentioned that the same activity coefficient is used to correct for non-ideality of a given mixture or solution in any type of experiment, and that activity coefficients can also be calculated theoretically in many cases from a knowledge of the forces operating between the particles. Activities should strictly be used in place of concentrations in all discussion of colligative properties and in all electrochemical formulae, as well as in connection with equilibria. These fields of chemistry provide many methods of obtaining experimental values of activity coefficients and the fact that the different methods yield values in close agreement with each other provides further confirmation of the validity of the concept of activity. On the other hand, the use of activities can often be avoided by the method mentioned on p. 410 of extrapolating values of a quantity calculated in terms of concentrations to the point of zero concentration.

Gibbs Free Energy Changes in Chemical Reactions

For the very general chemical reaction considered on p. 275, i.e.

$$aA + bB + \ldots = qQ + rR + \ldots, \qquad [7.22]$$

the total Gibbs energy of the initial state, i.e. of the reactants, is given by equation [7.81] in the form:

$$G \text{ (reactants)} = aG_A + bG_B + \ldots \qquad [7.96]$$

and that of the final state, i.e. of the products, is similarly given by:

$$G \text{ (products)} = qG_Q + rG_R + \ldots \qquad [7.97]$$

An equation of the following type applies to each reactant and product:

$$G_Q \equiv \mu_Q = \mu_Q^\ominus + RT \ln a_Q \qquad [7.87]$$

$$\therefore \; qG_Q \equiv q\mu_Q = q\mu_Q^\ominus + qRT \ln a_Q = q\mu_Q^\ominus + RT \ln a_Q^q \qquad [7.98]$$

The Gibbs energy change for the reaction is given by:

$$\Delta G = G \text{ (products)} - G \text{ (reactants)} \qquad [7.99]$$

$$\therefore \; \Delta G = q\mu_Q + r\mu_R + \ldots - a\mu_A - b\mu_B - \ldots \qquad [7.100]$$

In order for the reaction to proceed in a forward direction ΔG must be negative. Then $(-\Delta G)$ measures the 'affinity' of the reaction. When the reactants and products are in their standard states of unit activity, ΔG becomes $\Delta G^\ominus$, the standard Gibbs energy change, and the chemical potentials, μ_Q etc., become the standard chemical potentials $\mu_Q^\ominus$, etc. By analogy with equation [7.100]

$$\Delta G^\ominus = q\mu_Q^\ominus + r\mu_R^\ominus + \ldots - a\mu_A^\ominus - b\mu_B^\ominus - \ldots \qquad [7.101]$$

Introducing the expressions for chemical potential (e.g. equation [7.98]) into equation [7.100]):

$$\Delta G = q\mu_Q^\ominus + r\mu_R^\ominus + \ldots + RT \ln (a_Q^q) + RT \ln (a_R^r) + \ldots$$
$$- a\mu_A^\ominus - b\mu_B^\ominus - \ldots - RT \ln (a_A^a) - RT \ln (a_B^b) - \ldots$$

Introducing $\Delta G^\ominus$ (equation [7.101]) and uniting the logarithmic terms:

$$\Delta G = \Delta G^\ominus + RT \ln \left(\frac{a_Q^q \times q_R^r \times \ldots}{a_A^a \times a_B^b \times \ldots} \right)$$
$$= \Delta G^\ominus + RT \ln \frac{\{\text{products}\}}{\{\text{reactants}\}}. \qquad [7.102]$$

where {products} is the product of the activities of the products, each raised to the power of the stoichiometric number in the equation, and where {reactants} is the analogous product for the reactants.

Let us carry out the reaction in such a way that the reactants in the initial state proceed to the products in the final state through an intermediate equilibrium state (see Fig. 7.8a). If the reactants and the products are already in the equilibrium state, $\Delta G = 0$. In other words, no net work is done in the equilibrium state when the reactants are converted into products in small amounts so that the activities do not

Fig. 7.8 Variation of Gibbs energy G with extent of reaction
(a) a so-called 'reversible' reaction
(b) a so-called 'irreversible' reaction

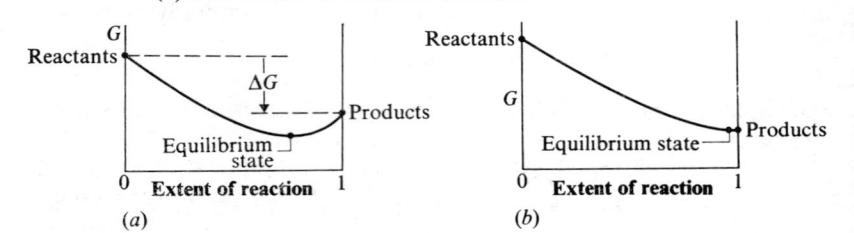

change from their equilibrium values at constant temperature and pressure. Thus, at equilibrium:

$$0 = \Delta G^{\ominus} + RT \ln \left(\frac{a_Q^q \times a_R^r \times \ldots}{a_A^a \times a_B^b \times \ldots} \right)_{eq.}, \quad [7.103]$$

where the subscript eq. indicates that the reactants and products are in chemical equilibrium. At constant temperature, T, and at constant pressure the $\mu^{\ominus}$ values for the substances involved are constant and therefore according equation [7.101] $\Delta G^{\ominus}$ is a constant. Equation [7.103] then leads to the result that

$$\left(\frac{a_Q^q \times a_R^r \times \ldots}{a_A^a \times a_B^b \times \ldots} \right)_{eq.} = K \quad [7.104]$$

is a constant. The constant, K, is known as the equilibrium constant and is always positive and dimensionless, because activities are always positive and dimensionless. Equation [7.103] is usually written in the form:

$$\Delta G^{\ominus} = -RT \ln K. \quad [7.103]$$

Equation [7.102] shows that, when the reactants at unit activity are converted into the products at unit activity, $\Delta G = \Delta G^{\ominus}$, which accords with the definition of $\Delta G^{\ominus}$.

If the reactants are allowed to react spontaneously, ΔG is negative and therefore G decreases, reaching a minimum value at the equilibrium state where all the reactants and products are present at their equilibrium activities or concentrations (Fig. 7.8a). Since G cannot decrease any more, no further reaction will take place unless G is increased by doing work on the system. If the products are allowed to react spontaneously, the reaction will at first proceed in reverse. G will then decrease to the same minimum value at the same equilibrium state. This state can therefore be approached from either direction.

If the equation symbolising the reaction is written in reverse, thus:

$$qQ+rR+ \ldots = aA+bB+ \ldots,$$

the reactants and products exchange roles, Fig. 7.8(a) is reversed, the signs of ΔG and $\Delta G^{\ominus}$ change, as in the case of other extensive quantities, and the equilibrium constant becomes its reciprocal. The equilibrium state is, however, unchanged.

If K is large, e.g. > 100, the equilibrium state contains such a small fraction of reactants as to be almost indistinguishable from the products (Fig. 7.8b). Such reactions are sometimes referred to as 'irreversible' and the sign $=$ in the chemical equation may be replaced by the sign $\rightarrow$. $\Delta G^{\ominus}$ is then relatively large and negative. If K is of the order unity, i.e. $K \approx 0.1$ to 10, Fig. 7.8(a) represents the situation quite well. Such reactions are sometimes called 'reversible' and the sign $=$ in the chemical equation may be replaced by the sign $\rightleftharpoons$. $\Delta G^{\ominus}$ is then relatively small, being positive when $K < 1$ and negative when $K > 1$. The words 'reversible' and 'irreversible' when used to describe thermodynamic processes have meanings different from, and more exact than, those implied above. The stricter meaning of these words, as explained on p. 282 et seq. should always be borne in mind when they are used in a thermodynamic sense.

Dynamical Equilibrium in Chemical Reactions

Although the bulk properties and thermodynamic functions of a system at equilibrium suggest that equilibrium is a static process, the molecules are continually undergoing change, which shows that equilibrium is a dynamic process at the molecular level. When radioactive atoms are incorporated in one of the reactants or products in an equilibrium mixture, the radioactivity of this component decreases with time even when radioactive decay has been accounted for. This shows that the labelled component is still being converted into products or reactants in spite of the overall state of equilibrium.

At equilibrium there is no net reaction because the rates of the forward and the backward reactions are the same. Each reactant is being regenerated by the reverse reaction as fast as it is destroyed by the forward reaction and the converse is true for each product. When the law of mass action (p. 145) is applied to the forward and reverse reactions, it can be shown that:

$$K = \frac{k_f}{k_b}, \qquad [7.105]$$

where K is the equilibrium constant, k_f is the rate constant of the forward reaction and k_b is the rate constant of the backward reaction. This equation always holds true at constant temperature and pressure no matter how complex is the stoichiometric equation or the mechanism of the forward or reverse reaction.

Le Chatelier proposed a very important principle which states that every change imposed on a system in equilibrium produces modifications in the system which tend to nullify the imposed change. Dynamic equilibrium accords with this principle. If more reactant A is added to the equilibrium mixture, the forward reaction will proceed faster so as to reduce the activity (or concentration) of A and increase the activities (or concentrations) of the products until the terms in the numerator of equation [7.104] once more balance the terms in the denominator to give the same value of the constant, K. If more product Q is added to the equilibrium mixture, the backward reaction will proceed faster so that the activity (or concentration) of Q decreases and that of each of the reactants increases until equilibrium is again restored as governed by the same value of K. For example, the reaction

$$Na_2CO_3 + CaCl_2 \rightleftharpoons CaCO_3 + 2NaCl$$

is known to proceed from left to right in the laboratory, showing that it has a large equilibrium constant. Berthelot noticed that in certain Egyptian lakes the reaction was proceeding in the reverse direction and sodium carbonate was crystallising out, probably because the water in the lakes contained a large excess of sodium chloride.

If one of the products is removed as it is formed, the reaction will attempt to restore equilibrium by producing more of this and other products. The product may be removed in a number of ways. It may disappear in a subsequent reaction. For example, Williamson found that the following equilibrium is attained at low temperatures:

$$C_2H_5OH + H_2SO_4 \rightleftharpoons C_2H_5O.SO_2.OH + H_2O.$$

If the temperature is raised so that ethyl hydrogen sulphate is removed by reaction with ethanol to form diethyl ether, continuous reaction takes place, thus:

$$C_2H_5O.SO_2.OH + C_2H_5OH = C_2H_5.O.C_2H_5 + H_2SO_4.$$

Other ways in which a product may be removed are by continual precipitation out of solution when its solubility limit has been exceeded or by liberation as a gas, e.g. carbon dioxide in respiration, or by diffusion through a membrane, e.g. waste products of metabolism from a cell.

A development of this idea which is used in the interpretation of the majority of biological reactions is that of interdependent equilibria. Here, one or more of the components of one equilibrium enters into further equilibria and any changes in the latter automatically disturb the first equilibrium. An example of this of pharmaceutical interest is provided by the effect of a drug on bacterial metabolism. If we take the view that a drug blocks the receptor groups on the bacterial cells, so that they are no longer available for normal metabolic functions, we must consider the two following equilibria, which are interdependent by virtue of the fact that the receptors appear in both:

$$\text{Receptors} + \text{Drug} \rightleftharpoons \text{Complex 1},$$
$$\text{Receptors} + \text{Metabolite} \rightleftharpoons \text{Complex 2}.$$

The destruction of bacteria by the drug depends on complex 1 being formed to such an extent that the second equilibrium is disturbed and complex 2 is broken down to provide the receptors required in the first equilibrium. The phenomenon of drug antagonism is the reverse of this. A drug antagonist is a substance which can reverse the inhibiting effect of a drug and it is probably an essential metabolite for the bacteria. In this way it was found that p-aminobenzoic acid is an essential metabolite because it is capable of reversing the effects of sulphonamide drugs. This was confirmed by its wide distribution and by the discovery of its role in the synthesis of the growth factor folic acid. A consideration of the first equilibrium also shows that if the concentration of the drug is reduced in any way in the patient, for example, by reaction with serum proteins, or by conversion to an inactive ionic form, it will not be so effective in blocking the receptors, and so may not result in the death of the bacteria. This emphasises the well-known principle that it is the concentration of the drug in the form in which it is required, and at the correct site, which is important, rather than the total dose given to the patient.

Equilibrium Constants

For the chemical reaction in the general form:

$$a\text{A} + b\text{B} + \ldots = q\text{Q} + r\text{R} + \ldots$$

the true thermodynamic equilibrium constant at constant temperature and pressure is dimensionless and is given by

$$K = \left(\frac{a_Q^q a_R^r \ldots}{a_A^a a_B^b \ldots} \right)_{\text{eq.}} \quad [7.104]$$

If one of the substances in the equilibrium mixture is present in its standard state, e.g. a liquid or a solid in excess of the solubility limit, or is very nearly in its standard state, e.g. the solvent, its activity is by convention put equal to unity in the expression for K. The amount of this component will in any case not be appreciably altered whatever the position of equilibrium.

If the equilibrium mixture behaves ideally, or nearly so, activities in terms of the pure substances as the standard state may be replaced by mole fractions, x. The equilibrium constant, K_x, in terms of mole fractions is then nearly identical to K and is still dimensionless, because mole fractions are dimensionless.

$$K_x = \left(\frac{x_Q^q x_R^r \cdots}{x_A^a x_B^b \cdots}\right)_{eq.} \qquad [7.106]$$

If the reaction involves only gases which behave ideally or nearly so, mole fractions may be replaced by partial pressures. The equilibrium constant, K_p, in terms of partial pressures is given by:

$$K_p = \left(\frac{p_Q^q p_R^r \cdots}{p_A^a p_B^b \cdots}\right)_{eq.} \qquad [7.107]$$

According to Dalton's law of partial pressures (p. 25), $p_Q = x_Q P$, $p_A = x_A P$, etc., where P is the total pressure of the system.

∴
i.e.
$$K_p = K_x P^{(q+r+\cdots - a-b-\cdots)}$$
$$K_p = K_x P^{\Delta v} \qquad [7.108]$$

where v is the stoichiometric number of a given reactant or product and Δv is the sum of the stoichiometric numbers of the products minus the sum of the stoichiometric numbers of the reactants, i.e.

$$\Delta v = q + r + \cdots - a - b - \cdots.$$

If the equilibrium mixture is sufficiently dilute to behave ideally or nearly so, activities may be replaced by concentrations, c, represented by square brackets [ ], or by molality, m. If the mixture is dilute, $a \propto c$ or $a \propto m$; the proportionality constant depends on the standard state at which the activity is unity. The equilibrium constant K_c expressed in terms of molar concentrations is given by:

$$K_c = \left(\frac{c_Q^q c_R^r \cdots}{c_A^a c_B^b \cdots}\right)_{eq.} = \left(\frac{[Q]^q [R]^r \cdots}{[A]^a [B]^b \cdots}\right)_{eq.} \qquad [7.109]$$

If the reaction involves only gases which behave ideally or nearly so, the ideal gas law may be applied.

For n mol of gas occupying a volume V, $p = \frac{n}{V} RT$.

Since $c = n/V$, in general $p = cRT$.

The partial pressures are given by $p_Q = c_Q RT$, $p_A = c_A RT$, etc.

∴ $K_p = K_c (RT)^{(q+r+\cdots - a - b - \cdots)}$,

i.e. $K_p = K_c (RT)^{\Delta v}$. $\qquad [7.110]$

Equilibrium Constants

Inspection of equations [7.108] and [7.110] shows that if $\Delta v = 0$, $a+b+\ldots = q+r+\ldots$, $K_c = K_p = K_x$, and K_p and K_c are dimensionless and independent of the units of partial pressure or concentration. If $\Delta v \neq 0$, i.e. if $q+r+\ldots \neq a+b+\ldots$, $K_c \neq K_p \neq K_x$ and K_c and K_p are not dimensionless but depend on the units of concentration and partial pressure respectively.

It has been shown theoretically, for a given chemical reaction at constant temperature and pressure, with different proportions of reactants and products, that the equilibrium constant is always constant. Table 7.3 illustrates this point using as an example the water gas reaction at 1259 K, thus

$$H_2 + CO_2 \rightleftharpoons H_2O + CO.$$

Table 7.3. Initial and equilibrium mole fractions in the water gas reaction at 1259 K

Initial mole fractions		Equilibrium mole fractions				$\dfrac{x(H_2O)\,x(CO)}{x(H_2)\,x(CO_2)}$
H_2	CO_2	H_2	CO_2	H_2O	CO	
0.899	0.101	0.8052	0.0069	0.0940	0.0940	1.59
0.699	0.301	0.4693	0.0715	0.2296	0.2296	1.57
0.519	0.491	0.2285	0.2144	0.2786	0.2786	1.58
0.297	0.703	0.0686	0.4751	0.2282	0.2282	1.60

Although the initial mole fractions of hydrogen and carbon dioxide and the equilibrium mole fractions of the components show quite wide variations,

$$K_x = \frac{x(H_2O)\,x(CO)}{x(H_2)\,x(CO_2)}$$

remains practically constant at about 1.59. If K had been determined instead of K_x, even better agreement would be expected.

Equilibrium constants usually vary considerably with temperature at constant pressure. The equation which describes this variation can be readily derived from equation [7.103],

$$\Delta G^\ominus = -RT \ln K,$$

by partial differentiation with respect to temperature, thus:

$$\left(\frac{\partial(\Delta G^\ominus)}{\partial T}\right)_p = -R \ln K - RT \left(\frac{\partial \ln K}{\partial T}\right)_p.$$

Substituting both expressions into the Gibbs–Helmholtz equation [7.71] in the form:

$$\Delta G^\ominus = \Delta H^\ominus + T \left(\frac{\partial(\Delta G^\ominus)}{\partial T}\right)_p$$

gives

$$-RT\ln K = \Delta H^\ominus - RT\ln K - RT^2\left(\frac{\partial \ln K}{\partial T}\right)_p,$$

whence

$$\left(\frac{\partial \ln K}{\partial T}\right)_p = \frac{\Delta H^\ominus}{RT^2}. \qquad [7.111]$$

This is the van't Hoff equation (or isochore) and may be integrated if $\Delta H^\ominus$, the standard enthalpy of reaction, is assumed not to vary with temperature over the range studied, thus:

$$\ln K = -\frac{\Delta H^\ominus}{RT} + \text{constant}'. \qquad [7.112]$$

Since $\ln x = 2.303 \log x$, we can write,

$$\log K = -\frac{\Delta H^\ominus}{2.303R} \cdot \frac{1}{T} + \text{constant}. \qquad [7.113]$$

This equation is analogous to the logarithmic form of the Arrhenius equation (5.55, p. 165) and to the integrated form of the Clausius equation (7.79, p. 305). By plotting a graph of $\log K$ against $1/T$, $\Delta H^\ominus$ may be calculated from the gradient or K may be obtained at any temperature within or slightly beyond the range studied. By means of the following equation (cf. equation [5.57], p. 165) $\Delta H^\ominus$ may be calculated if K is known at two temperatures or K may be calculated at a temperature T_2, if $\Delta H^\ominus$ and K at another temperature T_1 are known:

$$\log K_2 - \log K_1 = \log \frac{K_2}{K_1} = \frac{\Delta H^\ominus}{2.303R}\left(\frac{T_2-T_1}{T_1 T_2}\right). \qquad [7.114]$$

It can be seen from the form of this equation that the effect on K of varying the temperature conforms with Le Chatelier's principle. If $\Delta H^\ominus$ is positive (i.e. the forward reaction is endothermic) then K is increased by increasing the temperature. That is to say, the forward reaction which absorbs heat is favoured by an increase in temperature. Conversely, an exothermic reaction with negative $\Delta H^\ominus$ would not proceed as far and so would not have as large a value of K if the temperature were increased. For many reactions under normal conditions the enthalpy of reaction, ΔH, approximates closely to the standard value $\Delta H^\ominus$.

The equation which describes the variation of equilibrium constant with pressure at constant temperature can be readily derived from equation [7.103],

$$\Delta G^\ominus = -RT\ln K,$$

by partial differentiation with respect to pressure, thus:

$$\left(\frac{\partial (\Delta G^\ominus)}{\partial p}\right)_T = -RT\left(\frac{\partial \ln K}{\partial p}\right)_T.$$

Equation [7.73] can be written in the form

$$\left(\frac{\partial (\Delta G^\ominus)}{\partial p}\right)_T = \Delta V^\ominus.$$

Comparison with the previous equation shows that:

$$\left(\frac{\partial \ln K}{\partial p}\right)_T = -\frac{\Delta V^\ominus}{RT}, \qquad [7.115]$$

where $\Delta V^\ominus$ is the change in volume when the reactants in their standard states are converted to the products in their standard states in amounts corresponding to their stoichiometric numbers.

This equation accords with Le Chatelier's principle. If the forward reaction is associated with an increase in volume, i.e. if ΔV is positive, an increase in pressure will decrease K, and a decrease in pressure will increase K. The reverse reaction, which is associated with a decrease in volume, is favoured by an increase in pressure. Dissociation and decomposition reactions in the gas phase are examples of this type, e.g.

$$PCl_5 \rightleftharpoons PCl_3 + Cl_2.$$

Conversely, if the forward reaction is associated with a decrease in volume, i.e. if ΔV is negative, an increase in pressure will increase K and favour the forward reaction. Association reactions in the gas phase are of this type, e.g.

$$2\, CH_3COOH \rightleftharpoons (CH_3COOH)_2.$$

If the reaction is accompanied by no change in volume, i.e. if $\Delta V = 0$, K will be unaffected by pressure changes, e.g.

$$2\, HI \rightleftharpoons H_2 + I_2.$$

For reactions which do not involve gases, moderate changes in pressure usually have a negligible effect on K, because the volume changes in such reactions are usually very small.

The 'practical' equilibrium constants, K_p and K_c, for ideal gaseous reactions are always independent of changes in pressure and volume. The proof is based on equation [7.108] which, since $K = K_x$, may be written as:

$$K_p = K p^{\Delta v}.$$

Taking logarithms, $\qquad \ln K_p = \ln K + \Delta v \ln p$

Energy and Equilibrium

Differentiating with respect to the applied pressure, p, which in fact equals the total pressure, P, of the mixture, we have:

$$\left(\frac{\partial \ln K_p}{\partial P}\right)_T = \left(\frac{\partial \ln K}{\partial P}\right)_T + \frac{\Delta v}{P}.$$

Introducing equation [7.115] we have:

$$\left(\frac{\partial \ln K_p}{\partial P}\right)_T = -\frac{\Delta V^{\ominus}}{RT} + \frac{\Delta v}{P}. \qquad [7.116]$$

Applying the ideal gas equation to each product and reactant of the reaction mixture and adding, we have:

for the products $\qquad PV_2 = (q+r+\ldots)RT$
for the reactants $\qquad PV_1 = (a+b+\ldots)RT.$
Subtracting $\qquad P\Delta V = \Delta v . RT.\qquad$ [7.117]

If the reactants and products are in their standard states,

$$\frac{\Delta v}{P} = \frac{\Delta V^{\ominus}}{RT}.$$

Substitution into equation [7.116] gives

$$\left(\frac{\partial \ln K_p}{\partial P}\right)_T = 0. \qquad [7.118]$$

This result means that K_p is a constant independent of pressure or volume change. Further, since the equation [7.110] connecting K_p and K_c does not contain a pressure or volume change, K_c is also independent of these quantities.

If α is the fraction decomposed and if P is the total pressure at equilibrium, the amounts and partial pressures of the various components at equilibrium for the dissociation of 1 mole of phosphorus pentachloride are as follows:

$$\begin{array}{cccc} PCl_5 & \rightleftharpoons PCl_3 & + \; Cl_2; & \text{total} \\ 1-\alpha & \alpha & \alpha & 1+\alpha \\ \dfrac{(1-\alpha)P}{1+\alpha} & \dfrac{\alpha P}{1+\alpha} & \dfrac{\alpha P}{1+\alpha} & P \end{array}$$

$$K_p = \frac{p(PCl_3).p(Cl_2)}{p(PCl_5)}$$

$$= \frac{\alpha P.\alpha P.(1+\alpha)}{(1+\alpha)(1+\alpha)(1-\alpha)P}$$

$$= \frac{\alpha^2 P}{1-\alpha^2}.$$

Since K_p is independent of pressure at constant temperature, then if P is increased, α must decrease and the back reaction is favoured.

If n mol of PCl_5 are present initially and V is the total volume at equilibrium, the concentrations of the substances participating in the above reaction may be calculated as follows:

$$[PCl_3] = [Cl_2] = \alpha n/V; \quad [PCl_5] = (1-\alpha)n/V.$$

$$K_c = \frac{[PCl_3][Cl_2]}{[PCl_5]}$$

$$= \frac{\alpha^2 n^2/V^2}{(1-\alpha)n/V}$$

$$= \frac{\alpha^2 n}{(1-\alpha)V}.$$

At constant temperature K_c is independent of pressure. If the total pressure is increased, V will decrease. In order for V to decrease while K_c and n are constant, α must decrease and the back reaction is again favoured. The movement of equilibrium in favour of the back reaction as the total pressure is increased, or as the total volume is decreased, corresponds exactly to the decrease in the true thermodynamic equilibrium constant, K, according to equation [7.115], and accords with Le Chatelier's principle.

Problems

1 The heat capacity at constant pressure, C_p in $J\ K^{-1}\ mol^{-1}$, for steam is given by the equation

$$C_p = 36.8 - 7.9 \times 10^{-3}\ T + 9.2 \times 10^{-6}\ T^2$$

whereas C_p for water may be taken as $75.2\ J\ K^{-1}\ mol^{-1}$, independent of temperature. Given that ΔH for the vaporisation of water at 373.15 K is 40.6 kJ mol^{-1}, calculate ΔH at 310.6 K.

2 The enthalpies of combustion of liquid carbon disulphide, solid carbon and solid sulphur are -1076, -394, and -297 kJ mol^{-1}, respectively. Calculate the enthalpy of formation of carbon disulphide.

3 Use the enthalpy of formation of carbon disulphide calculated in the previous problem to obtain the mean bond energy of the C=S bond, given that the enthalpies of atomisation of solid carbon and solid sulphur are 719 and 223 kJ mol^{-1} respectively, and that the enthalpy of vaporisation of liquid carbon disulphide is 27 kJ mol^{-1}.

4 The value of ΔH for the reaction

$$CH_2=CH_2(g) + Cl_2(g) = CH_2Cl-CH_2Cl(g)$$

is -183 kJ. The values of ΔH of formation from gaseous atoms for methane, ethane, and ethene (ethylene) may be taken to be -1662, -2830, and -2257 kJ mol^{-1}. The ΔH of formation of chlorine from gaseous atoms is -243 kJ mol^{-1}. Calculate the mean bond energies of the bonds C—H, C—C, C=C and C—Cl.

5 The following are the enthalpies of hydrogenation (in kJ mol^{-1}) of propene and of compounds containing two double bonds:

propene	-126.1	hexa-1,5-diene	-253.3
buta-1,3-diene	-238.8	cyclohexa-1,3-diene	-231.7
penta-1,4-diene	-254.4	cyclopentadiene	-212.9

Calculate the resonance energy in each of the dienes, and interpret the results in terms of the possibilities of conjugation in each molecule.

6 Assuming that water vapour behaves as an ideal gas, calculate the maximum work done (a) when 1 mole of water is vapourised at 373.15 K and 101.3 kPa (neglecting the volume of the liquid) (b) when this vapour is expanded isothermally by reducing the pressure to 50 kPa.

7 Calculate the change in (a) entropy and (b) free energy when 0.3 moles of a gas A and 0.7 moles of a gas B are mixed at 373 K, assuming that the gas mixture behaves ideally.

8 From the expression for the heat capacity at constant pressure as a function of temperature for steam, given in problem 1, calculate the change of entropy per mole as steam is heated from 373.15 to 473.15 K at constant pressure.

9 The following table gives the boiling points and enthalpies of vaporisation of a number of liquids. Calculate the corresponding entropies of vaporisation and comment on the values obtained for water, methanol and ethanoic acid.

liquid	T_b/K	ΔH_{vap}/kJ mol^{-1}
water	373.2	40.6
mercury	630.1	59.1
hexane	341.9	28.6
tetrachloromethane	349.7	30.0
methanol	337.9	35.2
ethanoic (acetic) acid	391.1	23.6

10 The vapour pressure of ethyl acetate (relative molecular mass 88.11) in the region of the boiling point (350.3 K at 101.3 kPa) varies with temperature to the extent of 3.06 kPa K^{-1}. The densities of the liquid and the vapour at the boiling point are 815.4 and 3.23 kg m^{-3}, respectively. Calculate the enthalpy of vaporisation

using (a) the approximate form and (b) the exact form of the Clapeyron–Clausius equation.

11 A mixture of water and ethyl alcohol, in which the mole fraction of water is 0.4, has a density of 849.4 kg m^{-3} and the partial molar volume of alcohol in this mixture is 5.75×10^{-3} m^3 mol^{-1}. Calculate the partial molar volume of water in the mixture.

12 From the data given for the water gas reaction on p. 319 calculate $\Delta G^\ominus$ for this reaction at 1259 K.

13 For the water gas reaction at 1259 K, $\Delta H^\ominus$ has the value -31.89 kJ mol^{-1}. From this and the value of $\Delta G^\ominus$ calculated in the previous problem, calculate the value for the standard entropy change, $\Delta S^\ominus$, at this temperature.

14 The equilibrium constant for the reaction
$$\text{iso-citrate} \rightleftharpoons \text{cis-aconitate}$$
at 311 K is 0.5. Calculate the free energy change at 311 K when 1 mole of *iso*-citrate at a concentration of 0.2 mol dm^{-3} is converted into 1 mole of *cis*-aconitate at a concentration of 1×10^{-3} mol dm^{-3}. (Activities can be taken as being equal to the concentrations in mol dm^{-3}.)

15 The equilibrium constant for the reaction
uridine + inorganic phosphate $\rightleftharpoons$ uracil + D-ribose-1-phosphate
is 0.91 at 311 K and pH 7.0. Calculate (a) the standard free energy change, (b) the ratio of the product of the activities of products to the product of the activities of the reactants which would produce 4.0 kJ mol^{-1} of available energy.

16 The following table shows values of the equilibrium constant K at various temperatures T for the reaction
$$\text{fumarate} + H_2O \rightleftharpoons \text{L-malate}$$

T/K	293.4	298.2	307.8	317.6	322.8
K	4.46	3.98	3.27	2.75	2.43

Use a graphical method to obtain $\Delta H^\ominus$ and also calculate $\Delta G^\ominus$ and $\Delta S^\ominus$ for the reaction at 298.2 K.

8 Electrolytes

Strong Electrolytes

When an ionic solid dissolves in a polar solvent, such as water, the ions are separated from each other and become solvated (p. 203). This separation of the positive and negative ions enables the solution to conduct electricity, since the positive ions are attracted to a negative electrode placed in the solution, and the negative ions are attracted to a positive electrode. Any solution which is capable of conducting electricity is termed an electrolyte and the process of passing electricity through an electrolyte is called electrolysis.

Ionic solids, e.g. salts, have been shown by X-ray crystallographic methods (p. 36) to be completely ionised in the solid state and must therefore be completely ionised even in concentrated solution. Substances which are completely ionised in solution are called strong electrolytes and include not only salts, e.g. KCl, Na_2HPO_4, $FeSO_4 \cdot 7H_2O$, sodium salicylate, proflavine hemisulphate and quinine hydrocloride, but also strong acids and strong bases in polar solvents. Strong bases in aqueous solution include fully ionised hydroxides, such as those of the alkali metals (Group IA), e.g. NaOH, and of the alkaline earth metals (Group IIA), e.g. $Mg(OH)_2$, and the quaternary ammonium hydroxides (R_4N^+ OH^-). Strong acids in aqueous solution include HCl, HBr and HI, certain inorganic oxyacids, e.g. $HClO_4$, HNO_3, H_3PO_4 and H_2SO_4, the sulphonic acids (RSO_3H) and trichloroacetic acid (CCl_3COOH). These acids are not ionised in the pure state but react with water to give the hydrated proton (H_3O^+) and an anion, e.g.

$$H_2SO_4 + H_2O \rightarrow H_3O^+ + HSO_4^-$$
$$HSO_4^- + H_2O \rightarrow H_3O^+ + SO_4^{2-}.$$

The splitting of covalent or neutral molecules, such as these acids, into solvated ions in polar solvents is called electrolytic dissociation or ionisation.

Equilibria in Ionic Systems

If two solutions are mixed, each containing a fully ionised electrolyte, it is only possible to say that reaction takes place between them if either some non-ionised molecules or complex ions are formed, or some product is thrown out of solution as a precipitate or as a gas, etc. Then it is possible to describe the reaction in terms of the equilibrium between the non-ionic material, or the material which is thrown out of solution, and the particular ions from which it is formed. For example, when a strong acid and a strong base react to form a salt and water, the full equation might be of the form:

$$H^+ + Cl^- + Na^+ + OH^- \rightleftharpoons Na^+ + Cl^- + H_2O$$

but the reaction is really dependent upon the equilibrium:

$$H^+ + OH^- \rightleftharpoons H_2O.$$

Again, when two salts react in aqueous solution to give a precipitate, the full equation might be of the form:

$$Ag^+ + NO^- + Na^+ + Cl^- \rightleftharpoons AgCl + Na^+ + NO_3^-$$

but the reaction is really dependent upon the equilibrium:

$$Ag^+ + Cl^- \rightleftharpoons AgCl \text{ (solid)}.$$

In either of the two cases considered, the extent of the reaction can be affected by altering the concentrations and therefore the activities of any of the ions which enter into the significant equilibrium, and this is known as the common ion effect. The equilibrium between ions and an unionised molecule is usually considered in reverse as ionic dissociation. This will be discussed later in this chapter and it also forms the basis for the whole of the quantitative discussion of acids and bases in Chapter 9. The equilibrium between ions and a solid is usually considered in reverse in terms of the solubility product and it is worth pausing to discuss this further here.

Solubility Product

When a substance dissolves to give ions in solution, the extent of its solubility is determined by equilibrium being set up between the solid and the ions in solution. For the equilibrium: AB (solid) $\rightleftharpoons A^+ + B^-$, the expression for the equilibrium constant is:

$$K = \frac{a_{A^+} \times a_{B^-}}{a_{AB}}.$$

The term a_{AB}, which represents the activity of the excess solid, is taken as unity by convention (see p. 223), so the equilibrium constant may be written:

$$K_s = a_{A^+} \times a_{B^-} \approx m_{A^+} \times m_{B^-} \approx [A^+][B^-] \qquad [8.1]$$

K_s is called the solubility product of the substance AB. This is an alternative way of expressing the solubility of an electrolyte, and it can easily be related to the solubility if this is first expressed in mol kg^{-1} or in mol dm^{-3}. Thus, if the solubility of AB is s mol dm^{-3}, and if the electrolyte can be regarded as being completely dissociated, there must be s mol dm^{-3} of A$^+$ and s mol dm^{-3} of B$^-$ in the solution.

Hence, $\qquad\qquad K_s \approx s \times s = s^2,$

or, $\qquad\qquad\qquad s \approx \sqrt{K_s}.$ [8.1a]

(A precise equality can be stated if the activity coefficients of the ions are included, but this will be ignored in the present treatment, see p. 339.)

The advantage of expressing the solubility of an electrolyte as a solubility product is that it explains the way in which the solubility varies when the material is in the presence of other electrolytes containing an ion in common with the original substance. If a second electrolyte CB is added to a saturated solution of the original electrolyte AB, it is found that some of the substance AB is precipitated out of the solution because of the common ion B$^-$. The equilibrium between solid AB and the ions A$^+$ and B$^-$ must be maintained, by keeping the product of [A$^+$] and [B$^-$] constant, and equal to K_s. This can only be done by removing equivalent amounts of A$^+$ and B$^-$ from solution as solid AB, until the product is again equal to K_s. It should be noted that [A$^+$] is no longer equal to [B$^-$] when equilibrium has been established, nor has all the excess B$^-$ ion, introduced by adding CB, been removed – the decrease in [A$^+$] means that [B$^-$] is not reduced to its previous value, in satisfying the constancy of the solubility product.

Experiment 8.1 Determination of the solubility product of potassium hydrogen tartrate.

Potassium hydrogen tartrate, which may be represented as KHTa, is a sparingly soluble salt which breaks up in solution to give potassium ions and weakly acidic hydrogen-tartrate ions:

$$KHTa \rightleftharpoons K^+ + HTa^-$$

$\therefore K_s \approx [K^+][HTa^-]$

The hydrogen-tartrate ion is such a weak acid that its dissociation into hydrogen and tartrate ions may be neglected. The concentration of hydrogen–tartrate ions in solution may be determined readily by titration against standard alkali, using phenolphthalein as indicator.

Addition of another salt containing potassium ion reduces the solubility of potassium hydrogen tartrate, because of the common-ion effect, but it should be found that K_s remains constant.

About 1 g of finely powdered KHTa is placed in each of five test-tubes. To the first, distilled water is added until the tube is nearly full. To the others, the following four standard solutions of KCl are added: tube 2, 0.01 mol dm^{-3}; tube 3, 0.02 mol dm^{-3}; tube 4, 0.03 mol dm^{-3}; tube 5, 0.04 mol dm^{-3}. The five test-tubes are stoppered and well shaken for five minutes, and then allowed to stand, preferably in a thermostat, for a further twenty minutes, with occasional shaking. At the end of this period, 5 cm^3 portions are withdrawn from the tubes by means of a pipette fitted with a glass-wool filter. (Glass wool is contained in a rubber connection between the end of the pipette and an extra jet drawn from glass tubing. The filter is removed before delivering the liquid from the pipette.) Each portion is titrated with 0.02 mol dm^{-3} NaOH, or 0.01 mol dm^{-3} Ba(OH)$_2$. From each titre, the concentration of HTa$^-$ in the corresponding solution is calculated. The concentration of K$^+$ arising from the KHTa must be the same as the concentration of HTa$^-$. The total concentration of K$^+$ is therefore the sum of this and the added concentration of KCl. The solubility product is obtained by multiplying the total K$^+$ concentration in each solution by the HTa$^-$ concentration. It should be found that K_s is approximately the same in all the solutions.

The principle of reducing the concentration of a sparingly soluble substance by the common-ion effect has many applications, particularly in quantitative and qualitative analysis. Thus, in the gravimetric determination of sulphate ion by precipitation as the barium salt the precipitate is washed with a dilute sulphuric acid solution rather than with water.

Electrolysis

When an electric current passes along a metal conductor, such as a length of copper wire, the atoms of the metal do not move. The conduction arises because some of the electrons of the metal are not firmly attached to particular atoms but are free to move from atom to atom. They are therefore attracted towards the positive end of the wire, while, at the negative end, more electrons are fed in from the battery or some other source of electricity. However, conduction through an electrolyte is of a different nature, since it arises by the migration of ions through the solution to the electrode of opposite charge. In order that the electric current may flow through both the electrolyte and the external circuit, the ions must be discharged at the electrodes which dip into the solution, giving up electrons at the positive electrode, or receiving electrons at the negative electrode.

During this process of discharge the ions are converted into neutral atoms or molecules. There are a number of processes which may then occur, e.g. (a) a metallic film may be formed by the deposition of atoms, as in the process of electroplating; (b) a gas may be evolved, as when water is electrolysed; (c) the material of the electrode may be attacked, causing it to be dissolved or to be coated with an insoluble film; (d) the substances which have been discharged may react with the solvent to form various products. These electrode processes, as they are called, may be represented by chemical equations in which the symbol e^- represents an electron. The following may be quoted as examples of the above types of processes:

(a) $Ni^{2+} + 2e^- = Ni$ (deposited as a metal film).
(b) $2H^+ + 2e^- = H_2$ (given off as a gas).
(c) $Cl^- + Ag = AgCl + e^-$ (deposited as an insoluble film) or,
$NO_3^- + Ag = AgNO_3 + e^-$ (electrode dissolves).
(d) $2Na^+ + 2e^- + 2H_2O = 2NaOH + H_2$ (reaction with solvent).

In each case the amount of chemical change which takes place is determined by the quantity of electricity (or electric charge) which passes through the solution. Each ion which is discharged alters by z the number of electrons on the electrode, where z is the valency of the ion. The number of ions in one mole of any substance is given by the Avogadro constant, L. Hence for every mole discharged there must be a change of zL in the number of electrons on the electrode, i.e. a quantity which depends on the valency of the ions. Therefore, when $1/z$ mole of substance undergoes chemical change the number of electrons exchanged at the electrodes is L, which corresponds to 1 mole of electrons. The amount of substance associated with one mole of electrons (i.e. L electrons) is known as one equivalent, which is here abbreviated to eq. and is the same as $1/z$ mole. In the case of an ionic compound, e.g. a salt, z is the number of positive or negative electronic charges associated with the chemical formula of the compound, as written. Thus, for $AgNO_3$, $z = 1$, and for $CuSO_4$, $z = 2$. The atomic or molecular weight of an element, ion or compound when divided by z (i.e. A_r/z or M_r/z) is often referred to as the equivalent weight of the chemical species.

The constant quantity of electricity (or electric charge) associated with one equivalent of substance (and therefore with one mole of electrons) is called the faraday. This quantity is so named in honour of Faraday, who was a pioneer in the study of electrolysis and who enunciated the following laws:

1. The mass or amount of any substance liberated or dissolved is proportional to the quantity of electricity (or electric charge, Q) passed.

2. The masses of different substances liberated or dissolved by the same quantity of electricity are proportional to their chemical equivalent weights (i.e. M_r/z).

Algebraically these laws may be expressed thus:

1. For constant M_r, $m \propto Q$.
2. For constant Q, $m \propto M_r/z$.

Combining these, $m \propto Q.M_r/z$ or

$$Q = F\frac{mz}{M_r}, \quad (m \text{ is in g here}) \qquad [8.2]$$

where F is a constant known as the Faraday constant. But $m/M_r = n$, the amount of substance (number of moles) of the substance being liberated or deposited. Thus:

$$Q = Fzn. \qquad [8.2a]$$

By accurate experiments on electrolysis the Faraday constant, F, is found to have the value 96 487 C mol^{-1} (and the faraday is equal to 96 487 C) where C is the abbreviation for the coulomb, the SI unit of electric charge (see below).

It is interesting to note that it was a consequence of Faraday's laws of electrolysis that the idea of the fundamental particle of electricity came into being. It was argued that, since one equivalent ($1/z$ mol) of any substance is associated with one faraday of electricity, 1 mol of any ion must be associated with z faradays, where z is the valency. Therefore, one ion of any substance must be associated with z faradays/L, where L is the Avogadro constant. This quantity must be a simple multiple (z) of the quantity F/L, which is therefore the smallest quantity of electricity ever found associated with an atom in the form of an ion. As stated on p. 60, it was this fundamental unit of electricity which was first given the name 'electron', and it was some time later that particles of electricity having this charge were discovered and given the same name. Since $L = 6.0222 \times 10^{23}$ mol^{-1}, the charge of the electron (and of the proton, p. 64) is given by

$$e = \frac{F}{L} = \frac{96\,487 \text{ C mol}^{-1}}{6.0222 \times 10^{23} \text{ mol}^{-1}} = 1.6022 \times 10^{-19} \text{ C}$$

The *electric current* I (SI unit the ampere A) is the rate at which electric charge (quantity of electricity) is passing. The ampere is defined as that constant current which, if maintained in two straight parallel conductors of infinite length, of negligible circular cross section, and placed 1 metre apart in vacuum, would produce between these conductors a force equal to 2×10^{-7} newton per metre length. The *quantity of electricity* or *electric charge* Q (SI unit the coulomb C) is the electric current I passing multiplied by the time t (SI unit the second s) for which it passes, thus:

$$Q = It. \qquad [8.3]$$

One coulomb represents one ampere flowing for one second.

In any quantitative study of electrolysis it is necessary to know how much electricity has passed through the apparatus during the course of an experiment. Because of the possibility of variation of the current as the experiment progresses, the quantity of electricity is not conveniently measured by multiplying the current by the time for which it passes. Instead, a coulometer is included in the circuit. This is a piece of apparatus in which the passage of electricity is made to bring about some easily measurable chemical change. By determining the amount of this change, the quantity of electricity passed may then be calculated by applying Faraday's laws.

In one accurate form of coulometer silver nitrate is electrolysed between a silver anode and a platinum dish as cathode, and the quantity of electricity which has passed is calculated from the weight of silver deposited on the platinum dish. In another more convenient coulometer, iodine is liberated from potassium iodide solution at a platinum anode. At the same time, a solution of iodine in potassium iodide round a platinum cathode decreases in strength by conversion of some of the iodine into iodide ions. The amount of conversion of I^- to I_2, or vice versa, expressed in equivalents, gives the number of faradays which have passed through the solution.

Transport Numbers

During the passage of electricity through an electrolyte, there is a contribution to the total conductivity from the positive ions moving to the cathode and a contribution from the negative ions moving to the anode. These two contributions are not generally equal, since the positive ions move at a different speed from the negative ions. The fraction of the total amount of electricity which is carried by any particular type of ion is called the transport number of that ion. In studying systems in which the movement of ions takes place, it is often necessary to consider the movement of the various types of ions separately, and, in order to do this, their respective transport numbers must be known.

If the simplest type of situation is considered, in which neither ion is capable of attacking the electrode, and neither is capable of being reformed after deposition, it may be seen that concentration changes must take place in the solution during the course of the electrolysis. The positive and negative ions may be imagined to be evenly distributed throughout the solution, on the average, before the electrolysis. When the current is switched on, however, all the positive ions move towards the cathode, leaving the region round the anode lacking in positive ions. The excess negative ions in this region are deposited on the anode. Thus, there is a decrease in concentration of the total electrolyte in the region of the anode, and it will be shown below that this decrease is proportional to the transport number of the cation. There is also a decrease in concentration round the cathode, due to the movement of anions, and this is proportional to the transport number of the anion.

The relationship between the changes in concentration and the transport numbers may be derived by considering, in detail, the movements of the ions through the cell. The whole solution may be divided into three portions: the anode compartment, which includes all the region in which there is a change in concentration round the anode; the middle compartment, in which there is no change in concentration; and the cathode compartment, which includes all the region in which there is a change in concentration round the cathode. Let t_+ be the transport number of the positive cation, and t_- be the transport number of the negative anion. Suppose that, during the whole electrolysis, f faradays of electricity pass through the cell (as measured by a coulometer in series with the transport cell). Then, t_-f faradays are carried by movement of the anion, and t_+f faradays by movement of the cation. At the same time, f equivalents of anion are deposited on the cathode, and f equivalents of cation are deposited on the cathode (Faraday's laws). The movements of the ions may be represented diagramatically, as follows:

Cathode Compartment	Middle Compartment	Anode Compartment
t_-f eq. of anion move out	t_-f eq. of anion in and out	t_-f eq. of anion move in
t_+f eq. of cation move in	t_+f eq. of cation in and out	t_+f eq. of cation move out
f eq. of cation lost by desposition		f eq. of anion lost by deposition

Total Loss and Gain:

t_-f eq. of anion lost by migration	No change	t_+f eq. of cation lost by migration
$(f-t_+f)$ eq. of cation lost		$(f-t_-f)$ eq. of anion lost

But since the two fractions of electricity transported must equal the total quantity which passes during electrolysis:

$$t_+ f + t_- f = f, \quad (\text{or } t_+ + t_- = 1) \quad [8.4]$$

Hence, the $(f - t_+ f)$ eq. of cation lost from the cathode compartment may be represented by $t_- f$ eq. Also, the $(f - t_- f)$ eq. of anion lost from the anode compartment may be rewritten as $t_+ f$. Hence, totalling the losses in the two electrode compartments:

Cathode Compartment	Anode Compartment
$t_- f$ eq. of anion lost	$t_+ f$ eq. of cation lost
$t_- f$ eq. of cation lost	$t_+ f$ eq. of anion lost

Thus, $t_- f$ eq. of the total electrolyte are lost from the cathode compartment, and $t_+ f$ eq. of the total electrolyte are lost from the anode compartment. As a general rule, it may be said that the loss of electrolyte from the region round either electrode is proportional to the transport number of the ion leaving that region, the constant of proportionality being the number of faradays that have passed, provided that the loss is measured in equivalents.

In cases where an electrode is attacked by a liberated substance, or where a liberated substance reappears in the solution in ionic form, the calculation above must be modified. Thus, in the case of the electrolysis of silver nitrate between silver electrodes, where the nitrate ion, on being deposited, attacks the anode, and re-forms silver nitrate, we should have:

Cathode Compartment	Anode Compartment
$t_- f$ eq. of NO_3^- move out	$t_- f$ eq. of NO_3^- move in
$t_+ f$ eq. of Ag^+ move in	$t_+ f$ eq. of Ag^+ move out
f eq. of Ag^+ lost by deposition	f eq. of NO_3^- lost by deposition, followed by f eq. of $AgNO_3$ redissolving as f eq. of NO_3^- and f eq. of Ag^+

Total loss and gain:

$t_- f$ eq. of NO_3^- lost	$t_- f$ eq. of NO_3^- gained
$(f - t_+ f)$ eq. of Ag^+ lost	$(f - t_+ f)$ eq. of Ag^+ gained

Thus, there has been a gain of electrolyte in the anode compartment exactly equal to the loss in the cathode compartment. Rewriting the quantities lost or gained, they become t_-f eq. of Ag^+ and t_-f eq. of NO_3^-; i.e. t_-f eq. of the total electrolyte. From this experiment it would only be possible to determine directly the transport number of the nitrate ion. However, the transport number of the silver ion is readily obtainable from the relationship $t_+ + t_- = 1$.

Measurement of Transport Numbers

In order to apply the above relationships to the measurement of transport numbers, the electrolysis must be carried out in a cell which has facilities for separating the electrolyte into anode, middle and cathode compartments. A simple piece of apparatus in which this separation may conveniently be carried out is shown in Fig. 8.1.

Fig. 8.1 Cell for determining transport numbers

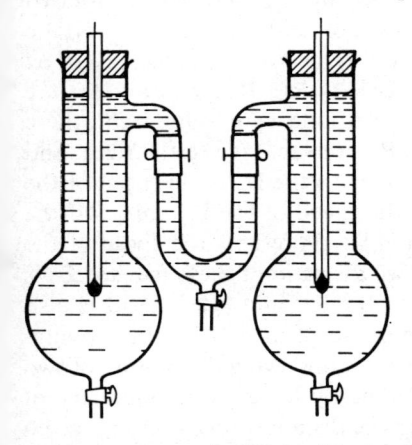

The apparatus is filled with the electrolyte at a known concentration, and, after the electrodes have been inserted, the cell is connected in series with a coulometer to a direct-current supply. After passing the current for about two hours, it is switched off, and the clips on the rubber connections are closed. The solutions in the three compartments are run out into separate flasks and analysed. The amount of chemical change which has taken place in the coulometer is also found. In calculating the transport number, the weight of water and the weight of solute in each of the three compartments is first found. These are compared with the weight of solute which was associated with the same weight of water in each compartment before electrolysis. The loss or gain in the amount of solute (in eq.) associated with the water in each compartment is thus found. This quantity is divided by the total number of faradays passed, as found from the coulometer, to obtain the transport number. There should be no concentration change in the middle compartment. If some change is observed, it is an indication that the electrolysis has been continued for too long, so that the regions in which the concentration has changed extend further than the electrode compartments. If this is so, the experiment must be repeated, the current being passed for a shorter time.

Since the transport number is a measure of the current-carrying power of one type of ion relative to another, it must be proportional to the relative speeds of the ions. Thus, if the positive ion moves to the cathode twice as quickly as the negative ion moves to the anode, the positive ion must be carrying two-thirds of the total charge. This principle forms the basis of another method of measuring transport numbers – the moving-boundary method. In this, the actual progress of the ions towards the electrodes may be observed. The ratio of the speeds of the two types of ions (positive and negative), measured under the same conditions, will then be equal to the ratio of their transport numbers. This ratio may be found directly by setting up a cell with two liquid boundaries, one indicating the motion of the positive ion and the other the motion of the negative ion. This would be achieved by having three electrolytes in the sequence: PQ, RQ, RS. If the anode were to the left of such a sequence, and the cathode were to the right, there would be a junction between P^+ and R^+ ions moving to the right, and another between Q^- and S^- ions moving to the left. The speed of the junction between P^+ and R^+ gives the speed of the R^+ ion, and the speed of the junction between Q^- and S^- represents the speed of the Q^- ion under the same conditions. So, the ratio of these two speeds is the ratio of the transport numbers of the R^+ and Q^- ions. It is also possible to measure the transport number of an ion by the moving-boundary method without the necessity of observing the motion of two boundaries simultaneously. This is done by having one boundary of the type PQ, RQ, and by comparing the distance moved after a given time with the total amount of electricity which has passed through the cell.

Some Factors Influencing Transport Numbers

The transport number of an ion is not a constant, irrespective of the electrolyte. It represents the fraction of the current carried by a particular type of ion, so that it is determined by the current-carrying powers of all the types of ion present in the solution. Thus, the transport number of the chloride ion in 0.1 mol dm^{-3} NaCl is 0.61, but in 0.1 mol dm^{-3} HCl it is only 0.17, because the hydrogen ion moves faster than the sodium ion under the same conditions (p. 351). The hydrogen ion therefore carries a greater proportion of the current in HCl than does the sodium ion in NaCl.

In the determination of transport numbers by the methods described, no account is taken of the possibility that water molecules may migrate, under the influence of the current, with the ions. This does in fact happen, since most ions are hydrated, i.e. there are water molecules more or less loosely attached to them. The transport numbers obtained from such experiments are therefore only 'apparent' transport numbers. To obtain a 'true' transport number, it is necessary to measure the movement of the ions relative to some added solute which will not migrate with the ions during the electrolysis. For this purpose, sugars are convenient substances to use as added solutes, because their concentrations can easily be determined by measuring optical rotations. Then, by finding the number of equivalents of electrolyte associated with the sugar (and not with the water) in the electrode compartments before and after electrolysis, the true transport numbers may be calculated. Further, by comparing the true and apparent transport numbers, the relative extents of hydration of the positive and negative ions may be determined.

Transport number studies are used in conjunction with conductance studies in the determination of molar conductivities of separate ions (p. 351). They also give some information about the nature of the conducting species, e.g. their relative sizes (ion + hydration shell), chemical composition, and tendency to occur as complex ions. A further application is in the detection of anomalous conduction mechanisms, such as the interchange of protons among stationary water molecules.

Deviations of Strong Electrolytes from Ideal Behaviour

Accurate experimental observation of the colligative properties of strong electrolytes shows that the van't Hoff correction factor, i, defined by equation [6.46], is usually less than the number of ions produced by one molecule (p. 246). These departures from ideality cannot be due to incomplete dissociation as was once thought, since strong electrolytes are completely dissociated in water and in other ionising solvents. These deviations by electrolytes are expressed in the divergence from unity of both the practical osmotic coefficients, ϕ, defined by equation [6.51], and of the activity coefficients (p. 311).

Since electrolytes consist of oppositely charged ions, it might be thought that the activity of individual ions could be determined. For example, equation [10.35] shows that the e.m.f. of a concentration cell with transport is related to the activity of one type of ion. There are, however, practical difficulties due to the inability to measure accurate liquid junction potentials (see p. 421). In view of this, the mean activity, $a_{\pm}$, of both types of ion, or the activity, a, of the electrolyte as a whole, are considered instead. An electrolyte $C_{\nu_+}A_{\nu_-}$, where C^{z+} is the cation and A^{z-} is the anion, will dissociate as follows:

$$C_{\nu_+}A_{\nu_-} = \nu_+C^{z+} + \nu_-C^{z-}$$

(e.g. $Fe_2(SO_4)_3 = 2\,Fe^{3+} + 3\,SO_4^{2-}$; $\nu_+ = 2$, $\nu_- = 3$, $z_+ = 3$, $z_- = 2$). The activity of the cation, a_+, and of the anion, a_-, are related to the experimentally accessible activities, a and $a_\pm$, by the equation:

$$a = (a_+)^{\nu_+}(a_-)^{\nu_-} = (a_\pm)^{\nu_+ + \nu_-}. \qquad [8.5]$$

The quantity a is known as the 'activity' of the electrolyte and $a_\pm$, which is the geometric mean of the ionic activities, is known as the 'mean activity' of the electrolyte. It can be seen that:

$$a_\pm = a^{1/(\nu_+ + \nu_-)}. \qquad [8.6]$$

The activity coefficient of the cation, γ_+, and of the anion, γ_-, are related to the respective activities, a_+ and a_-, and to the respective molalities, m_+ and m_-, by equations analogous to [7.94], thus:

$$a_+ = \gamma_+ m_+; \qquad a_- = \gamma_- m_-. \qquad [8.7]$$

Since the activities of individual ions cannot be measured, the activity coefficients γ_+ and γ_- cannot be determined. The experimentally accessible quantity is the geometric mean of these activity coefficients, known as the mean activity coefficient, $\gamma_\pm$, which is defined by an equation analogous to [8.5], thus:

$$(\gamma_+)^{\nu_+}(\gamma_-)^{\nu_-} = (\gamma_\pm)^{\nu_+ + \nu_-}. \qquad [8.8]$$

Introduction of equation [8.7] into [8.5] gives:

$$(\gamma_+ m_+)^{\nu_+}(\gamma_- m_-)^{\nu_-} = (a_\pm)^{\nu_+ + \nu_-}.$$

Dividing this equation by [8.8], we have:

$$(m_+)^{\nu_+}(m_-)^{\nu_-} = \left(\frac{a_\pm}{\gamma_\pm}\right)^{\nu_+ + \nu_-},$$

whence

$$a_\pm = \gamma_\pm (m_+^{\nu_+} m_-^{\nu_-})^{1/(\nu_+ + \nu_-)}. \qquad [8.9]$$

This equation is applicable to a single electrolyte or to a mixture of electrolytes. For a solution of a single electrolyte of molality m, the respective molalities of the cation and anion will be:

$$m_+ = \nu_+ m \quad \text{and} \quad m_- = \nu_- m. \qquad [8.10]$$

Introducing these ionic molalities into equation [8.9]:

$$a_\pm = \gamma_\pm [(m^{\nu_+} m^{\nu_-})(\nu_+^{\nu_+} \nu_-^{\nu_-})]^{1/(\nu_+ + \nu_-)},$$

whence

$$a_\pm = \gamma_\pm m (\nu_+^{\nu_+} \nu_-^{\nu_-})^{1/(\nu_+ + \nu_-)}. \qquad [8.11]$$

Deviations of Strong Electrolytes from Ideal Behaviour

The mean activity and the mean activity coefficient are related by an equation analagous to [8.7], thus:

$$a_\pm = \gamma_\pm m_\pm, \qquad [8.12]$$

where $m_\pm$ is the mean molality of the ions of the electrolyte. Comparing this equation with equations [8.9] and [8.11] we have:

$$m_\pm = (m_+^{\nu_+} m_-^{\nu_-})^{1/(\nu_+ + \nu_-)} \qquad [8.13]$$

and

$$m_\pm = m(\nu_+^{\nu_+} \nu_-^{\nu_-})^{1/(\nu_+ + \nu_-)}. \qquad [8.14]$$

Equations [8.6], [8.10] and [8.14] may be exemplified by $Fe_2(SO_4)_3$, in which case $a_\pm = a^{\frac{1}{5}}$, $m_+ = 2m$, $m_- = 3m$ and $m_\pm = 108^{\frac{1}{5}} m = 2.551\ m$. For a uni-univalent electrolyte, e.g. KCl, $a_\pm = a^{\frac{1}{2}}$, $m_+ = m_- = m$ and $m_\pm = m$.

An electrolyte $C_{\nu_+} A_{\nu_-}$ in saturated solution partakes in the following equilibrium:

$$C_{\nu_+} A_{\nu_-}\ (\text{solid}) \rightleftharpoons \nu_+ C + \nu_- A.$$

The solubility product of this more complex electrolyte is given by an equation analogous to the first part of equation [8.1], thus:

$$K_s = (a_+)^{\nu_+} (a_-)^{\nu_-}. \qquad [8.15]$$

Introduction of the expressions for the ionic activities (equations [8.7]) gives:

$$K_s = (m_+ \gamma_+)^{\nu_+} (m_- \gamma_-)^{\nu_-}. \qquad [8.16]$$

The activities, molalities, concentrations and activity coefficients refer, of course, to the saturated solution. Comparison of equations [8.15] and [8.5] shows that the activity of a saturated electrolyte as a whole is equal to the solubility product of the electrolyte. Introduction of the expressions for the ionic molalities (equations [8.10]) into equation [8.16] gives:

$$K_s = (\gamma_+ \nu_+ m)^{\nu_+} (\gamma_- \nu_- m)^{\nu_-}, \qquad [8.17]$$

where m is the molality of the electrolyte as a whole. If the electrolyte is so sparingly soluble that the activity coefficients may be assumed to be close to unity, equations [8.16] and [8.17] approximate to:

$$K_s \approx m_+^{\nu_+} m_-^{\nu_-} \qquad [8.18]$$

and

$$K_s \approx (\nu_+ m)^{\nu_+} (\nu_+ m)^{\nu_-}. \qquad [8.19]$$

These principles may be exemplified by CaF_2, a salt which is sparingly soluble in water and for which $\nu_+ = 1$ and $\nu_- = 2$. If the solubility of calcium flouride is m mol kg^{-1} the solubility product according to equation [8.19] is approximately equal to $4m^3$. For a uni-univalent electrolyte, AB, such as AgCl, $v_+ = v_- = 1$, and if its solubility is s mol dm^{-3} the solubility product is s^2 as previously stated on p. 328.

The deviations of strong electrolytes from ideality are caused by the electrostatic forces between the ions in solution. Debye and Hückel proposed a theory of strong electrolytes which enables these departures from ideality, the activity coefficients of individual ions and the conductance of strong electrolytes (p. 347), to be calculated from the forces between the ions. This theory considers the tendency for positive ions to group themselves around the negative ions and vice versa, thus preventing the free movement of the ions relative to each other. (Similarly, in a sodium chloride crystal each sodium is surrounded by chloride ions. In solution, however, the distribution of ions is less regular than in a crystal and the oppositely charged ionic atmosphere around each ion contains solvent molecules.) Debye and Hückel calculated the electric energy of interaction of each ion and related it to the chemical potential of the ion. They obtained the result

$$\log \gamma_\pm = -A|z_+ z_-| \sqrt{I}, \qquad [8.20]$$

where $\gamma_\pm$ is the mean activity coefficient of the electrolyte, z_+ and z_- are the number of protonic or electronic charges on each ion (the ionic valencies) and the modulus signs $||$ indicate that the positive value of the product $z_+ z_-$ should be taken. A is a constant which depends on the temperature and relative permittivity (dielectric constant) of the solvent ($A = 0.509$ kg$^{\frac{1}{2}}$ mol$^{-\frac{1}{2}}$ for water at 25° C) and I is the ionic strength of the solution. This is defined by

$$I = \tfrac{1}{2} \Sigma(mz^2) \qquad [8.21]$$

where m is the molality and z the valency of each ion in the solution and the summation, Σ, includes every type of ion in the solution.

For a solution of a strong uni-univalent electrolyte, e.g. KCl, $z_+ = z_- = 1$, so that:

$$I = m \quad \text{and, in water,} \quad \log \gamma_\pm = -0.509 \sqrt{m}.$$

For a solution of Na$_2$SO$_4$ in water, $z_+ = 1$ and $z_- = 2$, so that:

$$I = \tfrac{1}{2}(2m.1^2 + m.2^2) = 3m \quad \text{and} \quad \log \gamma_\pm = -1.018 \sqrt{3m}.$$

Mean activity coefficients calculated from the Debye–Hückel theory agree satisfactorily with those determined experimentally at ionic strengths below about 10^{-3} to 10^{-2} mol kg^{-1} because only in very dilute solutions can the ions be regarded as point charges. In extremely dilute solutions, the ions are on average so far apart that their electric interactions are negligible, i.e. as $I \to 0$, $\gamma \to 1.0$; this is not only required by the theory but is found experimentally (p. 312, Table 7.2).

Weak Electrolytes

Substances which are incompletely ionised in solution are called weak electrolytes, and include weak acids and bases. Weak acids in aqueous solution include certain inorganic oxy-acids, e.g. H_3BO_3, H_2SO_3 and H_2CO_3, the carboxylic acids (RCOOH, but not CCl_3COOH), phenols, e.g. C_6H_5OH, and thiols (RSH). Weak bases in aqueous solution include ammonia and the amines. The above substances exist mainly as neutral covalent molecules in the pure state and undergo partial electrolytic dissociation in polar solvents. The concentrations of the ions in weak electrolytes is so small that the interactions between the ions, described by Debye and Hückel, do not occur to any appreciable extent.

Consider a solute AB which partly dissociates into two ions, A^+ and B^-. Let α be the degree of dissociation, i.e. the fraction of each mol of AB which is converted into ions. Then, for each mol of AB originally dissolved, $(1-\alpha)$ mol will remain undissociated, and α mol of A^+ and a mol of B^- will be formed. This may be represented as follows:

$$AB \rightleftharpoons A^+ + B^-.$$
$$(1-\alpha) \quad \alpha \quad \alpha$$

The van't Hoff factor i (p. 246) is here equal to the ratio of the number of particles after dissociation to the number of particles before dissociation thus:

$$i = \frac{(1-\alpha)+\alpha+\alpha}{1}$$

$$\therefore \quad i = 1+\alpha \quad [8.22]$$

In a more general case, if each molecule dissociates to give m ions,

$$i = (1-\alpha)+m\alpha.$$

$$\therefore \quad i-1 = \alpha(m-1) \quad [8.23]$$

$$\therefore \quad \alpha = \frac{i-1}{m-1}. \quad [8.24]$$

By means of an accurate measurement of a colligative property (p. 246), i can in principle be determined and hence α may be calculated. This is only possible, however, if at moderate concentrations the solute is appreciably dissociated, i.e. if i differs appreciably from unity and α differs appreciably from zero. If α is very small, it may still be possible to determine α from conductance measurements discussed in the next section.

If the original solute AB is present at a concentration c mol dm^{-3} before dissociation, the concentrations after dissociation will at equilibrium be as follows:

$$[AB] = c(1-\alpha); \quad [A^+] = [B^-] = c\alpha \text{ mol dm}^{-3}.$$

Applying the law of mass action (p. 318) in the appropriate form using concentrations:

$$K_c = \frac{[A^+][B^-]}{[AB]}$$

$$= \frac{(c\alpha)(c\alpha)}{c(1-\alpha)} \quad [8.25]$$

$$K_c = \frac{c\alpha^2}{1-\alpha}. \quad [8.26]$$

The relationship between α and c in terms of the constant K_c is known as the Ostwald Dilution Law. If the solute is a very weak electrolyte, α is very small compared with unity, so that $(1-\alpha) \approx 1$, $\therefore K_c \approx c\alpha^2$. It can be seen that the lower the concentration of the solution, the higher is the degree of dissociation, because the chance of a pair of oppositely charged ions coming within the sphere of attraction of one another and recombining to form a molecule is smaller.

For weak electrolytes K_c calculated from values of α determined experimentally at various concentrations is constant or nearly so, particularly at low concentrations. If AB is a weak acid, K_c is the acid dissociation constant. Ostwald's dilution law is used to calculate dissociation constants from degrees of dissociation determined from conductance data. The law cannot be applied to strong electrolytes, since they are completely dissociated at all concentrations.

Resistance and Conductance

Electrical conduction of a body is the tendency of electrically charged particles (or current carriers) to flow when an electric potential difference is applied to the body and it is an important property of many materials, including electrolytes. The electric current measures the rate of flow of electrical charges and its SI unit is the ampere, A (p. 331). The *potential difference* measures the electrical pressure driving the charges through the body and its SI unit is the volt, V. Ohm's law states that the current I flowing through a conducting body is directly proportional to the potential difference V applied to the body, and may be stated as

$$I = V/R, \quad [8.27]$$

where R is the electrical *resistance* of the body, the SI unit being the ohm, Ω. The reciprocal of resistance is the *conductance G*, whose SI unit is the siemens, S, which is also known as reciprocal ohm, Ω^{-1}.

Resistivity and Conductivity

The resistance of a body depends partly on the material from which it is made and partly on the shape of the body. The problem is best considered by analogy with the passage of water through a pipe (cf. p. 395). The smaller the cross-sectional area of the pipe, the greater will be the resistance to flow. Also, the longer the pipe, the smaller will be the quantity of water flowing in a given time. Similarly, in electrical conduction, the resistance will increase as the cross-sectional area of the conductor decreases and as the length increases. In order to eliminate this effect of shape, and to obtain a resistance which is characteristic of a given material at a given temperature, a quantity known as *resistivity*, ϱ, is defined. This is the resistance between opposite faces of a unit cube of material (of side 1 m) and is related to the resistance of a particular block of material by the formula:

$$\varrho = \frac{RA}{l} \quad \text{or} \quad R = \frac{\varrho l}{A}, \qquad [8.28]$$

where R is the resistance of the block (in Ω),
  l is the length of the block (in m),
  A is the cross-sectional area of the block (in m^2)
and ϱ is the resistivity of the material (in Ω m).

The *conductivity*, κ, of a material is the reciprocal of the resistivity and its SI unit is Ω^{-1} m^{-1} or S m^{-1}.

$$\kappa = \frac{Gl}{A} \quad \text{or} \quad G = \frac{\kappa A}{l}. \qquad [8.29]$$

In the earlier c.g.s. system ϱ has the units Ω cm ($= 10^{-2}\,\Omega$ m) and κ has the units Ω^{-1} cm^{-1} ($= 10^2\,\Omega^{-1}$ m^{-1}).

In order to compare the resisting or conducting powers of different types of solid material it is necessary to compare their resistivities or conductivities at a given temperature, since these properties are temperature dependent. At 25° C *conductors*, which include highly conducting materials such as metals and graphite, have resistivities of the order 10^{-9} to $10^{-5}\,\Omega$ m, which increase with increasing temperature. The current carriers are electrons. As pointed out on p. 329 some electrons are not firmly held by particular atoms so they can move in an electric field and conduct electricity. These electrons occupy a set of energy levels, known as the conduction band, which can be regarded as a group of delocalised molecular orbitals embracing all the atoms of the crystal.

Materials known as *semiconductors* have resistivities at 25° C of the order 10^{-5} to $10^{+7}\,\Omega$ m which, unlike metals, decrease with increasing temperature and which are profoundly affected by impurities. The current carriers are also electrons, but these are more strongly held than in metals. Most electrons in semiconductors are held in the valency bonds and relatively few are present in the conduction band. The relationship between conductivity and temperature has the same form as that between reaction rate constant and temperature (p. 164). On increasing the temperature more electrons jump the energy gap between the valency band, comprising the bonds, and the conduction band. Impurities affect the conductivity by supplying or withdrawing electrons to or from the conduction band. Semiconductors include elements with properties of both metals and non-metals, e.g. germanium and silicon, and metal oxides with lattice defects. They are used in electronic devices, e.g. transistors, for amplification and switching, and thermistors, for temperature measurement and control.

Insulators or dielectrics include non-metals, e.g. sulphur, and plastics such as polyethylene, and gases. These have high resistivities of the order 10^7 to 10^{15} Ω m at 25° C. The electrons are even more strongly held than in semiconductors and virtually none are present in the conduction band. There is a large energy gap between the valency band and the conduction band. High temperatures and high voltages can, however, excite electrons so as to overcome this barrier, whereupon conduction takes place. This phenomenon is known as dielectric breakdown. When an increasing potential is applied across two metal electrodes in a gas at a low pressure, dielectric breakdown occurs, when a certain potential, known as the ionisation potential, is reached. The gas then conducts an electric current and emits a spectrum (p. 81). The *ionisation potential* is defined as the energy which must be absorbed by a molecule in order just to remove one electron from it to infinity, thereby producing a positive ion. Ionisation potentials are frequently measured in electronvolts (eV) per molecule and have the dimensions of charge multiplied by potential difference and therefore of energy. Since the charge associated with one electron is 1.6022×10^{-19} coulomb, and Avogadro's constant is 6.0222×10^{23} molecules mol^{-1}, 1 eV per molecule = 1.6022×10^{-19} J per molecule = 96 487 J mol^{-1} (= the Faraday constant multiplied by 1 V).

When a gas conducts electricity, the current carriers are not only the electrons but also the positive ions formed at the same time. In electrolytes the carriers of electric current are the positive and negative ions they contain and not electrons themselves.

Measurement of Conductance and Conductivity of Electrolytes

The conductance of an electrolyte is measured by determining the resistance of a cell in which the electrolyte is contained between two parallel electrodes. Essentially, the method is the same as that used to

Measurement of Conductance and Conductivity of Electrolytes

find the resistance of a solid conductor by means of a Wheatstone bridge. The circuit is set up as shown in Fig. 8.2 (*a*), the main modification for electrolytes being that the bridge is supplied with alternating current (a.c.) instead of direct current (d.c.), and this is detected by means of a telephone ear-piece, an a.c. meter or cathode ray tube, instead of a galvanometer.

If direct current were passed through the cell, the electrodes would become polarised by the liberation of gas, and a back electromotive force would be set up opposing the passage of current. With alternating

Fig. 8.2 (*a*) Wheatstone bridge circuit
(*b*) Conductance cell

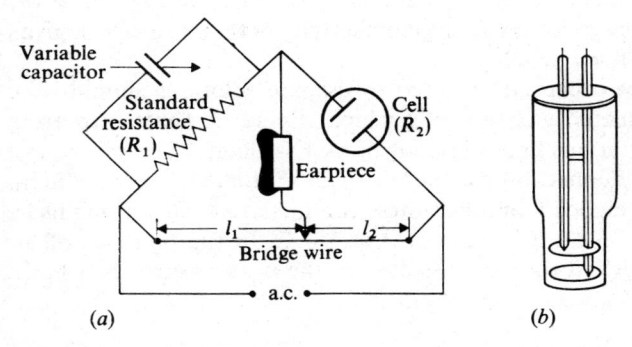

current, however, each electrode is alternately positive and negative, and any small amount of chemical change which takes place during one half-cycle is reversed in the next. With the standard resistance R_1 set to a value roughly equal to the estimated resistance R_2 of the cell, the slider on the bridge wire is moved until no sound is heard in the earpiece. If the distances of the slider from the two ends are then l_1 and l_2, and if the bridge wire has a uniform resistance throughout its length, the resistance R_2 may be calculated from the relationship, $R_1/l_1 = R_2/l_2$.

The cell and its connecting leads behave like a capacitor since they allow a small amount of a.c. to pass between them even when there are no ions between the plates. This *capacitance effect* is balanced out by including a variable capacitor in parallel with R_1 (Fig. 8.2*a*) and adjusting its capacitance until the detector gives a minimum response. Since the capacitance effect becomes greater with increasing frequency, whereas the polarisation effect increases with decreasing frequency, the frequency of the a.c. used to energise the bridge is a compromise and is usually of the order 1 kHz (1000 cycles per second). The ear is particularly sensitive to a sound of this frequency.

A typical conductance cell is shown in Fig. 8.2(*b*). The electrodes are two circular plates of platinum which are platinised (coated electrolytically with finely divided platinum) to reduce the possibility of polarisation. Short pieces of platinum wire welded to the electrodes are sealed into glass tubes, and are connected via copper wire leads to the external circuit.

It is not possible, by measurement of the electrode areas and separation, to convert the resistance accurately into resistivity or a conductivity. This is done, instead, by first making measurements on a substance of known conductivity, e.g. standard KCl solution, and using this to determine a cell constant. If the KCl solution has a conductivity, κ_0, and its measured resistance is R_0, then the cell constant K is given by $K = \kappa_0 . R_0$. (The conductivity of 0.100 mol dm^{-3} KCl at 25° C is 1.289 Ω^{-1} m^{-1}, i.e. 0.01289 Ω^{-1} cm^{-1}.) The KCl is rinsed out, and the electrolyte whose conductivity κ is required is placed in the cell. If the measured resistance is now R, the conductivity of the electrolyte is given by the equation $\kappa = K/R$.

The water or other solvent used to prepare solutions should have as low a conductivity as possible which should be subtracted from those of the solutions under investigation. Conductivity water is most conveniently obtained by passage through an ion exchange column. Air must be excluded from the purest conductivity water which has a conductivity of 3.8×10^{-6} Ω^{-1} m^{-1} at 18° C. In the presence of air atmospheric carbon dioxide dissolves in the water which then has a conductivity of about 10^{-4} Ω^{-1} m^{-1}.

Molar Conductivity of Electrolytes

For a given solid material the conductivity or resistivity depends only on the temperature. For an electrolyte, however, conductivity depends also on the number of ions in a given volume of solution and on the interactions between them. It therefore depends upon the concentration in a rather complex manner. If the electrolyte were fully dissociated and if the ions did not interact, the conductivity would be proportional to the concentration. In order to eliminate the latter effect the conductivity κ of the electrolyte is divided by the concentration c and the resulting quantity, which is known as the *molar conductivity* Λ is used, thus:

$$\Lambda = \kappa/c. \qquad [8.30]$$

Here the term molar means divided by concentration expressed in the SI units mol m^{-3}. Since κ has the SI unit Ω^{-1} m^{-1}, Λ has the SI unit Ω^{-1} m^2 mol^{-1}. The molar conductivity is actually the conductance of the solution if it were contained between parallel electrodes 1 m apart and having sufficient area to accommodate exactly 1 mol of the solute between the plates. This can be seen by substituting in equation [8.29] $G = \Lambda$, $l = 1$ and $Al = 1/c$, whereupon equation [8.30] arises. When molar conductivity is considered, the chemical formula unit whose concentration is c and to which 1 mol refers must be specified (see p. 349).

If the electrolyte were fully dissociated, and if the ions did not interact, the molar conductivity would be independent of concentration. Since neither of these conditions is fulfilled by actual electrolytes, Λ varies with c. Λ is plotted against $1/c$ in Fig. 8.3(a) and against $\sqrt{c}$ in Fig. 8.3(b) for a typical strong electrolyte, e.g. KCl in water, and for

Fig. 8.3 Equivalent conductivity, Λ, of strong and weak electrolytes as functions of concentration, c. (The dotted portions of the curves are not accessible experimentally.)

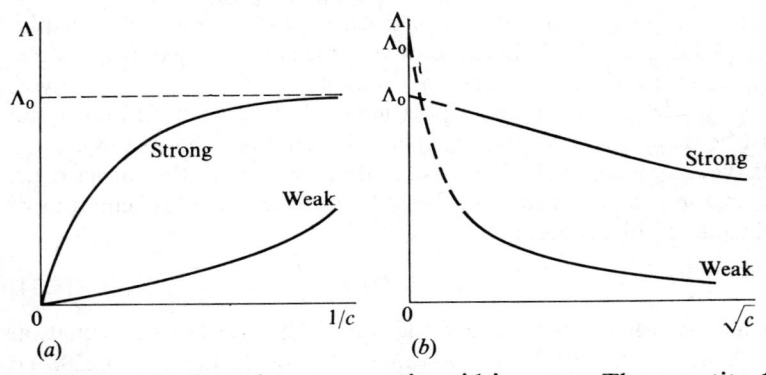

a typical weak electrolyte, e.g. acetic acid in water. The quantity $1/c$ is known as the dilution of the solution and is the volume which contains 1 mol of solute. The reason why $\sqrt{c}$ is chosen in preference to c will be apparent later. At infinite dilution ($1/c = \infty$), i.e. at zero concentration ($c = 0$), any electrolyte, whether strong or weak, is completely dissociated and its ions are on average so far apart that they do not interact. The value to which Λ tends as $1/c$ approaches infinity, or as c approaches zero, measures the current-carrying power of the completely dissociated electrolyte without ionic interactions and is represented by the symbol Λ_0.

For *strong electrolytes* plots of Λ against $\sqrt{c}$ are almost linear at low concentrations and may be readily extrapolated to zero concentration to obtain an accurate value of Λ_0. This linear relationship is expressed by the equation:

$$\Lambda = \Lambda_0 - a\sqrt{c}, \qquad [8.31]$$

where a is a constant for a given electrolyte at a given temperature. The decrease in Λ with increasing concentration expresses the increasing non-ideality introduced by increasing interactions between the ions as they get closer together. By experiment and by derivation from the Debye–Hückel theory Onsager found that:

$$a = A + B\Lambda_0, \qquad [8.32]$$

where A and B are functions only of certain properties of the solvent and the temperature. For water at 25° C and with Λ and Λ_0 in Ω^{-1} m² mol⁻¹ and c in mol m⁻³, $A = 1.90 \times 10^{-4}$ and $B = 7.23 \times 10^{-3}$.

For weak electrolytes Λ is a measure of the current carrying power of the incompletely dissociated electrolyte at a given concentration c, whereas Λ_0 measures that of the completely dissociated electrolyte. Thus, for a weak electrolyte the degree of dissociation α at a given concentration c is given by

$$\alpha = \Lambda/\Lambda_0. \qquad [8.33]$$

The relationship between Λ and c (or $1/c$ or $\sqrt{c}$) is more complex than that for strong electrolytes and is given by eliminating α from equations [8.33] and [8.26]. Because Λ continues to increase with decreasing concentration (or increasing dilution), Λ_0 for weak electrolytes cannot be determined by extrapolation of Λ to zero concentration and must be determined by another method (equation 8.35, below).

If the quantities in this discussion are expressed in the earlier c.g.s. system, the usual units are: $\kappa\, \Omega^{-1}\,\text{cm}^{-1}$, c mol litre⁻¹, $\Lambda\, \Omega^{-1}\,\text{cm}^2\,\text{mol}^{-1}$, and equation [8.30] becomes:

$$\Lambda = 1000\, \kappa/c, \qquad [8.34]$$

but all the other equations are the same. The constants in equation [8.32], however, now have the values $A = 60.2$ and $B = 0.229$. To convert the c.g.s. units to SI base units the numerical values must be treated as follows:

$$\kappa \times 10^2, \quad c \times 10^2, \quad \Lambda \times 10^{-4}.$$

Molar Conductivity and Electric Mobility of Ions

The molar conductivity of an electrolyte at zero concentration depends only on the ions, the solvent and the temperature. Since the current flowing through an electrolyte is carried by both cations and anions, the value of Λ_0 may be regarded as the sum of the contributions from both types of ion. Since Λ_0 refers to a state in which the electrolyte is completely dissociated and in which the ions are so far apart that they do not interact with each other, the contribution made by a particular type of ion to Λ_0 is a constant independent of the other type of ion present in the electrolyte. This conclusion is known as Kohlrausch's law of independent migration of ions. The contribution to Λ_0 from each type of ion is known as the molar conductivity of the ion (or the ionic conductivity) λ, thus:

$$\Lambda_0 = \lambda_+ + \lambda_-. \qquad [8.35]$$

Kohlrausch's law enables Λ_0 for any electrolyte, including a weak electrolyte, to be calculated by addition of the molar conductivities of the two types of ion involved or from the Λ_0 values for an appropriate series of strong electrolytes. For example, Λ_0 for acetic acid, HA, $[\Lambda_0(HA) = \lambda(H^+) + \lambda(A^-)]$ may be calculated from the following known values of Λ_0 in water at 25° C:

$$\Lambda_0(NaA) = \lambda(Na^+) + \lambda(A^-) = 9.10 \text{ k}\Omega^{-1} \text{ m}^2 \text{ mol}^{-1} \quad \text{(i)}$$

$$\Lambda_0(HCl) = \lambda(H^+) + \lambda(Cl^-) = 42.62 \text{ k}\Omega^{-1} \text{ m}^2 \text{ mol}^{-1} \quad \text{(ii)}$$

$$\Lambda_0(NaCl) = \lambda(Na^+) + \lambda(Cl^-) = 12.64 \text{ k}\Omega^{-1} \text{ m}^2 \text{ mol}^{-1} \quad \text{(iii)}$$

Then, adding (i) and (ii) and subtracting (iii):

$$\lambda(Na^+) + \lambda(A^-) + (H^+) + \lambda(Cl^-) - \lambda(Na^+) - \lambda(Cl^-) =$$

$$9.10 + 42.62 - 12.64 \text{ k}\Omega^{-1} \text{ m}^2 \text{ mol}^{-1}$$

$$\therefore \Lambda_0(HA) = \lambda(H^+) + \lambda(A^-) = 39.08 \text{ k}\Omega^{-1} \text{ m}^2 \text{ mol}^{-1}.$$

It can be seen that the molar conductivities of ions and of electrolytes at infinite dilution are additive and obey the usual chemical laws of stoichiometry. The following examples for the sulphate ion and for sodium sulphate in water at 25° C emphasise this and show that when molar conductivities are considered the chemical formula unit to which 1 mol refers must be specified (p. 347).

$$\lambda(SO_4^{2-}) = 2\lambda(\tfrac{1}{2}SO_4^{2-})$$
$$15.80 = 2 \times 7.90 \text{ k}\Omega^{-1} \text{ m}^2 \text{ mol}^{-1}$$

$$\Lambda_0(Na_2SO_4) = 2\lambda(Na^+) + \lambda(SO_4^{2-})$$
$$25.98 = 2 \times 5.09 + 15.80 \text{ k}\Omega^{-1} \text{ m}^2 \text{ mol}^{-1}$$

$$\Lambda_0(\tfrac{1}{2}Na_2SO_4) = \lambda(Na^+) + \lambda(\tfrac{1}{2}SO_4^{2-})$$
$$12.99 = 5.09 + 7.90 \text{ k}\Omega^{-1} \text{ m}^2 \text{ mol}^{-1}.$$

The value of λ for which the chemical formula unit has a charge equal to *one* protonic charge or to *one* electronic charge is sometimes known as the *equivalent* conductivity of the ion. Similarly, the value of Λ for which the chemical formula contains *one* protonic charge and *one* electronic charge is sometimes called the *equivalent* conductivity of the electrolyte. In these cases the amount of substance to which λ and Λ refer is not one mole but one equivalent (p. 330). For example, the equivalent conductivity of the sulphate ion in water at 25° C is 7.90 kΩ^{-1} m^2 eq.$^{-1}$ and the equivalent conductivity of sodium sulphate at infinite dilution is 12.99 kΩ^{-1} m^2 eq.$^{-1}$. For singly charged ions, e.g. Na$^+$ and Cl$^-$, and for uni-univalent electrolytes, e.g. KCl and acetic acid, the molar and equivalent conductivities are equal.

350   Electrolytes

The solubility s of a sparingly soluble salt can be determined from conductivity measurements, provided that the molar conductivities of the ions at the temperature of these measurements are known to enable Λ_0 to be calculated. Since the saturated solution is dilute, $\Lambda_0 \approx \Lambda = \kappa/c$, where $c = s$ (mol m^{-3}). The conductivity κ is the conductivity of a saturated solution of the salt in conductance water minus that of the water alone. The solubility s is calculated from:

$$s = \kappa/\Lambda_0. \qquad [8.36]$$

A particle which bears a charge, e.g. an ion, tends to move in the direction of an applied electric field with a velocity proportional to the field strength. The electric field strength between two parallel plates is the potential difference applied to the plates divided by the distance between them and its SI unit is V m^{-1}. The electric mobility u of a particle, e.g. an ion, is the velocity of motion of the particle (in m s^{-1}) divided by the electric field strength (in V m^{-1}) creating the motion and its SI unit is m^2 s^{-1} V^{-1}.

Consider an enclosed block of liquid solvent containing c mol m^{-3} of a uni-univalent electrolyte completely dissociated into ions which do not interact. This block is arranged to fit exactly between two plates each of area A m^2 and separated by l m. The block has a volume Al m^3 and therefore contains cAl mol of electrolyte.

Since the charge associated with 1 mol of a univalent ion (1 eq.) is F coulomb, the charge associated with all the cations or all the anions in the block is $FcAl$ coulomb, or for each unit length of the block FcA C m^{-1}. If a potential difference V volt is applied to the plates l m apart, the electric field strength between the plates is V/l V m^{-1}. If the electric mobility of the cation is u_+ and of the anion is u_- m^2 s^{-1} V^{-1}, the velocity of migration of the cation will be u_+V/l m s^{-1} and of the anion will be u_-V/l m s^{-1}. The rate of flow of charge across the block is the product of charge per unit length and velocity and is $FcAu_+V/l$ C s^{-1} due to the cation and $FcAu_-V/l$ C s^{-1} due to the anion; these are the respective currents in A. The total current I passing between the plates is the sum of these values, thus:

$$I = (u_+ + u_-)FcAV/l. \qquad [8.37]$$

The conductance of the block of electrolyte is given by

$$G = I/V \qquad [8.38]$$

∴ $$G = (u_+ + u_-)FcA/l. \qquad [8.39]$$

The conductivity κ of the electrolyte is given by substitution into equation [8.29], thus:

$$\kappa = (u_+ + u_-)Fc. \qquad [8.40]$$

The molar conductivity Λ of the electrolyte is given by substitution into equation [8.30], as follows:

$$\Lambda = (u_+ + u_-)F. \qquad [8.41]$$

Since the electrolyte is completely dissociated into ions which do not interact (equation 8.35) gives:

$$\Lambda = \Lambda_0 = \lambda_+ + \lambda_-.$$

Comparison of the last two equations shows that:

$$\lambda_+ = u_+ F, \quad \lambda_- = u_- F \quad \text{and} \quad \Lambda_0 = (u_+ + u_-)F \qquad [8.42]$$

$$u_+ = \frac{\lambda_+}{F}, \quad u_- = \frac{\lambda_-}{F} \quad \text{and} \quad (u_+ + u_-) = \frac{\Lambda_0}{F}. \qquad [8.43]$$

The fraction of the total current carried by a particular type of ion is the transport number t of that type of ion. It has been shown that the currents carried are as follows: by the cations, constant $\times u_+$; by the anions, constant $\times u_-$; the total current is, constant $\times (u_+ + u_-)$; where the constant $= Fc\Lambda V/l$. The transport number of the cation is therefore given by:

$$t_+ = \frac{u_+}{u_+ + u_-} = \frac{\lambda_+}{\Lambda_0}. \qquad [8.44]$$

The transport number of the anion is similarly given by:

$$t_- = \frac{u_-}{u_+ + u_-} = \frac{\lambda_-}{\Lambda_0}. \qquad [8.45]$$

We note here that $t_+ + t_- = 1$ as in equation [8.4]. The transport numbers are the values for the completely dissociated electrolytes whose ions do not interact, i.e. the values at infinite dilution. Knowing t and Λ_0 it is possible to obtain λ for any ion. Since molar conductivities and electric mobilities of ions increase by about 2 per cent for every degree increase, temperature must be accurately controlled in conductance work.

In water at 25° C the molar conductivities of most univalent ions are of the order 4 to 8 $k\Omega^{-1}$ m^2 mol^{-1} (or mS m^2 mol^{-1}), but those of H^+ and OH^- are 35.0 and 19.2 $k\Omega^{-1}$ m^2 mol^{-1} respectively. The larger values for H^+ (more properly H_3O^+) and OH^- reflect the greater electric mobility of these ions, which probably arises from the ability of the electrons in the water molecule to alternate rapidly between a covalent bond and a lone pair or hydrogen bond. This enables the following chain mechanisms to occur rapidly:

$$\overset{H}{\underset{+}{|}}\text{H—O—H} \quad \overset{H}{|}\text{O—H} \quad \overset{H}{|}\text{O—H} \quad \overset{H}{|}\text{O—H} \quad \overset{H}{|}\text{O—H}$$

$$\downarrow$$

$$\overset{H}{|}\text{H—O} \quad \overset{H}{|}\text{H—O} \quad \overset{H}{|}\text{H—O} \quad \overset{H}{|}\text{H—O} \quad \overset{H}{|}\text{H—O—H} \quad \text{and}$$
$$+$$

$$\overset{H}{\underset{-}{|}}\text{O} \quad \overset{H}{|}\text{H—O} \quad \overset{H}{|}\text{H—O} \quad \overset{H}{|}\text{H—O} \quad \overset{H}{|}\text{H—O}$$

$$\downarrow$$

$$\overset{H}{|}\text{O—H} \quad \overset{H}{|}\text{O—H} \quad \overset{H}{|}\text{O—H} \quad \overset{H}{|}\text{O—H} \quad \overset{H}{\underset{-}{|}}\text{O}$$

The high electric mobilities of H^+ and OH^- cause their transport numbers to be rather higher than those of the accompanying ions (p. 336).

Conductimetric Titrations

Since the conductivity κ of a dilute solution of a strong electrolyte is approximately proportional to the concentration (p. 346), changes in the concentrations of electrolytes which react with each other during titration can often be studied by measurements of the conductance of the solution. The comparatively high molar conductivities of H^+ and OH^- makes the conductimetric method very suitable for acid-base titrations (see Fig. 8.4). If a solution of a strong base, such as sodium hydroxide, is added from a burette to a solution of a strong acid, e.g.

Fig. 8.4 Conductimetric titration curves
(a) acids with a strong base (b) a weak acid with a weak base

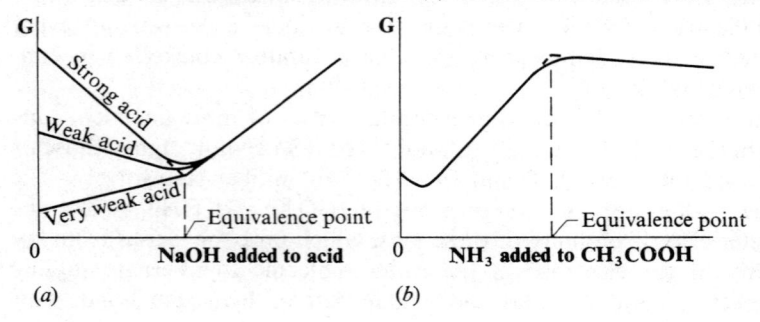

(a) NaOH added to acid

(b) NH₃ added to CH₃COOH

hydrochloric acid, in a beaker containing the two electrodes of a conductance cell, the conductance falls as hydrogen ions are removed and replaced by other cations, e.g. Na^+ (Fig. 8.4a), the equation for the reaction being:

$$Na^+ + OH^- + H^+ + Cl^- = Na^+ + Cl^- + H_2O.$$

After the equivalence point, free hydroxyl ions appear in the solution causing a rise of conductance. If the acid is weak, it produces only a small concentration of hydrogen ions, so that the slope before the endpoint is small. If the acid is very weak, e.g. acetic acid, HA, the concentration of hydrogen ions may be so small that the conductance rises as the salt is formed (Fig. 8.4a), the equation being:

$$Na^+ + OH^- + HA = Na^+ + A^- + H_2O.$$

Analogous behaviour is observed when a strong acid is added to a weak base.

A weak acid, e.g. acetic acid, CH_3COOH, may be accurately titrated conductimetrically with a weak base, e.g. ammonia, according to the following equation:

$$NH_3 + CH_3COOH = NH_4^+ + CH_3COO^-.$$

Very few other satisfactory methods are available for the titration of weak acids with weak bases. When a small quantity of aqueous ammonia is added to aqueous acetic acid, the small amount of salt which is formed suppresses the slight ionisation of acetic acid and the conductance falls slightly. As more ammonia is added, more salt is formed. The number of ions and the conductance of the solution then increase linearly and reach a maximum at the equivalence point (Fig. 8.4b). Excess of weakly ionised ammonia solution merely dilutes the ammonium acetate solution and causes a relatively small change in conductance.

Many types of titration involving ions can be carried out conductimetrically in water or in other solvents, e.g. glacial acetic acid. Such reactions include the formation of molecules or complexes which are sparingly soluble, e.g. silver halides, poorly ionised, e.g. water, or which carry more or less charge than the reactants, e.g. complex ions.

The increase in volume associated with the addition of the titrant will itself affect the conductance of the solution. A correction can be made by multiplying the conductance by the total volume of the solution in the beaker divided by the initial volume. Alternatively, the effect may be reduced to insignificance by titrating with a much stronger solution than the one to be titrated. The conductance is plotted against the volume of titrant added and straight lines are drawn through the points. The equivalence point is given by the point of intersection. Points near the equivalence point often deviate from the straight lines due to the hydrolysis or dissociation of a soluble reaction product or to the solubility of an insoluble reaction product.

Problems

1 At 25° C, silver chloride has a solubility of 0.195 mg in 100 cm^3 of water. Calculate the solubility product at 25° C assuming that activity coefficients are unity when concentrations are expressed in mol dm^{-3}. What would be the concentration of silver ions in a saturated solution of silver chloride at 25° C in an aqueous solution of hydrochloric acid containing 2.0 mol of HCl per dm^3 of solution?

2 During a transport number determination, an iodine coulometer was wired in series with the apparatus. The iodine liberated was equivalent to 30.8 cm^3 of sodium thiosulphate solution containing 0.1 mol dm^{-3} of $Na_2S_2O_3$. Calculate the number of faradays (moles of electrons) which passed through the transport apparatus.

3 In the transport number determination mentioned in problem 2, silver nitrate was electrolysed between silver electrodes. The analysis of the anode compartment after electrolysis showed that it contained 2.358 g of $AgNO_3$ and 99.533 g of water. In the original silver nitrate solution, 99.533 g of water contained 2.079 g of $AgNO_3$. Calculate the transport numbers of the Ag^+ and NO_3^- ions in this solution. Given that the molar conductivity at zero concentration for $AgNO_3$ is 1.333×10^{-2} Ω^{-1} m^2 mol^{-1}, calculate the separate molar conductivities at zero concentration for the Ag^+ and NO_3^- ions,

4 For the aqueous solutions (a) 0.001 mol kg^{-1} barium chloride and (b) 0.0003 mol kg^{-1} magnesium sulphate calculate for each total electrolyte at 25° C (i) the ionic strength, (ii) the mean activity coefficient from the Debye–Hückel equation, (iii) the mean molarity, (iv) the mean activity and (v) the activity.

5 The resistivities of solutions containing 0.01 mol dm^{-3} of HCl, KCl and CH_3COOH are 2.43, 7.10 and 61.80 Ω m, respectively. Calculate their molar conductivities, and give a qualitative explanation of their relative orders of magnitude.

6 The following table gives the conductivities of HCl and CH_3COOH solutions (in Ω^{-1} m^{-1}) at various concentrations c. Calculate the molar conductivities and plot graphs of molar conductivity versus the reciprocal of the concentration. Estimate the molar conductivity at zero concentration for HCl.

$c/\text{mol m}^{-3}$	100	50	20	10
HCl	3.91	2.00	8.14×10^{-1}	4.12×10^{-1}
CH_3COOH	5.20×10^{-2}	3.68×10^{-2}	2.31×10^{-2}	1.62×10^{-2}

$c/\text{mol m}^{-3}$	5	2	1
HCl	2.08×10^{-1}	8.40×10^{-2}	4.21×10^{-2}
CH_3COOH	1.14×10^{-2}	7.12×10^{-3}	4.86×10^{-3}

7 Plot the molar conductivities given in problem 6 for HCl against an appropriate function of the concentration to obtain an approximately linear graph from which to obtain a more reliable value for the molar conductivity at zero concentration and the constant a for the electrolyte. Compare this value of a with that predicted by equation [8.32].

8 The molar conductivities at zero concentration for the ions H^+ and CH_3COO^- are 3.50×10^{-2} and $4.09 \times 10^{-3}\, \Omega^{-1}\,\text{m}^2\,\text{mol}^{-1}$ respectively. Use this information and the molar conductivities at various dilutions for CH_3COOH calculated in problem 6 to calculate the conductivity ratio Λ/Λ_0 at each dilution. From the approximate form of the Ostwald dilution law, in which $(1-\alpha)$ is taken as being always 1, devise and carry out an appropriate linear plot of α against a function of the concentration from which to obtain the dissociation constant from the slope.

9 The conductivity of a saturated aqueous solution of calcium fluoride at $25°$ C is $5.04 \times 10^{-5}\, \Omega^{-1}\,\text{cm}^{-1}$ and that of the sample of water from which it was prepared was $7.0 \times 10^{-7}\, \Omega^{-1}\,\text{cm}^{-1}$ at $25°$ C. At this temperature the molar conductivities of the ions are:

$$\lambda(\tfrac{1}{2}Ca^{2+}) = 59.5\, \Omega^{-1}\,\text{cm}^2\,\text{mol}^{-1}; \quad \lambda(F^-) = 55.4\, \Omega^{-1}\,\text{cm}^2\,\text{mol}^{-1}.$$

Calculate the solubility of calcium fluoride in mass concentration units. Also calculate the solubility product, assuming activity coefficients to be unity.

10 The molar conductivities extrapolated to zero concentration for Na^+ and Cl^- are 4.28×10^{-3} and $6.60 \times 10^{-3}\, \Omega^{-1}\,\text{m}^2\,\text{mol}^{-1}$ respectively at $18°$ C. Calculate the transport numbers and electric mobilities of the ions in a very dilute solution of sodium chloride at $18°$ C.

9 Acids and Bases

The Definitions of Acid and Base

Acids and bases constitute a group of electrolytes of such great importance to biology and pharmacy that a separate chapter is devoted to them. Before any detailed information was available as to the chemical composition of acids and bases, these substances were characterised by the fact that acids possess a sour taste, and cause vegetable dyes, such as litmus, to turn from blue to red, whereas bases are capable of reversing the action of acids, and, in particular, of reacting with acids to form salts. Davy was among the first to realise that hydrogen is the essential constituent of acids.

When Arrhenius studied the dissociation of electrolytes (see p. 341), he found that it was those hydrogen atoms of the molecule which were capable of dissociating in solution to form hydrogen ions, which conferred the characteristic properties on an acid. The higher the degree of dissociation, the stronger was the acid. Moreover, he was able to correlate this measure of acid strength with another property which had come to be regarded as typical of acids, namely, catalytic power (see p. 182). The electrical conductance, which was used to determine the degree of dissociation, was found to be roughly proportional to the catalytic power for a series of acids. This point is discussed further on p. 387.

According to the dissociation theory of acids and bases, an acid is a substance which ionises to form hydrogen ions in solution. This process was represented by the equation:

$$HA \rightleftharpoons H^+ + A^-.$$

And a base is a substance which ionises to give hydroxyl ions in solution:

$$MOH \rightleftharpoons M^+ + OH^-.$$

The process of neutralisation is therefore a reaction between hydrogen ions and hydroxyl ions to form water molecules. This view is supported by the fact that the enthalpies of neutralisation of all strong acids with strong bases are approximately the same (p. 375).

Various difficulties arose, however, on further consideration of the dissociation theory. It was not clear whether a covalent molecule, such as hydrogen chloride, should be regarded as an acid, or whether it only became an acid when it ionised in solution. Again, many typical bases, such as ammonia and the amines, were known to contain no hydroxyl groups, and could only produce hydroxyl ions by reaction with water. It was suggested that undissociated ammonium hydroxide was first formed when ammonia was dissolved in water, and that this ionised to give hydroxyl ions:

$$NH_3 + H_2O \rightleftharpoons NH_4OH \rightleftharpoons NH_4^+ + OH^-.$$

Quite apart from the fact that there is now some doubt as to the existence of undissociated ammonium hydroxide the difficulty as to whether anhydrous ammonia should be regarded as a base was not overcome by this suggeston.

Such difficulties are increased if acids and bases are studied in solvents other than water. Ammonia dissolved in alcohol does not give rise to hydroxyl ions, yet it still exhibits typically basic properties in that it neutralises acids. In fact, the typical basic ion in ethanol solutions, which takes the place of the hydroxyl ion in water, is EtO^-. This may be seen by the fact that sodium ethoxide is a stronger base in ethanol than is sodium hydroxide. In order to extend the dissocation theory to cover non-aqueous solvents, a base must be defined as a substance which ionises to give anions characteristic of the solvent (in the sense that EtO^- is characteristic of EtOH, and OH^- is characteristic of H_2O). A similar modification must be made in the definition of acids, but here the change is not as fundamental. An acid must be defined as a substance which ionises to give cations characteristic of the solvent. In each case, the cations are hydrogen ions. The earlier view that the hydrogen ion is a proton, H^+, has been superseded, because it is found that the proton is always solvated in solution. The hydrogen ion is therefore characteristic of the solvent in that it consists of a proton attached to a solvent molecule. Thus, in water, acids give rise to H_3O^+ ions $(H_2O + H^+)$, and in ethanol they give $EtOH_2^+$ ions $(EtOH + H^+)$.

The modified definitions of acid and base are applicable in most cases, but they still leave difficulties in connection with acidic and basic substances in solvents in which ionisation does not occur, or when reactions take place in the absence of solvents. It is also inconvenient to have to modify one's view of what constitutes an acid or a base for every different solvent. A more general definition of acids and bases was proposed simultaneously by Brönsted and Lowry, and their view has been very widely adopted. It states that an acid is a species which tends to lose protons, and a base is a species which tends to accept protons. Thus, the fundamental acid-base equilibrium may be written:

$$A \rightleftharpoons B + H^+.$$

A is the acid which tends to lose protons by the forward reaction, and B is the base which tends to accept protons in the reverse reaction. An acid and a base which differ in their structure by a proton, and are related by an equilibrium such as the one above, are known as a 'conjugate' acid-base pair.

The Brönsted-Lowry view of acids and bases is the one which is adhered to in this book, and its main implications are discussed in this chapter. It will be seen that it provides an excellent basis for the quantitative description of acid-base equilibria, and, though it will be applied here only to aqueous systems, it is equally applicable to any solvent system.

In the fundamental equilibrium above, A and B may be molecules or ions of any sign or charge, provided only that A is one unit more positive in charge than B. Thus, the electrically neutral acid molecule HCl has as its conjugate base the negatively charged chloride ion. (The chloride ion is a base, because it tends to accept a proton to form the molecule HCl.) Similarly, the acid salt ion HSO_4^- has as its conjugate base the sulphate ion SO_4^{2-}. On the other hand, the ammonium ion NH_4^+ may equally well be regarded as an acid, in that it tends to lose a proton to form its conjugate base NH_3.

The Brönsted-Lowry definition of an acid does not differ much from that given by the dissociation theory. The significant modification is that the acid need only *tend* to lose *protons*, rather than ionise to form hydrogen ions in solution. The definition therefore applies in any solvent, and even to substances which do not ionise, since they can still donate protons directly to the base in an acid-base reaction. The modification in the definition of a base is more fundamental, since the capacity for accepting protons is emphasised, and not the ability to give hydroxyl ions. In fact, a substance which owes its basic properties to the formation of hydroxyl ions in solution is a special case of the Brönsted-Lowry definition, since it is the hydroxyl ion itself which is the base in such cases. The hydroxyl ion is able to accept protons to form water, and is therefore a base. A substance such as ammonia is a base whether it is in the gaseous or liquid state, or whether it is in solution, since it is capable of accepting protons to form ammonium ions.

The Strengths of Acids and Bases

The fundamental equilibrium relating the conjugate acid-base pair is not one which can occur in practice in solution. As the acid loses protons, there must be some other substance present to accept them, even if this is only the solvent of the acid solution. This other substance must necessarily be a base, since it is accepting protons, and the substance which is formed by the acceptance of the protons must be the conjugate acid of this base. In practice, therefore, every acid-base equilibrium must involve two conjugate acid-base pairs:

$$A_1 + B_2 \rightleftharpoons B_1 + A_2,$$

where A_1 and B_1 are one conjugate pair, and A_2 and B_2 are the other.

The strength of an acid is its power to provide protons, and this would theoretically be measured by determining how far to the right is the equilibrium $A \rightleftharpoons B + H^+$. In practice, the tendency to provide protons must be measured in a way which unavoidably includes the tendency of some base to accept the protons provided. Thus, if the strength of the acid A_1, above, is measured by the extent of its reaction with the base B_2, this clearly depends also on the power of B_2 to accept protons. A strong base would cause the equilibrium to be further over to the right than would a weak base for a given acid strength. In order to compare strengths of a series of acids or a series of bases, it is necessary to have a standard base or acid, respectively, against which to measure the strengths. The solvent is chosen as the standard base or acid in each case.

Thus, in aqueous systems, the strength of an acid HA is measured by the forward tendency (left to right) of the equilibrium:

$$HA + H_2O \rightleftharpoons H_3O^+ + A^-.$$

In this equilibrium, water is acting as a base, because it is accepting protons from the acid. The hydrogen ion, H_3O^+, is the conjugate acid of H_2O, and the anion A^- is the conjugate base of HA. The forward tendency of the equilibrium is measured by its equilibrium constant, which may be written (in its correct form, involving activities instead of concentrations) thus:

$$K = \frac{a_{H_3O^+} \times a_{A^-}}{a_{HA} \times a_{H_2O}}. \qquad [9.1]$$

The higher the value of K, the further to the right is the equilibrium and the stronger is the acid. The solvent, water, is usually present in great excess, and the small amount which is used up in the dissociation of the acid leaves its concentration and activity virtually unaffected. The activity of water is therefore that of pure water in its standard state (i.e. under the conditions in which it usually exists at room temperature), and, by convention, this is made equal to unity (see p. 223). The dissociation constant of the acid is therefore usually written:

$$K_a = \frac{a_{H_3O^+} \times a_{A^-}}{a_{HA}}. \qquad [9.2]$$

360 Acids and Bases

The approximate form of the dissociation constant, involving concentrations, is often used, and may be distinguished from the correct form by writing it as k_a, thus:

$$k_a = \frac{[H_3O^+][A^-]}{[HA]}. \qquad [9.3]$$

Apart from writing the hydrogen ion in its correct form, this dissociation constant is identical with that which would be given by the dissociation theory.

Polybasic acids, such as sulphuric and phosphoric acids, have a separate dissociation constant for the ionisation of each hydrogen ion. Thus, in the case of phosphoric acid:

$$H_3PO_4 + H_2O \rightleftharpoons H_3O^+ + H_2PO_4^- \qquad K_1 = \frac{a_{H_3O^+} \times a_{H_2PO_4^-}}{a_{H_3PO_4}}.$$

$$H_2PO_4^- + H_2O \rightleftharpoons H_3O^+ + HPO_4^{2-} \qquad K_2 = \frac{a_{H_3O^+} \times a_{HPO_4^{2-}}}{a_{H_2PO_4^-}}.$$

$$HPO_4^{2-} + H_2O \rightleftharpoons H_3O^+ + PO_4^{3-} \qquad K_3 = \frac{a_{H_3O^+} \times a_{PO_4^{3-}}}{a_{HPO_4^{2-}}}.$$

The successive ionisations become increasingly difficult, because the protons must be removed from residues which bear increasingly greater negative charges. The higher the negative charge on the residue, the greater is its attraction for protons. This is reflected in the magnitudes of the dissociation constants, which are: $K_1 = 1.1 \times 10^{-2}$; $K_2 = 7.5 \times 10^{-8}$; $K_3 = 4.8 \times 10^{-13}$.

The strength of a base in aqueous solution is measured relative to water by the equilibrium:

$$B + H_2O \rightleftharpoons BH^+ + OH^-.$$

Here, water is acting as an acid because it is giving protons to the base. The dissociation constant of the base is derived from this equilibrium, with the omission of a term for the activity of the solvent, for the same reason as before:

$$K_b = \frac{a_{BH^+} \times a_{OH^-}}{a_B}. \qquad [9.4]$$

It should be noted that the hydroxyl ions in this equilibrium come from the water molecules and are not produced by the breakdown of the base B. Thus in the definition of basic strength, there is a fundamental difference between the Brönsted–Lowry and the dissociation theories. According to the latter, the strength of the base would be judged by the tendency for the molecules of the base to ionise. The equilibrium above would be regarded as giving the hydrolysis constant for the base.

An interesting fact emerges if the strength of an acid is compared with that of its conjugate base. Thus, it has already been shown above that the strength of an acid HA is given by:

$$K_a = \frac{a_{H_3O^+} \times a_{A^-}}{a_{HA}}. \qquad [9.2]$$

The conjugate base is A^-, and its strength is determined by the equilibrium:

$$A^- + H_2O \rightleftharpoons HA + OH^-.$$

The basic dissociation constant of A^- is therefore:

$$K_b = \frac{a_{HA} \times a_{OH^-}}{a_{A^-}}. \qquad [9.5]$$

If K_a and K_b are multiplied together, we have:

$$K_a \times K_b = \frac{a_{H_3O^+} \times a_{A^-} \times a_{HA} \times a_{OH^-}}{a_{HA} \times a_{A^-}}$$

$$= a_{H_3O^+} \times a_{OH^-}. \qquad [9.6]$$

The product of the activities of hydrogen and hydroxyl ion in aqueous solutions is a constant at a given temperature, called the ionic product for water, and denoted by K_w. It is a measure of the dissociation of water according to the equilibrium:

$$H_2O + H_2O \rightleftharpoons H_3O^+ + OH^-.$$

The equilibrium constant, putting $a_{H_2O} = 1$ as before, is:

$$K_w = a_{H_3O} \times a_{OH^-}. \qquad [9.7]$$

Thus, the strength of an acid and its conjugate base are related by the expression:

$$K_a \times K_b = K_w. \qquad [9.8]$$

This relationship is a useful one in that it enables one to say that, if an acid is weak, its conjugate base must be strong, and vice versa. It also enables one to calculate the dissociation constant of a base if only the dissociation constant of its conjugate acid is known, and vice versa. The strength of a base is often expressed by quoting K_a for its conjugate acid.

Many dissociation constants are very small numbers, and, like hydrogen-ion activities, they are more conveniently written in logarithmic form. By analogy with the hydrogen-ion exponent, pH, the dissociation constant exponents are written as 'pK', thus:

$$\text{pH} = -\log_{10} a_{H^+}, \qquad [9.9]$$

$$\text{p}K = -\log_{10} K. \qquad [9.10]$$

On this notation, fairly strong acids have small pK_a values, and weak acids have large values. Water itself could be regarded as a very weak acid, of $pK_a = 14$, since K_w is known to be 10^{-14} at room temperature. Strong bases have small pK_b values, but, if they are quoted in terms of the pK_a values of their conjugate acids, these are large numbers. This follows by taking logarithms of the expression $K_a \times K_b = K_w$, giving:

$$\log K_a + \log K_b = \log K_w.$$

Multiplying throughout by -1:

$$-\log K_a - \log K_b = -\log K_w.$$

Hence:

$$pK_a + pK_b = pK_w. \qquad [9.11]$$

Thus, if water is regarded as a very weak base, its pK_b value is 14, but, since pK_w is 14, the basic dissociation constant of water may be expressed by the pK_a value of its conjugate acid, the hydrogen ion, for which $pK_a = 0$.

The dissociation constants of acids and bases may be measured by a number of methods. The conductance method was referred to above. This depends upon the comparison of the molar conductivity at a known dilution with that at infinite dilution, to obtain the degree of dissociation, and hence the dissociation constant, by Ostwald's dilution law (see pp. 341, 342). The most accurate method employs a concentration cell and an extrapolation of a function involving the measured e.m.f. The theory of this method is beyond the scope of this book. A simpler method, involving the measurement of the e.m.f. of cells, is one in which the pH of solutions containing the acid or base and its salt is determined by means of a hydrogen or glass electrode (see pp. 396, 414). The pH may also be determined by an indicator method, such as that described in experiment 9.2 p. 381. The pK value is then obtained from the pH by employing one of the important relationships derived as follows:

(a) The pH of a solution of a weak acid.

Suppose the solution were prepared by making up c mol of HA to 1 dm³ in water. If α is the degree of dissociation, then αc mol of hydrogen ion, and αc mol of A⁻ are formed by the dissociation, and $c(1-\alpha)$ mol of the acid remain as undissociated HA (cf. p. 342). This may be written:

$$\underset{c(1-\alpha)}{HA} + H_2O \rightleftharpoons \underset{\alpha c}{H_3O^+} + \underset{\alpha c}{A^-}$$

The dissociation constant may now be calculated in terms of α and c, if the approximation is made of replacing activities by concentrations (the dissociation constant being now written as k_a). Thus:

$$k_a = \frac{\alpha c \times \alpha c}{c(1-\alpha)}$$

$$= \frac{\alpha^2 c}{(1-\alpha)}. \qquad [9.12]$$

It would be possible to solve this equation exactly for α. More frequently, an approximate solution is obtained by assuming that α is sufficiently small for it to be negligible in comparison with unity.

Then, $\qquad (1-\alpha) \approx 1$

and $\qquad k_a = \alpha^2 c$

∴ $\qquad \alpha^2 = k_a/c$

and $\qquad \alpha = \sqrt{k_a/c}. \qquad [9.13]$

Now, $\qquad [H_3O^+] = \alpha c, \qquad [9.14]$

∴ $\qquad [H_3O^+] = c\sqrt{k_a/c}$

$\qquad = \sqrt{k_a \cdot c,} \qquad [9.15]$

∴ $\qquad \log[H_3O^+] = \tfrac{1}{2}\log k_a + \tfrac{1}{2}\log c,$

∴ $\qquad -\log[H_3O^+] = -\tfrac{1}{2}\log k_a - \tfrac{1}{2}\log c,$

∴ $\qquad \text{pH} = \tfrac{1}{2}\text{p}k_a - \tfrac{1}{2}\log c. \qquad [9.16]$

(b) *The pH of a solution of a weak acid and its salt.*

Whatever the relative proportions of the acid and its salt in the solution, their activities must always be related to that of the hydrogen ion by the equilibrium constant. Expressing this in the approximate form, involving concentrations:

$$k_a = \frac{[H_3O^+][A^-]}{[HA]}. \qquad [9.3]$$

Taking logarithms of both sides, and separating the term involving the hydrogen ion concentration on the right-hand side:

$$\log k_a = \log[H_3O^+] + \log\left(\frac{[A^-]}{[HA]}\right),$$

∴ $\qquad -\log k_a = -\log[H_3O^+] - \log\left(\frac{[A^-]}{[HA]}\right),$

∴ $\qquad \text{p}k_a = \text{pH} - \log\left(\frac{[A^-]}{[HA]}\right).$

Finally, adding the log term to both sides.

$$\text{pH} = \text{p}k_a + \log\left(\frac{[A^-]}{[HA]}\right).$$

If the acid is fairly weak, the amount of HA converted into A^- is small, and this ionisation is further inhibited by the addition of the fully dissociated salt of the acid, which provides a relatively large concentration of A^- ions (the common ion effect; cf. p. 328). The amount of HA in the solution, after dissociation has taken place, may therefore be taken to be the same as the concentration of acid before dissociation. Also, the concentration of A^- may be taken as being entirely due to the added salt of the acid. The equation may then be written:

$$\mathrm{pH} = \mathrm{p}k_a + \log\left(\frac{[\text{salt}]}{[\text{acid}]}\right). \qquad [9.17]$$

Alternatively, since A^- is the conjugate base of the acid HA, it is sometimes expressed as:

$$\mathrm{pH} = \mathrm{p}k_a + \log\left(\frac{[\text{base}]}{[\text{acid}]}\right). \qquad [9.18]$$

This is known as the Henderson equation, and it is used in the calculation of the pH of buffer solutions (see p. 367). A more accurate equation, [9.29], is derived below which does not involve the approximations of the concentrations of acid and base.

(c) *The pH of a solution of a weak base.*

Suppose the solution were prepared by making up c mol of base to 1 dm³ in water. If the degree of dissociation were α, the equilibrium and the concentrations after dissociation would be:

$$\underset{c(1-\alpha)}{B} + H_2O \rightleftharpoons \underset{\alpha c}{BH^+} + \underset{\alpha c}{OH^-}.$$

The dissociation constant will be given, as in the case of the acid dissociation, by:

$$k_b = \frac{\alpha^2 c}{(1-\alpha)}. \qquad [9.19]$$

As before, when $(1-\alpha)$ is made equal to unity

$$\alpha = \sqrt{k_b/c}. \qquad [9.20]$$

Now $\qquad [OH^-] = c\alpha \qquad [9.21]$

$$= \sqrt{k_b \cdot c}. \qquad [9.22]$$

In order to obtain the pH, it is necessary to relate the concentration of hydroxyl ion to that of hydrogen ion by means of the ionic product:

$$[H_3O^+] \times [OH^-] = k_w,$$

$$\therefore \qquad [H_3O^+] = k_w/[OH^-]. \qquad [9.23]$$

Hence, substituting the value of the hydroxyl ion concentration derived above:

$$[H_3O^+] = k_w/\sqrt{k_b \cdot c},$$
$$\log [H_3O^+] = \log k_w - \tfrac{1}{2}\log k_b - \tfrac{1}{2}\log c.$$
$$\therefore \quad pH = pk_w - \tfrac{1}{2}pk_b + \tfrac{1}{2}\log c. \qquad [9.24]$$

(d) *The pH of a solution of a weak base and its salt.*

The salt of the base B provides a relatively large concentration of the ions BH^+, and the concentration of OH^- ions is related to the concentrations of these two species by the approximate equilibrium constant:

$$k_b = \frac{[BH^+][OH^-]}{[B]},$$

$$\therefore \quad [OH^-] = k_b \cdot \frac{[B]}{[BH^+]}.$$

But $\quad [H_3O^+] \times [OH^-] = k_w,$

$$\therefore \quad [H_3O]^+ = k_w/[OH^-]$$
$$= \frac{k_w \times [BH^+]}{k_b \times [B]},$$

$$\therefore \quad \log [H_3O^+] = \log k_w - \log k_b + \log\left(\frac{[BH^+]}{[B]}\right),$$

$$\therefore \quad pH = pk_w - pk_b - \log\left(\frac{[BH^+]}{[B]}\right). \qquad [9.25]$$

As in the case of the acid dissociation, the concentration of BH^+ may be taken as being approximately that of the salt which was added, and the concentration of B may be taken as being approximately that of the base originally present. Thus:

$$pH = pk_w - pk_b - \log\left(\frac{[salt]}{[base]}\right). \qquad [9.26]$$

This equation is actually identical with the Henderson equation, derived above, for the pH of mixtures of a weak acid and its salt. From the relationship between the dissociation constants of an acid and its conjugate base, it is seen that $pk_w - pk_b$ may be replaced by pk_a. Then, the salt is the conjugate acid of the weak base, so the equation just derived may be expressed as:

$$pH = pk_a - \log\left(\frac{[acid]}{[base]}\right),$$

or, $\quad pH = pk_a + \log\left(\frac{[base]}{[acid]}\right), \qquad [9.18]$

which is identical with the previous equation. This emphasises the fact that problems of dissociation of acids and bases do not require separate treatment on the Brönsted–Lowry view. One set of equations suffices for both if the conjugate acid-base relationships are borne in mind.

A pH equation which is more accurate than the Henderson equation will now be derived by considering the general case of an acid A and its conjugate base B. They are related by the equilibrium:

$$A + H_2O \rightleftharpoons B + H_3O^+.$$

(No charges are shown for A and B since this is a general case, but it must be remembered that A has one hydrogen atom and one unit of positive charge more than B.)

Let the stoichiometric concentrations of A and B as the solution is made up be c_A and c_B respectively. The actual concentrations [A] and [B] will differ from these according to the position of equilibrium reached in the process mentioned above and also by the fact that the base B is capable of removing protons from water according to the equilibrium:

$$B + H_2O \rightleftharpoons A + OH^-.$$

(The position of equilibrium in this second reaction will, of course, be governed by that in the first reaction, as described on p. 317.)

The first equilibrium reduces the concentration of A by an amount equal to the concentration of H_3O^+ formed (cf. p. 362) and also increases the concentration of B by this same amount. The second equilibrium increases the concentration of A by an amount equal to the concentration of OH^- present at equilibrium and reduces the concentration of B by the same amount. Expressed algebraically:

$$[A] = c_A - [H_3O^+] + [OH^-] \qquad [9.27]$$
$$[B] = c_B + [H_3O^+] - [OH^-]. \qquad [9.28]$$

We have already seen that:

$$\mathrm{pH} = \mathrm{p}k_a + \log\left(\frac{[\text{base}]}{[\text{acid}]}\right), \qquad [9.18]$$

where the concentrations of base and acid were taken approximately as being the stoichiometric concentrations. If this approximation is not made, the correct concentrations for substitution into this relationship are the actual concentrations of acid and base present at equilibrium, namely [A] and [B]. The relationship then becomes:

$$\mathrm{pH} = \mathrm{p}k_a + \log\left(\frac{c_B + [H_3O^+] - [OH^-]}{c_A - [H_3O^+] + [OH^-]}\right). \qquad [9.29]$$

This equation gives better values of pH over a wider range of conditions of concentration and acid strength than the Henderson equation previously derived. Both equations have an important application in the calculation of the pH values of buffer solutions (see below). They also permit the calculation of pk_a or pk_b from measurements of pH as in experiment 9.2, p. 381. Still more accurate equations involve the use of activities instead of concentrations, but this is not necessary for most purposes.

Buffer Solutions

A buffer solution is one which maintains a constant pH even when the amount of acid or alkali in the solution varies slightly. A constant pH is required in certain laboratory experiments, but it is even more necessary in biological systems where the life processes can only take place within a limited range of pH. The buffer action in such systems is discussed on p. 369.

It is found that a reasonably concentrated solution of a weak acid and its salt, or of a weak base and its salt, is capable of acting as a buffer solution. The reason for this is that the pH is governed by the ratio of two fairly large concentrations, and a small amount of acid or alkali added to the solution can only produce negligible changes in these fairly large concentrations.

Thus, if a buffer consists of a weak acid and its salt, the pH is governed by the equation [9.17] deduced in (b) above, i.e. it depends upon the pk_a value of the acid and the logarithm of the ratio of the concentrations of salt and acid. The addition of a small quantity of any acid material would convert a little of the salt into the acid, but, if the concentrations of A^- and HA are both reasonably large, these small changes would be negligible and the pH would remain constant. If the buffer consists of a weak base and its salt, the pH is governed by the equation [9.26] deduced in (d) above, i.e. it depends upon pk_b and the logarithm of the ratio of the concentrations of salt and base. Addition of a small quantity of any acid would convert a little of the base into salt, and addition of a small quantity of alkali would convert a little of the salt into the base, but again, if the concentrations of salt and base are reasonably large, these changes would be negligible.

From the form of the equation it may be seen that the pH of a buffer solution does not depend on the absolute values of the concentrations of the two components, but only on their ratio. However, it is clear that the absolute concentrations must govern whether the small changes brought about by the addition of small amounts of acid or alkali are negligible or not. The resistance to change of pH of a buffer solution is given by the buffer value or buffer capacity of the solution. This is the concentration (usually in $mol\ dm^{-3}$) of strong monobasic acid, e.g. HCl, or strong monoacid base, e.g. NaOH, necessary to produce a change of one unit in pH in the solution. The higher the concentrations of the two components of the buffer solution, the higher is the buffer capacity. Also, the buffer capacity is a maximum when the two components are present with equal concentrations, and falls off as the ratio of the concentrations becomes progressively less than, or greater than, unity. The following calculation illustrates these points.

When a buffer solution is made up of equal concentrations of acid and salt, the pH is equal to the pk_a value of the acid. Thus:

$$\text{pH} = pk_a + \log\left(\frac{[\text{salt}]}{[\text{acid}]}\right). \qquad [9.17]$$

When $\qquad [\text{salt}] = [\text{acid}]$,

$$\text{pH} = pk_a + \log(1),$$

and, since $\qquad \log(1) = 0$,

$$\text{pH} = pk_a.$$

In order to make the pH increase by one unit, the logarithm of the ratio of the concentrations must be made equal to unity. The ratio of [salt] to [acid] must therefore be made equal to 10 (because $\log 10 = 1$):

$$\text{pH} = pk_a + \log\left(\tfrac{10}{1}\right).$$

If the salt and acid were both present originally to the extent of 0.1 mol dm^{-3}, to alter the ratio to 10 : 1, the final concentrations would have to be $\tfrac{10}{11} \times 0.2$ mol dm^{-3} and $\tfrac{1}{11} \times 0.2$ mol dm^{-3}, respectively, i.e. 0.182 mol dm^{-3} and 0.018 mol dm^{-3} approximately. Such an alteration of the concentrations would be brought about by the addition of 0.082 mol dm^{-3} of a strong base. The buffer capacity is therefore 0.082 mol dm^{-3} towards strong base. However, if the salt and acid were both originally present to the extent of 0.01 mol dm^{-3}, the change to a 10:1 ratio would be brought about by the addition of 0.0082 mol dm^{-3} of strong base (giving 0.0182 mol dm^{-3} salt and 0.0018 mol dm^{-3} acid). The buffer capacity is then only 0.0082 mol dm^{-3} towards strong base.

If the salt and acid were originally present in a 10 : 1 ratio, giving a pH of pk_a+1, a change of pH of one unit, to the value pk_a+2, is obtained by changing the concentration ratio from 10 : 1 to 100 : 1 (because $\log 100 = 2$). If the salt concentration were 0.182 mol dm^{-3} and the acid 0.018 mol dm^{-3} as above, the change to a 100 : 1 ratio is brought about by the addition of approximately 0.016 mol dm^{-3} of strong base. This would produce final concentrations of 0.198 mol dm^{-3} salt and 0.002 mol dm^{-3} acid. The buffer capacity is therefore 0.016 mol dm^{-3} towards strong base, i.e. less than one-quarter of the capacity which the solution possessed when the concentrations of salt and acid were the same. It should be noted that the buffer capacities towards strong acid and strong base are not the same when the original concentrations of acid and salt in the buffer are unequal. For example, to reduce the pH of the solution just considered from pk_a+1 to pk_a, instead of increasing it to pk_a+2, 0.082 mol dm^{-3} of strong acid would be required (since this is the reverse of the process considered in the first calculation). The buffer capacity towards strong acid is therefore 0.082 mol dm^{-3}, even though it is only 0.016 mol dm^{-3} towards strong base.

As a general rule, effective buffer solutions can only be made up, from a given acid, to work within a pH range of one unit on either side of the pk_a value. Outside this range, the buffer capacity is too small. Similarly, for weak bases, the buffer solution only functions efficiently within one unit of pH from the value ($pk_w - pk_b$).

Buffers in Biological Systems

In blood plasma, a uniform pH is maintained by three acids and their salts (mostly sodium salts). These are carbonic acid (H_2CO_3), the dihydrogen-phosphate ion ($H_2PO_4^-$), and protein. Protein is a class of substance which exists in a zwitterion form (see p. 377), but which functions mainly as an acid in equilibrium with its salt at pH values which lie above its isoelectric point (see p. 512). The phosphate and protein buffer systems are present in relatively small concentrations, so the carbonate buffer is mainly responsible for regulating the pH. The bicarbonate ions are present to an extent of about 0.025 mol dm^{-3} and carbonic acid to about 0.00125 mol dm^{-3}. The ratio of salt to acid is therefore about 20 : 1. Hence:

$$pH = pk_a + \log 20.$$

Since the pk_a value, for dissociation of the first hydrogen ion of carbonic acid, is about 6.1:

$$pH = 6.1 + \log 20$$
$$= 6.1 + 1.3$$
$$= 7.4.$$

The bicarbonate buffer has a special property which increases its capacity towards acids. The carbonic acid, which is produced in the neutralisation of added acids, splits up, after a certain concentration is reached, to form CO_2 and H_2O. The carbon dioxide is then readily passed out of the system through the lungs.

Buffering is also essential in plants. Quite large quantities of acid may be produced, particularly in fruits, yet the pH must be maintained within the range which is appropriate for the proper functioning of the various enzyme systems. Again, this is done partly by bicarbonate and phosphate buffers, though the acids produced and their salts (e.g. malate, citrate, oxalate, etc.) also make a considerable contribution to the buffering action.

Hydrolysis of Salts

When a salt of a strong acid and strong base is dissolved in water, it does not undergo any reaction with the water, i.e. it is not hydrolysed. This is because neither of the ions of the salt have any appreciable acid or basic properties. For example, sodium chloride solution contains sodium ions and chloride ions. Of these, only the chloride ions have any tendency to react with water:

$$Cl^- + H_2O \rightleftharpoons HCl + OH^-.$$

But the chloride ion is the conjugate base of the strong acid HCl, so it must be a very weak base. The reaction with water is therefore quite negligible. Solutions of such salts therefore remain neutral. The activities of hydrogen and hydroxyl ions are equal, as they are in pure water. The product of the two activities is 10^{-14}, so each must be 10^{-7}. The pH of a neutral solution is therefore 7.

When one of the ions of a salt is derived from either a weak acid or a weak base, the ion must be a strong conjugate base or a strong conjugate acid respectively, and it has a marked tendency to undergo reaction with water. The salt is then said to be hydrolysed in solution, and the pH differs from 7. Thus, the acetate ion in sodium acetate is the conjugate base of acetic acid. It is therefore hydrolysed thus:

$$CH_3COO^- + H_2O \rightleftharpoons CH_3COOH + OH^-.$$

The acetate ion is a fairly strong base, and the equilibrium favours the production of excess hydroxyl ions in the solution. The product of the activities of the hydroxyl and hydrogen ions must remain at 10^{-14}, so the activity of hydrogen ions must be less than 10^{-7}. The pH of the solution is therefore greater than 7. The hydrolysis constant for the salt (i.e. for the equilibrium above) must be the same as the dissociation constant of the acetate ion. The actual pH may be easily calculated, therefore, if the basic dissociation constant of the acetate ion is known, by employing the same relationship as that derived for pH in terms of pk_b for a weak base. If pk_b for the ion is not known, it may be replaced by $pk_w - pk_a$, where pk_a is the value for the conjugate acid. The equation for the pH then becomes:

$$pH = \tfrac{1}{2}pk_w + \tfrac{1}{2}pk_a + \tfrac{1}{2}\log c, \qquad [9.30]$$

where c is the concentration of the salt in mol dm^{-3}.

As an example of a salt of a weak base and a strong acid, ammonium chloride, may be considered. The ammonium ion tends to lose protons to the water, and the equilibrium:

$$NH_4^+ + H_2O \rightleftharpoons NH_3 + H_3O^+$$

favours the production of hydrogen ions, because NH_4^+ is a fairly strong acid (being conjugate with the weak base NH_3). The hydrogen ion activity in the solution of the salt is therefore greater than 10^{-7}, and the pH is less than 7. The hydrolysis constant of the salt is the same as the dissociation constant of the ammonium ion, so the actual pH is given by the same equation as that for the dissociation of a weak acid above. Again, if the pk_a value for NH_4^+ is not known, it is obtained by taking the difference between pk_w and pk_b for the weak base. The equation then becomes:

$$pH = \tfrac{1}{2}pk_w - \tfrac{1}{2}pk_b - \tfrac{1}{2}\log c, \qquad [9.31]$$

where c is the concentration of the salt in mol dm^{-3}.

When the salt is that of a weak acid and a weak base, both ions undergo acid-base reactions with the water. In the case of ammonium acetate, for example:

$$NH_4^+ + H_2O \rightleftharpoons NH_3 + H_3O^+$$
$$CH_3COO^- + H_2O \rightleftharpoons CH_3COOH + OH^-.$$

The total equilibrium is therefore:

$$NH^+ + CH_3COO^- + 2H_2O \rightleftharpoons NH_3 + CH_3COOH + H_3O^+ + OH^-.$$

However, part of this equilibrium is that which exists between water and hydrogen and hydroxyl ions, irrespective of whether the salt is present or not:

$$2H_2O \rightleftharpoons H_3O^+ + OH^-.$$

The concentrations of H_3O^+ and OH^- are not generally equal in the presence of salts, but their values will be determinable from the concentrations of the other species taking part in the equilibrium. The terms for the ionisation of water may therefore be subtracted from the total equilibrium, to give:

$$NH_4^+ + CH_3COO^- \rightleftharpoons NH_3 + CH_3COOH.$$

This simplification of the equilibrium makes it appear that the concentrations of the salt ions must always be equal, since they are equal before hydrolysis, and occur in equimolecular proportions in the hydrolysis equation. It must be remembered that this is not generally true, but it is approximately correct if the dissociation constants of the acid and the base from which the salt is derived are not widely different.

The hydrolysis constant, k_h, is therefore given by:

$$k_h = \frac{[NH_3][CH_3COOH]}{[NH_4^+][CH_3COO^-]}.$$

The same ratio of concentration terms is given by $\dfrac{k_w}{k_a k_b}$, where k_a is the dissociation constant of acetic acid and k_b is the dissociation constant of ammonia, thus:

$$k_a k_b = \frac{[H_3O^+][CH_3COO^-]}{[CH_3COOH]} \times \frac{[NH_4^+][OH^-]}{[NH_3]},$$

and $\quad k_w = [H_3O^+][OH^-],$

$$\therefore \quad \frac{k_w}{k_a k_b} = \frac{[CH_3COOH][NH_3]}{[CH_3COO^-][NH_4^+]} = k_h. \qquad [9.32]$$

Now, if the solution of the salt were made up to contain c mol dm^{-3} and if the degree of hydrolysis is assumed to be approximately the same for each ion and equal to x, the concentrations of the various species after hydrolysis would be:

$$\underset{c(1-x)}{NH_4^+} + \underset{c(1-x)}{CH_3COO^-} \rightleftharpoons \underset{cx}{NH_3} + \underset{cx}{CH_3COOH}$$

The value of k_h would then be:

$$k_h = \frac{cx \cdot cx}{c(1-x) \cdot c(1-x)}$$

$$= \frac{x^2}{(1-x)^2}. \qquad [9.33]$$

If x is small compared with unity, the approximate form of this equation is:

$$k_h \approx x^2$$

$$\therefore \quad x \approx \sqrt{k_h}. \qquad [9.34]$$

The concentration of both ammonia and acetic acid is therefore $c\sqrt{k_h}$. To obtain the pH of the solution, it is necessary to find the concentration of hydrogen ions in equilibrium with either the concentration of ammonia or the concentration of acetic acid in the solution. It is immaterial which of these is considered, since the equilibria adjust themselves in such a way that the concentrations of both these species are in equilibrium with the same concentration of hydrogen ions. Considering the dissociation of acetic acid, the concentration of hydrogen ions is given by:

$$[H_3O^+] = \frac{k_a \cdot [CH_3COOH]}{[CH_3COO^-]}.$$

Substituting cx for $[CH_3COOH]$, and $c(1-x)$ for $[CH_3COO^-]$:

$$[H_3O^+] = \frac{k_a \cdot cx}{c(1-x)}$$

$$= \frac{k_a \cdot x}{(1-x)}. \qquad [9.35]$$

But $\dfrac{x}{(1-x)} = \sqrt{k_h},$

$\therefore \quad [H_3O^+] = k_a \sqrt{k_h}$ [9.36]

$\qquad\qquad\quad = k_a \sqrt{k_w/k_a k_b}$

$\qquad\qquad\quad = \sqrt{k_w k_a / k_b}.$ [9.37]

Taking logarithms:

$\log [H_3O^+] = \tfrac{1}{2} \log k_w + \tfrac{1}{2} \log k_a - \tfrac{1}{2} \log k_b,$

$\therefore \quad pH = \tfrac{1}{2} pk_w + \tfrac{1}{2} pk_a - \tfrac{1}{2} pk_b.$ [9.38]

Referring back to the assumption, which was made above, that the concentrations of the two salt ions would remain approximately equal after hydrolysis, it is seen that this equation is only strictly true when pk_a and pk_b are approximately the same. Nevertheless, it is found that the equation holds reasonably well, even when pk_a and pk_b differ by several units, provided that the solutions are not very dilute. If pk_a and pk_b are equal, the pH is $\tfrac{1}{2} pk_w$, i.e. 7. The solution remains neutral in spite of hydrolysis. In the case of the example considered, namely ammonium acetate, pk_a and pk_b are both approximately 4.76. A solution of this salt is therefore a very effective neutral buffer solution.

Many salt solutions which are used in pharmacy are subject to hydrolysis. Therefore, when the salt solution is prepared by neutralisation of an acid with a base, the addition of base must be continued until the pH is that of the hydrolysed salt. The *British Pharmacopœia* defines the permissible range of pH within which the salt solution of a given dilution must lie. In some cases the pH of a solution may be affected by alkali dissolving from glass containers, and this may prejudice the keeping qualities of the solution. The glass used for the containers (e.g. ampoules for injections) is therefore subjected to tests for alkalinity before use.

Neutralisation Curves

During a neutralisation reaction the pH may be followed (e.g. by inserting a hydrogen or glass electrode into the solution) and plotted on a graph against the amount of acid or alkali added. The result is called a neutralisation or titration curve for the acid-base system. From such a curve it is often possible to derive the dissociation constant (or constants) for a weak acid or a weak base. The interpretation of such curves is more easily understood, however, by calculating the form of typical curves from known dissociation constants.

In a solution of a strong acid, the concentration of hydrogen ions is the same as that of the acid, since it is completely dissociated. The pH of the acid solution is readily obtained, therefore, by taking the negative logarithm of this concentration. In a solution of a strong base, the concentration of the hydroxyl ions is the same as that of the base. Dividing this into the ionic product for water, the hydrogen ion concentration, and hence the pH, is readily found. Formulae have been derived above for finding the pH in a solution of a weak acid or a weak base.

At the equivalence point of the acid and the base, the pH is that given by the hydrolysis of the salt according to the equations derived above. When a weak acid is half neutralised, i.e. when the concentration

Fig. 9.1 Neutralisation curves

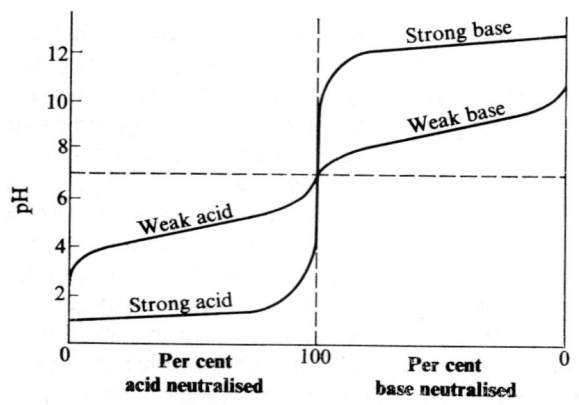

of salt is equal to the concentration of acid, the pH is equal to pk_a (as seen from the Henderson equation). Similarly, when a weak base is half neutralised, the Henderson equation shows that the pH is equal to $pk_w - pk_b$. Between these special points the curve follows a logarithmic form, because of the logarithmic relationship between pH and hydrogen ion concentration. Figure 9.1 is a composite graph covering all the typical cases for the neutralisation of a monobasic acid (weak or strong) with a weak or strong base. In each case the pH changes most rapidly at the end-point, and the sharpness of the change increases with the strengths of the acid and the base. It should be noticed that the pH changes most slowly, in the weak acid or the weak base curve, at the point of half neutralisation. This is another way of demonstrating that the buffer capacity is greatest for such solutions when there are equal concentrations of acid or base and salt.

The pH range over which the rapid change takes place varies according to the strength of the acid and base. For a weak acid and a strong base it is on the alkaline side of pH = 7, whereas for a strong acid and a weak base it is on the acid side. The position of this range may be deduced from the fact that its mid-point corresponds to the pH given by the hydrolysis of the salt. The mid-point is only at pH = 7 for the neutralisation of a strong acid with a strong base, or for a weak acid with a weak base if both have equal dissociation constants ($k_a = k_b$).

If a polybasic acid is titrated with a strong base, there should be one sharp change in pH for each hydrogen atom of the molecule which is ionised. This is only true if the dissociation constants for the various stages of ionisation are well separated in pk value. If two of the successive pk values are similar, there is found to be no sharp change in pH between the two points where pH = pk. The pH for half neutralisation of a dibasic acid with acid dissociation constants k_1 and k_2 is $\frac{1}{2}(pk_1+pk_2)$. This can be seen by analogy with the pH for the hydrolysis of a salt of a weak acid and a weak base. When a dibasic acid is half neutralised the stronger acid group is in its anionic or basic form except for the equilibrium which is set up between this and the weaker acid group. Putting $\frac{1}{2}pk_1$ in place of $\frac{1}{2}pk_w - \frac{1}{2}pk_b$ for the basic group, and $\frac{1}{2}pk_2$ for $\frac{1}{2}pk_a$ for the weaker acid group in the pH equation of p. 373, we obtain the relationship quoted above.

Enthalpy of Neutralisation

The enthalpy of neutralisation, ΔH_n, is the increase in enthalpy when 1 equivalent of an acid and 1 equivalent of a base neutralise each other. One equivalent of an acid is that amount of the acid which can donate one mole of hydrogen ions to a base, i.e. one mole of the acid divided by its basicity. One equivalent of a base is that amount of the base which can accept one mole of hydrogen ions from an acid, i.e. one mole of the base divided by its acidity. The term equivalent has here a meaning different from that in Chapter 8 (see p. 330). Enthalpy change ΔH is discussed in Chapter 7 (see p. 275).

Enthalpies of neutralisation may conveniently be considered from the point of view of the dissociation theory of acids and bases according to Arrhenius (p. 356). For a completely ionised (i.e. strong) acid and base, e.g. HCl, HNO$_3$, NaOH, KOH, R$_4$NOH, ΔH_n is simply equal to ΔH for the reaction:

$$H^+(aq) + OH^-(aq) = H_2O; \qquad \Delta H^\ominus_{298} = -56.7 \text{ kJ},$$

and is independent of the anion of the acid and the cation of the base. These ions constitute the salt and therefore appear on both sides of the equation. If either the acid or the base or both are incompletely ionised (i.e. weak), e.g. CH$_3$COOH, NH$_3$, ΔH_n is made less negative by any heat absorbed in the ionisation reaction(s) or by the fact that the neutralisation reaction is incomplete.

Experiment 9.1 Determination of the enthalpy of neutralisation of various acid-base pairs.

The calorimeter consists of a 500 cm³ vacuum flask fitted with a cork through which passes a wire stirrer and a thermometer. The water equivalent, e kg, of the apparatus is determined as follows. Into the calorimeter is pipetted 100 cm³ of water and, after stirring, the temperature t_1 is noted. Into a 250 cm³ flask previously rinsed with water is pipetted 100 cm³ of water. The flask is heated to about 40° C, swirled and the temperature t_2 is noted. The water in the flask is poured into the calorimeter which is stirred rapidly; the temperature is noted every 10 seconds for the first 30 seconds, then every 30 seconds until it is constant. The temperature is plotted against the time and the instantaneous temperature t_3 attained on mixing is determined by extrapolation of the nearly horizontal portion to zero time.

Heat changes may be calculated from the equation:

$$q = mc\,\Delta T \qquad [9.39]$$

where q is the heat exchanged (in J) by a material,

c is the specific heat capacity of the material (in J K⁻¹ kg⁻¹; 4187 J K⁻¹ kg⁻¹ for water);

ΔT is the change in temperature (in K or °C).

Heat lost by the warm water =
Heat gained by the cold water + Heat gained by the calorimeter

∴ $\quad 0.100 \times 4187 \times (t_2 - t_3) =$
$[e \times 4187 \times (t_3 - t_1)] + [0.100 \times 4187 \times (t_3 - t_1)]$.

Hence e is calculated.

ΔH_n is determined in aqueous solution for any of the following acids: HCl, HNO₃, CH₃COOH, with any of the following bases: NaOH, KOH, NH₃. Each determination is carried out in duplicate according to the following procedure. Into the clean, dry calorimeter is pipetted 100 cm³ of a 2 mol dm⁻³ solution of the base. Into a 250 cm³ beaker, previously rinsed with the solution, is pipetted 100 cm³ of a 2 mol dm⁻³ solution of the acid and the temperature is adjusted to within ±0.2° C of that of the solution in the calorimeter. Both solutions are stirred and the mean temperature t_4 is noted. The acid is poured into the base in the calorimeter. The mixed solution is rapidly and continuously stirred and the temperature is recorded as before until it is constant. The instantaneous temperature t_5 attained on mixing is determined as before.

The total mass of each solution may be taken to be 0.10 kg, and the specific heat capacity of the mixed solution may be taken to be that of pure water. If the acidic and the basic solutions have different concentrations, the weaker solution controls the number of equivalents (moles in these cases) of the acid and base neutralised. Let f be the factor for the weaker solution. Each solution is diluted by a further factor of 2 on mixing with the other.

Using equation [9.39] the heat evolved in the neutralisation of 100 cm³ (0.10 kg) of $f \times 1$ mol dm⁻³ solution

$$= (0.10+e) \times 4187 \times (t_5-t_4) \text{ J}$$

∴ heat evolved in the neutralisation of 1 dm³ of a 1 mol dm⁻³ solution (1 equivalent)

$$= (0.10+e) \times 4187 \times (t_5-t_4) \times \frac{1000}{100f} \text{ J eq.}^{-1}$$

$$= -\Delta H_n \text{ (in J eq.}^{-1}\text{)}$$

Hence ΔH_n is calculated.

Amphoteric Electrolytes or Ampholytes

Substances which are capable of acting both as acids and bases are termed amphoteric electrolytes or ampholytes. Amino acids are typical of this class of substances. They combine, in the same molecule, the basic —NH₂ group and the acidic —COOH group. When they are dissolved in water, the pH is determined by the relative strengths of the acidic and basic portions of the molecule. In practice, the normal form of the molecule is that in which the acidic and the basic groups have already undergone reaction with each other to form —NH₃⁺ and COO⁻. In this state the molecule is said to exist as a zwitterion, which may be regarded as an internal salt.

If a solution of such an ampholyte is acidified, the —COO⁻ groups of some molecules are converted into —COOH groups. If alkali is added, the —NH₃⁺ groups of some molecules are converted into —NH₂ groups. If the pH is adjusted so that the average number of —COO⁻ groups is equal to the average number of —NH₃⁺ groups, the ampholyte is said to be at its isoelectric point. The properties of proteins depend a great deal on whether they are at the isoelectric pH or on the acid or alkaline side of it, and the matter is discussed further, in this connection, on p. 512. At its isoelectric point, the solubility of an ampholyte is a minimum. The pH corresponding to the isoelectric point is most easily obtained theoretically by considering the equilibrium of the ampholyte with water in terms of the hydrolysis of the sat- of a weak acid and a weak base, for which it was shown above that:

$$\text{pH} = \tfrac{1}{2}\text{p}k_w + \tfrac{1}{2}\text{p}k_a - \tfrac{1}{2}\text{p}k_b, \qquad [9.38]$$

where $\text{p}k_a$ is the value for the dissociation of the weak acid (in this case the —COOH group) and $\text{p}k_b$ is the value for the dissociation of the weak base (in this case the —NH₂ group). Alternatively, one may consider the acid dissociation of the —NH₃⁺ group and the basic dissociation of the —COO⁻ group. Representing the pk_a value for —NH$_3^+$ as pk_a' and the pk_b value for —COO⁻ as pk_b', we have:

$$pk_a = pk_w - pk_b'$$

and
$$pk_b = pk_w - pk_a',$$

∴
$$\begin{aligned}
pH &= \tfrac{1}{2}pk_w + \tfrac{1}{2}(pk_w - pk_b') - \tfrac{1}{2}(pk_w - pk_b') \\
&= \tfrac{1}{2}pk_w + \tfrac{1}{2}pk_w - \tfrac{1}{2}pk_b' + \tfrac{1}{2}pk_w + \tfrac{1}{2}pk_a' \\
&= \tfrac{1}{2}pk_w + \tfrac{1}{2}pk_a' - \tfrac{1}{2}pk_b'.
\end{aligned}$$

This is the same equation as before, so it is immaterial whether pk_a is taken as referring to the dissociation of —COOH or —NH$_3^+$, and whether pk_b is taken as referring to —NH$_2$ or —COO⁻.

It was mentioned on p. 361 that the dissociation constant for a base may be quoted as the acid dissociation constant of its conjugate acid. A description of the dissociation of amino acids is simplified if only acid dissociation constants are used. Thus, if NH$_2$RCOOH is taken as representing an amino-acid molecule, and NH$_3^+$RCOO⁻ is its zwitterion form, its basic dissociation is described from the point of view of the acid dissociation of the —NH$_3^+$ group:

$$NH_3^+RCOO^- + H_2O \rightleftharpoons NH_2RCOO^- + H_3O^+.$$

The acid dissociation of the —COOH group is represented by the equilibrium:

$$NH_3^+RCOOH + H_2O \rightleftharpoons NH_3^+RCOO^- + H_3O^+.$$

The equilibrium constant for the acid dissociation of NH$_3^+$RCOOH is given by:

$$k_1 = \frac{[NH_3^+RCOO^-][H_3O^+]}{[NH_3^+RCOOH]}.$$

The equilibrium constant for the acid dissociation of NH$_3^+$RCOO⁻ is given by:

$$k_2 = \frac{[NH_2RCOO^-][H_3O^+]}{[NH_3^+RCOO^-]}.$$

The constant k_2 really represents the acid dissociation constant of the zwitterion, k_a', and k_1 must be related to the basic dissociation constant of the zwitterion, k_b', by $k_1 k_b' = k_w$ (since the zwitterion is the conjugate base of the acid NH$_3^+$RCOOH). The isoelectric pH may be obtained in terms of k_1 and k_2, by substituting pk_2 for pk_a' and $(pk_w - pk_1)$ for pk_b':

$$\begin{aligned}
pH &= \tfrac{1}{2}pk_w + \tfrac{1}{2}pk_2 - \tfrac{1}{2}(pk_w - pk_1) \\
&= \tfrac{1}{2}pk_2 + \tfrac{1}{2}pk_1,
\end{aligned} \qquad [9.40]$$

i.e. the isoelectric pH is the mean of pk_1 and pk_2.

Indicators

Although various electrochemical methods are available for the accurate determination of the pH of solutions, rough determinations are made, and changes in pH are usually detected, by the use of indicators. These substances are themselves weak acids or bases, and possess different colours in their conjugate acidic and basic forms, or are colourless in one form. If an indicator is added in very small quantity to a solution, it does not disturb the acid-base equilibrium of the solution, but it adjusts the ratio of its conjugate acidic and basic forms according to the pH of the solution. The relative amounts of the acidic and basic forms present at a given pH are determined by the Henderson equation:

$$pH = pk_a + \log\left(\frac{[\text{base}]}{[\text{acid}]}\right), \qquad [9.18]$$

where pk_a refers to the acid dissociation constant of the indicator. If the ratio of the concentrations of the two forms of the indicator can be found by observing the colour (in the case of a two-colour indicator) or the depth of colour (in the case of an indicator which is colourless in one form), and if the pk value for the indicator is known, the pH of the solution may be calculated.

There are various ways in which the relative amounts of the two forms of the indicator may be estimated. One method is to have a long, oblong, glass-sided tank, which is divided diagonally by another piece of glass to give two wedge-shaped tanks. In one wedge is placed a dilute solution of the acid form of the indicator, and in the other is placed a solution of the same concentration of the basic form. The colour of the solution, seen through both wedges, shows a gradation from the acid colour to the basic colour. The amount of each form contributing to the colour at any point is given by the relative thicknesses of the two solutions at this point in the tank. The colour of the indicator in the solution whose pH is required is matched with the colour of the wedges at a certain point. The ratio of the thicknesses of the two solutions in the wedges at this point gives the ratio of the concentrations of the two forms for substitution in the pH equation.

More frequently, a set of buffer solutions of successively increasing pH is made up in similar test-tubes and the same number of drops of indicator solution is added to each. These give the range of colours to be compared with that produced by adding the same number of drops of indicator to the same volume of the solution whose pH is required. This is the basis of the comparator method, and it is described in more detail in Experiment 9.2. A similar determination may be carried out, using smaller volumes of liquid, by the capillator method. Colour standards sealed in capillary tubes are commercially available, and standard quantities of the solution to be examined, and the indicator, are mixed and drawn into a similar capillary tube. This is compared with the set of colour standards, and the pH may be read directly when the match has been found.

In these methods for the determination of pH, each indicator has a useful range which extends only about one unit on either side of the pH which is equal to pk_a. This is because the eye cannot distinguish between the colour for the acid form of the indicator and the colour given by a ratio of acid to basic forms greater than 10 : 1. The logarithm of this ratio is unity, so the pH is (pk_a+1). Similarly, a ratio of acid to basic forms any greater than 1 : 10 would be indistinguishable from the pure basic form. The logarithm of this ratio is -1, so the pH is (pk_a-1). Therefore, in order to cover a wide range of pH values, a number of indicators must be used of successive, appropriate pk_a values.

If a sufficiently wide variety of indicators is available, the pH of a solution may be estimated to within a fairly narrow range simply by discovering which indicators are caused to be in their acid form and which are caused to be in their basic form by the solution under test. This is the basis of pH determination by indicator test-papers. It is also used, in conjunction with the microscope, to determine the pH in cells or sections of biological material. In such systems, care must be exercised in the choice of indicators, since many of them are affected by the neutral salts which are also present in the cells.

In analytical work, indicators are not used to determine pH but to detect the sharp change in pH which occurs at the end-point of an acid-base titration (see Fig. 9.1, p. 374). The most appropriate indicator to use for a given acid and base is one which has a pk_a value close to the pH given by hydrolysis of the salt of the acid and the base. The pH at the equivalence point is that due to hydrolysis only, and this occurs

Table 9.1. pk_a Values of Some Common Indicators

Indicator	pk_a	Acid Colour	Alkaline Colour
Methyl orange	3.6	Red	Orange-yellow
Bromphenol blue	4.0	Yellow	Blue
Methyl red	5.1	Red	Yellow
Bromcresol purple	6.3	Yellow	Purple
Bromthymol blue	7.0	Yellow	Blue
Phenol red	7.9	Yellow	Red
Thymol blue	8.9	Yellow	Blue
Phenolphthalein	9.6	Colourless	Red
Thymolphthalein	9.9	Yellow	Blue

in the middle of the range of rapid change of pH at the end-point. An indicator having this pk_a value is altered from almost entirely acid form to almost entirely basic form (or vice versa) during the sharp change in pH. The colour change therefore takes place completely over a very narrow range of the amount of titrant added at the end-point. Table 9.1 gives the pk_a values of some common indicators.

Experiment 9.2 To determine the hydrolysis constant of ammonium chloride (i.e. the acid dissociation constant of the ammonium ion) by an indicator method.

Ammonium chloride is the salt of a weak base and a strong acid, so it hydrolyses to give an acid solution:

$$NH_4^+Cl^- + H_2O \rightleftharpoons NH_3 + H_3O^+ + Cl^-.$$

The hydrolysis constant k_h is therefore the same as the acid dissociation constant of the ammonium ion:

$$k_h = k_a = \frac{[NH_3][H_3O^+]}{[NH_4^+]}.$$

It was deduced above that the dissociation constant of an acid is related to the pH of the solution by:

$$pH = \tfrac{1}{2}pk_a - \tfrac{1}{2}\log c, \qquad [9.16]$$

where c is the concentration of the acid in mol dm^{-3}. Applying this to the present problem, we have:

$$pH = \tfrac{1}{2}pk_h - \tfrac{1}{2}\log c,$$

where c is the concentration of the ammonium ion, and therefore equal to the total concentration of the salt in mol dm^{-3}. The value of pk_h is therefore found by determining the pH of a solution of known concentration of the salt. The pH is to be determined by comparing the colour of methyl-red indicator in the solution with the colours in a series of buffer solutions of known pH.

A series of buffer solutions covering the pH range required for this experiment may be made up after calculation of appropriate concentrations for acetic acid and sodium acetate mixtures. This is done by mixing equal proportions of 0.2 mol dm^{-3} NaOH solutions with different amounts of excess 0.2 mol dm^{-3} acetic acid solution, and finally making up to the same total volume in each case. Since the same quantity of NaOH is present in each solution, and it is all neutralised by the acid, there is a constant quantity of sodium acetate in each buffer solution. This is advantageous, because pk values (e.g. for indicators) vary if the total salt concentration in the solution varies. (The dissociation constant K, in terms of activities, is a constant, but the approximate value k, in terms of concentrations, depends upon the extent to which the concentrations deviate from the activities.) Each buffer solution is therefore made up as follows. 10 cm³ of 0.2 mol dm⁻³ NaOH solution are pipetted into a 100 cm³ graduated cylinder. A volume $(10+x)$ cm³ of 0.2 mol dm⁻³ acetic acid solution is added from a burette – where x cm³ is the volume calculated as explained below – and the mixture is made up to 70 cm³ with distilled water. The volume x cm³ represents the concentration of acetic acid in the solution, compared with 10 cm³ of salt. The pH of the buffer is therefore related to x by the equation (cf. p. 364):

$$pH = pk_a + \log\left(\frac{10}{x}\right),$$
$$= pk_a + \log 10 - \log x,$$
$$= pk_a + 1 - \log x.$$

The value of pk_a for acetic acid at the final salt concentration in each buffer solution may be taken as 4.63.

Hence

$$pH = 5.63 - \log x.$$

From this equation, values of x should be calculated which will give pH values of 5.6, 5.5, 5.4, ... 5.0. Then 10 cm³ of each buffer solution are placed in each of a set of uniform test-tubes, and the same number of drops of methyl-red indicator is added to each tube. (The amount of methyl red added should be the minimum necessary for the colour gradation to be clearly visible; approximately 4 drops of a 0.04 per cent solution.)

An ammonium chloride solution of concentration 1 mol dm⁻³ is then made up in fresh distilled water, by weighing out 5.35 g of the salt and making up to 100 cm³ in a standard flask. 10 cm³ of this solution are placed in a tube uniform with those containing the buffer solutions, and the same number of drops of methyl red is added as was placed in each buffer solution. The colour of the ammonium chloride tube is then matched with that of one of the buffer tubes. The pH in the ammonium chloride solution is then known to be within ±0.05 units of that of the buffer solution (neglecting salt errors). Hence, the hydrolysis constant of the salt may be calculated from the formula quoted above, substituting 1.0 for c. The value obtained is liable to inaccuracy due to impurities in the water and the salt and to neglecting salt errors.

pH and Drug Action

A number of drugs are either weak acids or weak bases, such as phenols, sulphonamides, alkaloids, e.g. codeine, and barbiturates, e.g. phenobarbitone. The extent to which such drugs are ionised depends upon their pk_a values and the pH of the medium in which they are used. The relative amounts of the ionised and non-ionised forms present in a given solution are related to the pH by the Henderson equation (p. 364). For a drug which dissociates as a base, the basic form is un-ionised and the acid form ionised. A drug which dissociates as an acid, however, is un-ionised in its acid form and ionised in its basic form.

When the effect of pH on the toxic power of a drug is investigated, the results often suggest that the effective agent is either (*a*) the non-ionised form, or (*b*) the ionised form or (*c*) both forms of the drug. Thus, the inhibiting effect of salicylic acid on cell division in echinoderm eggs varies with pH in such a way that it is clear that the concentration of neutral molecules is the factor determining the toxic effect. Over the range of pH 8 to 5, the total minimum concentration of salicylic acid necessary to stop the cell division decreases. If the concentration of neutral molecules at each pH is calculated, it is found to be approximately constant. Thus, the ionic form is apparently without effect in this case, and a fixed concentration of neutral molecules must be attained at each pH.

In the case of the acridines, a similar study indicates that the concentration of the cationic form is the main factor in determining the bacteriostatic power. The situation is complicated, however, by the effect of the pH on the electrical charge at the external surface of the cell membrane, which appears to be the site of action of these drugs (p. 513). Furthermore, active molecules must possess a planar structure of sufficient area, in order to form van der Waals interactions (p. 30) with the receptor groups of the bacteria.

A number of drugs, e.g. phenols, are active against certain organisms in both the ionised and the non-ionised forms, although the neutral molecule is often much more active than the ion. The antifungal activity of 2,4-dinitrophenol is a case in point. On the other hand, bacteriostatic power does not correlate in any simple way with pK_a among the sulphonamide group but rather with electronic and other structural features of the molecule. The activities of the sulphonamides are thought to be related to their structural similarity to the natural bacterial metabolite, *p*-aminobenzoic acid. This similarity is sufficient to ensure that the drug is absorbed by the receptor groups for the metabolite but is insufficient for the drug to perform any metabolic function. The biologically active form of *p*-aminobenzoic acid, pK_a 4.8, is probably the anion.

Acids and Bases

p-aminobenzoate   sulphonamide

Neutral molecules usually penetrate cell membranes much more readily than do ions, because ions, being hydrated, have a greater size and, depending on their charge, are either repelled by or adsorbed by the charged protein in the membrane. Neutral molecules usually penetrate cell membranes easily, provided that they have no more than about three hydrophilic groups and are not too large. For example, barbiturates enter the eggs and larvae of sea urchins in the unionised form only and the resulting inhibitions are attributable only to neutral molecules. Furthermore, in the stomach and duodenum where the pH is between 1 and 3, a basic or ampholytic drug, such as an alkaloid or amino acid, will acquire a proton and therefore a positive charge and will consequently not be absorbed through the gut. On the other hand, a weakly acidic drug, such as acetylsalicylic acid or a barbiturate, will be unionised and may therefore be absorbed. It should be noted that a non-penetrating ion can sometimes be made to penetrate a cell membrane by introducing a lipophilic group, e.g. a chloro- group in the case of an antimalarial drug, such as chloroquine and mepacrine. Moreover, there are specific mechanisms for the transport of biologically important ions, e.g. Na^+, across membranes (pp. 428 and 429), but the importance of such mechanisms in the membrane transport of drugs remains to be elucidated.

The relative concentrations of the ionised form of a drug on the two sides of a cell membrane can be readily calculated from the pH of both the external solution and the internal solution. We usually assume that the cell membrane is impermeable to the ionised form, is freely permeable to the unionised form and that the concentration of the latter is the same on both sides of the membrane. In the case of a basic drug, the non-ionised form is the free base, B, and the ionised form is the conjugate acid, BH^+. The situation can then be represented as follows:

Applying the Henderson equation [9.18] to the external solution:

$$pH_e = pk_a + \log\left(\frac{[B]_e}{[BH^+]_e}\right),$$

pH and Drug Action

Fig. 9.2 Distribution of a basic drug, B, across a membrane impermeable to ions

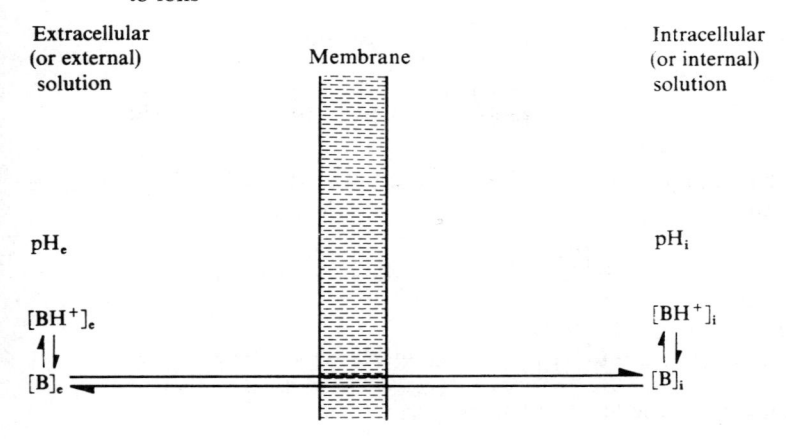

| Extracellular (or external) solution | Membrane | Intracellular (or internal) solution |

pH_e pH_i

$[BH^+]_e \updownarrow$ $[BH^+]_i \updownarrow$

$[B]_e \longleftrightarrow [B]_i$

and to the internal solution:

$$pH_i = pk_a + \log\left(\frac{[B]_i}{[BH^+]_i}\right).$$

Subtracting the first equation from the second:

$$pH_i - pH_e = \log\left(\frac{[B]_i}{[BH^+]_i}\right) - \log\left(\frac{[B]_e}{[BH^+]_e}\right).$$

$$= \log\left(\frac{[B]_i\,[BH^+]_e}{[BH^+]_i\,[B]_e}\right)$$

But $[B]_i = [B]_e$, therefore:

$$pH_i - pH_e = \log\left(\frac{[BH^+]_e}{[BH^+]_i}\right), \qquad [9.41]$$

which is the equation required for a basic drug.

In the case of an acidic drug, the unionised form is the free acid, HA, and the ionised form is the conjugate base, A^-. Analogous reasoning, in which $[HA]_e = [HA]_i$, leads to the following equation:

$$pH_i - pH_e = \log\left(\frac{[A^-]_i}{[A^-]_e}\right), \qquad [9.42]$$

where the subscript i refers to the intracellular solution and the subscript e to the external solution as before.

The ratio of the total concentration of a drug (both ions and neutral molecules) on one side of the membrane to that on the other side of the membrane may also be expressed in terms of an equation. In the case of a basic drug this may be derived as follows. The two forms of the

Henderson equation [9.18] stated above are written in the following alternative ways:

$$\log\left(\frac{[BH^+]_i}{[B]_i}\right) = pk_a - pH_i$$

$$\log\left(\frac{[BH^+]_e}{[B]_e}\right) = pk_a - pH_e.$$

Adding unity to the antilogarithms of both sides of each equation, we obtain:

$$1 + \frac{[BH^+]_i}{[B]_i} = 1 + \text{antilog}\,(pk_a - pH_i)$$

$$1 + \frac{[BH^+]_e}{[B]_e} = 1 + \text{antilog}\,(pk_a - pH_e).$$

Rearranging the left-hand sides gives:

$$\frac{[B]_i + [BH^+]_i}{[B]_i} = 1 + \text{antilog}\,(pk_a - pH_i)$$

$$\frac{[B]_e + [BH^+]_e}{[B]_e} = 1 + \text{antilog}\,(pk_a - pH_e).$$

When one equation is divided by the other, cancelling $[B]_e$ and $[B]_i$ which are equal, we obtain the following result:

$$\frac{[B]_i + [BH^+]_i}{[B]_e + [BH^+]_e} = \frac{1 + \text{antilog}\,(pk_a - pH_i)}{1 + \text{antilog}\,(pk_a - pH_e)}$$

$$= \frac{\text{total concentration of drug in the internal solution}}{\text{total concentration of drug in the external solution}}. \quad [9.43]$$

This is the ratio required. In the case of an acidic drug analogous reasoning, in which $[HA]_i = [HA]_e$, gives the following equation for the ratio of total concentration of drug:

$$\frac{[HA]_i + [A^-]_i}{[HA]_e + [A^-]_e} = \frac{1 + \text{antilog}\,(pH_i - pk_a)}{1 + \text{antilog}\,(pH_e - pk_a)}. \quad [9.44]$$

Intracellular pH may be determined using an indicator whose non-ionised and ionised forms conform to the conditions mentioned in either of the last two paragraphs. A suitable indicator is 5,5-dimethyl-2,4-oxazolidinedione which is of the type HA, A$^-$ and for which pK_a is 6.13 at 37° C. The pH of the external medium, pH_e, is measured and the extracellular and intracellular concentrations of the drug may be determined using a radioactively labelled form of the indicator or a spectrophotometric method. The data are substituted into the appropriate equation, e.g. [9.42] or [9.44] in the case of the indicator mentioned above, and the pH inside the cells, pH_i, is calculated.

Receptors for drugs may be on the outer surfaces of membranes (extracellular), e.g. bacterial receptors for aminoacridines, or inside the cell (intracellular), e.g. receptors for barbiturates in the eggs and larvae of the sea urchin. The receptor may be ionised, like drugs (see above), to an extent which depends on their pK_a values and on the pH of the extracellular or intracellular medium, as appropriate. Drugs active as cations combine with anionic receptors, e.g. —COO$^-$, —S$^-$ (from SH) and ionised phosphate groups, whereas drugs active as anions combine with cationic receptors, e.g. —NH$_3^+$, $>$NH$^+$. The pH of the external medium will not only affect the extent of ionisation of the drugs but also of the receptors, and even the viability and biochemical properties of the organism. There is some evidence that the external pH affects the ionisation of the intracellular as well as the extracellular receptors.

Acid-Base Catalysis

It has already been mentioned that the catalytic power of acids has long been recognised as one of their characteristic properties. The mechanism of the catalysis has also been indicated on p. 182. The acid donates a proton to the reacting species which is then more capable of undergoing reaction. After the reaction, the proton is lost again by the product of the reaction. The essential feature of this type of catalysis is a reversible acid-base reaction between the catalyst and the reactant and product molecules.

It was recognised at an early date that many reactions which are catalysed by acids are also catalysed by bases. For example, the hydrolysis of esters, the mutarotation of glucose and the enolisation of ketones proceed more rapidly in the presence of either acids or alkalis than they do in neutral solution. In the acid range the rate of reaction is approximately proportional to the concentration of hydrogen ions in the solution, and in the alkaline range the rate is approximately proportional to the concentration of hydroxyl ions.

A further study of such catalysed reactions, coupled with the extension of the definitions of acid and base, eventually led to the realisation that all the acidic and basic species present in the solution might contribute to the catalysis. Thus, not only hydrogen ions but also the undissociated acid can bring about acid catalysis. The anion of the acid, which is its conjugate base, also exerts some influence as a basic catalyst. Each catalytic species which is present increases the rate of the reaction by an amount proportional to its activity (or approximately proportional to its concentration). At a constant temperature the rate constant k of a catalysed reaction in the presence of an acid HA, which is partly dissociated, would therefore be given approximately by:

$$k = k_0 + k_{H_3O^+}[H_3O^+] + k_{OH^-}[OH^-] + k_{HA}[HA] + k_{A^-}[A^-]. \quad [9.45]$$

The constant of proportionality for each concentration term is called its 'catalytic coefficient', e.g. $k_{H_3O^+}$ is the catalytic coefficient of H_3O^+. It must be emphasised that k in this expression denotes a contribution to the total rate constant for the reaction (and does not represent a dissociation constant as in the preceding part of this chapter). k_0 is the rate constant for the 'spontaneous' reaction, i.e. in the presence of the solvent only. Catalytic coefficients are measured by determining the rate of the reaction under various conditions of different concentrations of the catalytic species. It has been established that there is an approximate relationship between the catalytic coefficient of a given species and its acid or base strength, as given by its dissociation constant. This relationship is given by Brönsted as:

$$k_A = G_A K_A^\alpha, \quad [9.46]$$

where k_A is the catalytic coefficient of the species A,
  G_A and α are constants for the particular reaction in a given solvent at a given temperature,
and K_A is the dissociation constant of the species A.

This relationship may be used to determine acid or base strengths in an approximate manner. Equation [9.46] can be written in the following alternative logarithmic form:

$$\log k_A = \log G_A + \alpha \log K_A. \quad [9.47]$$

A plot of $\log k_A$ against $\log K_A$ gives a linear graph whose gradient is α and which intercepts the ordinate at $\log G_A$.

The following mechanisms, which have been proposed for the tautomerisation of a ketone to its enol form in the presence of an acid or a basic catalyst respectively, serve to illustrate the part played by the catalyst in such reactions:

Acid catalysis.

$$\underset{H}{\overset{R_2}{\underset{R_1}{\rangle}}}C-C\overset{R_3}{\underset{O}{\langle}} + A \rightleftharpoons \underset{H}{\overset{R_2}{\underset{R_1}{\rangle}}}C-C\overset{R_3}{\underset{OH^+}{\langle}} + B \rightleftharpoons \overset{R_2}{\underset{R_1}{\rangle}}C=C\overset{R_3}{\underset{OH}{\langle}} + A.$$

Basic catalysis.

$$\underset{H}{\overset{R_2}{\underset{R_1}{\rangle}}}C-C\overset{R_3}{\underset{O}{\langle}} + B \rightleftharpoons \overset{R_2}{\underset{R_1^-}{\rangle}}C-C\overset{R_3}{\underset{O}{\langle}}$$

$$\updownarrow$$

$$\overset{R_2}{\underset{R_1}{\rangle}}C=C\overset{R_3}{\underset{O^-}{\langle}} + A \rightleftharpoons \overset{R_2}{\underset{R_1}{\rangle}}C=C\overset{R_3}{\underset{OH}{\langle}} + B.$$

In these reactions, A may be any sort of acid species and B may be any sort of basic species.

The recognition of catalytic power as one of the fundamental properties of an acid or a base served as one of the starting-points for the development, by Lewis, of an alternative theory of acids and bases. This theory succeeds in correlating more information than the Brönsted–Lowry theory, but it is not as convenient for the discussion of quantitative relationships. Lewis defined a base as a donor of a lone pair of electrons and an acid as a lone pair acceptor. A reaction between a Lewis base and a Lewis acid therefore results in the formation of a co-ordination compound in which the two reactant molecules are linked by a dative bond. The reaction is exemplified by the reaction between ammonia, a Lewis base, and boron trifluoride, a Lewis acid. When a metal atom or ion forms a co-ordination compound or complex with another molecule or ion, which is called a ligand, e.g. NH_3, H_2O, Cl^-, OH^-, ethylenediaminetetraacetate, the metal is the Lewis acid and the ligand is the Lewis base. Since Lewis acids and bases can form co-ordination compounds with organic compounds, they are important catalysts in organic chemistry. Since the ability to donate a lone pair of electrons is both a necessary and sufficient property of a proton acceptor, every Lewis base is a Brönsted–Lowry base and vice versa. There is, however, a fundamental difference between a Lewis acid (a lone pair acceptor) and a Brönsted–Lowry acid (a proton donor). For example, boron trifluoride and aluminium chloride are acids only in the Lewis sense, while hydrogen chloride and acetic acid are acids only in the Brönsted–Lowry sense. Nevertheless, Lewis acids often react with water and other protonic solvents to produce Brönsted-Lowry acids, e.g.

$$AlCl_3 + 3H_2O = Al(OH)_3 + 3HCl.$$

Problems

1 Find the hydrogen ion concentrations and hence the approximate pH (assuming activities to be equal to concentrations) in the following solutions: hydrochloric acid of concentration 0.1 mol dm^{-3} (assumed completely dissociated), formic acid of concentration 0.16 mol dm^{-3} (degree of dissociation 0.033), chloracetic acid of concentration 0.01 mol dm^{-3} (degree of dissociation 0.367), phenol of concentration 1.0 mol dm^{-3} (degree of dissociation 0.000 011), ammonia of concentration 0.022 mol dm^{-3} (degree of dissociation 0.029), sodium hydroxide of concentration 0.1 mol dm^{-3} (assumed completely dissociated).

2 The degrees of dissociation of solutions of formic acid, chloracetic acid and phenol, each of concentration 1.0 mol dm^{-3}, are 0.0133,

0.0367, and 0.000 011 respectively. Calculate the dissociation constants and pk_a values for the three acids.

3 Taking pk_a for acetic acid to be 4.756, calculate how many moles of solid sodium acetate must be added to 1 dm³ of a solution of acetic acid of concentration 0.1 mol dm⁻³ to obtain a buffer solution of pH = 4.50.

4 The bicarbonate buffer in blood plasma consists of about 0.025 mol of bicarbonate ion and 0.001 25 mol of carbonic acid of $pk_a = 6.10$ in each dm³ of plasma. Find the buffer capacity towards acid and alkali.

5 Taking pk_b for ammonia to be 4.76, calculate pk_a for the NH_4^+ ion, the hydrolysis constant of ammonium chloride, the pH of an ammonia solution of concentration 0.1 mol dm⁻³, and the pH of of an ammonium chloride solution of concentration 0.1 mol dm⁻³.

6 A 50 cm³ portion of a solution of a crystalline organic acid containing 9.00 g dm⁻³ was titrated with NaOH solution of concentration 0.1 mol dm⁻³. The pH at various stages in the titration is shown in the following table:

cm³ NaOH added	0	10	20	30	40	45	48	50
pH	1.15	1.21	1.29	1.38	1.46	1.68	2.08	2.80
cm³ NaOH added	52	55	60	70	80	90	95	98
pH	3.30	3.60	3.80	4.12	4.43	4.82	5.32	5.88
cm³ NaOH added	99	100	101	102	105	110	120	
pH	6.30	8.41	10.78	11.01	11.52	11.83	12.12	

Plot the titration curve, and deduce the basicity of the acid and its molecular weight (relative molecular mass).

7 Glycine has values for pK_1 and pK_2, for its two dissociation constants, of 2.35 and 9.78 respectively. Calculate the isoelectric pH. Sketch roughly the form of the titration curve for glycine by first plotting the isoelectric pH and the points of half-neutralisation of the acidic and basic groups.

8 The total concentration c (in mol dm⁻³) of salicylic acid required to inhibit cell division in an echinoderm egg varies with pH in the following manner:

log c	−1.4	−1.1	−0.8	−0.5	−0.2	+0.1	+0.4	+0.7
pH	5.3	5.7	6.0	6.2	6.6	6.95	7.3	7.5

Taking pk_a for salicylic acid to be 3.0, calculate, at each pH, the concentration of neutral molecules in the solution. Plot graphs of the logarithm of the total concentration, and of the logarithm of the concentration of neutral molecules, versus pH, and hence confirm that the inhibition is due to the neutral molecules present in each solution.

9 Two aqueous solutions of pH 5.40 and 8.45 respectively are separated by a membrane permeable to neutral molecules but impermeable to ions. Morphine, a basic drug of pk_a 7.87, distributes itself between the two solutions. Calculate the ratio of the concentration of cationic form of the drug in the more acidic solution to that in the more basic solution. Calculate the corresponding ratio for the total concentration of drug (cations and neutral molecules).

10 The rate constants k for the mutarotation of glucose catalysed by mandelic acid and mandelate ion are given in the following table for various concentrations (in mol dm^{-3}) of these two species.

mandelic acid	Na mandelate	$k \times 10^4 /\text{s}^{-1}$
0.001	0.050	0.966
0.001	0.100	1.063
0.001	0.125	1.111
0.050	0.100	1.118
0.100	0.100	1.172

Assuming that the catalytic effects of hydrogen and hydroxyl ions are negligible over this range of concentrations, plot appropriate graphs and determine the catalytic coefficients for mandelic acid and mandelate ions (in the units mol^{-1} dm^3 s^{-1}) and the rate constant for the spontaneous reaction.

10 Electrochemical Cells

Introduction

In an electrochemical cell the energy released by a chemical reaction is converted into electrical energy. During this process the formation and destruction of ions results in the development of charges on the electrodes. Each electrode can be regarded as having a certain electrode potential and this depends upon the concentration of ions in its vicinity. The generation of electrical energy in a cell is therefore the opposite of the process of electrolysis described in Chapter 8 because, in the latter, ions are formed or destroyed in a cell by the action of electrical energy fed in from the external circuit.

There are two main reasons why the study of electrical cells is important from the chemical point of view. Firstly, since the potential of an electrode varies according to the concentrations of the ions in equilibrium with it, electrical measurements on cells can give valuable information about ionic equilibria in solution. The most important application of this type, in the measurement of pH, is described later (p. 414). Secondly, from such measurements it is possible to calculate with great accuracy the energy changes associated with a chemical reaction taking place in a cell.

Transformation of Chemical into Electrical Energy

It has been seen in Chapter 7 that the amount of energy released during a chemical reaction which is capable of being transformed into work or into forms of energy other than heat is not usually equal to the heat content change ΔH. Allowance must be made for a change in entropy ΔS (i.e. a change in the amount of disorder) in the system. If, in the course of the change, the system becomes more ordered and the entropy decreases, an amount of chemical energy $T \Delta S$ is used to bring this about. This appears as heat which raises the temperature of the cell unless it is kept in a thermostat bath. It is not available for conversion into electrical energy. The maximum amount of energy which is available for transformation into electrical energy is, in fact,

Transformation of Chemical into Electrical Energy

equal to the decrease in free energy during the reaction. It may also happen that the system becomes more disordered during the chemical change (especially if the reactants become split into a large number of small molecules or ions). If that is the case, ΔS is positive and an extra amount of energy $T\Delta S$ is made available for conversion into electrical energy over and above the amount provided by the decrease in heat content. This extra amount of energy is obtained by the cell absorbing heat from its surroundings as it functions. Again, the maximum amount of energy available for transformation into electrical energy is equal to $-\Delta G$. Both situations are summarised by the equation (cf. p. 300):

$$\text{Maximum electrical energy} = -\Delta G = -\Delta H + T\Delta S. \qquad [10.1]$$

The condition which must be fulfilled in order that a transformation of chemical to electrical energy can take place is that the reaction must involve the transfer of electrons to and from materials which can be made up in the form of electrodes. A typical example is provided by the reaction which takes place in the Daniell cell, namely:

$$Zn + CuSO_4 \rightarrow Cu + ZnSO_4.$$

Actually the two sulphates are fully ionised in solution and the sulphate ion undergoes no change, so the reaction can be represented as:

$$Zn + Cu^{2+} \rightarrow Cu + Zn^{2+}.$$

Written in this form it is clear that the reaction involves the transfer of two electrons from each zinc atom to each Cu^{2+} ion. In the cell, this transfer is made to take place in two stages, at the two electrodes. One electrode is a zinc rod dipping into zinc sulphate solution and during the reaction zinc atoms leave the rod to become Zn^{2+} ions in solution. In doing so, each zinc atom must leave behind on the rod two electrons and these create a negative potential at this electrode. The other electrode consists of a copper plate dipping into copper sulphate solution. During the reaction each Cu^{2+} ion takes two electrons from the plate as the ion is deposited on it. This drain of electrons from the plate gives it a positive electrical potential. If the two electrodes are now connected through an external circuit, electrons flow through this circuit from the negative (zinc) electrode to the positive (copper) electrode, so the reaction is accompanied by the production of electrical energy.

The condition for obtaining the maximum amount of electrical energy from a cell it that is should operate reversibly (see p. 285). This means that the processes occurring at the two electrodes must be cabable of being reversed by an infinitesimally small change in the external conditions. For this to be possible, the tendency for the cell to generate electrical energy must be almost exactly balanced by an external circuit which is trying to cause the electrons to flow in the opposite direction. The fact that this condition can be satisfied in practice, as described below, means that it is possible to measure the maximum electrical energy which a cell can provide and hence obtain the free energy change in the corresponding chemical reaction. This facility of measuring directly values of ΔG for reactions in solution provides one of the reasons why the study of electrochemical cells is so important in physical chemistry.

Reversible Electrode Processes

The concept of reversibility as it applies to the processes which take place at the electrodes of a cell must now be discussed in more detail. During the working of a cell there are normally chemical changes taking place at each electrode and associated with this is the gain of electrons at the cathode and the loss of electrons at the anode. (It is this gain and loss of electrons at the cathode and anode respectively which enables the cell to maintain a flow of current in the external circuit.) In order to be reversible, the electrode materials must only take part in chemical changes which correlate exactly with the electrical changes. No energy or material must be transformed in any other way and a very small increase in the potential difference in the external circuit opposing that of the cell must exactly reverse the chemical changes.

The simplest type of reversible electrode is a metal dipping into a solution containing ions of the same substance. Two opposing processes take place simultaneously and reach an equilibrium state depending on the conditions. Atoms of the metal tend to go into solution forming positive ions, leaving a corresponding number of electrons on the metal electrode. At the same time, ions in solution tend to become attached to the electrode and to be converted into atoms by accepting electrons from the electrode. The more negative the electrode becomes by losing atoms, the more will it attract the positive ions from the solution, until equilibrium is established. On the other hand, the more electropositive is the metal the greater is the tendency to form positive ions in solution and the more negative will the electrode be when equilibrium is established. The equilibrium itself can be represented by the equation:

$$M^{z+} + ze^- \rightleftharpoons M$$

(where z is the valency of the ion of the metal M and e^- represents an electron). If more ions M^{z+} are formed by chemical reaction they become converted into metal atoms at the electrode by accepting a proportionate number of electrons. Conversely, if electrons are removed from the electrode during electrolysis, metal atoms become ions in an attempt to restore the equilibrium. In this way, the chemical change and the flow of electric current to and from the electrode are seen to be closely correlated and to be completely reversible. This is only true if no other process happens simultaneously whereby M^{z+} and M are interconverted without affecting the charge on the electrode. (An example of irreversibility is provided by the Daniell cell as it is usually set up. Impurities in the zinc and the acidity of the solution cause some of the zinc to dissolve with the evolution of hydrogen. This does not cause an equivalent amount of electrical change at the electrode. Moreover, if the cell were run in reverse by electrolysis this side reaction would not take place in reverse.)

Another important type of reversible electrode consists of an inert metal (usually platinum or gold) dipping into a solution containing both the reduced and oxidised forms of a molecule or ion. Under these conditions the metal acts as a reservoir of electrons which are used up if some of the oxidised form is converted into the reduced form, or are liberated if the reaction happens in reverse. A simple example of this type of oxidation-reduction or redox system is a solution containing both ferric and ferrous ions. The equilibrium is:

$$Fe^{3+} + e^- \rightleftharpoons Fe^{2+}.$$

The position of equilibrium, which determines the charge on the electrode, depends in this case on the tendencies of Fe^{3+} and Fe^{2+} to take and give electrons. That is, it depends on how strongly reducing or oxidising the system is. It also depends on the relative concentrations of the oxidised and reduced forms. Again, both the chemical state of the system and the electrical state of the electrode are interrelated and changes in either will affect the other in a reversible manner.

Electrode Potentials

For a cell to be able to supply electric current to an external circuit one electrode must be at a higher electrical potential than the other. Electric charge then flows from the higher to the lower potential in an attempt to equalise the two potentials. It is helpful to consider this process from the point of view of the analogy of water flowing through a tube (cf. p. 343). The force which causes the flow of current is equal to the difference in potential at the two ends of the circuit. In the electrical case it is called the electromotive force (often abbreviated to e.m.f.) and it is the counterpart of the difference in pressure between the two ends of the tube in the water analogy. Just as the rate of flow of water through a pipe offering a given resistance is proportional to the pressure, so also the flow of charge, which is the electric current, is proportional to the e.m.f. of the source, provided the circuit has a constant resistance. The e.m.f. and the potentials of the two electrodes are measured in volts. Only under reversible conditions, i.e. when the e.m.f. of the cell is balanced by an external e.m.f., is the e.m.f. of the cell exactly equal to the difference between the potentials of the two electrodes.

It is convenient to be able to discuss the potentials of separate electrodes, but since it is only possible to measure a potential relative to another it is necessary to choose one reproducible electrode as defining arbitrarily a zero of potential. The electrode which is chosen for this is the standard hydrogen electrode. This is a reversible elec-

Fig. 10.1 The hydrogen electrode

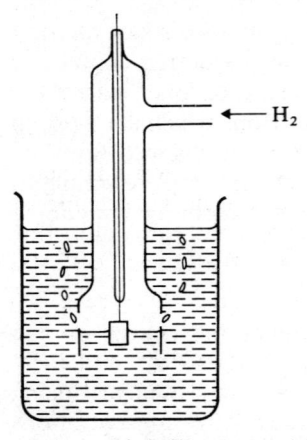

trode and it is illustrated in Fig. 10.1. It consists of hydrogen bubbling over a plate of platinised platinum which also dips into a solution containing hydrogen ions. The platinised platinum plate responds to the equilibrium between hydrogen ions and hydrogen molecules:

$$H^+ + e^- \rightleftharpoons \tfrac{1}{2} H_2.$$

In the standard hydrogen electrode the hydrogen ions and the hydrogen gas are each present at unit activity. This condition is fulfilled by a 1.2 mol kg^{-1} solution of hydrochloric acid with the pressure of hydrogen gas equal to 101 325 Pa (1 atm). The potential of any electrode measured relative to this is therefore taken as being its electrode potential E_h (sometimes stated as being on the hydrogen scale).

Energy Changes from Measurements on Cells

It has already been stated (on p. 393) that if a reaction were allowed to take place in a cell under reversible conditions (which also implies that it must take place infinitely slowly) the electrical energy produced would be equal to the decrease in free energy for the reaction. The electrical energy is equal to the quantity of charge transferred multiplied by the potential difference across which it is transferred. It is therefore equal to the amount of charge which flows multiplied by the e.m.f. of the cell. By Faraday's laws, each equivalent of chemical change is accompanied by the transfer of charge equal to the Faraday constant, F. The amount of chemical change in moles represented by the chemical equation (the usual unit in thermochemistry; see pp. 276 and 313), is therefore accompanied by the transfer of nF units of charge, where n is a small whole number equal to the number of moles of electrons exchanged between the chemical species represented by the chemical equation. This corresponds to the performance of electrical work nFE, where E is the e.m.f. of the cell. If the cell is operating under reversible conditions, the work done by the cell is a maximum and is given by:

$$-\Delta G = nFE. \qquad [10.2]$$

In order to find the change in Gibbs energy for the reaction it is therefore only necessary to measure the e.m.f. of the appropriate cell and to know, from the equation and the form of the cell, the value of n. For example, in the case of the reaction occurring in the Daniell cell, the chemical equation is

$$Zn + CuSO_4 = ZnSO_4 + Cu,$$

which is here taken to mean that one mole of zinc and one mole of copper sulphate react to yield one mole of zinc sulphate and one mole of copper and for which $n = 2$ mols. Under standard conditions, the e.m.f. of this cell is 1.103 V. Substituting these values and $F = 96\,487$ C mol^{-1} into equation [10.2] gives $\Delta G^\ominus$ as follows:

$$-\Delta G^\ominus = 2 \text{ mol} \times 96\,487 \text{ C mol}^{-1} \times 1.103 \text{ V}.$$

We note that $1\text{V} \times 1\text{C} = 1\text{J}$, so that

$$\Delta G^\ominus = -212\,850 \text{ J for the above reaction.}$$

The entropy change and enthalpy change of a cell reaction can also be obtained entirely from e.m.f. measurements by making use of the equations leading to the Gibbs–Helmholtz equation (p. 302), namely

$$\Delta S = -\left(\frac{\partial(\Delta G)}{\partial T}\right)_p \qquad [7.70]$$

and
$$\Delta G = \Delta H - T\Delta S. \qquad [7.62]$$

Differentiation of equation [10.2] with respect to temperature at constant pressure gives:

$$-\left(\frac{\partial(\Delta G)}{\partial T}\right)_p = nF\left(\frac{\partial E}{\partial T}\right)_p. \qquad [10.3]$$

Substituting into equation [7.70] we have:

$$\Delta S = nF\left(\frac{\partial E}{\partial T}\right)_p. \quad [10.4]$$

Introducing the expressions for ΔS (equation [10.4]) and ΔG (equation [10.2]) into equation [7.62] and rearranging to obtain ΔH:

$$\Delta H = -nFE + nFT\left(\frac{\partial E}{\partial T}\right)_p. \quad [10.5]$$

By measuring the e.m.f. of a cell at a number of different temperatures at constant pressure (usually atmospheric) and plotting a graph of E versus T, the value of E at a given temperature T can be read off and the value of $\left(\dfrac{\partial E}{\partial T}\right)_p$ is the gradient of the graph at this point. Substitution of the appropriate values into equations [10.4] and [10.5] enables ΔS and ΔH to be calculated for the cell reaction at the particular temperature chosen. Enthalpy changes measured by this method are in good

Fig. 10.2 Potentiometer circuit

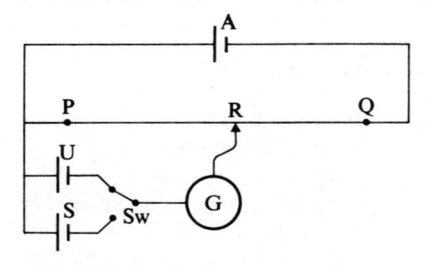

agreement with those determined calorimetrically and it is generally acknowledged that, where it is possible to measure energy changes by an electrochemical method, the greatest accuracy can be obtained.

The method of measuring the e.m.f. of a cell under reversible conditions, by opposing it with an equal and opposite e.m.f., is illustrated in Fig. 10.2. This is a potentiometer circuit, and the principle on which it works is that of opposing the cell U of unknown e.m.f., or a standard cell S, by a fraction of the e.m.f. of the accumulator A. In the symbol for each cell, the longer of the two lines represents the positive electrode. PQ represents a bridge wire with a sliding contact R. G is the galvanometer which may be connected between R and either of the two cells by adjusting the switch marked Sw. Between the points P and Q there is the full e.m.f. of the accumulator A with P positive and Q negative. Along the bridge wire between P and Q there will be a constant voltage drop per unit length. Hence, between P and R there will be a fraction PR/PQ of the e.m.f. of A, with R negative relative to P. From these two points there will be a tendency for electrons to flow from R through G, and through either U or S to P. But, at the same time, U (or S) will tend to send electrons in the opposite direction round the same circuit. When the deflection on G is zero, these two opposing tendencies are equal, and the fraction of the e.m.f. PR/PQ is equal to the e.m.f. of U (or S). Two null positions of the sliding contact are found, in turn, for U and S respectively. Then,

$$\frac{\text{e.m.f. of U}}{\text{e.m.f. of S}} = \frac{\text{PR for U}}{\text{PR for S}}.\qquad [10.6]$$

Since the e.m.f. of S is known, the e.m.f. of U may be calculated.

Fig. 10.3 Weston standard cell

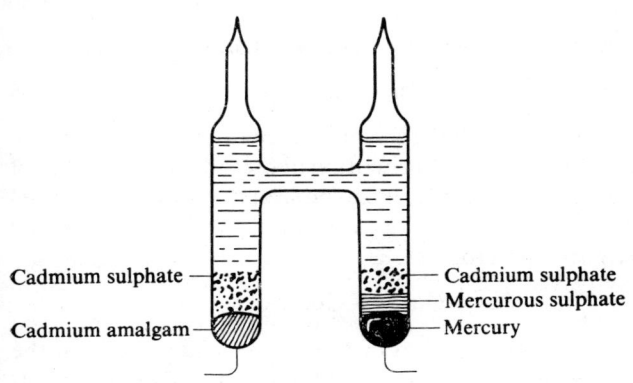

Cadmium sulphate — Cadmium sulphate
— Mercurous sulphate
Cadmium amalgam — Mercury

The Weston cell (see Fig. 10.3) is commonly used as the standard cell S. In this, the positive electrode is mercury in contact with solid mercurous sulphate, and the negative electrode is a solution (or amalgam) of cadmium in mercury (12 per cent, w/w). The electrolyte is a saturated solution of cadmium sulphate, and excess of the salt is present to ensure that the solution is maintained in a saturated condition. The Weston cell is easily made up to give a reproducible e.m.f. of 1.018 37 V at 25° C. The temperature coefficient $(\partial E/\partial T)_p$ of the cell is $-40\ \mu\text{V K}^{-1}$.

Sign Convention

Because of the relative nature of the potentials of electrodes, it is necessary to adopt a convention, firstly as to what constitutes a positive and what a negative e.m.f. of a cell and secondly as to whether a given electrode potential is positive or negative. The definition of the second depends on the first, and both refer to the symbolic method of drawing a cell so as to represent the reaction which takes place within it when it is allowed to supply current. The cell is represented in such a way that the symbols for the electrolytes appear between those for the electrode materials and separated from them by vertical lines. A boundary between two electrolytes is represented by a double vertical line, if the liquid junction potential is eliminated (pp. 408, 421) or is insignificant, which we shall now assume. A boundary between two electrolytes is represented by a single vertical line, if the liquid junction potential is significant (p. 422). The relationship between the representation of a cell and the reaction which it implies is that, as the forward reaction takes place, positive charges move through the cell from left to right and/or negative charges move through the cell from right to left and electrons flow through the external circuit from the left-hand electrode to the right-hand electrode. Thus, the Daniell cell written as:

$$Zn \mid Zn^{2+} \mid\mid Cu^{2+} \mid Cu$$

implies that the reaction taking place at the left-hand electrode is:

$$Zn \rightarrow Zn^{2+} + 2e^-.$$

This corresponds to positive charges on the zinc ions moving away from the left-hand side in the cell as written above and to the release of electrons at the left-hand electrode. It can be seen that oxidation occurs at this electrode. The Daniell cell as written above implies that the reaction at the right-hand electrode is:

$$2e^- + Cu^{2+} \rightarrow Cu.$$

This corresponds to positive charges on the copper ions arriving at the right-hand electrode and to the withdrawal of electron from the right-hand electrode by these ions. It can be seen that reduction occurs at this electrode. When the reactions of the two electrodes (or half-cells) are added together so as to eliminate the electrons, the overall reaction taking place in the cell is obtained:

$$Zn + Cu^{2+} \rightarrow Cu + Zn^{2+}.$$

The e.m.f. of the cell, E, is always taken as the potential of the electrode written on the right-hand side, E_r, relative to that of the left-hand electrode, E_l, thus:

$$E = E_r - E_l. \qquad [10.7]$$

The convention of sign of the e.m.f. of a cell can now be stated quite simply. If the cell is written in the way indicated above (with positive charges moving from left to right within the cell, etc.), and the reaction which this implies is the direction of the spontaneous reaction as the cell supplies current, then the e.m.f. is positive. If this way of writing the cell implies the reverse of the spontaneous reaction the e.m.f. is negative. In the example of the Daniell cell the representation shown above does imply the spontaneous reaction because zinc does tend to displace copper from solution. The cell as written would therefore have a positive e.m.f. which means that the copper electrode would be positive relative to the zinc electrode.

It might be thought to be unnecessary to establish a convention of the sign of the e.m.f. of a cell in this way when the sign can so easily be measured by, say, a voltmeter. It must be remembered, however, that the labelling of the terminals + and − on a voltmeter depends on an equivalent sign convention to the one above. Also, in electrochemistry, cells are often considered which could not readily be set up and measured experimentally. It is mainly for this reason that the convention as stated above is necessary.

The conventions defining the sign of the e.m.f. of a cell discussed above and the potential of a separate electrode described below have been recommended for some years by IUPAC. The potential of an electrode on the hydrogen scale should be visualised in terms of a complete cell, with a standard hydrogen electrode to the left and the electrode in question to the right. The potential of the right-hand electrode is then the total e.m.f. of the cell. The question as to whether the total e.m.f. is positive or negative with the electrodes arranged in this way is decided by the convention already given. If this e.m.f. is positive, then the electrode potential is also positive. If the e.m.f. for the cell written in this way is negative, then the electrode potential is negative.

These two cases can be illustrated by considering a copper and a zinc electrode separately. A cell combining a standard hydrogen electrode with a copper electrode would be written:

$$\text{Pt}, \text{H}_2 | \text{H}^+ || \text{Cu}^{2+} | \text{Cu}.$$

Two electrode materials which constitute different phases, e.g. Pt and H_2 in this cell, are separated by a comma in the representation of a cell or half-cell. The above cell corresponds to the cell reaction:

$$H_2 + Cu^{2+} \rightarrow 2H^+ + Cu.$$

Actually copper is less electropositive than hydrogen because it does tend to be displaced from its salts by hydrogen gas. The reaction as written above does therefore represent the spontaneous direction of the reaction as the cell supplies current. According to the convention, therefore, the e.m.f. of the cell is positive and so the potential of the copper electrode is positive.

In the case of zinc the cell is written:

$$\text{Pt}, \text{H}_2 | \text{H}^+ || \text{Zn}^{2+} | \text{Zn}$$

and this corresponds to the reaction:

$$H_2 + Zn^{2+} \rightarrow 2H^+ + Zn.$$

In this case, however, the spontaneous reaction is the reverse of this because zinc is more electropositive than hydrogen and tends to displace it from solution. The e.m.f. of the cell is therefore negative and the electrode potential of zinc is negative.

The signs of the potentials of oxidation-reduction electrodes is defined in exactly the same way. In fact, there is no fundamental difference between this sort of electrode where a molecule or ion receives electrons to become reduced and the type of electrode considered above in which a metal ion receives electrons to become a metal atom. It simplifies the electrochemical theory to treat both situations together as reduction processes (and the reverse as oxidation processes) and this corresponds to normal terminology since salts of metals are often spoken of as being reduced to the corresponding metals.

The ferric-ferrous system can be taken again to illustrate the definition of sign of an electrode where both the oxidised and the reduced forms are in solution. The cell would be written:

$$\text{Pt, } H_2 | H^+ || Fe^{3+}, Fe^{2+} | \text{Pt}.$$

Two substances present in the same solution, e.g. Fe^{3+} and Fe^{2+} in this cell, are also separated by a comma in the representation of the cell or half-cell. The cell represented here corresponds to the reaction:

$$\tfrac{1}{2} H_2 + Fe^{3+} \rightarrow H^+ + Fe^{2+}.$$

It so happens that this is the direction of the spontaneous reaction so the e.m.f. of the cell is positive. The redox potential of the ferric-ferrous system is therefore positive.

Just as the convention of sign of the e.m.f. of a cell can often be replaced by the simple experiment of measuring which electrode is positive with the aid of a voltmeter or a potentiometer circuit, so the convention about single electrode potentials can be put in a simple form if the appropriate measurements can be made. If an electrode is measured as being positive relative to a standard hydrogen electrode (after eliminating liquid-junction potentials, as mentioned on p. 408), then it has a positive electrode potential.

Some earlier texts express electrode potentials in terms of a sign convention which is the opposite of that recommended by IUPAC. To obtain this sign convention one must imagine such a measurement being made and then thought of in terms of the potential of the standard hydrogen electrode relative to the electrode in question. This is less direct but it does have one advantage over the recommended convention, namely, that the more electropositive metals then have the more positive electrode potentials (cf. Table 10.1, p. 405). A glance at a table of standard electrode potentials shows in an easy way which sign convention is adopted in a particular book, and this fact should always be ascertained before reading any electrochemical discussion, if confusion is to be avoided.

Standard Electrode (or Oxidation-Reduction) Potentials

From the discussion above on electrode processes it is seen that at constant temperature and pressure the potential of a reversible electrode depends upon two factors: (a) the nature of the materials present, and (b) their concentrations, or more correctly, their activities. The form of the dependence on the second factor must be the same as the dependence of change in Gibbs energy, ΔG. The relationship between ΔG and activity has already been derived (equation [7.102], p. 313) and can be written as:

$$\Delta G = \Delta G^{\ominus} + RT \ln \frac{\{\text{products}\}}{\{\text{reactants}\}}. \qquad [7.102]$$

In this equation, braces { } are used to signify activities. The electrode potential must correspond to the decrease in Gibbs energy for the reaction taking place in a cell formed by combining the electrode in question with a standard hydrogen electrode in the manner described above. The electrode reaction in question may be written in the general form:

$$ox + ne^- = red,$$

where ox signifies the oxidised form and red the reduced form. The electrode potential (or oxidation-reduction potential) E_h of this electrode on the hydrogen scale is the e.m.f. E of the cell:

$$\text{Pt}, \text{H}_2 | \text{H}^+ \, || \, ox, red | \text{Pt}.$$

The cell reaction corresponding to this cell is:

$$\frac{n}{2} \text{H}_2 + ox \rightarrow n\text{H}^+ + red.$$

We note that the IUPAC sign convention requires that molecular hydrogen and the oxidised form are the reactants and that hydrogen ions and the reduced form are the products. When this information is substituted into equation [7.102] above we have:

$$\Delta G = \Delta G^{\ominus} + RT \ln \frac{\{\text{H}^+\}^n \{red\}}{\{\text{H}_2\}^{n/2} \{ox\}}.$$

Since both hydrogen ions and hydrogen gas in the standard hydrogen electrode are in their standard states of unit activity, this equation reduces to:

$$\Delta G = \Delta G^{\ominus} + RT \ln \frac{\{red\}}{\{ox\}}. \qquad [10.8]$$

The amount of chemical change represented by the chemical equation is accompanied by the transfer of charge nF corresponding to the performance of electrical work nFE (equation [10.1]), so that:

$$-\Delta G = nFE = nFE_h. \qquad [10.9]$$

In the same way a standard electrode potential (or standard oxidation–reduction potential) $E_h^\ominus$ of the electrode in question is related to the standard Gibbs energy change $\Delta G^\ominus$ by the analogous equation:

$$-\Delta G^\ominus = nFE^\ominus = nFE_h^\ominus. \qquad [10.10]$$

$E_h^\ominus$ is the potential of the electrode on the hydrogen scale when all the substances which partake in the electrode process are present in their standard states. For a metal half-cell this means that the metal in a pure, stable form is dipping into a solution in which its ions are present at unit activity. For a redox half-cell it usually means that the inert electrode, e.g. bright Pt, is dipping into a solution containing both the oxidised and reduced forms of the system at unit activity. Like $\Delta G^\ominus$ and equilibrium constants, $E_h^\ominus$ is a constant for a given electrode system at a particular temperature and at a pressure of 101 325 Pa (1 atm).

Eliminating ΔG and $\Delta G^\ominus$ from equations [10.8], [10.9] and [10.10], we have:

$$-nFE_h = -nFE_h^\ominus + RT \ln \frac{\{red\}}{\{ox\}}.$$

Driving throughout by $-nF$ we obtain the general electrode potential equation:

$$E_h = E_h^\ominus - \frac{RT}{nF} \ln \frac{\{red\}}{\{ox\}}. \qquad [10.11]$$

This equation is sometimes called the Nernst equation and can be written in the following alternative forms [10.12 to 10.14]:

$$E_h = E_h^\ominus + \frac{RT}{nF} \ln \frac{\{ox\}}{\{red\}} \qquad [10.12]$$

$$E_h = E_h^\ominus + \frac{2.303\ RT}{nF} \log \frac{\{ox\}}{\{red\}}. \qquad [10.13]$$

Since $R = 8.314$ J K^{-1} mol^{-1}, $F = 96\ 487$ C mol^{-1} and 1 J = 1 V × 1 C,

$$2.303\ R/F = 1.985 \times 10^{-4}\ \text{V K}^{-1}$$

$$\therefore \qquad E_h = E_h^\ominus + (1.985 \times 10^{-4}) \cdot \frac{T}{n} \log \frac{\{ox\}}{\{red\}}. \qquad [10.14]$$

The factor $2.303\ RT/F$ ($= 1.985 \times 10^{-4}\ T$) has the value 0.0578 V at 18° C, 0.0582 V at 20° C, 0.05916 V at 25° C and 0.0616 V at 37° C.

If the chemical equation of the cell reaction is multiplied by a constant factor and the derivation of the Nernst equation is repeated, it can

Standard Electrode (or Oxidation-Reduction) Potentials

be seen that the e.m.f. of the cell and the electrode potential E_h are independent of the multiplying factor and are therefore intensive quantities (p. 272). On the other hand, ΔG (and also ΔH and ΔS) are found to be multiplied by the factor, since they are extensive quantities.

The Nernst equation shows how the electrode potential depends upon the two factors mentioned at the beginning of this section. $E_h^\ominus$, which depends only on the nature of the materials present, is a measure, under standard conditions, of the relative tendencies of both the oxidised and the reduced forms to attract electrons. The more positive is $E_h^\ominus$ the more strongly oxidising is the system, i.e. the greater is the tendency for the oxidised form to be converted into the reduced form if it can take electrons from some other system. In terms of a metal electrode and its ions, the more positive is $E_h^\ominus$ the greater is the tendency for the ions to become metal atoms if they can accept electrons from some other system.

In quoting values for $E_h^\ominus$ it is usual to replace the subscript h by the oxidised and reduced forms of the electrode material, preferably in parentheses. If they are quoted in this order they imply correctly the order of the components in the cell used to define the electrode potential. To be quite specific, the electrode reaction is sometimes quoted. This is the part of the total cell reaction which takes place at the electrode under consideration and, since the terms involving H_2

Table 10.1. Standard Electrode Potentials of Metals at 298.15 K

Ion and electrode	$E_h^\ominus/V$
K^+, K	−2.924
Ca^{2+}, Ca	−2.76
Na^+, Na	−2.711
Mg^{2+}, Mg	−2.375
Zn^{2+}, Zn	−0.763
Fe^{2+}, Fe	−0.409
Co^{2+}, Co	−0.28
Ni^{2+}, Ni	−0.23
Sn^{2+}, Sn	−0.136
Pb^{2+}, Pb	−0.126
H^+, H_2	0.000
Cu^{2+}, Cu	+0.340
Hg_2^{2+}, Hg	+0.796
Ag^+, Ag	+0.7996

and H^+ have disappeared from the Gibbs energy expression, the electrode reaction includes all the terms which should appear in the electrode potential formula. The correct convention is maintained if the electrons (which are normally included when writing the reaction at one electrode) always appear on the left hand side of the equilibrium. Table 10.1 gives some standard electrode potentials for metals and Table 10.2 gives some standard oxidation-reduction potentials. In the second case the corresponding electrode reaction is quoted, in the form required by the sign convention.

Table 10.2. Oxidation-Reduction Potentials at 298.15 K

System	Reaction	$E_h^\ominus$/V
MnO_4^-, H^+, Mn^{2+}	$MnO_4^- + 8H^+ + 5e^- \rightleftharpoons Mn^{2+} + 4H_2O$	+1.491*
Ce^{4+}, Ce^{3+}	$Ce^{4+} + e^- \rightleftharpoons Ce^{3+}$	+1.443
$Cr_2O_7^{2-}$, H^+, Cr^{3+}	$Cr_2O_7^{2-} + 14H^+ + 6e^- \rightleftharpoons 2Cr^{3+} + 7H_2O$	+1.33*
O_2, H, H_2O	$O_2 + 4H^+ + 4e^- \rightleftharpoons 2H_2O$	+1.229*
Hg^{2+}, Hg_2^{2+}	$2Hg^{2+} + 2e^- \rightleftharpoons Hg_2^{2+}$	+0.905
Fe^{3+}, Fe^{2+}	$Fe^{3+} + e^- \rightleftharpoons Fe^{2+}$	+0.770
Cu^{2+}, Cu^+	$Cu^{2+} + e^- \rightleftharpoons Cu^+$	+0.158
Sn^{4+}, Sn^{2+}	$Sn^{4+} + 2e^- \rightleftharpoons Sn^{2+}$	+0.15
Cr^{3+}, Cr^{2+}	$Cr^{3+} + e^- \rightleftharpoons Cr^{2+}$	−0.41

* Since hydrogen ions enter into this equilibrium, the value for $E_h^\ominus$ depends upon pH (cf. p. 418). The value quoted corresponds to unit activity of hydrogen ions, pH = 0.

In order to show how the general equation of p. 404 gives the variation of electrode potentials with activities in particular cases, three typical inorganic examples will now be given. A typical organic redox system is discussed in detail on p. 417.

(a) A metal M and its ions M^{z+}.

The electrode reaction is:

$$M^{z+} + ze^- \rightleftharpoons M.$$

When equation [10.12] is applied, we have:

$$E(M^{z+}, M) = E^\ominus(M^{z+}, M) + \frac{RT}{zF} \ln \frac{\{M^{z+}\}}{\{M\}}. \qquad [10.15]$$

Because the metal is in excess, its activity is constant. If, as is usual, the metal is in its pure stable form, it is in its standard state, so that

$\{M\} = 1$ and the final equation is:

$$E(M^{z+}, M) = E^{\ominus}(M^{z+}, M) + \frac{RT}{zF} \ln \{M^{z+}\} \qquad [10.16]$$

(b) A metal M in contact with its insoluble salt MX.

This consists of a metal wire, plate or rod which has been electrolytically coated with a layer of its insoluble salt MX. It is reversible with respect to X^- ions and the electrode reaction is

$$MX + e^- \rightleftharpoons M + X^-.$$

Application of equation [10.11] gives:

$$E(MX, M, X^-) + E^{\ominus}(MX, M, X^-) - \frac{RT}{F} \ln \frac{\{M\}\{X^-\}}{\{MX\}}.$$

Putting the activities of the two solids equal to unity as before:

$$E(MX, M, X^-) = E^{\ominus}(MX, M, X^-) - \frac{RT}{F} \ln \{X^-\}. \qquad [10.17]$$

Systems of this type form the basis of electrodes specific to anions. For example, the activity of chloride, bromide or iodide ion in a solution can be determined by measuring the electrode potential of a silver electrode coated with the appropriate silver halide and dipping into the solution.

(c) The ferric–ferrous redox system.

A bright platinum electrode dips into a solution containing ferric ions at activities $\{Fe^{3+}\}$ and $\{Fe^{2+}\}$ respectively. The electrode reaction is:

$$Fe^{3+} + e^- \rightleftharpoons Fe^{2+}.$$

When equation [10.12] is applied, we have:

$$E(Fe^{3+}, Fe^{2+}) = E^{\ominus}(Fe^{3+}, Fe^{2+}) + \frac{RT}{F} \ln \frac{\{Fe^{3+}\}}{\{Fe^{2+}\}} \qquad [10.18]$$

Formal Electrode Potentials

Since activity is equal to the concentration multiplied by the activity coefficient y (equation [7.95]), the Nernst equation [10.12] can be written as:

$$E_h = E_h^{\ominus} + \frac{RT}{nF} \ln \frac{[ox]y_{ox}}{[red]y_{red}}. \qquad [10.18]$$

Whence:

$$E_h = E_h^\ominus + \frac{RT}{nF} \ln \frac{y_{ox}}{y_{red}} + \frac{RT}{nF} \ln \frac{[ox]}{[red]}$$

$\therefore \qquad E_h = E_h^{\ominus\prime} + \dfrac{RT}{nF} \ln \dfrac{[ox]}{[red]} \qquad\qquad$ [10.19]

where $\qquad E_h^{\ominus\prime} = E_h^\ominus + \dfrac{RT}{nF} \ln \dfrac{y_{ox}}{y_{red}}.\qquad\qquad$ [10.20]

Comparison of equation [10.12] with [10.19] shows that activities may be replaced by the corresponding concentrations provided that the *standard* electrode potential $E_h^\ominus$ is replaced by $E_h^{\ominus\prime}$, which is sometimes called the *formal* electrode potential. $E_h^{\ominus\prime}$ depends not only on $E_h^\ominus$ but also on the activity coefficients of the species *ox* and *red*. The latter quantities are affected by the various substances present in the solution particularly ions, including hydrogen and hydroxyl ions, as a result of interionic and intermolecular interactions. Consequently, whenever a formal electrode potential is quoted, the appropriate conditions to which it refers should also be stated. For example, $E^{\ominus\prime}(Fe^{3+}, Fe^{2+})$ is 0.67 V in the presence of 1 mol dm^{-3} sulphuric acid at 20° C. This value differs appreciably from $E^\ominus(Fe^{3+}, Fe^{2+})$, which is 0.770 V at 25° C.

Measurement of Electrode Potentials

The principles of the measurement of electrode potentials have already been discussed, but attention must be drawn to one or two practical points. If the solutions which contain the ions in equilibrium with the electrodes are different either in nature or in concentration, there is likely to be an unwanted further potential difference at the junction of the two liquids. This liquid–junction potential arises because of the transport of ions across the boundary between the two liquids. If the positive and negative ions have different transport numbers, there is a separation of positive and negative charge, and this constitutes the potential difference. The liquid-junction potential can usually be calculated, and an appropriate correction made to the measured e.m.f. of the cell, or alternatively the potential can be eliminated by the use of a 'salt bridge'. A salt bridge consists of a saturated solution of potassium chloride, ammonium nitrate or potassium nitrate which is placed between the two other solutions and in contact with both. One way of bringing this about is to allow side arms from the two electrode vessels to dip into a beaker containing the salt bridge solution. Another method, which is sometimes more convenient is to have the salt bridge in the form of a connecting tube. The bridge is then made up by warming the saturated solution with a little agar so that it cools in the tube to a solid gel. The salts used in making up the bridge are found experimentally to give a zero liquid-junction potential, and this is attributed to the fact that, for each of the two salts, the positive and negative ions diffuse at about the same rate. An example of a cell with a salt bridge, set up for the measurement of the electrode potential of silver, is shown in Fig. 10.4.

Fig. 10.4 Cell for measurement of the electrode potential of silver

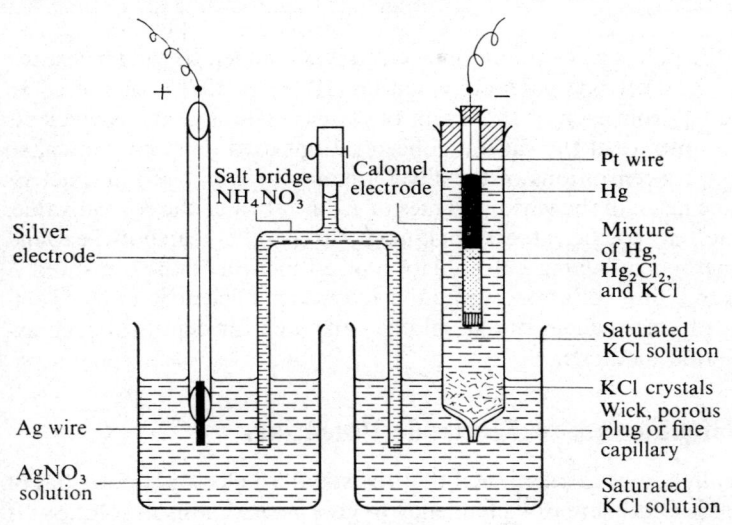

The calomel electrode is often used instead of the hydrogen electrode as a reference half-cell and is available in various forms. The calomel electrode is easily set up to give a reproducible potential depending on the strength of the potassium chloride solution it contains, as shown in Table 10.3. Figure 10.4 illustrates a common and versatile type of saturated calomel electrode. The potassium chloride solution makes liquid contact at the base of the tube with the solution in the other half-cell through a porous plug, wick or a cap with a small grove. These devices slow down the diffusion of the potassium chloride solution into the other solution. The 1.0 mol dm^{-3} and 0.1 mol dm^{-3} calomel electrodes do not contain crystals of potassium chloride. Calomel is mercurous chloride, Hg_2Cl_2. The half-cell reaction in the calomel electrode is:

$$\tfrac{1}{2} Hg_2Cl_2(s) + e^- = \tfrac{1}{2} Hg(l) + Cl^-(aq).$$

To obtain the standard electrode potential $E^\ominus$ for a silver electrode, measurements are made of the e.m.f. of the cell shown in Fig. 10.4 at a number of different concentrations of the silver nitrate solution. Each measured value of the e.m.f. gives the potential of the silver electrode relative to the calomel electrode. To obtain values on the hydrogen

Electrochemical Cells

Table 10.3. Potential of The Calomel Electrode on the Hydrogen Scale

Concentration of KCl	Potential at 298.15 K	Temperature coefficient
0.1 mol dm^{-3}	0.3337 V,	-0.09 mV K^{-1}
1.0 mol dm^{-3}	0.2807 V,	-0.28 mV K^{-1}
saturated	0.2415 V,	-0.66 mV K^{-1}

scale, the potential of the calomel electrode is added to each measured e.m.f. The electrode potential equation (10.16, p. 407) is then used to obtain $E_h^{\ominus}$ from each of these sets of values. The activity coefficients of silver nitrate at the various concentrations used must be known, so that the concentrations of the silver ion may be convtered into activities. The mean of the various values of $E_h^{\ominus}$ is taken as the correct value. If the activity coefficients are not known, they, and $E_h^{\ominus}$ can both be found by a method involving extrapolation of a function of the measured E values to infinite dilution, where the activity coefficient is unity. Textbooks of electrochemistry should be consulted for details of such extrapolation methods.

The Significance of Electrode Potentials

It has already been mentioned that the standard electrode potential of a metal is a measure of its tendency to give positive ions in solution. If two elements are considered, the first having a higher negative $E^{\ominus}$ value than the second, the first will have a greater tendency than the second to go into solution as positive ions. It will therefore displace the second element from its salts. So, when the metallic elements are arranged in a series, with the one having the highest negative standard electrode potential at the top, and the one with the highest positive standard electrode potential at the bottom, the series will give the order in which the metals tend to displace each other from their salts. Such an arrangement as shown in Table 10.1 is called the electrochemical, or electromotive series.

It can easily be shown how the tendency for one element to displace another can be put on a quantitative basis. Consider the case of zinc and copper once more. The electrode potential of zinc is given by:

$$E(Zn^{2+}, Zn) = E^{\ominus}(Zn^{2+}, Zn) + \frac{RT}{2F} . \ln \{Zn^{2+}\}.$$

Similarly for copper:

$$E(Cu^{2+}, Cu) = E^{\ominus}(Cu^{2+}, Cu) + \frac{RT}{2F} . \ln \{Cu^{2+}\}$$

It is the tendency for zinc to displace copper from its salts which causes the difference of potential between the electrodes in a Daniell cell. When the displacement has gone to completion so that there is equilibrium between Zn^{2+}, Zn, Cu^{2+} and Cu, there is no longer a difference in potential between the two electrodes, and the cell ceases to supply current. That is, $E(Zn^{2+}, Zn) = E(Cu^{2+}, Cu)$.
It follows that:

$$E^\ominus(Zn^{2+}, Zn) + \frac{RT}{2F}.\ln\{Zn^{2+}\} = E^\ominus(Cu^{2+}, Cu) + \frac{RT}{2F}.\ln\{Cu^{2+}\}.$$

Subtracting $E^\ominus(Zn^{+2}, Zn)$ and $\frac{RT}{2F}.\ln\{Cu^{2+}\}$ from both sides of the equation, we have:

$$\frac{RT}{2F}.\ln\{Zn^{2+}\} - \frac{RT}{2F}.\ln\{Cu^{2+}\} = E^\ominus(Cu^{2+}, Cu) - E^\ominus(Zn^{2+}, Zn).$$

Writing the difference between the two logarithms as the logarithm of the quotient:

$$\frac{RT}{2F}.\ln\frac{\{Zn^{2+}\}}{\{Cu^2\}} = E^\ominus(Cu^{2+}, Cu) - E^\ominus(Zn^{2+}, Zn).$$

If we substitute the numerical values at 298.15 K, we have:

$$\frac{0.0592}{2}.\log\frac{\{Zn^{2+}\}}{\{Cu^{2+}\}} = 0.340 - (-0.763),$$

$$\therefore \quad \log\frac{\{Zn^{2+}\}}{\{Cu^{2+}\}} = \frac{1.103}{0.0296},$$

$$\therefore \quad \log\frac{\{Zn^{2+}\}}{\{Cu^{2+}\}} = 37.3$$

$$\therefore \quad \frac{\{Zn^{2+}\}}{\{Cu^{2+}\}} = 2 \times 10^{37}.$$

From this it follows that, when equilibrium has been reached, the zinc has displaced the copper ions from solution to such an extent that the zinc ions are present to about 10^{37} times the amount of the copper ions. This is virtually complete displacement.

The standard oxidation-reduction potential for a redox system is the value of the potential when the oxidised and reduced forms are present to the same extent (i.e. when the ratio of the activities is 1). It therefore provides a measure of the oxidising or reducing power of the system under standard conditions. The greater the tendency for the oxidised form to be converted into the reduced form, the greater is the tendency to take electrons from the electrode, thus making it more positive. This means that the more positive is the value of $E^\ominus$ for an oxidation-reduction system the more powerful are its oxidising properties. Corresponding to the electrochemical series based on electrode potentials, it is possible to draw up a table of standard oxidation-reduction potentials, indicating the order of oxidising or reducing powers of the various systems. If the most positive value of $E_h^\ominus$ is placed at the top of the table, and the most negative value at the bottom, a system will oxidise another if it lies above it in the table, and reduce it if it lies below. Some standard oxidation-reduction potentials have been shown in this way in Table 10.2. Again, as in the case of standard electrode potentials, this table is of quantitative as well as of qualitative significance. The difference between two $E_h^\ominus$ values may be used to calculate the state of equilibrium which is attained when two systems are mixed, and one is oxidised by the other. It is found, in fact, that one system will oxidise another virtually to completion if the standard oxidation–reduction potentials differ by at least 0.3 volt.

Potentiometric Titrations

Since the potential of a reversible electrode depends upon the activity of the ions with which it is in equilibrium, it is possible to follow activity changes during titrations by measuring the potential of an appropriate electrode relative to a calomel or any other standard electrode. If the electrode is reversible to both the system being titrated and the system with which it is being titrated, then there is usually a large change in the potential of the electrode at the equivalence point. The two most important types of reaction in which this principle is applied are pH titrations and oxidation-reduction titrations. An example of the latter which illustrates the principle quite well is the oxidation of a ferrous salt by permanganate in acid solution.

In order to follow this reaction electrochemically, the procedure is as follows. The solution of the ferrous salt is placed in the titration flask, from which it is preferable to exclude air by passing a stream of nitrogen. A salt bridge connects the liquid in the flask with a potassium chloride solution into which a calomel electrode dips. A platinum electrode is placed in the solution in the titration flask, and its potential is measured relative to the calomel electrode by means of a potentiometer. (If desired, the potential on the hydrogen scale may be obtained by adding the potential of the calomel electrode as has been done in Fig. 10.5a).

As the first few cm³ of permanganate solution are added from the burette, the oxidation-reduction potential will be that of a ferric–ferrous system in which there is excess ferrous ion, i.e. somewhat less than $E^\ominus(Fe^{3+}, Fe^{2+})$. When the ferrous salt has been half oxidised, the activities of ferrous and ferric ion are equal, and the potential is $E^\ominus(Fe^{3+}, Fe^{2+})$, i.e. $+0.770$ V on the hydrogen scale. Addition of further permanganate produces excess ferric ion, so, according to the equation for the ferric–ferrous system on p. 407, the potential is somewhat greater than 0.770 V. As long as some ferrous ion remains, the amount of permanganate ion in the system is vanishingly small, so the contribution to the measured potential from the permanganate–manganous system is negligible. From the form of the equation it can be seen that the oxidation-reduction potential for the ferric–ferrous system will not differ much from the $E_h^\ominus$ value of 0.770 V until the ratio $\{Fe^{3+}\}/\{Fe^{2+}\}$ has very extreme values. Thus, until the end-point has almost been reached, the total potential will always be in the region of 0.770 V. However, if the titration is continued to the end-point, there is a rapid change in the measured potential, from around this value, to a figure not far removed from 1.491 V, the $E_h^\ominus$ value for the permanganate–manganous system. This is because the permanganate ion added after the end-point is in equilibrium with the manganous ion which has already been formed, and there is now an insignificant amount of ferrous ion present. The contribution to the total potential from the ferric-ferrous system is now negligible and the permanganate–manganous system quickly gives potentials not far removed from its $E_h^\ominus$ value. The sudden rise in potential at the end-point is due therefore to the large difference in the $E_h^\ominus$ values for the two oxidation-reduction systems involved in the reaction. The condition for accurate determination of the end-point by the potentiometric method is the same as that given on p. 412 for complete oxidation of one system by another, namely, that the $E_h^\ominus$ values should differ by at least 0.3 V.

An end-point which cannot be located accurately because the change in E_h is not sharp enough can often be pin-pointed by means of a differential plot shown in Fig. 10.5(b). If ΔE is the change in E_h which results from the addition of a small volume of titrant (ΔV) from the burette, $\Delta E/\Delta V$ approximates to the gradient of the titration curve, and is plotted against the volume V of titrant added. The peak in this graph marks the point at which the gradient of the titration curve is steepest; which is the end-point. For precise results the volume scale in this region should be expanded.

In an acid-base titration, it is possible to use a hydrogen electrode to respond to the change in the activity of hydrogen ions in solution according to the Nernst equation. For the hydrogen electrode the standard electrode potential on the hydrogen scale $E_h^\ominus$ is by definition zero (p. 396). If the pressure of the hydrogen gas is 101 325 Pa (1 atm), $\{red\} = \{H_2\} = 1$, and equation [10.14] can be written:

$$E(H^+, H_2) = 1.985 \times 10^{-4} \times T \log \{H^+\}. \qquad [10.21]$$

Now, pH $= -\log\{H^+\}$, so that:

$$E(H^+, H_2) = -1.985 \times 10^{-4} \times T \times pH \qquad [10.22]$$

Fig. 10.5 Typical potentiometric oxidation-reduction titration curves

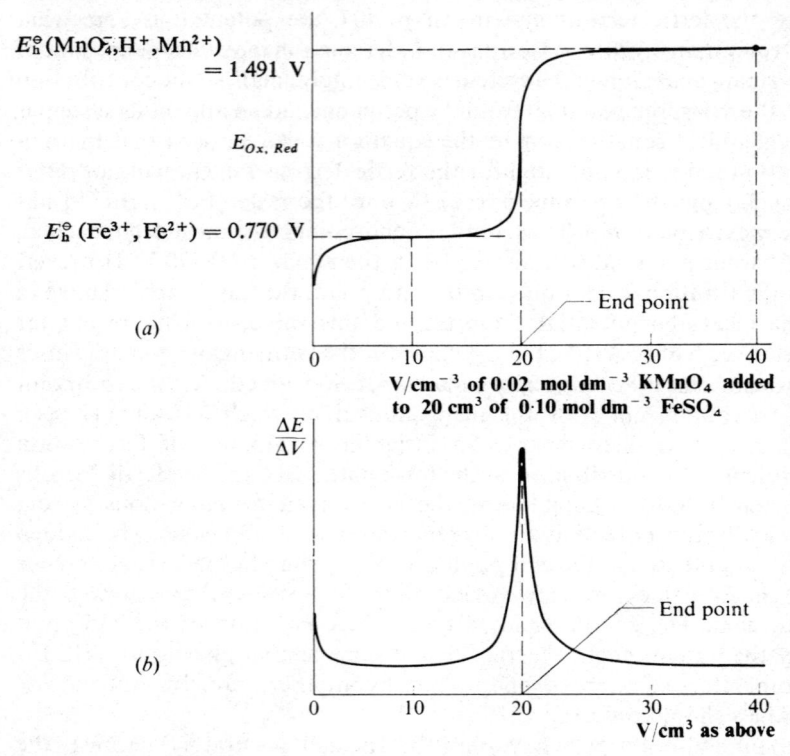

$E_h^\ominus(MnO_4^-,H^+,Mn^{2+}) = 1.491$ V

$E_{Ox, Red}$

$E_h^\ominus(Fe^{3+},Fe^{2+}) = 0.770$ V

(a)

End point

0   10   20   30   40
V/cm^{-3} of 0·02 mol dm^{-3} KMnO$_4$ added to 20 cm^3 of 0·10 mol dm^{-3} FeSO$_4$

$\dfrac{\Delta E}{\Delta V}$

(b)

End point

0   10   20   30   40
V/cm^3 as above

The potential of a hydrogen electrode during an acid-base titration, being proportional to pH, changes according to the form of the neutralisation curves given in Fig. 9.1 (p. 374). Providing the pK_a values for the acid and the base are not too close together, there is an abrupt change in pH and therefore in the hydrogen electrode potential at the equivalence point. Greater precision in locating the equivalence point can be obtained by a differential plot similar to that above. The change in pH (Δ pH) which results from the addition of a small volume ΔV of titrant is determined during the titration. A sharp peak in the plot of Δ pH/ΔV against the volume of titrant added gives the end-point. In practice, the glass electrode is used instead of the hydrogen electrode for potentiometric acid-base titrations as well as for pH measurements, but the basic principles of the titration are unchanged.

Glass Electrode

There are a number of difficulties associated with the use of the hydrogen electrode: the gas must be carefully purified, it may be necessary to apply corrections for the fact that the partial pressure of hydrogen in the electrode vessel is not exactly 101 325 Pa and the electrode cannot be used in the presence of oxidising agents. Because of these difficulties, a glass electrode is most commonly used for pH measurement and for following pH titrations.

If a thin glass membrance separates two solutions containing hydrogen ions at different activities, there is found to be a potential difference between the two surfaces of the glass. This potential difference is called a membrane potential (p. 425) and varies with the activi-

Fig. 10.6 Cell with a glass electrode

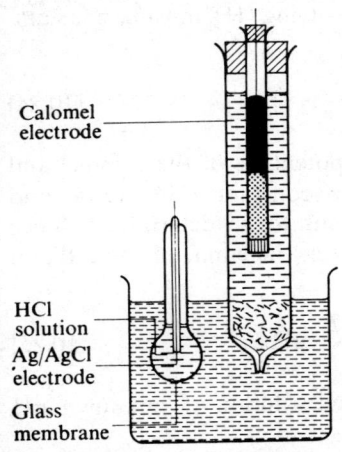

ties of the hydrogen ions in the same way in which the potential difference between two hydrogen electrodes dipping into these two solutions would vary. To measure this potential difference, the glass membrane must be made part of a cell with a sub-standard electrode on either side of it. Such a cell is illustrated in Fig. 10.6. The thin glass membrane is in the form of a bulb which contains hydrochloric acid solution. The sub-standard electrode inside the bulb is a silver-silver chloride electrode, which is reversible to chloride ions (see p. 407).

Outside the bulb of the glass electrode is the external solution of which the hydrogen-ion activity $\{H^+\}_e$ is to be determined, and dipping into this is a saturated calomel electrode. There is no liquid-junction potential where the calomel electrode dips into the solution, because one of the liquids is a saturated potassium chloride solution. The whole cell illustrated in Fig. 10.6 may be written as:

$$Hg(l), Hg_2Cl_2(s) | KCl(saturated) || H^+\{H^+\}_e | HCl\{H^+\}_i | AgCl(s), Ag(s)$$

within calomel electrode.   glass membrane.   within glass electrode.

The e.m.f. of the whole cell is given by the difference in the potentials of the calomel and silver-silver chloride electrodes, plus the difference between the potentials of the two glass surfaces, the latter being:

$$E = \frac{RT}{F} \ln \frac{\{H^+\}_e}{\{H^+\}_i} = -E_m. \qquad [10.23]$$

E_m is the membrane potential (p. 426) of the glass membrane with respect to hydrogen ions. This term can be split into two, thus:

$$E = \frac{RT}{F} \ln \{H^+\}_e - \frac{RT}{F} \ln \{H^+\}_i.$$

If $\{H\}_i$ represents the activity of hydrogen ions in the hydrochloric acid solution inside the bulb, the term involving $\{H\}_i$ must be constant. The total e.m.f. is therefore given by:

$$E = E_G^\ominus + \frac{RT}{F} \ln \{H^+\}_e \qquad [10.24]$$

where $E_G^\ominus$ is the constant involving the potentials of the calomel and silver–silver chloride electrodes and the constant $\{H^+\}_i$ term, and $\{H^+\}_e$ is the activity of the hydrogen ions to be determined. Since $pH = -\log \{H^+\}$, the relationship between the e.m.f. of the cell and the pH of the external solution is:

$$E = E_G^\ominus - \frac{2.303\, RT}{F} \cdot pH. \qquad [10.25]$$

If the electrodes are placed in an external solution of unknown pH (pH_x), the e.m.f. of the cell is given by:

$$E_x = E_G^\ominus - \frac{2.303\, RT}{F} \cdot pH_x. \qquad [10.26]$$

If the electrodes are placed in a standard solution of known pH (pH_s), the e.m.f. of the cell is given by:

$$E_s = E_G^\ominus - \frac{2.303\, RT}{F} \cdot pH_s. \qquad [10.27]$$

Subtracting equation [10.26] from [10.27] we have:

$$E_s - E_x = \frac{2.303\, RT}{F} (pH_x - pH_s). \qquad [10.28]$$

The constant $E_G^\ominus$ varies slightly with time for a given glass electrode and differs from one glass electrode to another. It may be found by measuring the e.m.f. for a standard buffer solution of known pH or may be eliminated as indicated in equation [10.28]. Values of pH for standard buffer solutions at various temperatures can be looked up in tables. For example, 0.05 mol kg^{-1} potassium hydrogen phthalate has a pH equal to 4.00 at 15° C and a solution of 0.025 mol kg^{-1} KH_2PO_4 and 0.025 mol kg^{-1} Na_2HPO_4 has a pH of 6.84 at 38° C.

Owing to the high resistance of the glass membrane, the e.m.f. of a cell containing the glass electrode must be measured using a voltmeter which takes a vanishingly small current, such as a high resistance voltmeter with a d.c. amplifier. Although a potentiometer of the type shown in Fig. 10.2 takes no current at the nul point, it takes far too great a current on either side of this point, when a normal galvanometer is in the circuit. Modern pH meters usually consist of a glass electrode, a saturated calomel electrode and a high resistance voltmeter which is calibrated directly in both pH and e.m.f. The voltmeter often has a temperature adjustment or compensator to allow for or to compensate for the variation of the factor 2.303 RT/F in equation [10.28] with temperature. A second standard buffer of very different pH from the first enables the calibrations to be checked. The valve voltmeter of a pH meter may be used instead of a potentiometer to determine the e.m.f. of any electrochemical cell.

Oxidation-Reduction Indicators

There are not many organic oxidation-reduction processes which can be made to take place under reversible conditions so that they can be studied electrochemically. However, some such processes are known of which the most important are oxidation-reduction (or redox) indicators and electron carriers in living cells.

Substances which change colour in passing from the reduced to the oxidised form or *vice versa* may be used as redox indicators, since a small quantity of such a substance, when added to a solution, quickly adjusts itself so that it has the same redox potential as the rest of the solution. The oxidation-reduction equilibrium in many such systems corresponds to the equation:

$$In + 2H^+ + 2e^- \rightleftharpoons InH_2,$$

where In is the oxidised form and InH_2 is the reduced form of the indicator. The change in Gibbs energy for this reaction is given by substitution into equation [10.8], thus:

$$\Delta G = \Delta G^\ominus + RT \ln \frac{\{InH_2\}}{\{In\}\{H^+\}^2}. \qquad [10.29]$$

The oxidation-reduction potential for the system is given by the Nernst equation (e.g. 10.12), thus:

$$E_h = E_h^\ominus + \frac{RT}{2F} \ln \frac{\{In\}\{H^+\}^2}{\{InH_2\}}. \qquad [10.30]$$

After separating the term in $\{H^+\}$ this equation becomes:

$$E_h = E_h^\ominus + \frac{RT}{2F} \ln \frac{\{In\}}{\{InH_2\}} + \frac{RT}{2F} \ln \{H^+\}^2.$$

Next, putting $2\ln\{H^+\}$ in place of $\ln\{H^+\}^2$ and cancelling the 2 with the 2 in the denominator of that term:

$$E_h = E_h^\ominus + \frac{RT}{F} \ln\{H^+\} + \frac{RT}{2F} \ln \frac{\{In\}}{\{InH_2\}}.$$

Finally, writing 2.303 log for ln and $-\text{pH}$ for $\log\{H^+\}$, we have:

$$E_h = E_h^\ominus - \frac{2.303 RT}{F} \cdot \text{pH} + \frac{2.303 RT}{2F} \log \frac{\{In\}}{\{InH_2\}}. \quad [10.31]$$

The oxidation-reduction potential depends not only on the ratio of the activities of the oxidised and reduced forms but also on the pH. When redox indicators are used to measure an oxidation potential, the solution is usually buffered to a known pH, so that:

$$E_h^\ominus - \frac{2.303 RT}{F} \cdot \text{pH} = E_h^{\ominus\prime}, \quad \text{a constant.} \quad [10.32]$$

The standard oxidation-reduction potential of an indicator or biological electron carrier is usually quoted at a definite temperature and pH as $E_h^{\ominus\prime}$. Equation [10.31] is then written as:

$$E_h = E_h^{\ominus\prime} + \frac{2.303 RT}{2F} \log \frac{\{In\}}{\{InH_2\}}. \quad [10.33]$$

The standard Gibbs energy change corresponding to $E_h^{\ominus\prime}$ is defined by an equation analogous to [10.10] thus:

$$-\Delta G^{\ominus\prime} = nFE_h^{\ominus\prime}. \quad [10.34]$$

Methylene blue is an indicator of the above type, its redox equilibrium being:

$$\text{Me}_2\text{N}-\text{[structure]}-\text{NMe}_2 \; \text{(Blue)} \;\; +2\text{H}^+ + 2e^- \rightleftharpoons \;\; \text{Me}_2\text{N}-\text{[structure]}-\text{NMe}_2 \; \text{(Colourless)}$$

For methylene blue $E_h^{\ominus\prime}$ is $+0.011$ volt at $25°$ C and at pH 7.00. Since its reduced form is rapidly oxidised by molecular oxygen, methylene blue can only be used as a redox indicator under anaerobic conditions. Some other indicators do not suffer from this disadvantage.

If an indicator is placed in a solution whose oxidation-reduction potential is above the $E_h^{\ominus\prime}$ value, the indicator adjusts itself so that it is present mostly in the oxidised form. On the other hand, if the potential of the solution is below the $E_h^{\ominus\prime}$ value for the indicator, it is present mostly in the reduced form. The solution may therefore be tested with a number of indicators of descending $E_h^{\ominus\prime}$ values, until the last indicator to be reduced by the solution is found. It is then known that the oxidation-reduction potential of the solution must be below the $E_h^{\ominus\prime}$ value for this indicator, and above the $E_h^{\ominus\prime}$ value for the next in the sequence.

This method is of considerable importance in the determination of the oxidation-reduction potentials of biological fluids, where only a small quantity of the material may be available. These potentials are generally small positive or negative quantities, and a large number of indicators is available for the study of such materials, covering the range $+0.3$ to -0.5 volt at pH 7.

Another use of oxidation-reduction indicators is in the determination of end-points in oxidation-reduction titrations. We have already seen (p. 414) that there is a rapid change in potential at the end-point, from values characteristic of the system originally present to values characteristic of the system being added. If an indicator is present whose $E_h^{\ominus\prime}$ value at the pH of the titration falls within the region of rapid change, it is converted, at the end-point, from its reduced to its oxidised form (or vice versa). There is therefore an abrupt colour-change which marks the end-point of the titration. The use of diphenylamine in the titration of ferrous salts by dichromate is an example of the application of this type of indicator.

Oxidation-Reduction Processes in Biological Systems

The metabolic processes which take place in living cells convert nutrients and oxygen into substances of lower Gibbs energy. Much of the Gibbs energy which is liberated by these processes is either used directly or stored for future use, generally through the formation and utilisation of phosphate bond energy (p. 281). The overall process, known as respiration, is the oxidation of the nutrient substances and the reduction of oxygen to water. This proceeds by a series of intermediate redox steps or electron (or hydrogen) transfers (shown in Table 10.4) each of which is catalysed and made reversible by its own specific enzyme, known as an oxido-reductase. The whole series of processes is known as the electron transport (or respiratory) chain which in animal cells is located in the membranes of subcellular particles (or organelles) known as mitochondria.

Table 10.4 shows that the oxidation of the reduced form of NAD^+ or $NADP^+$ (i.e. NADH or NADPH, see below) by O_2 ($E_h^{\ominus\prime} = 1.14$ V; $\Delta G^{\ominus\prime} = -220$ kJ mol^{-1}) proceeds in a stepwise fashion with the formation of 3 mol of ATP from ADP and inorganic phosphate ($\Delta G^{\ominus\prime} = 3 \times 31 = 93$ kJ mol^{-1}). Comparison of the $\Delta G^{\ominus\prime}$ values quoted above in parentheses shows that the utilisation of the Gibbs energy is less than 50 per cent efficient. The remainder of the Gibbs energy ($\Delta G^{\ominus\prime} = -220+93 = -127$ kJ mol^{-1}) is not harnessed but is liberated as heat.

Nicotinamide adenine dinucleotide (NAD^+), nicotinamide adenine dinucleotide phosphate ($NADP^+$), flavine adenine dinucleotide (FAD) and flavine mononucleotide (FMN) are examples of coenzymes (or cofactors) which function as electron (or hydrogen) carriers. The reductions may be exemplified by:

$$NAD^+ + H^+ + 2e^- = NADH$$
$$(E_h^{\ominus\prime} = -0.32 \text{ V} \quad \text{at} \quad \text{pH 7 and 25°C});$$
$$FAD + 2H^+ + 2e^- = FADH_2$$
$$(E_h^{\ominus\prime} = -0.22 \text{ V} \quad \text{at} \quad \text{pH 7 and 25°C}).$$

A redox indicator can act as an electron carrier in place of a coenzyme provided that its $E_h^{\ominus\prime}$ value lies between the $E_h^{\ominus\prime}$ values of the redox systems above or below that of the coenzyme. By short-circuiting the electron transport system oxidation-reduction indicators can act as inhibitors. For example, the oxidised form of methylene blue ($E_h^{\ominus\prime}$ = 0.011 V, under the conditions in Table 10.4) is reduced by NADH and the reduced form is oxidised by O_2.

Table 10.4. Standard Oxidation-Reduction Potentials of the Various Stages of the Electron Transport Chain and the Corresponding Standard Free Energy Changes at pH 7 and 25° C

Redox stages	$E_h^{\ominus\prime}$ / volt	$-\Delta G^{\ominus\prime}$ / kJ mol^{-1}	Mode of utilisation
substrate			
$NAD(P)^+$, $NAD(P)H$	−0.32	50	ADP $+P_i$ ↓ ATP
flavoprotein (FAD or FMN cofactor)	−0.06		
ubiquinone	+0.08 to	52	ADP $+P_i$ ↓ ATP
cytochrome b	+0.10		
cytochrome c_1	+0.21		
cytochrome c	+0.23 to +0.26	118	ADP $+P_i$ ↓ ATP
cytochrome $\begin{matrix}a\\|\\a_3\end{matrix}$	+0.29		
O_2, H^+, H_2O	+0.82		

The measured redox potential of a bacterial culture is not that of the actual cells but of the nutrients, metabolites and waste products present in the medium. Nevertheless, the metabolism of bacterial cells is related to the oxidation-reduction potential of the media, e.g. strictly anaerobic bacteria (those which flourish only in the absence of O_2) must have a redox potential below -0.2 V before growth is possible. Oxygen inhibits growth by raising the oxidation-reduction potential of the medium.

Concentration Cells and Junction Potential

A cell consisting of two half-cells of the same type but with different concentrations of electrolyte is known as a concentration cell. Let us for convenience consider a concentration cell consisting of two electrodes of a univalent metal M in contact with a solution of one of its uni-univalent salts M^+X^-. The spontaneous reaction as the cell supplies current will tend to equalise the activities (or concentrations) of the two solutions. M^+ ions from the more concentrated solution will be discharged at its electrode and metal M from the other electrode will 'dissolve', introducing more ions into the more dilute solution. The positive pole of the cell will therefore be the electrode in contact with the more concentrated solution. Electrons will flow in the external circuit from the metal in contact with the more dilute solution to that in contact with the more concentrated solution. Chemical change and current flow will continue until the activity of the metal ion is the same in the two solutions and will then stop. In the simple cells to be considered transference of anions accompanies that of the cations.

If the junction potential is eliminated by means of a salt bridge, e.g. in Fig. 10.7(a), the cell can be represented with a double line separating the electrolytes as follows:

$$M\,|\,MX(a_{\pm l})\,|\,|\,MX(a_{\pm r})\,|\,M.$$

The e.m.f. of this cell is given by combining equation [10.7] with [10.16], thus:

$$E = \left(E^\ominus + \frac{RT}{F}\ln a_{+r}\right) - \left(E^\ominus + \frac{RT}{F}\ln a_{-l}\right)$$

$$\therefore \quad E = \frac{RT}{F}\ln\frac{a_{+r}}{a_{+l}} = \frac{RT}{F}\ln\frac{a_{\pm r}}{a_{\pm l}}, \qquad [10.35]$$

where a_{+l} and a_{+r} are activities of the cation, M^+, in the right-hand and left-hand half-cell respectively. These activities may be replaced by the corresponding mean ionic activities (p. 337) of the electrolyte, MX, in the respective half-cells.

Fig. 10.7 Concentration cells with $a_{\pm r} > a_{\pm l}$ and $n = 1$
(a) concentration cell with the liquid junction potential eliminated by means of a salt bridge
(b) concentration cell with transport
(c) concentration cell without transport

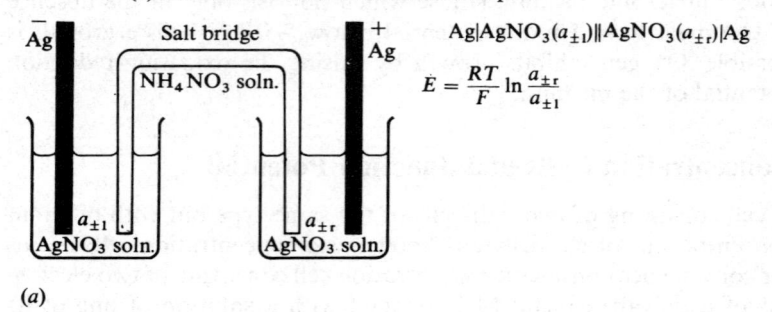

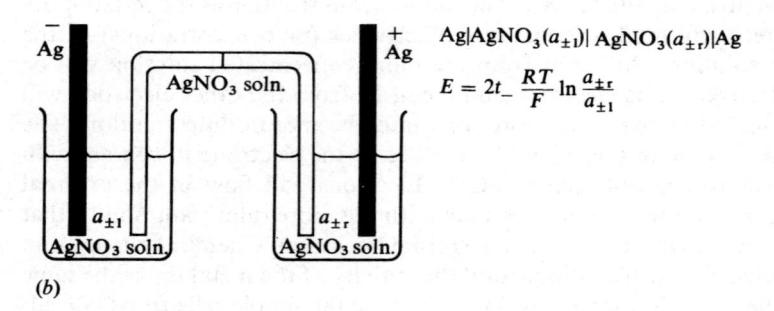

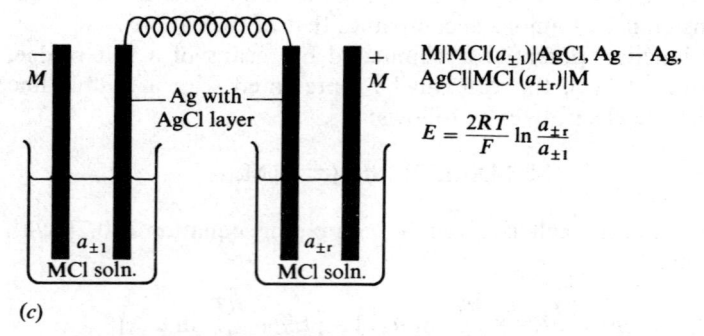

(a) Ag|AgNO$_3(a_{\pm l})$‖AgNO$_3(a_{\pm r})$|Ag

$$E = \frac{RT}{F} \ln \frac{a_{\pm r}}{a_{\pm l}}$$

(b) Ag|AgNO$_3(a_{\pm l})$| AgNO$_3(a_{\pm r})$|Ag

$$E = 2t_- \frac{RT}{F} \ln \frac{a_{\pm r}}{a_{\pm l}}$$

(c) M|MCl$(a_{\pm l})$|AgCl, Ag − Ag, AgCl|MCl$(a_{\pm r})$|M

$$E = \frac{2RT}{F} \ln \frac{a_{\pm r}}{a_{\pm l}}$$

If a salt bridge is not placed between the two solutions as in Fig. 10.7(b), the electrolyte can pass from one half-cell to the other. The cell is then said to be a concentration cell with transport and may have a liquid junction potential. In representing the cell, the boundary between the two electrolytes is indicated by a single vertical line thus:

$$M \,|\, MX(a_{\pm l}) \,|\, MX(a_{\pm r}) \,|\, M$$

$$M \to M^+ \xrightleftharpoons[t_-\text{mol X}^-]{t_+\text{mol M}^+} M^+ \to M.$$

Concentration Cells and Junction Potential

In order to calculate the e.m.f. of such a cell we must imagine that it is large enough to enable the following changes to take place without any alteration in the activities of the ions. Let the cell supply an electric charge equal to the Faraday constant, F. 1 mol of M^+ will be discharged at the electrode in contact with the more concentrated solution (on the right in this cell) and t_+ mol of M^+ will flow in from the more dilute solution (on the left in this cell), where t_+ is the transport number of M^+. The overall loss of M^+ from the more concentration solution will be $(1-t_+)$ mol which is equal to t_-, the transport number of X^-. During this process t_- mol of X^- will flow from the more concentrated to the more dilute solution. The overall result will be a transfer of t_- mol of MX from the more concentrated solution where the mean ionic activity is $a_{\pm r}$, to the more dilute solution where the mean ionic activity is $a_{\pm l}$. The decrease in Gibbs energy, $-\Delta G$, associated with this process is the number of moles of MX transferred times the difference in the partial molar Gibbs energy (chemical potential, μ) of MX between the two solutions, thus:

$$-\Delta G = t_-(\mu_r - \mu_l), \qquad [10.36]$$

where μ_r and μ_l are the chemical potentials of the electrolyte MX in the right-hand and left-hand solution respectively. Each ionic species contributes to these chemical potentials as follows:

$$\mu_r = \mu_{+r} + \mu_{-r} \quad \text{and} \quad \mu_l = \mu_{+l} + \mu_{-l}.$$

The chemical potential of each species is given by equation [7.87] in the form:

$$\mu_+ = \mu_+^\ominus + RT \ln a_+ \quad \text{and} \quad \mu_- = \mu_-^\ominus + RT \ln a_-$$

whence $\quad \mu_r = \mu_+^\ominus + \mu_-^\ominus + RT \ln a_{+r} + RT \ln a_{-r}$

and $\quad \mu_l = \mu_+^\ominus + \mu_-^\ominus + RT \ln a_{+l} + RT \ln a_{-l}.$

As before, the activities of the individual ions, a_+ and a_-, are replaced by the mean activities, $a_\pm$ (p. 337), so that:

$$\mu_r - \mu_l = 2RT \ln a_{\pm r} - 2RT \ln a_{\pm l}$$

or $\quad\quad \mu_r - \mu_l = 2RT \ln \dfrac{a_{\pm r}}{a_{\pm l}}. \qquad [10.37]$

Introducing this difference in chemical potentials into equation [10.36] we have:

$$-\Delta G = 2t_- RT \ln \dfrac{a_{\pm r}}{a_{\pm l}}. \qquad [10.38]$$

The e.m.f. of the cell is given by introducing equation [10.2] where $n = 1$, thus:

$$E = 2t_- \dfrac{RT}{F} \ln \dfrac{a_{\pm r}}{a_{\pm l}}. \qquad [10.39]$$

This equation enables transport numbers to be determined from the e.m.f. of concentration cells with transport provided that the activities of the electrolytes are known.

The difference between equations [10.35] and [10.39] is due to the liquid junction potential E_j, which is defined as the e.m.f. of the cell without a junction potential minus the e.m.f. of the corresponding cell with a junction potential. Therefore,

$$E_j = \frac{RT}{F} \ln \frac{a_{\pm r}}{a_{\pm l}} - 2t_- \frac{RT}{F} \ln \frac{a_{\pm r}}{a_{\pm l}}$$

$$E_j = (1-2t_-)\frac{RT}{F} \ln \frac{a_{\pm r}}{a_{\pm l}}. \qquad [10.40]$$

Since $t_- = 1-t_+$ from equation [8.4],

$$(1-2t_-) = (t_+ - t_-) = (2t_+ - 1). \qquad [10.41]$$

The junction potential as defined above decreases the e.m.f. of a cell with transport if positive and increases it if negative. If the anion migrates faster than the cation, $t_- > t_+$, i.e. $t_- > 0.5$, and E_j is negative and increases the e.m.f. of the cell. Consequently, the more dilute side of the liquid junction is more negative than the more concentrated side. A useful rule is that the more dilute side of the liquid junction acquires the polarity of the faster moving ion. If the cation and the anion at the liquid junction have approximately equal transport numbers, e.g. are monovalent ions with the same molar conductivities or electric mobilities, E_j will be small. An electrolyte of this type, e.g. KCl, NH_4NO_3 or KNO_3, in concentrated solution is often used as a salt bridge (see p. 408) to reduce E_j to a negligible value.

The liquid junction can also be eliminated by joining the two solutions with an identical reversible electrode, e.g. Fig 10.7(c), to give two cells which are connected in opposition, thus:

$$M \,|\, MCl(a_{\pm r}) \,|\, AgCl, Ag - Ag, AgCl \,|\, MCl(a_{\pm l}) \,|\, M$$

Such an arrangement is called a concentration cell without transport. When the cell supplies a charge equal to F, 1 mol of M^+ ions from the more concentrated solution on the right will be discharged at the M electrode and 1 mol of Cl^- ions will be discharged at the Ag/AgCl electrode to form AgCl and L electrons. These electrons will travel to the left hand AgCl/Ag electrode where they will cause 1 mol of Cl^- to go into the more dilute solution. 1 mol of M^+ will also dissolve into this solution from the M electrode. In this way 1 mol of MCl will appear to pass from the more concentrated to the more dilute solution. The e.m.f of this cell in which 1 mol of MCl is transferred can be calculated in the same way as that for the cell with transport, in which only t_- mol of MX was transferred. For the present cell the result is:

$$E = \frac{2RT}{F} \ln \frac{a_{\pm r}}{a_{\pm l}}. \qquad [10.42]$$

The e.m.f. of the cell without transport is always higher than the corresponding cell with transport and twice that of the cell in which the junction potential has been eliminated by a salt bridge. These differences relate to the fact that a concentration cell without transport has 4 electrodes where chemical change can take place, whereas the other types of concentration cell have only 2 electrodes each.

Membrane Potentials

A membrane (or demarcation) potential has the same origin as a liquid junction potential (equation [10.40]) and may be defined as the potential difference between two sides of a membrane as a result of an unequal distribution of ionic material on the two sides (cf. discussion of the glass electrode, p. 415). We shall first consider the simplest case in which the membrane separates two solutions of the same uni-univalent salt, MX, at different concentrations. The permeability of the membrane may greatly influence the transport numbers of the ions across it. If the membrane is freely permeable to both ions, the transport numbers across the membrane are the same as those in solution. If the permeability of the membrane is reduced towards one type of ion only, the transport number of that ion will fall, whereas that of the oppositely charged ion will rise, and the membrane potential, which is given by equation [10.40], will change accordingly. The potential difference has, in fact, been measured by Michaelis across collodion membranes of different pore sizes separating two solutions of KCl, one of molality 0.1 mol kg^{-1} (activity coefficient 0.770) and the other of molality 0.01 mol kg^{-1} (activity coefficient 0.901). At these concentrations in free solution at 25° C the transport numbers are: $t(K^+) = 0.490$, $t(Cl^-) = 0.510$. Substitution of the transport number into equation [10.40] indicates that the liquid junction potential is very small. In accordance with this, the membrane potential for membranes of high porosity was zero. As the porosity of the membrane was reduced, the membrane potential increased from zero to a limiting value of 55 mV. The more dilute solution was positive with respect to the more concentrated solution, showing that the Cl$^-$ ion was slowed down relative to the K$^+$ ion. The limiting values of the transport numbers can be calculated as follows from these data using equation [10.40] in the form:

$$(2t_+ - 1) = E_j F / 2.303 RT \log \frac{a_{\pm r}}{a_{\pm l}}.$$

Now $E_j = 55$ mV, $2.303RT/F = 59.2$ mV at $25°$ C,

$$a_\pm = m\gamma_\pm, \quad a_{\pm r} = 0.1 \times 0.770 \text{ and } a_{\pm l} = 0.01 \times 0.901,$$

whence $(2t_+ - 1) = 0.9972$

∴ $\quad t_+ = 0.9986 \quad$ and $\quad t_- = 0.0014.$

It can be seen that the transport number of the Cl^- ion has been reduced almost to zero by the collodion membranes of low porosity, and that the transport number of the K^+ ion is almost unity. In general, if the membrane is impermeable to either type of ion of a given strong uni-univalent electrolyte, equation [10.40] leads to the following equation for the membrane potential, E_m:

$$E_m = \pm \frac{RT}{F} \ln \frac{a_{\pm r}}{a_{\pm l}}, \qquad [10.43]$$

where $a_{\pm r}$ refers to the solution which is placed in the right-hand half-cell in the representation of the cell and where $a_{\pm l}$ refers to the solution in the left-hand half-cell. The $+$ sign is applicable if $t_+ = 1.0$ ($t_- = 0$, i.e. membrane impermeable to the anion), whereas the $-$ sign is applicable if $t_- = 1.0$ ($t_+ = 0$, i.e. membrane impermeable to the cation).

If the membrane is permeable to only one type of ion, it is more convenient to replace the mean activities, $a_{\pm r}$ and $a_{\pm l}$, by the activities of this ion. For example, the thin glass membrane of the glass electrode on p. 415 is usually permeable only to hydrogen ions even when the external solution, for which $\{H^+\}_e$ is being measured, contains many different types of ion. We can then write, $a_{\pm r} = \{H^+\}_i$ and $a_{\pm l} = \{H^+\}_e$. The transport number of the hydrogen ion is unity (i.e. $t_+ = 1.0$) and that of every other ion, including the anions, is zero (i.e. $t_- = 0$). Substitution into equation [10.40] or [10.43] leads directly to equation [10.23] for the membrane potential of the glass electrode.

If the membrane is permeable to more than one type of ion, the membrane potential may be expressed in terms of the activity of any one of the permeating ions. A membrane might, for example, be impermeable to an ion of high molecular weight, such as a protein anion, P^-, but might be permeable to accompanying K^+ and Cl^- ions. The presence of the non-permeating ion on one side of the membrane will cause other ions of the same charge, e.g. Cl^-, to be present at a greater concentration on the other side of the membrane. The permeating ions of opposite charge, e.g. K^+, will distribute themselves so as to maintain electrical neutrality. As a result there will be less potassium chloride on the same side of the membrane as the macromolecule than on the opposite side. The equilibria which are established in such cases as these are known as the Donnan membrane equilibria. The example considered above may be represented by:

```
        l      Membrane    r
     {P⁻}_l  ⎫         ⎧   —
     {K⁺}_l  ⎪         ⎪ {K⁺}_r
     {Cl⁻}_l ⎬    ⇌    ⎨ {Cl⁻}_r
     {H₂O}_l ⎭         ⎩ {H₂O}_r
```

where the subscripts l and r are used to denote activity on the left and right of the membrane respectively.

In the above example K^+, Cl^- and H_2O can diffuse through the membrane. The free energy of transfer of these species from left to right is given by equation [7.102] in the form:

$$\Delta G = \Delta G^\ominus + RT \ln \frac{\{K^+\}_r \{Cl^-\}_r \{H_2O\}_r}{\{K^+\}_l \{Cl^-\}_l \{H_2O\}_l}.$$

Since the amount of water is large compared with the amounts of the other species present, the inequalities in the activities of these species on either side of the membrane will not significantly change the activities of water on either side of the membrane. Consequently, $\{H_2O\}_r \approx \{H_2O\}_l$ and the last equation becomes:

$$\Delta G = \Delta G^\ominus + RT \ln \frac{\{K^+\}_r \{Cl^-\}_r}{\{K^+\}_l \{Cl^-\}_l}. \qquad [10.44]$$

When the equilibrium distribution of diffusible ions has been attained, there is no free energy change on transferring ions in either direction, so $\Delta G = 0$. Also $\Delta G^\ominus = 0$, because it represents the difference between the standard free energies of the ions on the left and the ions on the right. Each standard free energy depends on the type of ion and refers to the standard state of unit activity, so it is independent of the actual activities of the ion. Since the diffusing ions are the same on either side of the membrane, the standard free energies of those on the left are the same as those on the right. Substituting these zero terms we have:

$$0 = RT \ln \frac{\{K^+\}_r \{Cl^-\}_r}{\{K^+\}_l \{Cl^-\}_l}.$$

Since $T \neq 0$ and $\ln 1 = 0$,

$$1 = \frac{\{K^+\}_r \{Cl^-\}_r}{\{K^+\}_l \{Cl^-\}_l} \quad \text{or} \quad \frac{\{K^+\}_r}{\{K^+\}_l} = \frac{\{Cl^-\}_l}{\{Cl^-\}_r}. \qquad [10.45]$$

If the solutions of the diffusible ions are dilute, activities may be replaced by concentrations, so that

$$1 = \frac{[K^+]_r [Cl^-]_r}{[K^+]_l [Cl^-]_l} \quad \text{or} \quad \frac{[K^+]_r}{[K^+]_l} = \frac{[Cl^-]_l}{[Cl^-]_r}. \qquad [10.46]$$

The equation for the membrane potential in such cases, known as the Donnan membrane potential, may be derived by appropriate substitution into equation [10.43]. In the particular case considered, the membrane is impermeable to P^- and permeable to K^+, so that the membrane potential is given by:

$$E_m = \frac{RT}{R} \ln \frac{\{K^+\}_r}{\{K^+\}_l}. \qquad [10.47]$$

Expressing the activity ratio in the alternative form suggested by equation [10.45] we have

$$E_m = \frac{RT}{F} \ln \frac{\{Cl^-\}_l}{\{Cl^-\}_r}. \qquad [10.48]$$

Membrane potentials occur throughout all living organisms. In the higher animals and plants the most abundant ions are Na^+, K^+ and Cl^-. Inside many living cells $[K^+]$ is much higher than $[Na^+]$, whereas in the external solution $[Na^+]$ is higher than $[K^+]$. This unequal distribution of cations is in certain cases believed to be maintained by 'active transport' which functions as an 'ion pump' and requires a constant supply of Gibbs energy (p. 297). Leakage in the opposite direction is hindered by a somewhat lower permeability of the membrane to Na^+ compared with K^+ ions. Membrane potentials contribute towards nerve action potentials and muscle action potentials. Nerve and muscle action depends on the transmission of electrical impulses along a nerve fibre to another nerve cell or to a muscle cell. The electrical impulse consists of a temporary reversal of the sign of the membrane potential due to a change of permeability of the membrane.

Since membrane potentials are liquid junction potentials, they can be measured by making the membrane part of a concentration cell with transport. A simple method is to introduce two identical calomel electrodes or silver–silver chloride electrodes into the two solutions separated by the membrane, in which case the e.m.f. of the cell is equal to the membrane potential. Biological cell membrane potentials are now measured by inserting a micro-electrode into the solution inside the cell, a difficult task which requires great patience. The other electrode is introduced into the external solution.

A biological cell membrane potential, E_m, is usually taken to be the potential of the internal solution with respect to the external solution. If the solutions are dilute, equation [10.43] can be written:

$$E_m \approx \pm \frac{2.303 RT}{F} \log \frac{c_i}{c_e}, \qquad [10.49]$$

where c_i and c_e are the concentrations of a diffusible ion in the internal and external solutions respectively. In order to maintain the correct sign convention, a positive sign is used when the concentrations refer to a diffusible ion of the same sign as that of the non-diffusible ion and the non-diffusible ion is present only in the external solution. If one of these conditions is reversed the sign is reversed. The membrane potentials of muscle and nerve fibres are given by equations of the type [10.49]. Thus, if the non-diffusible ion is positive, e.g. Na^+, and is present only in the external solution, then at 20° C for which $2.303\, RT/F = 58$ mV, we have:

$$E_m/mV \approx 58 \log \frac{[K^+]_i}{[K^+]_e} \approx 58 \log \frac{[Cl^-]_e}{[Cl^-]_i}, \quad [10.50]$$

for the diffusible ions K^+ and Cl^- respectively.

Irreversible Electrodes, Overvoltage and Polarisation

Irreversible electrodes will be introduced by comparing them with a reversible electrode. Let us consider the following reversible cell consisting of a standard hydrogen electrode (p. 396) and a saturated calomel electrode (p. 409):

Pt, H_2 (1 atm) | HCl ($a_\pm = 1.0$) | | KCl (saturated) | Hg_2Cl_2, Hg.

The presence of saturated potassium chloride solution ensures that the liquid junction potential is eliminated. The cell reaction corresponding to the above cell as it supplies current is as follows:

$$\tfrac{1}{2} H_2(g) + \tfrac{1}{2} Hg_2Cl_2(s) = H^+(aq) + Cl^-(aq) + Hg(l),$$

and is spontaneous, the e.m.f. being $+0.2415$ V at 25° C (p. 410). Let the cell be connected in series with a current-measuring galvanometer to an opposing potential difference, i.e. of the same sign as that of the cell, provided by the potentiometer circuit shown in Fig. 10.2. If the opposing potential difference is increased steadily from zero by moving point R from P to Q the following changes occur. At first hydrogen gas is converted into hydrogen ions at the hydrogen electrode and the cell supplies current. This corresponds to point A in Fig. 10.8. The current falls off as the applied potential is increased and becomes zero when the applied potential is $+0.2415$ V, the e.m.f. of the cell (E in Fig. 10.8). A further increase in the applied potential causes current to flow in the reverse direction and electrolysis to occur. This corresponds to points between E and B in Fig. 10.8. The cell reaction then proceeds in reverse and hydrogen gas is now produced from the hydrogen electrode. The relationship between the current and the applied potential difference is approximately linear and is illustrated by the line AEB in Fig. 10.8.

The half-cell reaction in the standard hydrogen electrode (p. 396) takes place on the surface of a platinum plate coated with platinum black. If the supply of hydrogen gas which bubbles over this plate is turned off, the spontaneous reaction cannot occur. No current there-

Fig. 10.8 Current-voltage diagrams for a cell comprising a saturated calomel electrode and a reversible or irreversible hydrogen electrode

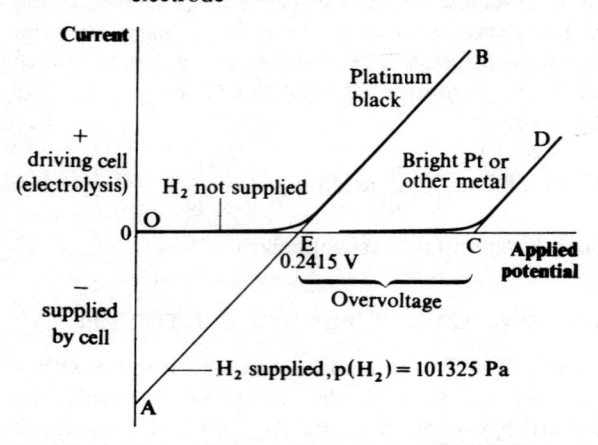

fore flows through the cell until the applied potential equals the e.mf. of the cell. Further increase in the applied voltage causes a gradual increase in current from zero and bubbles of hydrogen begin to appear on the platinum plate showing that electrolysis is taking place. This behaviour is illustrated by the current–voltage curve OEB in Fig. 10.8.

If the platinum electrode is bright and uncoated, a potential difference higher than for the platinum black electrode must be applied to the cell in order for the rise in current to occur and for hydrogen bubbles to appear. This behaviour is illustrated by OCD in Fig. 10.8. The increase in potential difference, EC, which is necessary to cause electrolysis, is called *hydrogen overpotential* or *hydrogen overvoltage* and indicates that the half-cell reaction which takes place on the bright platinum electrode is thermodynamically irreversible (p. 285). This reaction is:

$$2H^+ + 2e^- \rightarrow (2H) \rightarrow H_2$$

and proceeds through the formation and association of hydrogen atoms. Platinum black gives zero hydrogen overvoltage, because it actively catalyses the reaction. Consequently the activation energy is lowered practically to zero (p. 166), with the result that the reaction is sufficiently rapid to be always virtually in equilibrium on the metallic surface and the electrode is reversible (p. 283). Bright platinum and other metals are less effective catalysts on which the association of hydrogen atoms has an appreciable activation energy. The hydrogen overvoltage gives a measure of the energy required by the system to overcome this activation energy barrier. When the barrier is overcome, hydrogen gas appears and electrolysis takes place. On bright platinum and other metals the half-cell reaction is therefore not in equilibrium and is termed irreversible (p. 285). Such electrodes are said to be *polarised* when the applied potential difference lies between the e.m.f. of the cell and the e.m.f. plus overvoltage. The hydrogen overvoltage in 1 mol dm^{-3} hydrochloric acid is zero for platinum black, 0.17 V for bright platinum, rather higher for most metals and 1.04 V for mercury.

Polarography

When present in the solution, certain substances, which are termed depolarising or electroreducible substances, reduce the voltage at which current begins to flow (point C in Fig. 10.8) by being themselves reduced instead of hydrogen ions at the surface of the metal. Depolarising substances may be analysed qualitatively and quantitatively by

Fig. 10.9 Polarograph

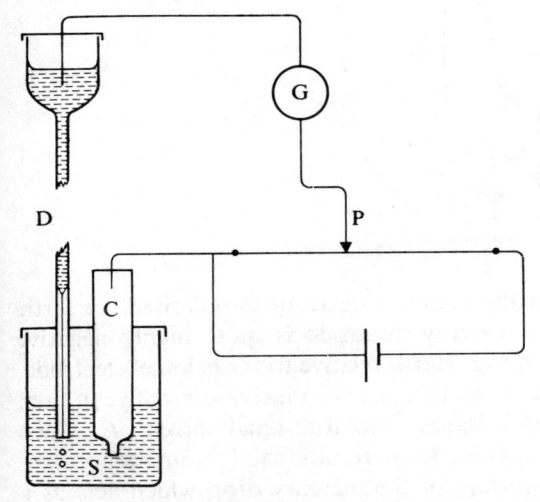

means of a valuable technique known as polarography which is based upon these principles and was developed by Heyrovsky. Mercury is the ideal electrode material for this purpose owing to its high hydrogen overvoltage and to the fact that it is a liquid whose surface can be readily renewed.

A common type of polarograph is illustrated in Fig. 10.9, and consists of a dropping mercury electrode D and a reference electrode, usually a calomel electrode C, dipping into the solution S containing a depolarising substance to be analysed. This solution must be a good electrolytic conductor and must therefore contain an inert salt, called a supporting electrolyte, e.g. 0.1 mol dm^{-3} potassium nitrate or potassium chloride. In less exact work the reference electrode may be substituted by a mercury pool anode of large surface area. The electrodes are connected to a potentiometer P for gradually increasing the applied voltage E. The galvanometer G measures the current I flowing through the cell. The plot of I against E is known as a polarogram (Fig. 10.10).

Analysis of depolarising electroreducible substances, e.g. metal ions, oxidising agents and dissolved oxygen, in the solution is carried out by making the mercury electrode the cathode, i.e. negative with respect to the calomel electrode. In the absence of depolarising substance but in the presence of inert salt the high hydrogen overvoltage

Fig. 10.10 Polarogram showing polarographic waves of substances A and B in the presence of a supporting electrolyte, S

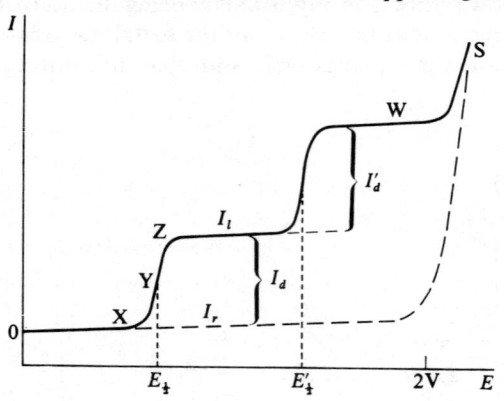

of mercury ensures that the mercury electrode is polarised, i.e. little current flows, until the mercury electrode is quite highly negative (about 2 V, dotted line in Fig. 10.10) relative to the calomel electrode. Higher applied voltages cause the current to increase rapidly. In fact, there is at lower applied voltages a small residual current, I_r, which increases with the voltage. This current results mainly from the accumulation of charge at the surface of the mercury drop which acts as a capacitor and partly from impurities in the solution.

If a depolariser is present in the solution, a rapid increase in current occurs at a lower applied voltage (XY in Fig. 10.10). This current depends on the rate at which the depolariser reaches the mercury cathode, where it is reduced according to the usual equation:

$$\text{oxidant} + ne^- \rightleftharpoons \text{reductant}.$$

This reaction reduces the concentration of depolariser in the immediate vicinity of the cathode. Since the depolariser is present at a low concentration 1 μmol dm^{-3} to 1 mmol dm^{-3}), a stage known as concentration polarisation is reached at Z, at which the vicinity of the cathode is depleted of depolariser. The current cannot now increase further and has reached its limiting value, I_l. The difference between the limiting current, I_l, and the residual current, I_r, is known as the diffusion current, I_d. The diffusion current is proportional to the limiting rate of reduction, which is equal to the rate at which the depolariser reaches the cathode by diffusion from the bulk of the solution across the depleted region. The rate of diffusion is itself proportional to the concentration of depolariser in the bulk of the solution. Thus, the diffusion current is proportional to the concentration of depolariser in the bulk of the solution and is independent of the applied voltage. The proportional plot of I_d against concentration of the depolariser enables the concentration of the depolariser in an unknown solution to be read from the measured value of I_d. This is the basis of quantitative analysis by polarography. The exact relationship is given by the Ilkovic equation:

$$I_d = knD^{\frac{1}{2}}cm^{\frac{2}{3}}t^{\frac{1}{6}}, \qquad [10.51]$$

where I_d is the diffusion current in A,
k is a constant equal to 706 or 607 (see later),
n is the number of electrons exchanged per molecule of depolariser in the electrode process,
D is the diffusion coefficient of the depolariser in $m^2\ s^{-1}$,
c is the concentration of the depolariser in mol m^{-3} (mmol dm^{-3}),
m is the rate of flow of mercury in kg s^{-1},
t is the drop time interval of the mercury in s.

The step-shaped current–voltage curve for a given depolarising substance is known as a *polarographic wave* (Fig. 10.10). If the applied voltage is steadily increased beyond the first wave, the current remains more or less constant until a further depolarising substance, if present, begins to be reduced, whereupon a second wave appears whose height gives the diffusion current I_d' for this substance. Eventually the beginning of the wave, S, for the supporting electrolyte appears (Fig. 10.10). The applied voltage corresponding to the mid-point of each wave is called the half-wave potential, $E_{\frac{1}{2}}$, which is a constant characteristic of the substance causing the wave. This is the basis of qualitative analysis by polarography. $E_{\frac{1}{2}}$, if expressed on the hydrogen scale, is often approximately equal to $E_h^{\ominus}$, the standard redox potential for the corresponding oxidation–reduction reaction. Strictly speaking, a polarographic wave is characteristic of each process of reduction. A substance, such as dissolved oxygen gas, gives two polarographic waves, if it undergoes two consecutive redox processes, e.g. a and b in acid solution or b and c in alkaline solution, thus:

(a) $\quad O_2 + 2H^+ + 2e^- = H_2O_2$
(b) $\quad H_2O_2 + 2H^+ + 2e^- = 2H_2O$

(c) $O_2 + 2H_2O + 2e^- = H_2O_2 + 2OH^-$
(d) $H_2O_2 + 2e^- = 2OH^-$

These waves enable dissolved oxygen to be determined polarographically. If, however, these waves overlap with the waves of another substance to be determined, dissolved oxygen must be displaced by bubbling nitrogen or hydrogen gas through the solution. A disadvantage of polarography is that it is prone to interference from substances in solution.

Provided that the mercury is pure and the capillary is clean, the surface of the dropping mercury electrode is chemically pure and reproducible because it is regularly renewed. Other electrodes which may be moved through the solution, such as a rotating platinum electrode, may be used instead. As the drop at the dropping mercury electrode is formed, grows and falls off, the current through the cell fluctuates. If the maximum value of I_d is used, the constant k in the Ilkovic equation [10.51] is equal to 706. If the average value of I_d is used, k equals 607. The fluctuations may be damped electronically or mechanically by the use of a long-period galvanometer.

The polarogram may be plotted automatically. The applied voltage is then increased steadily by turning the moving contact of a circular potentiometer with a motor and the current flowing through the cell is amplified to operate a recorder.

The analysis of electro-oxidisable substances, e.g. anions and reducing agents, may be carried out polarographically by making the mercury electrode the anode, i.e. positive with respect to the calomel electrode. In polarographic determinations the current flowing through the solution is so small that a negligible amount of material is removed from the solution. The solution can therefore be repeatedly polarographed and then sometimes used for other purposes. Polarography can be used to determine ions of most metals, e.g. lead, and many electro-reducible organic compounds, e.g. aldehydes, ketones, disulphides, lactones, peroxides and azo-, diazo-, nitro-, nitroso- and halogen compounds. Since these substances include many metabolites and drugs, polarography has wide applications in biochemical and pharmaceutical analysis.

Clark Oxygen Cell

The Clark oxygen cell, illustrated in Fig. 10.11, is a type of polarographic cell for measuring continuously the concentration of dissolved oxygen in biological fluids. The stirred biological solution is separated from a strong solution of potassium chloride as supporting electrolyte by a membrane of Teflon (polytetrafluoroethylene) which is permeable to O_2 but not to the solutions. The membrane therefore prevents one solution from contaminating the other, while allowing O_2 in the biological solution to come into equilibrium with the electrolyte. This causes negligible loss of O_2 from the biological solution, because its volume greatly exceeds that of the supporting electrolyte. The concentration of O_2 near the cathode is therefore equal to that in the biological solution.

Fig. 10.11 Clark oxygen cell with a polarising circuit

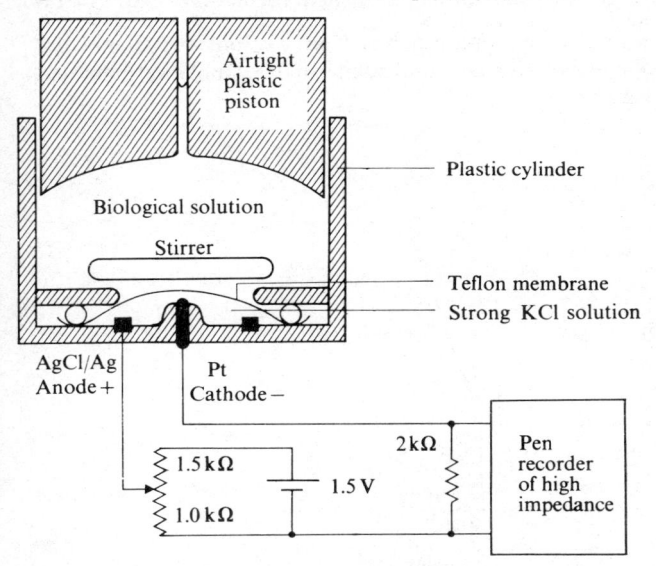

The cathode, which in this case is a platinum stud, is polarised in the absence of dissolved oxygen but depolarised in its presence. Since O_2 gives two polarographic waves (p. 433), the applied voltage is adjusted so that the cell is in a state corresponding to the limiting current plateau of the second wave (W in Fig. 10.10). The cathode has a potential of -0.6 V relative to the anode, which is a silver–silver chloride electrode in the form of a ring. Since the diffusion current is proportional to the concentration of the depolariser, dissolved oxygen, the limiting current is a linear function of the concentration of dissolved oxygen. The limiting current is fed into the y axis of a recorder in which it is plotted against time as the x axis. The y axis is readily calibrated in terms of the concentration of dissolved oxygen, by adjusting it to read 0 per cent when O_2 has been removed from the solution by sodium dithionite ($Na_2S_2O_4$) and to read 100 per cent when the solution is saturated with O_2 at the temperature of the experiment. From the continuous plot of concentration of O_2 (y) against time (x) respiration of cells and uptake or evolution of O_2 by their components can be followed with greater sensitivity, speed and flexibility than by Warburg manometry.

Amperometric (or Polarometric) Titrations

If the potential difference applied to a polarographic cell is kept constant, corresponding to the limiting current plateau of an electro-reducible substance A, the diffusion current is proportional to the concentration of A in the solution. Hence, changes in [A] during titration can be followed by repeatedly measuring the diffusion current. In practice, the limiting current, I_l, is measured, which is equal to the

Fig. 10.12 Amperometric titration curves. V is the volume of 0.01 mol dm^{-3} K$_2$Cr$_2$O$_7$ solution added to 20 cm^3 of 0.001 mol dm^{-3} Pb(NO$_3$)$_2$ solution

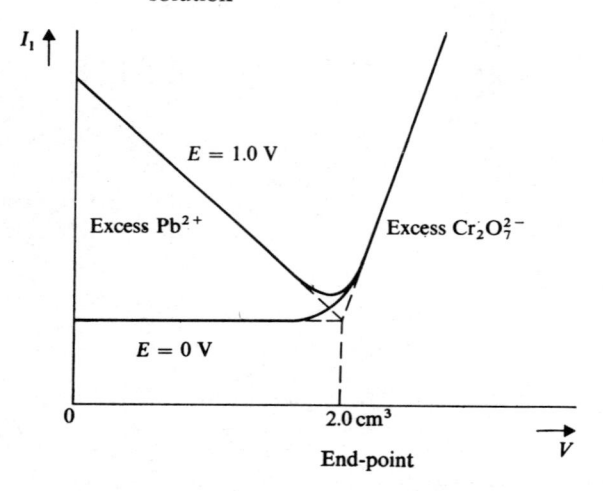

diffusion current plus a small constant residual current. Consider the effect of a substance B which reacts with A to give a polarographically inert substance, such as an insoluble precipitate. I_l is then plotted against the volume of a solution of B added from a burette. As B is added to A, [A] falls and I_l due to A decreases. At the equivalence point free B appears in the solution and [B] rises. If B is not electro-reducible, I_l remains low and constant so that the titration graph is shaped \＿. If B is electro-reducible, or in other words, if the applied voltage is set at or near its diffusion current plateau, I_l due to B increases as [B] rises after the end-point, so that the titration graph is V shaped. If B is electro-reducible and A is not, I_l is low and constant while A is in excess and increases when B accumulates in the solution, so that the graph is shaped ＿／.

These principles are illustrated in Fig. 10.12 by the titration of Pb(NO$_3$)$_2$ (= A) with K$_2$Cr$_2$O$_7$ (= B) using the polarographic apparatus in Fig. 10.9. The dropping mercury electrode dips into the deoxygenated solution in a titration flask, containing Pb(NO$_3$)$_2$ and 0.01 mol dm^{-3} KNO$_3$ as inert salt. The calomel electrode must not dip into this solution, because chloride ions precipitate lead ions as $PbCl_2$. Instead, the calomel electrode dips into saturated potassium chloride solution which is connected to the solution in the flask by a salt bridge of saturated potassium nitrate solution. Yellow $PbCr_2O_7$ is precipitated during the titration and the soluble inert salt KNO_3 is also formed, the ionic reaction being:

$$Pb^{2+} + Cr_2O_7^{2-} = PbCr_2O_7 \downarrow.$$

When the applied voltage is adjusted to 1.0 volt, both Pb^{2+} and $Cr_2O_7^{2-}$ are electro-reducible and the graph is V-shaped. When the applied voltage is set at 0 V, $Cr_2O_7^{2-}$ is electroreducible, but Pb^{2+} is not, so the graph is shaped __/. The end-point is given by the point of intersection of straight lines drawn through the points before and after the end-point. The shapes of the graphs, the deviations in the region of the end-point and the necessity for minimising or correcting for volume changes during titration are similar to those for conductimetric titrations (p. 352).

Amperometric (or polarometric) titrations are used to titrate various substances such as metabolites and drugs, with metal ions with which they form precipitates or complexes. For example, compounds containing thiol groups (—SH) may be titrated with silver nitrate or mercuric chloride.

Dead-Stop End-Point Titrations

This type of titration is based on the same principles as an amperometric titration in which substance A is not electro-reducible, i.e. not a depolariser, while substance B is. Dead-stop end-point titrations, however, require simpler apparatus (Fig. 10.13). Two bright platinum plates, each about 0.5 cm square and about 1.5 cm apart, dip into the titration vessel containing a solution of A. One or both electrodes are polarised by applying a small potential (about 50 mV) which is measured by the voltmeter, V, and adjusted by the potentiometer circuit, P. The current flowing through the electrolyte is measured by the galvanometer, G, which should give full scale deflection at about 1 μA and should have a number of shunts. These should enable the sensitivity to be decreased by factors of 10, 100 and 1000 to protect the galvanometer while the potentiometer is being adjusted. When the circuit is polarised, the current flowing through the electrolyte should be about 0.2 μA. To the stirred solution of A in the titration vessel a solution of B is added from a burette. With each addition of B the galvanometer gives only a temporary deflection. Just after the end-point a minute excess of B will permanently depolarise the electrodes and give a permanent deflection of the galvanometer.

Since iodine depolarises the electrodes, the method may be used

Fig. 10.13 Apparatus for a dead stop end-point titration

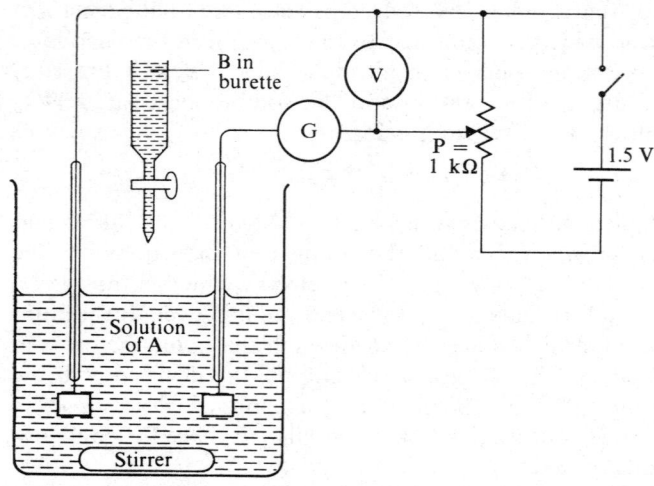

to titrate thiosulphate (A) with iodine (B) in the determination of the iodine value of lipids. The method may also be used to determine water (A) by titration with the Karl Fischer reagent (B) which contains free iodine. Since nitrous acid also depolarises the electrodes, the dead-stop end-point may be used to titrate the sulphonamide drugs and other primary aromatic amines (A) in acid solution with sodium nitrite (B). The reaction is one of quantitative diazotisation:

$$NaNO_2 + HCl = HNO_2 + NaCl$$
$$ArNH_2 + HNO_2 + HCl = ArN_2Cl + 2H_2O.$$

The end-point is the point at which a minute excess of free nitrous acid depolarises the electrodes.

Experiment 10.1 The quantitative determination of sulphonamide drugs using a dead-stop end-point titration.

The apparatus for the dead-stop end-point is set up as described (Fig. 10.13). The electrodes are cleaned by immersion for about 30 seconds in boiling concentrated nitric acid containing a little ferric chloride and are then thoroughly rinsed with water. The titration vessel is a 250 cm³ beaker to which is added 10 cm³ of concentrated hydrochloric acid and 75 cm³ of water. An accurately measured sample (2–3 mmol) of the sulphonamide or amine is dissolved in the solution, by warming if necessary, and then cooling to room temperature. A 50 cm³ burette is filled with approximately 0.1 mol dm^{-3} sodium nitrite. After setting up the apparatus, polarising the electrodes and switching on the magnetic stirrer, the solution in the beaker is titrated with the sodium nitrite solution at the rate of 1 drop per second.

Although each drop causes a deflection of the galvanometer, the spot or needle soon returns to its original position. As the end-point is approached, the temporary deflections on the galvanometer become larger and the spot or needle returns more slowly to rest. The rate of addition of sodium nitrite should then be much reduced. At the end-point one drop of sodium nitrite solution produces a larger permanent deflection.

The electrodes should be cleaned, as described, before each titration. The sodium nitrite solution should first be standardised with an accurately weighed sample of pure anhydrous sulphanilic acid (about 0.4 g). The experiment described accords with the official method in the British Pharmacopoeia for the assay of sulphacetamide sodium, sulphadiazine, benzocaine, suramin and a number of other similar drugs.

Problems

1 When a silver–silver chloride and a calomel electrode are incorporated in the same cell, the reaction that takes place as the cell supplies current is:

$$Ag(s) + \tfrac{1}{2}Hg_2Cl_2(s) = Hg(l) + AgCl(s).$$

The e.m.f. of the cell is 0.0455 V at 298 K and the temperature coefficient $(\partial E/\partial T)_p$ is 0.000338 V K^{-1}. Calculate ΔG, ΔS and ΔH for the reaction. What would be the values for the reaction:

$$2\,Ag(s) + Hg_2Cl_2(s) = 2\,Hg(l) + 2\,AgCl(s)?$$

2 Draw the cell diagram corresponding to the spontaneous cell reaction:

$$Zn(s) + CuSO_4(aq) = ZnSO_4(aq) + Cu(s).$$

Calculate the e.m.f. of the cell in which the zinc electrode dips into 10^{-4} mol dm^{-3} zinc sulphate and the copper electrode dips into 10^{-3} mol dm^{-3} copper sulphate at 25° C, given that $E_h^{\ominus}(Zn^{2+}, Zn) = -0.763$ V and $E_h^{\ominus}(Cu^{2+}, Cu) = +0.340$ V at this temperature. Assume that the liquid junction potential is negligible and that activity coefficients are unity if concentration is expressed in mol dm^{-3}.

3 Calculate the e.m.f. at 25° C of the cell:

$$Pt, H_2 \left| \begin{array}{c} 0.2 \text{ mol dm}^{-3} \\ \text{formic acid} \end{array} \right\| \left. \begin{array}{c} 0.5 \text{ mol dm}^{-3} \\ \text{acetic acid} \end{array} \right| H_2, Pt$$

given that the dissociation constant K_a of formic acid and acetic acid are 1.77×10^{-4} and 1.75×10^{-5} respectively at 25° C. Assume that the liquid junction potential has been eliminated, and that activity coefficients are unity.

4 The potential of a zinc electrode is measured at 298 K relative to a saturated calomel electrode at various concentrations of zinc chloride, zinc being the negative electrode. The following table gives the e.m.f. of the cell E, and the corresponding activities of the zinc ion, a, in the various solutions:

E/V	-1.066	-1.051	-1.037	-1.031
a	0.0076	0.0234	0.0589	0.1103

Calculate the standard electrode potential $E_h^\ominus$ of the zinc electrode for each value of the e.m.f., and find the mean value. The potential of the saturated calomel electrode on the hydrogen scale at 298 K is 0.242 V.

5 What would be the potential of a hydrogen electrode at 291 K (H_2 pressure = 1 atm = 101 325 Pa) on the hydrogen scale dipping into each of the solutions mentioned in problem 1, p. 389?

6 Given that the standard oxidation-reduction potentials at 298 K for the systems Fe^{3+}, Fe^{2+} and Sn^{4+}, Sn^{2+} are 0.77 and 0.15 V respectively, deduce the oxidation-reduction potentials E_h at the points suggested below during the titration of 10 cm³ of a solution of $SnCl_2$ of concentration 0.05 mol dm⁻³ with a solution of $Fe_2(SO_4)_3$ of concentration 0.05 mol dm⁻³.

$Fe_2(SO_4)_3$ added/cm³	1	2	4	5	6	8	9
$[Sn^{4+}]/[Sn^{2+}]$	$\frac{1}{9}$	$\frac{2}{8}$	$\frac{4}{6}$	$\frac{5}{5}$	$\frac{6}{4}$	$\frac{8}{2}$	$\frac{9}{1}$
$Fe_2(SO_4)_3$ added/cm³	11	14	18	20	22	24	
$[Fe^{3+}]/[Fe^{2+}]$	$\frac{1}{10}$	$\frac{4}{10}$	$\frac{8}{10}$	$\frac{10}{10}$	$\frac{12}{10}$	$\frac{14}{10}$	

Before the end-point, the contribution to the potential from the Fe^{3+}, Fe^{2+} system may be neglected, and after the end-point the contribution from the Sn^{4+}, Sn^{2+} system may be neglected. Concentrations may be used in place of activities. Plot the titration curve as E_h versus cm³ of $Fe_2(SO_4)_3$ added.

7 An oxidation-reduction indicator has a standard potential $E_h^{\ominus\prime}$ at 25° C and pH 7.0 of -0.296 V. What is the oxidation-reduction potential in a system in which the indicator is present as 50 per cent oxidised form and 50 per cent reduced form at pH 4.5? What is the oxidation-reduction potential in a system in which the indicator is present as 20 per cent oxidised form and 80 per cent reduced form at pH 7.0?

8 Calculate the e.m.f. of the cell corresponding to the reaction

$$2NADH + O_2 + 2H^+ = 2NAD^+ + 2H_2O$$

using the data in Table 10.4. Hence calculate the standard Gibbs energy and equilibrium constant for this reaction at pH 7 at 25° C.

9 A concentration cell consisting of two electrodes of a metal M, one dipping into a 0.05 mol dm^{-3} solution of one of its salts and the other into a 0.002 mol dm^{-3} solution of the same salt has an e.m.f. of 0.401 V at 25° C. State which solution is in contact with the positive pole of the cell. Assuming that there is no liquid junction potential, calculate the valency of M in the salt. Account for the fact that the calculated value is not exactly a whole number.

10 At 298 K the e.m.f. of the cell: −

Ag, AgI(s)|KI (0.001 mol kg^{-1})|KI (0.1 mol kg^{-1})|AgI(s), Ag

is −0.0838 V. Calculate the transport number for the K$^+$ ion assuming the mean ionic activity coefficients of KI to be 0.965 and 0.755 for the more dilute and the stronger solutions, respectively.

11 Assuming the mean activity coefficients and concentrations stated in the previous problem, calculate the e.m.f. of the corresponding cell in which the liquid junction potential is eliminated by a salt bridge.

12 A cell membrane ionises so as to become negatively charged. It is therefore impermeable to low concentrations of anions but permeable to cations. The surrounding fluid, 0.005 mol dm^{-3} aqueous potassium chloride, is positive by 74 mV with respect to the interior of the cell at 37° C. Calculate the concentration of electrolyte, which is here assumed to be potassium chloride, in the cell interior. What would be the cell membrane potential if the external solution consisted of 0.002 mol dm^{-3} aqueous potassium chloride? It may be assumed that activity coefficients are unity.

11 Surface Chemistry

Introduction

The study of phenomena at surfaces is important for pharmacy and biology, not only because the results are of interest in themselves but also because an understanding of the colloidal state depends upon a knowledge of surface properties. In biological systems, the transfer of materials into and out of cells must proceed by adsorption on to the cell membrane, penetration of the membrane and desorption at the opposite surface of the membrane. Many pharmaceuticals are administered in the form of a suspension (a solid dispersed in a liquid). The preparation, the stability and properties of these materials depend on the structure and properties of the surfaces of the particles dispersed. Besides such physical processes, it is probable that many chemical reactions take place at surfaces.

The most important types of surface are those formed by gas/solid, liquid/solid, gas (air)/liquid and liquid/liquid. Interest in the solid/solid interface is restricted mainly to solid state physics, and to the science of adhesion. A true gas/gas interface is never formed because gases always diffuse into each other.

Adsorption at all types of interface has several important features in common. Firstly, adsorption is highly selective. The amount depends greatly on the nature and previous treatment of the *adsorbent* surface and on the nature of the adsorbed substance, which is often known as the *adsorbate*. Secondly, adsorption is a rapid process whose rate increases with increasing temperature but decreases with increasing amount adsorbed. Thirdly, adsorption is a spontaneous process, i.e. ΔG is negative, and is usually associated with an increase in order of the adsorbate, i.e. ΔS is negative, and according to equation [7.62] is therefore usually exothermic, i.e. ΔH is negative. The change in enthalpy when one mole of adsorbate is adsorbed by the appropriate quantity of adsorbent is known as the enthalpy of adsorption or as the heat of adsorption at constant pressure. It may be determined by calorimetry or by application of the van't Hoff isochore ([7.113] p. 320) or the Clausius equation ([7.79] p. 305). The equilibrium mass],

x, or volume, V, of adsorbed substance or its pressure, p, or concentration, c, in the bulk phase at equilibrium is determined at various temperatures, T. When $\log x$ or $\log V$ is plotted against $1/T$, the gradient is equal to $-\Delta H/2.303R$, where ΔH is the enthalpy of adsorption. If $\log p$ or $\log c$ is plotted against $1/T$, the gradient is also equal to $-\Delta H/2.303R$, but ΔH is now the enthalpy of desorption of the adsorbate from the absorbent. This enthalpy change is opposite in sign but equal in magnitude to the enthalpy of adsorption. If the enthalpy of adsorption is negative, then as the temperature increases the quantity adsorbed, x or V, decreases, and the pressure, p, or the concentration, c, in the bulk phase increases in accordance with Le Chatelier's principle.

Two main types of adsorption have been recognised depending on the forces of interaction between adsorbent and adsorbate molecules. When these forces are dipole-dipole, dipole-induced dipole or dispersion forces (p. 30), the term *physical adsorption*, physisorption or van der Waals adsorption is used. When the forces are covalent or dative bonds, the term chemical adsorption or *chemisorption* is applied. Physical adsorption is associated with an enthalpy of adsorption numerically less than about -40 kJ mol^{-1}, whereas values numerically greater than about -80 kJ mol^{-1} characterise chemisorption. The enthalpy of physical adsorption is comparable to the enthalpy of condensation, ΔH_g^1, whereas the enthalpy of chemisorption is comparable to the enthalpy of chemical reaction. Physical adsorption is more common, whereas chemisorption is found only when the adsorbent and adsorbate tend to form a compound. Physical adsorption can usually be reversed by reducing the pressure surrounding the adsorbent unless the adsorbed substance condenses in capillaries, whereas chemisorption is difficult to reverse and usually takes place more slowly. A chemisorbed substance can only be removed by heating at a high temperature under vacuum, and may be liberated as a compound with the adsorbent rather than in its original form. For example, oxygen which has been chemisorbed on carbon is liberated as oxides of carbon when heated *in vacuo*. Heat treatment can alter the properties of the adsorbent. The adsorbed layer in physical adsorption may vary in thickness from one molecule to many molecules because van der Waals forces can extend from one layer of molecules to another as in the liquid (and solid) state. On the other hand, chemisorption cannot itself give rise to a layer more than one molecule thick owing to the specificity of bonding between the adsorbent and the adsorbate. However, subsequent layers may be physically adsorbed on top of the first layer. Thus, a single layer of adsorbed molecules, known as a 'monolayer', is given by both physical adsorption and by chemisorption, whereas a layer more than one molecule thick, known as a 'multilayer', is always associated with physical adsorption. Physical adsorption usually occurs more readily than chemisorption at lower temperatures but when the temperature is raised it gives way to chemisorption, showing that there is an activation energy barrier (p. 166) between physical and chemical adsorption. A plot of the quantity of the adsorbed substance against temperature at constant

Fig. 11.1 Adsorption isobar showing how the quantity adsorbed may vary with temperature

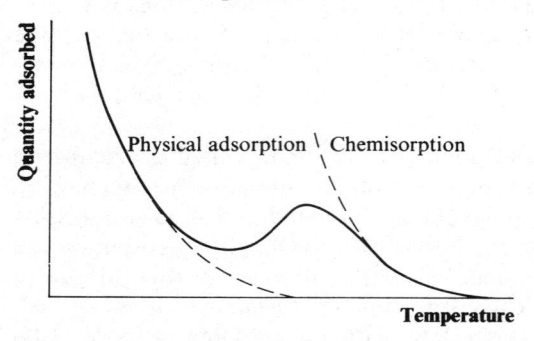

Fig. 11.2 Potential energy-intermolecular distance curves for chemical and physical adsorption

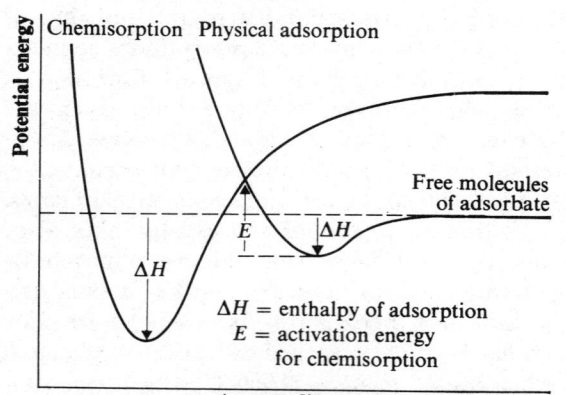

ΔH = enthalpy of adsorption
E = activation energy for chemisorption

pressure (Fig. 11.1), known as an adsorption isobar, shows the transition region over which the change takes place. A plot of the potential energy against the mean distance between adsorbate and adsorbent molecules (Fig. 11.2) shows that as the adsorbate molecule approaches the adsorbent physical adsorption first takes place. If the temperature is high enough for the activation energy barrier to be readily overcome, the adsorbate molecule approaches more closely and chemisorption takes place. The enthalpy of adsorption, ΔH, is seen to be numerically greater for chemisorption than for physical adsorption, showing that once formed chemisorption is not easily reversed. The existence of a significant activation energy barrier is referred to as 'activated adsorption'. For certain pairs of adsorbates and adsorbents, the potential energy curves for physical and chemical adsorption may overlap in such a way that the activation energy barrier is small or zero. The adsorbed molecules do not then remain in the physically adsorbed state for a significant time, and chemisorption occurs at all temperatures.

Gas-Solid Interface

Many of the fundamental principles of adsorption discussed above were discovered during work on the gas–solid interface. The properties of this interface govern the kinetics of heterogeneous gas reactions in which the solid adsorbent acts as the catalyst (p. 182). Adsorption of gases and vapours on solids provides a method of determining the specific surface area of solids and hence the mean particle size of powders in pharmacy.

Prior to any experimental work on the gas–solid interface, the solid adsorbent should be freed from previously adsorbed gases and vapours. This process has been mentioned above and is known as 'outgassing' or 'degassing'. The adsorbate gas or vapour is then allowed into the evacuated apparatus which contains the outgassed adsorbent of mass m. The quantity of gas adsorbed may be determined volumetrically or gravimetrically. In the volumetric method (Fig. 11.3a) the outgassed adsorbent is contained in a thermostatted tube and the gas is introduced from a gas burette. The equilibrium pressure, p, of the gas is measured with the manometer. The volume, V, of gas adsorbed is calculated from the total volume of the apparatus and from p and is corrected to NTP. The gravimetric method frequently makes use of the McBain–Bakr balance (Fig. 11.3b). The outgassed adsorbent is here contained in a small cup suspended from a quartz helical spring. After admission of the gas into the evacuated apparatus the equilibrium pressure, p, is measured on the manometer and the weight, x, of adsorbed gas is determined from the extension of the quartz spring. The relationship between the mass, x, the volume, V, and the amount, n mol, adsorbed by a given mass, m, of adsorbent is:

$$n = \frac{x}{M} = \frac{V}{V_m}, \quad [11.1]$$

where M is the molar mass of the adsorbate and V_m is the molar volume of the adsorbate (0.0224 m^3 mol^{-1} at NTP if the gas behaves ideally).

Plots of x/m or V/m ($= v$) against p at constant temperature are

Fig. 11.3 Apparatus for studying adsorption of gases by solids
(*a*) volumetric method (*b*) gravimetric method

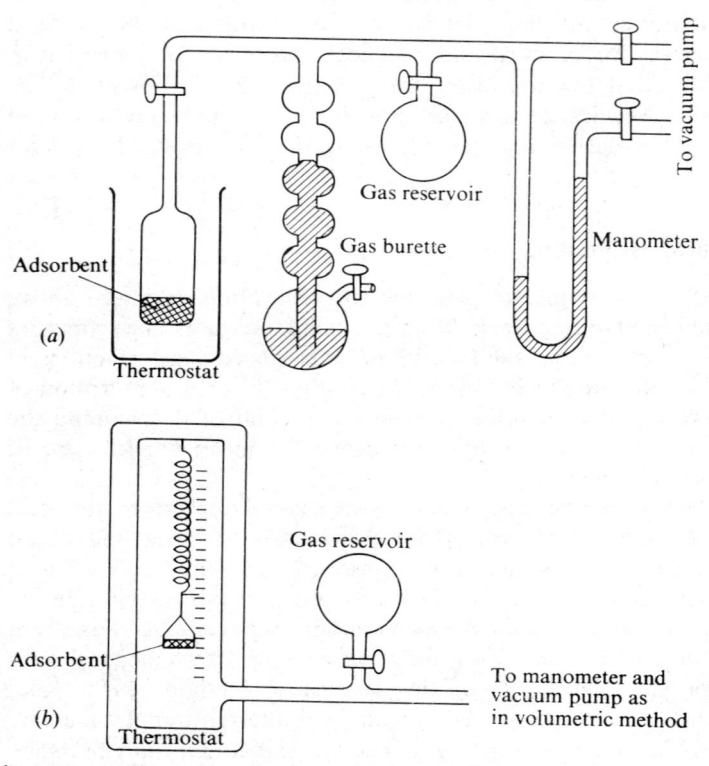

known as adsorption isotherms. Some gas–solid adsorption isotherms encountered in practice are shown in Fig. 11.4. A few other types may also be encountered. Type I is the Langmuir isotherm (p. 185) which corresponds to the formation of a monomolecular layer of adsorbate on the surface of the adsorbent. This isotherm is therefore only given by adsorbates which are chemisorbed or physically adsorbed on to solids with **very fine pores.** The value of v which corresponds to a complete monomolecular layer is indicated by v_m.

Isotherms of Type II, IIA, III, IV and V describe adsorption in multimolecular layers. When the pressure, p, of the adsorbate in the bulk phase is equal to the saturated vapour pressure, $p^{\ominus}$, the adsorbate gas or vapour can condense on the surface of the adsorbent to form a liquid. This can only happen if the temperature is below the critical temperature of the adsorbate. With isotherms of type II, IIA and III the curve approaches $p^{\ominus}$ asymptotically because there is no limit to the number of layers of adsorbate molecules. The horizontal approach of the curves of Type IV and V towards $p^{\ominus}$ indicates that there is a limit to the number of layers which can be accommodated in the

Fig. 11.4 Adsorption isotherms classified according to Brunauer (except IIA)
I N_2 on charcoal at 90 K II N_2 on silica gel at 77 K
III Br_2 on silica gel at 352 K IIA krypton on carbon black at 90 K
IV benzene on ferric oxide gel at 323 K V water vapour on charcoal at 373 K

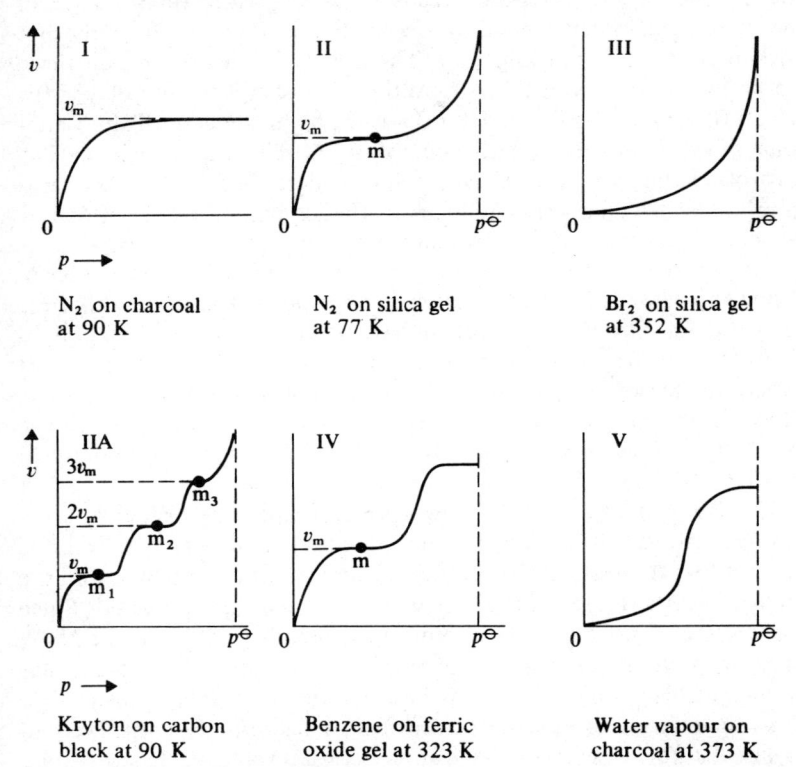

N_2 on charcoal at 90 K

N_2 on silica gel at 77 K

Br_2 on silica gel at 352 K

Kryton on carbon black at 90 K

Benzene on ferric oxide gel at 323 K

Water vapour on charcoal at 373 K

capillaries of the adsorbent. In stepwise isotherms, of which IIA is an example, each step corresponds to the completion of a layer of adsorbate molecules.

One of the most successful theoretical treatments of multilayer adsorption was made by Brunauer, Emmett and Teller who derived the BET equation:

$$\frac{p}{v(p^\ominus - p)} = \frac{1}{v_m c} + \frac{c-1}{v_m c} \cdot \frac{p}{p^\ominus}, \qquad [11.2]$$

where v, v_m, $p^\ominus$ and p are defined above and $c \approx \exp[(\Delta H_g^1 - \Delta H_I)/RT]$ where ΔH_I is the molar enthalpy of adsorption for the first molecular layer and ΔH_g^1 is the molar enthalpy of condensation which is assumed to apply to subsequent layers. For the adsorption of gases by solids ΔG and ΔS are always negative, so enthalpies of adsorption,

ΔH_I and ΔH_g^1 are also always negative. The BET equation is derived by equating the rates of adsorption and desorption for various adsorbed molecular layers, rather than for one layer as in the case of the Langmuir isotherm (p. 184). When the formation of the first layer of molecules is more exothermic than the formation of subsequent layers, i.e. when ΔH_I is more negative than ΔH_g^1, as is usual, c is greater than unity, and the BET equation corresponds to the sigmoid isotherm II and to the first part of isotherm IV in Fig. 11.4. The point of inflexion, m, on these isotherms results from the formation of a complete monolayer for which the corresponding value of v is v_m as in the case of the Langmuir (Type I) isotherm. The value of v_m enables the specific surface area of a solid to be calculated, as will be described in the next paragraph. The BET equation reduces to the Langmuir isotherm at low pressures or when ΔH_g^1 is insignificant compared with ΔH_I, and is particularly successful in explaining the very common Type II isotherm. When the formation of the first layer of molecules is less exothermic than the formation of subsequent layers, i.e. when ΔH_I is less exothermic than ΔH_g^1, c is less than unity and the BET equation corresponds to the isotherm III and to the first part of isotherm V in Fig. 11.4. These isotherms are comparatively rare and arise when the forces of adsorption which give rise to the first molecular layer are relatively small.

The BET equation [11.2] gives a linear graph when $p/v(p^{\ominus}-p)$ is plotted against $p/p^{\ominus}$. The gradient is equal to $(c-1)/v_m c$ and the intercept on the vertical axis is $1/v_m c$. The sum of the gradient and the intercept is therefore $1/v_m$, from which v_m may be calculated. Since the volume of gas adsorbed by unit mass of adsorbent is v_m at NTP, the corresponding amount n mol is equal to v_m divided by the molar volume at NTP which, if the gas behaves ideally, is 0.0224 m³ mol⁻¹. Thus the number of molecules adsorbed by 1 kg of adsorbent is equal to $v_m L/0.0224$, where L is Avogadro's constant and v_m has the units m³ kg⁻¹. Now the specific surface area of the adsorbent, a m² kg⁻¹ is equal to the number of molecules adsorbed per kg of adsorbent to form a complete monolayer multiplied by the area of each molecule, A m². Thus:

$$a = v_m LA/0.0224. \quad [11.3]$$

This adsorption method is particularly valuable for determining specific surface area, since gases readily penetrate aggregates (or 'clumps') of solid particles. The gas must be capable of forming a closely packed monolayer, which can only be achieved if the gas is adsorbed physically. The gases frequently chosen are nitrogen at 77 K, for which a is 1.62×10^{-19} m², and the noble gases, e.g. krypton. Many powerful adsorbents, including porous solids, e.g. charcoal and silica gel, and finely divided powders, have specific surface areas of the order 10^5 or $10^6\,\mathrm{m^2\,kg^{-1}}$. If the solid adsorbent is in the form of a powder whose particles may be assumed to be spherical, the mean particle radius, r m, may be calculated from a and the density ϱ kg m^{-3} ($= 1$/specific volume), thus:

$$\frac{\text{specific volume}}{\text{mean particle volume}} = \frac{\text{specific surface area}}{\text{mean particle surface area}} = \frac{\text{number of}}{\text{particles per kg}}$$

$$\therefore \quad \frac{1/\varrho}{\frac{4}{3}\pi r^3} = \frac{a}{4\pi r^2}$$

$$\therefore \quad r = 3/\varrho a. \qquad [11.4]$$

Introducing the expression for a (equation [11.3])

$$r = 3 \times 0.0224/\varrho v_\mathrm{m} LA. \qquad [11.5]$$

In the isotherms shown in Fig. 11.4 adsorption and desorption follow the same curve; that is to say, the curve of v versus p is the same when p is increased as when it is decreased. In certain cases,

Fig. 11.5 Adsorption isotherm with a hysteresis loop

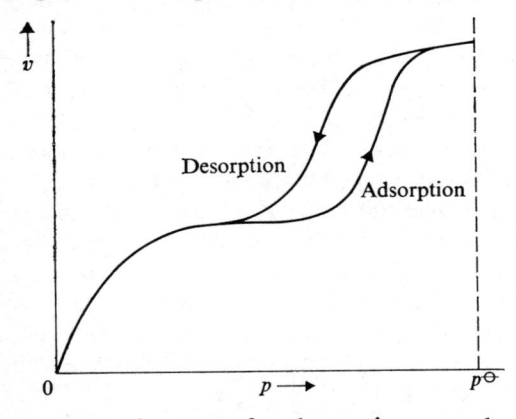

however, the curve for desorption may be displaced to lower bulk pressures than the curve for adsorption, thereby giving rise to a 'hysteresis loop' as shown in Fig. 11.5. This means that adsorption is not truly reversible, desorption being more difficult than adsorption. Hysteresis loops arise when the adsorbent has pores of diameter of the same order as molecular dimensions and when the adsorbate is below its critical temperature. On adsorption the adsorbate enters the capillaries and condenses to a liquid. On desorption this liquid must evaporate before leaving the the capillaries. Since the vapour pressure of the liquid within the capillaries is smaller than in the bulk phase, the external pressure necessary to cause desorption is less than that on adsorption. The presence of pores or capillaries of small size indicates that there is a limit to the quantity of adsorbent which they can accommodate. Consequently, hysteresis loops are most commonly found in isotherms of Types II and V in Fig. 11.4.

When the adsorbate is above its critical temperature capillary condensation and hysteresis loops do not usually arise, even though the pores may have diameters of the same order as the dimensions of the molecules of adsorbate. Under these circumstances those adsorbate molecules of smaller size than the pores will enter them and will be adsorbed, whereas those molecules larger than the pores will not be adsorbed and will remain in the bulk phase. The ease of adsorption will decrease with increasing molecular dimensions. The porous adsorbent is then acting as a 'molecular sieve'. Unlike a household sieve, a molecular sieve retains small particles while allowing larger particles to pass by. For example, natural zeolites are aluminosilicates with cavities and access windows of diameter about 1 nm and 0.4 nm respectively, which allow noble and diatomic gases, water and the normal alkanes to enter but not larger molecules. The rate of adsorption is found to decrease with increasing molecular weight. Molecular sieves are therefore useful in separating molecules of different sizes. In biology and pharmacy the most useful molecular sieves are those, such as Sephadex, which separate solutes in solution (see p. 455).

Closely related to the process of adsorption within the cavities of molecular sieves is the formation of the so-called 'clathrate compounds'. This is the process whereby small molecules of appropriate size (the guest molecules) may become trapped or 'clathrated' within an open framework formed by the linking together of the host molecules. Such entrapment takes place when the host substance is crystallised in the presence of the guest and often the guest molecules can only be released again by melting or dissolving the complex, when the framework of the host breaks down. Water and quinol are the best known substances that can form the requisite open framework structures by hydrogen bonding. They can give rise to clathrates of stoichiometric composition such as $3\,C_6H_4(OH)_2 \cdot CH_3CN$, $Xe \cdot 5\frac{3}{4}\,H_2O$ and $CHCl_3 \cdot 17\,H_2O$ and it is for this reason that such complexes are sometimes referred to as clathrate compounds. This term is rather misleading, however, since no bonds are formed between the host and guest molecules, the latter being merely entrapped and unable to pass between the host molecules forming the framework. It is because of this lack of dependence on chemical interaction that clathrates can readily be formed by the noble gases as well as by a wide variety of less inert substances such as halogens, hydrocarbons and their halogen derivatives. Clathrate formation may, however, be aided by van der Waals forces between the host and guest molecules (see p. 30). It is probable that transient clathrate structures are formed in appropriate solutions, e.g. anaesthetic gases in water. Pauling has proposed a theory of anaesthesia which is based on the formation of such clathrates, e.g.

$CHCl_3 \cdot 17 H_2O$ and $N_2O \cdot 5\frac{3}{4} H_2O$, in the synaptic regions in the brain. The unstable clathrates in microcrystalline form are believed to be stabilised by the side chains of the protein molecules and certain solutes in the encephalonic fluid. The microcrystals may increase the impedance of the neural network and thereby decrease the energy of the electrical oscillations of the brain or may inhibit enzyme reactions in the brain. In either case anaesthesia may result.

Liquid-Solid Interface

Solids adsorb not only gases but also one or more components of a liquid mixture or alternatively a solute or the solvent of a solution. Adsorption of solutes is exemplified by the uptake of drugs and substrates by the receptors of cells and by the removal of impurities from solution. It is the basis of adsorption chromatography and it is illustrated by Experiment 11.1.

It is sometimes possible to determine the composition of the adsorbed layer directly; for example, by introducing a radioactive form of the adsorbate into the bulk of the solution and then measuring the radioactivity of the adsorbent after its removal from the solution. For substances in solution an adsorption isotherm obtained from direct measurements of the composition of the adsorbed layer is known as an 'individual (or true) isotherm'. Individual isotherms may be shaped like isotherms I, II and III in Fig. 11.4; the ordinate may be x/m or n/m, where x is the mass and n the amount (number of moles) of adsorbate adsorbed by mass m of adsorbent, and the abscissa is the concentration, c, or the mole fraction of adsorbate in the bulk of the solution. The approach of the curve to the horizontal or vertical axes or to lines parallel to them may or may not be asymptotic.

Usually the composition of the adsorbed layer is not measured directly. Instead the concentration of adsorbate in the bulk of the solution is often determined before and after equilibration with the adsorbent and an apparent value of x/m (or n/m) is calcuated from the change in concentration. Let us consider the common case in which the substance being determined is the solute from a solution, and in which its concentration in the bulk of solution is decreased by equilibration with the adsorbent. The only conclusion which we can draw is that more solute than solvent has been adsorbed. We cannot ignore the adsorption of the solvent. The adsorption isotherm which we determine by analysing the bulk of the solution is a combination of the individual isotherms of the solute and solvent and is known as the 'composite (or apparent) isotherm' of the substance being determined. The same is true when considering mixtures or solutions consisting of more than two components.

Figure 11.6 shows four types of composite adsorption isotherm,

Fig. 11.6 Apparent adsorption isotherms for a substance in solution or in a liquid mixture

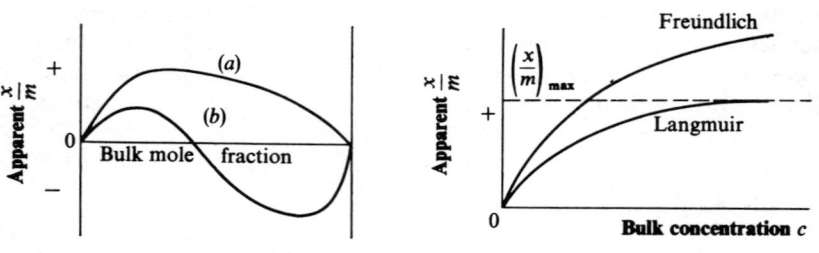

where x/m is the mass adsorbed per unit mass of adsorbent. Types (a) or (b) may apply to a component of a binary liquid mixture. Type (a) is exemplified by carbon tetrachloride adsorbed by charcoal from mixtures of carbon tetrachloride and chloroform. Type (b) is exemplified by benzene adsorbed by charcoal from benzene/methanol mixtures. Isotherms of the Freundlich and Langmuir types are also found (see Experiment 11.1). A component of a mixture or solution is said to be positively adsorbed if it is present at the solid surface in a greater proportion than that in the bulk. The component is said to be negatively adsorbed if its proportion at the surface is less than in the bulk. We note that negatively adsorbed components may still be present at the interface though in relatively small amounts. Zero adsorption of a component means that its proportion is the same at the interface as in the bulk and will always occur when the liquid component in question is in the pure state (i.e. at unit mole fraction). Adsorption also approaches zero as mole fraction or concentration approaches zero.

The composite adsorption isotherm of a component of a solution or mixture may be calculated from the individual isotherms of all the components of a mixture or solution, but the reverse is impossible without further information. However, for very dilute solutions, i.e. when the mole fraction of the solute is less than about 0.01, the mole fraction of the solvent in the bulk of the solution is virtually constant at a value very close to unity. Consequently, the amount of solvent adsorbed is virtually constant, although the amount of solute adsorbed, though small, will vary widely throughout the whole concentration range in the bulk. The composite isotherm of the solute is then almost identical with the individual isotherm of the solute. This approximation applies to aqueous solutions of many organic compounds, for example surface active agents and oxalic acid, which is considered in Experiment 11.1.

Experiment 11.1 The adsorption of oxalic acid from aqueous solutions by charcoal.

One dm³ of a stock solution of 0.25 mol dm⁻³ oxalic acid is first prepared, and 10 cm³ portions of this are warmed with sulphuric acid to 80° C and used to standardise an approximately 0.02 mol dm⁻³ solution of potassium permanganate. From the stock solution, 250 cm³ of each of the following oxalic acid solutions are prepared by dilution: 0.20, 0.15, 0.10, 0.05, 0.025 and 0.01 mol dm⁻³.

About 2 g of finely divided active charcoal are weighed accurately and placed in each of eight stoppered reagent bottles. A preliminary experiment is carried out with the 0.15 mol dm⁻³ oxalic acid solution. 100 cm³ of the solution are added to one of the reagent bottles, which is then shaken for successive periods of fifteen minutes (in a shaking machine, of possible). After each period of shaking, a 10 cm³ portion of the liquid is removed with a pipette, avoiding the charcoal as far as possible, and the portion is titrated with the permanganate solution. In this way it is possible to find the time required for the concentration to become approximately constant (i.e. for equilibrium to be established between the oxalic acid in solution and that adsorbed on the charcoal). In the experiment, the main reagent bottles should be shaken for at least one and a half times this period.

To the charcoal in each of the other seven bottles, 100 cm³ of one of the seven oxalic acid solutions is added in turn. The bottles are then shaken for the appropriate time determined by the preliminary experiment. The contents are filtered through small filter papers and the first 10 cm³ of each filtrate are rejected. (These two precautions minimise spurious results due to adsorption by the paper.) 25 cm³ portions of the filtrates are titrated with permanganate, and the concentration of oxalic acid, c, in each is calculated. These concentrations are compared with the original concentration, in each case, to obtain the mass of oxalic acid adsorbed by the charcoal. Hence the factor x/m, the mass adsorbed per unit mass of charcoal, is found.

In order to investigate whether the adsorption follows the Freundlich equation, it is necessary to plot a graph of log (x/m) against log c (where c is the concentration of oxalic acid in the solution at equilibrium, calculated above). This follows by consideration of the form of the equation applicable to adsorption from solution (cf. p. 184):

$$\frac{x}{m} = k \cdot c^{1/n}.$$

Whence, by taking logarithms of both sides:

$$\log \frac{x}{m} = \log k + \frac{1}{n} \cdot \log c.$$

This has the form of a straight-line graph with gradient $1/n$.

To test whether the adsorption follows the Langmuir equation (cf. p. 185):

$$\frac{x}{m} = \frac{k'kc}{(1+kc)}$$

(m/x) may be plotted against $1/c$. The reason for this procedure may be seen by multiplying both sides of the equation by $(1+kc)$ and by (m/xc):

$$\frac{1}{c} + k = k'k \cdot \frac{m}{x}.$$

Dividing both sides by $k'k$ we have:

$$\frac{1}{k'k} \cdot \frac{1}{c} + \frac{1}{k'} = \frac{m}{x}.$$

From this it is seen that, when the graph is plotted as indicated above, it should be a straight line of gradient $1/k'k$ and intercept at $c = 0$ of $1/k'$. At high concentrations kc is large compared with unity, i.e. $(1+kc) \approx kc$, and (x/m) approaches asymptotically a limiting value $(x/m)_{max}$ equal to k'.

It is instructive to plot a graph of the fraction, f, of adsorbate adsorbed against the original concentration, d, of adsorbate, where

$$f = (d-c)/d.$$

It will be seen that a greater fraction of oxalic acid is adsorbed from dilute solutions than from concentrated solutions. This fact forms the basis of decolorisation and purification of organic compounds in solution by shaking with charcoal or another adsorbent. The impurities, being at low concentrations, are preferentially removed, whereas the substance required, which is at a high concentration, suffers proportionally only a minor loss.

Adsorption from solution often obeys the same qualitative rules as solubility (p. 203). Adsorption on to polar adsorbents, e.g. silica gel or alumina, tends to increase with increasing polar character of the adsorbate. For example, a fatty acid dissolved in the non-polar solvent, toluene, will be adsorbed by the polar adsorbent, silica gel, more strongly than the toluene. Since increasing chain length of the fatty acid increases the non-polar character and reduces the polar character, adsorption of the fatty acid, as measured by x/m, in this system decreases in the order: acetic > propionic > n-butyric. Conversely, adsorption on to non-polar adsorbents, e.g. charcoal, tends to increase with decreasing polar character of the adsorbate. Thus, charcoal adsorbs fatty acids more strongly than the water used as the solvent, and the order of adsorption is the reverse of that above, namely: n-butyric > propionic > acetic. This type of simple, qualitative reasoning is used in choosing an adsorbent and an eluting solvent in adsorption chromatography. The common solvents may be placed in order of increasing eluting power from common polar adsorbents. The order in this elutropic series is approximately the order of increasing polarity, as expected.

Adsorption Chromatography

If a mixture of organic substances, in a suitable solvent, is placed on a column of alumina, or kieselguhr, or one of a variety of other adsorbents, and washed down with the solvent, the substances will pass down the column at different rates. Separation may be enhanced by using a number of different eluting solvents in turn. The technique of thin layer chromatography (TLC, p. 233) may also be applied.

As in partition chromatography (p. 233), the rate may be expressed by an R_F value, and it will be larger the smaller is the force of attraction between the adsorbent and the substance being adsorbed. The difference between adsorption and partition chromatography is that the substances to be fractionated apportion themselves between the solid of the column or layer and the developing liquid in the first case, but between the stationary and the moving liquids in the second case. The separation of the components along the column or layer may be detected visibly if they are coloured. Alternatively, they may be caused to fluoresce in ultra-violet light, or they may be caused to give colour reactions by streaking the extruded column or spraying the TLC plate with an appropriate reagent. The components in the various zones on the column or TLC plate may then be isolated by separating or scraping off each zone or spot and extracting the material. Instead, each component in turn may be washed completely through the column, and the successive fractions identified by the change in the properties of the eluate.

Separations by means of molecular sieves, such as Sephadex, is an important example of adsorption chromatography. Sephadex is a polysaccharide containing glucose subunits bonded together in the form of a cross-linked dextran. This structure swells to give a three dimensional gel network which acts as a molecular sieve (see p. 450). Small molecules may be removed from solutions of macromolecules by filtration through such a material. Macromolecules of different sizes, such as enzymes and other proteins, can be separated chromatographically by passing an aqueous solution containing them down a column of Sephadex. The volume of solvent which has flowed through the column before a given protein appears, known as the elution volume of that substance, is a linear function of the negative logarithm of the molecular weight of the protein. Calibration of the column with proteins of known molecular weight enables the molecular weights of other proteins to be determined.

Ion Exchange

Another process depending on somewhat similar principles is that of ion exchange on columns. Essentially, the process is one of double decomposition in which the material of the column is able to provide the type of ion required in place of one adsorbed from the solution. For example, natural zeolites and other aluminosilicates will replace calcium ions in hard water by sodium ions. Alternatively, they may be made to replace sodium ions by hydrogen ions. They are cation exchangers, and their reaction may be represented thus:

$$Z^-H^+ + Na^+ \rightleftharpoons Z^-Na^+ + H^+.$$
Exchanger In soln. Exchanger In soln.

The reaction may proceed in either direction, and the exchanger may be regenerated by passing through the column a fairly concentrated solution containing the ion which was removed.

A number of synthetic resins of different acidities are commercially available for use in ion-exchange columns. Those containing the sulphonic-acid group ($-SO_3H$) are strongly acid, and will exchange hydrogen ions for other cations, even when the latter are accompanied in solution by anions which are the conjugate base of a strong acid (see p. 358). Thus, even when Na^+ is accompanied by Cl^-, which is the conjugate base of the strong acid HCl, it can be replaced by H^+, thus:

$$R.SO_3^-H^+ + Na^+Cl^- \rightleftharpoons R.SO_3^-Na^+ + H^+Cl^-.$$
Resin In soln. Resin In soln.

Resins which are weaker acids will only replace cations which are associated with anions which are themselves conjugate bases of weaker acids. Thus, resins containing $-COOH$ groups will only replace Na^+ by H^+ in solutions of acetates, for example, and resins containing $-OH$ will only replace Na^+ by H^+ in alkaline solution.

Anion exchangers are usually synthetic resins containing amino or quaternary ammonium groups. The anion exchangers are also selective in their action, since their functional groups vary in basic strength. Strong anion exchangers are derived from the strongly basic quaternary ammonium hydroxides and their reactions are exemplified by:

$$R.N(CH_3)_3^+OH^- + Cl^- \rightleftharpoons R.N(CH_3)_3^+Cl^- + OH^-$$
Resin In soln. Resin In soln.

Weak anion exchangers are derived from the more weakly basic amines and may react as follows:

$$R.NH_2 + HCl \rightleftharpoons R.NH_3^+Cl^-$$

(which is not a true ion exchange, but an adsorption of acid), and

$$2\,R\,.\,NH_3^+Cl^- + SO_4^{2-} \rightleftharpoons (R\,.\,NH_3^+)_2\,.\,SO_4^{2-} + 2\,Cl^-.$$

During ion exchange the resins often undergo changes in density and therefore volume, in which case mechanical work is done. Thermodynamic considerations of Gibbs energy indicate that small ions are adsorbed in preference to large ions and ions of higher valency are taken up more readily than ions of lower valency.

Ion-exchange columns have many applications. If an aqueous solution is passed through a cationic and an anionic exchanger, it is possible to remove all the ions. In this way large quantities of water are readily prepared containing smaller amounts of ionic components than in distilled water as usually prepared. Separation of substances in chromatographic analysis may be brought about on an ion-exchange column by virtue of differences in acid or base strength. Moore and Stein have developed ion exchange chromatographic methods for separating and determining amino acids in the hydrolysates of proteins. It is also possible to use ion exchangers to recover small quantities of ionic substances from large volumes of otherwise nonionic solutions. This method is used in the recovery of aneurin from waste products of fermentation processes, for the removal of toxic or radioactive ions from waste and of small ions from solutions of macromolecules, a process known as 'desalting'. Ion exchange has a number of applications in pharmaceutical analysis, e.g. the purification and removal of alkaloids for assay and the removal of carbonate and silicate ions before hydroxide determinations.

Gas-Liquid and Liquid-Liquid Interfaces

These interfaces are most important in pharmacy and biology and are discussed under various headings in the remainder of this chapter. The most important gas–liquid interface is the air–liquid interface.

Surface Tension and Interfacial Tension

In the discussion of the liquid state in Chapter 2 it was mentioned that a characteristic property of all liquid surfaces is the surface tension, γ. This is the force acting along the liquid surface at right angles to any line of unit length on the surface, and is given by

$$\gamma = F/l, \qquad [11.6]$$

where F is the force acting at right angles to a line of length, l. The SI unit of surface tension is $N\,m^{-1}\,(=\,kg\,s^{-2})$ and the c.g.s. unit is dyne $cm^{-1}\,(=\,g\,s^{-2})$. Since 1 dyne $= 10^{-5}\,N$ and 1 cm $= 10^{-2}\,m$, 1 dyne $cm^{-1} = 10^{-3}\,N\,m^{-1} = 1\,mN\,m^{-1}$.

The surface free energy of a liquid surface is the work which must be done on the surface isothermally and reversibly to increase its area by unity, usually at constant pressure. In other words the surface free energy is equal to $w'_{rev}/\Delta A$ where w'_{rev} is the reversible work done and ΔA is the change in area. Since work is defined as the force, F, multipled by the distance, d, moved in the direction of the force, $w'_{rev} = Fd$. But $\Delta A = ld$, where l is defined above. The surface free energy is therefore equal to Fd/ld or F/l, which according to equation [11.6] is equal to γ, the surface tension. Thus, the surface free energy and surface tension of a given liquid surface are identical quantities and are usually represented by the same symbol γ. The SI unit of surface free energy is J m^{-2}, which is identical with N m^{-1} used for surface tension, since 1 J = 1 Nm. The c.g.s. unit of surface free energy is 1 erg cm^{-2} (= 1 dyne cm^{-1}).

Since the surface free energy, γ, is given by $w'_{rev}/\Delta A$ and since, according to equation [7.65], $\Delta G = w'_{rev}$, we can write

$$\gamma = \frac{\Delta G}{\Delta A} \quad \text{or} \quad \Delta G = \gamma \Delta A. \quad [11.7]$$

This equation only applies when γ is constant for a change in area ΔA. In general

$$\gamma = \left(\frac{\partial G}{\partial A}\right)_{T,p} \quad [11.7a]$$

The Gibbs energy due to the presence of a surface of area A is given by the analogous equation

$$G = \gamma A. \quad [11.8]$$

Just as a liquid surface has a surface tension (or surface free energy), so an interface or boundary layer between two phases possesses an interfacial tension (or interfacial free energy). The interface may be liquid–liquid, solid–liquid or gas–liquid. Since surface tension can be regarded as the interfacial tension of an air–liquid interface, interfacial tension is a more general concept than surface tension. The interfacial tension at the interface between two liquids, which have negligible mutual solubility usually has a value between the surface tensions of the individual liquids. On the other hand, the interfacial tension between two partially miscible liquids usually approximates to the difference in their surface tensions. The more similar are the intermolecular forces in the two liquids, the closer are the surface tensions. Surface tensions of liquids increase with increasing intermolecular forces and are greater for polar liquids than for non-polar liquids. At 20° C $\gamma/(\text{mN m}^{-1})$ is 72.8 for water, 22.3 for ethanol and 17.1 for diethyl ether. The surface tension of a liquid usually decreases with increasing temperature almost linearly and is very small near the critical temperature.

Surface tensions are commonly determined by the capillary rise method, by detachment methods using a Wilhelmy plate or a ring or by the drop weight or drop volume method. Liquid–liquid interfacial tensions can be determined by all these procedures with the exception of the capillary rise method. In determining surface tensions by this method the height, h, is measured by which the meniscus of the liquid rises in the capillary tube above the level of the liquid (Fig. 11.7) as a

Fig. 11.7 Enlarged view of a capillary rise

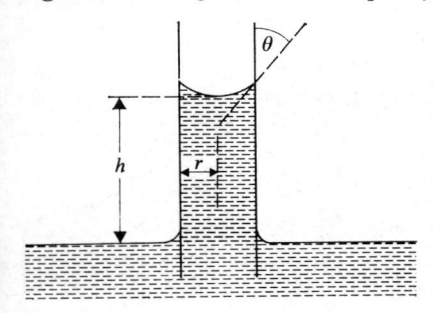

result of the upward pull due to surface tension. If the internal radius of the capillary is r, l in equation [11.6] is equal to the circumference, $2\pi r$, and

$$F = 2\pi r . \gamma. \qquad [11.9]$$

This force acts along the tangent to the meniscus at the circumference of the capillary tube. The angle between this tangent and the vertical is known as the contact angle, θ (see Fig. 11.7). The vertical upward component of the force F has the value $F \cos \theta$ and is balanced by the downward force, w, corresponding to the weight of the liquid raised, i.e.

$$F \cos \theta = w. \qquad [11.10]$$

Now
$$w = \pi r^2 h . \varrho . g, \qquad [11.11]$$

where $\pi r^2 h$ is the volume of the cylinder of liquid raised by the upward force, ϱ is the density of the liquid (so that $\pi r^2 h . \varrho$ is the mass of the liquid cylinder) and g is the acceleration due to gravity (9.81 m s^{-2}). Introducing the expressions for w (equation [11.11]) and F (equation [11.9]) into equation [11.10], we have

$$2\pi r \gamma \cos \theta = \pi r^2 h \varrho g.$$

Dividing both sides by $2\pi r \cos \theta$, we have the following general expression for the surface tension:

$$\gamma = \frac{h \varrho g r}{2 \cos \theta}. \qquad [11.12]$$

For aqueous solutions and many other liquids in a clean glass capillary,

Fig. 11.8 Plate detachment from a liquid surface

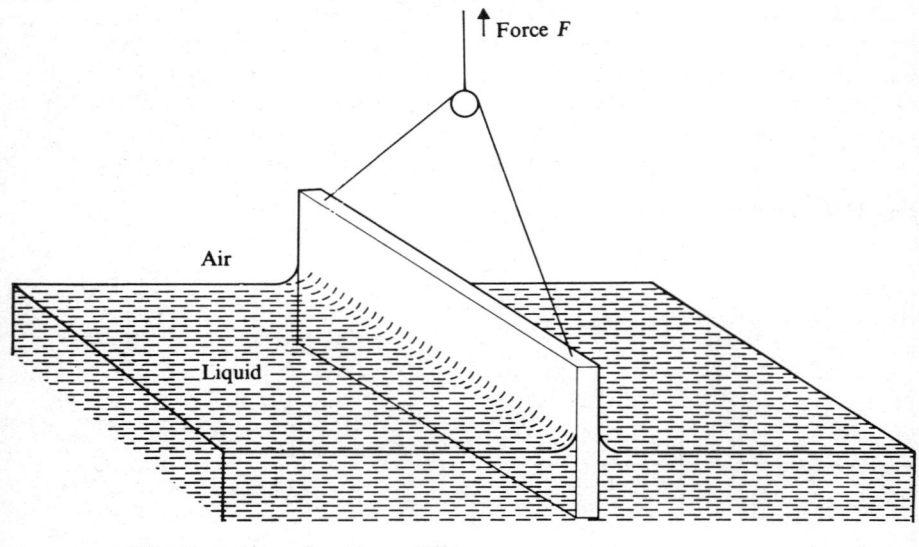

$\theta = 0$ and $\cos \theta = 1$, so that

$$\gamma = \tfrac{1}{2} h \varrho g r. \qquad [11.13]$$

For accurate determinations the capillary rise, h, is measured using a travelling microscope (or cathetometer) and the radius, r, at the meniscus must be very accurately known. For dilute solutions ϱ is close to 1000 kg m^{-3} at normal temperatures.

In the Wilhelmy plate method the minimum force, F, is measured which will just pull a plate of mica or glass out of the interface in question (Fig. 11.8). The plate is invariably rectangular so that its perimeter, l, is equal to $2 \times$ (length+thickness). The surface (or interfacial) tension is now given by

$$\gamma = F/2 \text{ (length+thickness)}. \qquad [11.14]$$

In the ring detachment method the minimum force, F, is measured which will just pull a platinum ring of circumference L out of the interface in question (Fig. 11.9). The surface (or interfacial) tension is now given by

$$\gamma = \frac{\beta F}{2L}. \qquad [11.15]$$

The divisor 2 allows for the fact that both the internal and external perimeter of the ring are detached from the interface. The factor β is to correct for the distortion of the liquid surface just before the ring is detached. The value of this correction factor depends on the dimensions of the ring and on the densities of the media on both sides

Fig. 11.9 Ring detachment from a liquid-liquid interface

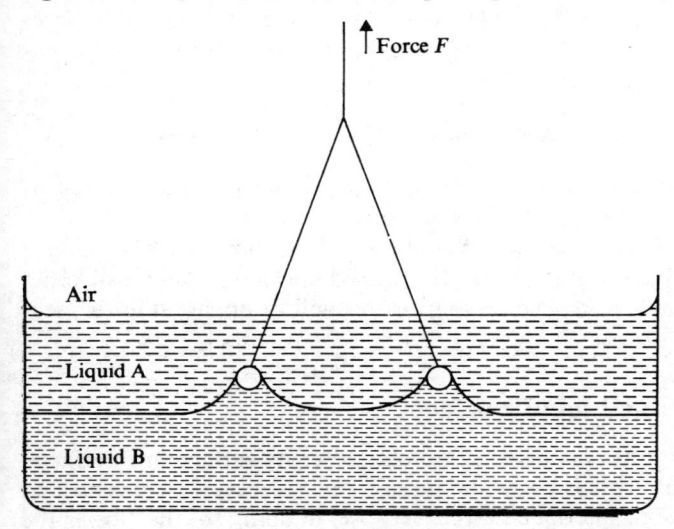

of the interface. β may be calculated or can be looked up in tables. The Wilhelmy plate or the ring must be wetted by the liquid in surface tension measurements or by the lower liquid in measurements of interfacial tension. The detachment force, F, is measured by increasing the upward pull very gradually until detachment is achieved. This is commonly carried out using a calibrated torsion balance in the Du Noüy tensiometer or may be done by hanging the plate or ring from an arm of a conventional mass balance. The clean dry plate or ring is first balanced below the surface of the liquid. The interface is moved downwards until the plate or ring is at the interface and wetted by the lower phase. The additional force or mass, m, required just to pull the plate or ring away from the interface gives F directly or as mg, where g is the acceleration due to gravity (9.81 m s^{-2}).

In the drop weight or drop volume method, the liquid whose surface tension is to be determined is allowed to drop very slowly from the tip of vertical narrow tube. The mass, m, or the volume, V, of one drop is determined by measurements on a convenient number of drops. The force, F, acting downwards at the point of detachment is given by

$$F = mg = V\varrho g.$$

The tip of the capillary tube of external radius r is either ground so that it is wetted completely or it may have a sharp edge. The surface tension of the liquid is then given by

$$\gamma = \frac{\phi F}{2\pi r},$$

where ϕ is a correction factor which depends on the ratio $r/V^{\frac{1}{3}}$. Satisfactory values of $r/V^{\frac{1}{3}}$ lie within the range 0.5 to 1.2, corresponding to values of ϕ between 1.54 and 1.67. Introducing the expression for F, we have:

$$\gamma = \frac{\phi mg}{2\pi r} = \frac{\phi V \varrho g}{2\pi r}. \qquad [11.16]$$

A micrometer syringe, i.e. a narrow plunger moved by a fine screw, may be used to control the flow of liquid down the tube.

All parts of the apparatus in contact with the liquids whose surface or interfacial tensions are to be determined must be scrupulously clean and free from grease. The reason for this will be apparent in the next section.

Adsorption at Liquid Surfaces

In general, solutes dissolved in a liquid alter its surface or interfacial tension. Those which lower the tension tend to become concentrated in the surface or interfacial layer, because, in doing so, they lower the Gibbs energy of the whole system. Surface active agents (surfactants or tensides) are substances which exhibit this effect even at low concentrations. Molecules of surface active agent always possess a polar *hydrophilic* group, hydrophilic meaning 'water-liking', and a non-poar *hydrophobic* or *lipophilic* group, so named because it is repelled by water and has an affinity for lipid or oil. Since both types of group are present, surface active agents are called amphiphilic agents, or *amphiphiles*, and their molecules tend to concentrate at or be adsorbed at the oil–water, air–water and air–oil interfaces (see Fig. 11.10). Each adsorbed molecule of surface active agent is so oriented that its polar hydrophilic head group, represented by a circle, is in the water layer and the non-polar lipophilic tail group, represented by a line, is in

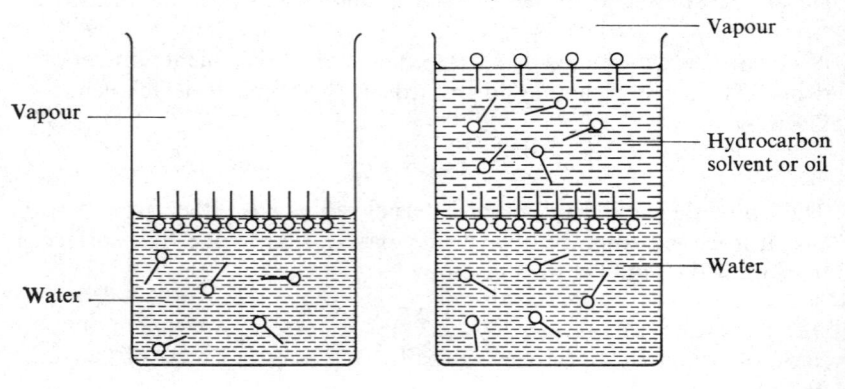

Fig. 11.10 Adsorption of molecules of surface active agent as a monomolecular layer at air–water, oil–water and air–oil interfaces

the oily layer. In practice lipophilic tail groups are usually hydrocarbon chains or occasionally fluorocarbon chains. The hydrophilic head group may be *anionic*, e.g. a carboxylate, sulphate or sulphonate group, *cationic*, e.g. a substituted ammonium group, or non-ionic, e.g. ester, hydroxy and polyoxyethylene groups, which are present in sorbitan esters and polyoxyethylene sorbitan esters.

Figure 11.10 should not be taken to imply that the molecules are static, but to represent an instant in time. The system is really in a state of dynamic equilibrium in which the solute molecules are continually undergoing exchange between each solution and the interface. Furthermore, after mixing, it may take time, perhaps several seconds, for equilibrium to be reached.

Fig. 11.11 An interface in discussion of the Gibbs adsorption isotherm

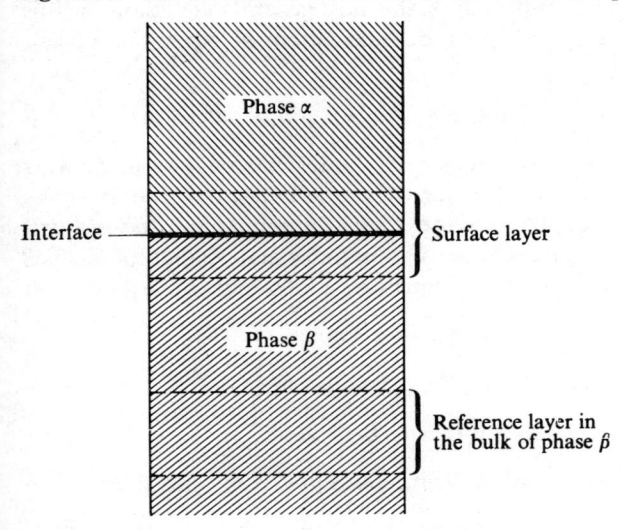

The relationship between the equilibrium amount of solute adsorbed at the interfacial layer and the extent to which the surface or interfacial tension is changed is given by the Gibbs adsorption isotherm which will now be derived. The interface between the two phases α and β is not sharp as indicated in Fig. 11.11 but has a composition which varies over a thickness of several molecules. Nevertheless it is convenient to recognise the quantities Γ_A and Γ_B, known as the surface excess concentrations of component A and component B respectively. These quantities are defined by:

$$\Gamma_A = \frac{n_A}{A} \quad \Gamma_B = \frac{n_B}{A}, \quad [11.17]$$

where A is the area of the interface and n_A and n_B are the amounts (mol) of A and B respectively in the surface layer in excess of that in the bulk of solution, which is here taken to be phase β. Since the interface is not sharp, the exact position of the surface layer is arbitrary and will be defined later. The Gibbs energy of the surface layer will be greater than that in the bulk of the solution by an amount G_σ. The excess amounts of the respective components, n_A and n_B, defined above and the interfacial tension, γ, both contribute to G_σ which is therefore given by adding together equations [7.81] and [11.8], thus:

$$G_\sigma = n_A \mu_A + n_B \mu_B + \gamma A. \qquad [11.18]$$

The chemical potentials, μ_A and μ_B, of the components are constant throughout all the phases and interfaces of the system, because the system is in equilibrium at constant temperature and pressure (see p. 310). If the system is made to undergo any infinitesimal change at constant temperature and pressure, the most general expression for this change is obtained by complete differentiation of equation [11.18], thus:

$$dG_\sigma = n_A\, d\mu_A + \mu_A\, dn_A + n_B\, d\mu_B + \mu_B\, dn_B + \gamma\, dA + A\, d\gamma.$$

If the interface only undergoes an infinitesimal change in area at constant temperature and pressure, the amounts of A and B at the interface will change and so the excess amounts n_A and n_B will undergo a corresponding change. The infinitesimal change in surface free energy is then given by (cf. equations [7.82] and [11.7a]):

$$dG_\sigma = \mu_A\, dn_A + \mu_B\, dn_B + \gamma\, dA.$$

Subtracting this equation from the previous one:

$$0 = n_A\, d\mu_A + n_B\, d\mu_B + A\, d\gamma.$$

Dividing throughout by A and subtracting $d\gamma$ from both sides:

$$-d\gamma = \frac{n_A}{A} d\mu_A + \frac{n_B}{A} d\mu_B$$

$$\therefore \qquad -d\gamma = \Gamma_A\, d\mu_A + \Gamma_B\, d\mu_B.$$

The position of the arbitrary surface layer is conveniently and sensibly defined such that the surface excess of the solvent, component A, is zero, i.e. $\Gamma_A = 0$.

$$\therefore \qquad -d\gamma = \Gamma_B\, d\mu_B. \qquad [11.19]$$

According to equation [7.89] $d\mu_B = RT\, d\ln a_B$, where a_B is the activity of the solute in the bulk phase. Introducing this expression into equation [11.19] and rearranging:

$$\Gamma_B = -\frac{1}{RT} \cdot \frac{d\gamma}{d\ln a_B} = -\frac{1}{2.303 RT} \cdot \frac{d\gamma}{d\log a_B}, \qquad [11.20]$$

(using the fact that $\ln x = 2.303 \log x$). Since $d \ln x = dx/x$

$$\Gamma_B = -\frac{a_B}{RT} \cdot \frac{d\gamma}{da_B}. \qquad [11.21]$$

The last two equations are exact forms of the Gibbs adsorption isotherm. For dilute solutions, $a_B = \text{constant} \times c_B$, where c_B is the concentration of B, so that $\ln a_B = \log \text{constant} + \ln c_B$, and $d \ln a_B = d \ln c_B$. Hence for dilute solutions a_B may be replaced by c_B expressed in any units. A useful form of the Gibbs adsorption isotherm is

$$\Gamma_B = -\frac{1}{2.303RT} \cdot \frac{d\gamma}{d \log c_B} \qquad [11.22]$$

where Γ_B is the surface excess of the solute B (SI unit mol m^{-2}), R is the gas constant (8.314 J K^{-1} mol^{-1}), T is the absolute temperature (in K), γ is the surface tension (SI unit N m^{-1}) and c_B is the concentration of B in the bulk of the solution (in any units).

Fig. 11.12 Relationship between the surface tension, γ, the concentration of solute, c_B, and the sign of the surface excess Γ_B

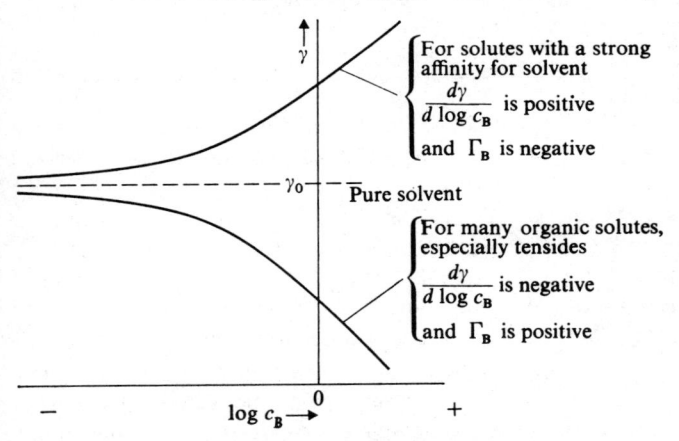

For solutes with a strong affinity for solvent
$\dfrac{d\gamma}{d \log c_B}$ is positive
and Γ_B is negative

Pure solvent

For many organic solutes, especially tensides
$\dfrac{d\gamma}{d \log c_B}$ is negative
and Γ_B is positive

In order to calculate Γ_B, measurements must be made of surface tension at a number of different concentrations of solute. A graph is then plotted of γ against $\log c_B$, as shown in Fig. 11.12, and the slope of the graph at any concentration is $d\gamma/d \log c_B$. The negative sign in the Gibbs adsorption isotherm implies that an excess of solute is present at the interface (Γ_B is positive) when the surface tension decreases with increasing concentration. Many systems behave in this way, expecially those containing surface active agents. Other examples are the surfaces of aqueous solutions of many organic solutes which do not have a strong affinity for the solvent, e.g. aqueous solutions of carboxylic acids, alcohols or phenols whose molecules have a high hydrocarbon content. By slicing surface layers about 0.1 mm thick off these solutions using a fast moving microtome blade and analysing the material collected, McBain and coworkers measured the surface excess directly. The values they obtained agreed well with those calculated as described above using the Gibbs adsorption isotherm. Surface concentrations may also be measured by labelling the solute with a radioactive isotope which emits β-particles and measuring the activity of the radiation immediately above the surface of the solution using a Geiger counter. Since β-particles generally have low penetrating powers, the radiation measured comes only from the surface layer and perhaps a thin layer of solution immediately below it.

Certain solutes which have a great affinity for the solvent, e.g. sugars or electrolytes in water, tend to leave the air–liquid interface and concentrate in the bulk of the solution. Such solutes are said to be negatively adsorbed. In these systems the surface excess of solute, Γ_B, is negative. This sign and the negative sign in the Gibbs adsorption isotherm indicates that the surface tension increases with increasing concentration of solute, as shown in Fig. 11.12. It can be seen that the surface tension, γ, of a solution approaches the value for the solvent, γ_0, as the concentration of the solute, c_B, tends or zero and therefore as log c_B tends towards $-\infty$.

Experiment 11.2 To show the effect of increasing number of CH_2 groups on the surface tension of aqueous solutions of alcohols and the determination of surface excess.

Solutions containing 0.2 mol dm^{-3} of straight chain aliphatic alcohols are prepared by making the following volumes up to 250 cm^3 with distilled water: methanol 2.02 cm^3; ethanol 2.92 cm^3; *n*-propanol 3.86 cm^3; *n*-butanol 4.58 cm^3; *n*-pentanol 5.43 cm^3. The *n*-pentanol must be pure; inferior samples containing appreciable quantities of isomers and higher homologues may not completely dissolve. The surface tension of each of these solutions is measured and plotted against the number of carbon atoms in the alcohol. The shape of the graph may be discussed.

The solution containing 0.2 mol dm^{-3} *n*-pentanol is diluted quantitatively with distilled water in 100 cm^3 graduated flasks, so as to give concentrations, c, equal to 0.15, 0.10, 0.07, 0.05, 0.032, 0.020, 0.014 and 0.010 mol dm^{-3}, respectively. The surface tension, γ, of all the solutions (0.01 to 0.2 mol dm^{-3}) of *n*-pentanol is determined and γ is plotted against log c. The maximum value of the surface excess, Γ_{max}, is calculated from the limiting value of the slope, dγ/d log c, at high values of c. The quantity Γ_{max} corresponds to closest approach of the alcohol molecules in the surface film. The minimum area, A_{min}, of each molecule in the surface film is L/Γ_{max}, where L is Avogadro's constant (6.022 × 10^{23} mol^{-1}). A_{min} is slightly greater than the actual area of the molecules because the molecules may be solvated and because there are still some solvent molecules in the surface layer at the highest concentrations used here.

The surface tensions of the solutions can be determined by any of the methods described above. The most convenient are the detachment methods, using the Du Noüy tensiometer, and the capillary rise method. The radius, r, of the capillary tube should be approximately 0.1 mm, determined by measurements on a liquid of known surface tension, e.g. distilled water, or by weighing a mercury thread of length, l, and of density 13 546 kg m^{-3} at 20° C. The formula used to calculate r is

$$\pi r^2 l = (\text{volume}) = \text{mass/density}.$$

Soluble and Insoluble Surface Films

When a soluble substance is adsorbed at the surface of the solvent, it may be said to form a surface film. In such cases the film is not generally complete, i.e. the whole of the surface does not consist of adsorbed molecules. Moreover, a particular molecule may spend part of the time in the bulk of the solution and part of the time in the surface layer. However, by application of the Gibbs adsorption isotherm it has been seen that the surface excess, and the average area of surface occupied per molecule, may be calculated.

Many substances which are insoluble in the bulk of a liquid, such as water, are nevertheless capable of spreading on its surface to form a surface film. For this to be possible, each molecule must be largely non-polar, to make the material insoluble, but it must also have a polar group which is attracted into the water surface. A completely non-polar substance would not spread on water, but would remain in the form of insoluble droplets. The insoluble surface films most commonly studied have been those of long-chain acids, alcohols or esters which have a non-polar paraffin chain and a polar head group. For such films the average surface area occupied per molecule is easy to calculate, if a known amount of the spreading material is placed on a known surface area. This is done by making a standard solution of the material in a volatile solvent, and by allowing a known small volume of this solution to spread on the surface. The volatile solvent soon evaporates, leaving the solute as an insoluble surface film containing a known number of molecules. Dividing this number into the total area of the film gives the average area occupied per molecule.

The spreading tendency of surface films may be regarded as being due to a surface pressure. This surface pressure may be measured, for an insoluble film, by confining it between barriers and a movable float and by measuring the pressure which it exerts on the float, as described below. However, another equally valid interpretation of the force exerted on the float is that it is due to the greater surface tension of the clean water surface beyond the float pulling more strongly than that of the surface bearing the film. Thus the surface pressure, Π (SI unit N m^{-1}), is given by

$$\Pi = \gamma_0 - \gamma, \qquad [11.23]$$

where γ_0 is the surface tension of the clean surface and γ is that of the surface bearing the film. The surface pressure may be determined by measuring γ_0 and γ using the Wilhelmy plate method for insoluble or soluble films, but is usually determined directly in the case of insoluble films. Surface pressure varies with the concentration of molecules in the surface, and this variation is usually illustrated by plotting graphs of surface pressure against the average surface area per molecule, A (SI unit m^2). For a fairly insoluble substance, such as lauric acid, the Π versus A curve may be obtained both directly from measurements of surface pressure and indirectly from surface tension measurements and the Gibbs adsorption isotherm. The validity of the Gibbs equation is confirmed by the fact that the results obtained by the two methods are in good agreement.

The Study of Insoluble Surface Films

Langmuir studied insoluble surface films spread on water in a shallow trough, and the basic features of his apparatus, as illustrated in Fig. 11.13, have been retained in similar apparatus designed by other workers.

Fig. 11.13 Langmuir trough for insoluble surface films

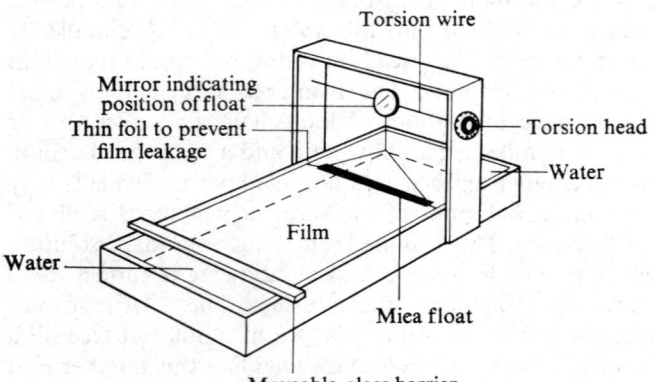

To prevent leakage of the film, the trough, moveable barrier and float must be hydrophobic. The trough and barrier may be made of polytetrafluoroethylene or alternatively of glass or metal coated with paraffin wax. The level of the liquid in the trough must be slightly above the brim.

The film is confined between a glass barrier and a mica float suspended from a torsion wire. The pressure of the film on the float causes a small rotation of the suspension, which is magnified by a beam of light reflecting from a mirror attached to the suspension. By rotation of the torsion head, the float and the reflected beam of light are brought back to their original positions. The amount of rotation of the head necessary to do this is a measure of the pressure on the float. (The torsion wire is previously calibrated by placing weights on a balance pan attached to the suspension but not shown in the diagram.) The total surface area of the film is known from the distance between the glass barrier and the float. By moving the glass barrier, the surface pressure Π can be measured at a number of a surface areas per molecule A, and the Π versus A curve can be plotted for the material of the film.

Types of Insoluble Surface Films

The results of studies of insoluble surface films have revealed important analogies between three-dimensional solids, liquids and gases, and two-dimensional surface films of molecules. The types of film have been labelled 'condensed', 'liquid expanded' and 'gaseous', to emphasise these analogies.

(a) *Gaseous or vapour films*

At very low surface pressures and large surface areas, the molecules of the film move about independently in the surface, and so behave as a two-dimensional gas (cf. the discussion of the gaseous state, p. 22). The Π versus A curves are found to approximate to the equation

$$\Pi A = kT, \qquad [11.24]$$

corresponding to the ideal gas equation for 1 mole, $PV = RT$. The former equation involves k, the gas constant per molecule, known as the Boltzmann constant, since A is the area of surface occupied per molecule. ($k = 1.381 \times 10^{-23}$ J K^{-1}). Moreover, the deviations from the equation $\Pi A = kT$ are similar to the deviations of the behaviour of gases from the ideal gas equation. The liquid expanded films discussed later represent extreme deviations from ideal behaviour. The ideal equation for gaseous films is only obeyed if the lateral forces between the molecules are negligible and the area of the molecules themselves is negligible compared with the area available to them for motion. When cetyltrimethylammonium bromide is spread on water it gives gaseous films because the cetyltrimethylammonium ions repel each other causing Π and A to be relatively large.

An interesting application of measurements of surface pressures of gaseous films is in the determination of the molecular weights, M_r, of protein molecules which form such films. A known mass of protein is spread on a known area, so the surface mass concentration κ (SI unit kg m^{-2}) is known. If the molar mass of the protein is $M_r/1000$ (SI unit kg mol^{-1}), the surface concentration is $1000\ \kappa/M_r$ (SI unit mol m^{-2}). The surface area per molecule (SI unit m^2 mol^{-1}) is given by

$$A = M_r/1000\ \kappa L, \qquad [11.25]$$

where L is Avogadro's constant (6.022×10^{23} mol^{-1}). The deviations of real gaseous films from ideal behaviour shown by equation [11.24] become smaller the further the molecules are apart, i.e. as κ is reduced. Eliminating A from equation [11.24] and [11.25] we have:

$$\frac{\Pi M_r}{1000 L \kappa} = kT \quad \text{as} \quad \kappa \to 0$$

Multiplying both sides by $1000\ L/M_r$ and replacing kL by the gas constant, R, we obtain

$$\left(\frac{\Pi}{\kappa}\right)_{\kappa \to 0} = \frac{1000\ RT}{M_r}. \qquad [11.26]$$

To eliminate the effects of non-ideality, surface pressures, Π, are measured over a wide range of surface mass concentrations, κ, and Π/κ is extrapolated to zero concentration (cf. p. 245). This extrapolated value is equal to $1000\ RT/M_r$, from which the molecular weight is calculated. If is found that although proteins are denatured on forming surface films, the molecular weights are not generally altered. (The denaturing usually involves only a change of structure and not decomposition of the protein.)

(b) *Condensed films*

A film becomes condensed when the molecules are closely packed together and almost vertically oriented, e.g. straight chain fatty acids, above and including stearic acid. In this state the arrangement is almost completely ordered, as in a crystal lattice. Once this section of the curve is reached, the surface pressure increases rapidly for small decreases in surface area, because the molecules are hardly compressible at all. In fact, if the surface area is reduced much at this stage, the film ceases to be a monomolecular layer of molecules and breaks up, with layers of molecules sliding over one another.

Two types of Π versus A curve for condensed films may be obtained. Both are illustrated in Fig. 11.14. The simpler type, which is given by stearic acid on water, has one linear portion BCD extending down to low surface pressures. When this linear portion is extrapolated to zero surface pressure, the area A gives the cross-sectional area of the molecule of the film under conditions of zero compression, i.e. when the molecules are just touching in a closely packed arrangement but

Fig. 11.14 Surface pressure versus area curves for condensed films (e.g. stearic acid on water, BCD, or on dilute hydrochloric acid, BCE), and for an expanded film of the oleic acid type (FG)

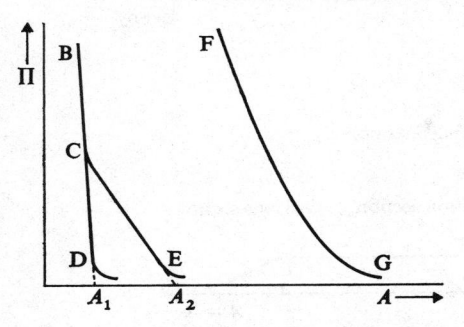

are not deformed by being squashed together. This cross-sectional area, amounting to 0.205 nm², is very similar to the corresponding value measured by X-ray diffraction of the material in the crystalline form, 0.185 nm². The second type of curve, which results when stearic acid is spread on dilute hydrochloric acid, has two linear portions BC and CE which extrapolate to areas A_1 and A_2 respectively at zero surface pressure. The area A_2 is thought to corresponds to incomplete close packing of the polar head groups of the molecules. On compression the molecules show normal behaviour along CB, owing to more efficient packing perhaps resulting from staggering of the head groups and interlocking of the alkyl chains.

(c) *Expanded films*

Introduction of a hydrophilic double bond into the stearic acid molecule to form oleic acid reduces the hydrophobic attraction between the vertical hydrocarbon chains at small areas and causes the molecules to lie flat on the aqueous surface at large areas. Consequently, Π for oleic acid is always greater than Π for stearic acid at any given value of A as shown by Fig. 11.14. Since stearic acid forms a condensed film, oleic acid is said to form an expanded film.

At higher surface pressures and smaller surface areas than are necessary to produce the gaseous or vapour films, the Π versus A curves for many substances, e.g. myristic acid spread on dilute hydrochloric acid, have the form shown in Fig. 11.15. Over the section marked 'liquid-expanded' the curve follows an equation of the form:

$$(\Pi - \Pi_0)(A - A_0) = kT. \qquad [11.27]$$

This form of equation is consistent with the polar head groups moving independently in the water surface with not much attraction between them while the hydrocarbon chains are more strongly interacting, forming a structure which is like a liquid.

Fig. 11.15 Surface pressure versus area curve for a liquid-expanded film

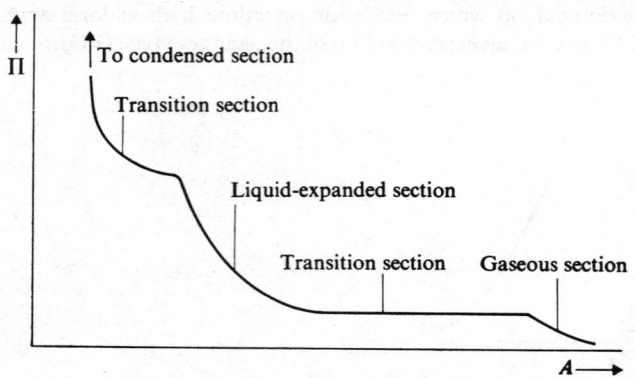

[Figure 11.15: Graph of Π versus A showing curve with labels: To condensed section, Transition section, Liquid-expanded section, Transition section, Gaseous section]

The above equation for liquid–expanded films resembles the van der Waals equation of state for real gases and liquids (p. 27). Films of long-chain esters and myristic acid on hydrochloric acid (Fig. 11.15) give Π versus A curves bearing a strong resemblance to Andrew's isothermals for carbon dioxide (see Fig. 2.3, p. 27), with horizontal portions analogous to the saturated vapour portions of the isothermals. Over the horizontal regions of the curves the surface films can be shown to consist of islands of condensed or liquid-expanded film with regions in between in which the molecules are moving about independently as in a gaseous film. This corresponds to the co-existence of liquid and vapour over the horizontal portions of the Andrews' isothermals.

Reactions in Surface Films

Information about the nature of surface films has been of value in interpreting the behaviour of emulsions, which are discussed in Chapter 12. We have already seen that the study of surface films of complex molecules may also be of assistance in determining their molecular weights and in obtaining some information about their structures. However, there is another application which is worthy of further mention here, namely, the study of reactions taking place in surface films.

Any reaction which takes place between materials in two different phases must of necessity be a surface reaction, and is therefore most appropriately studied by surface film techniques. An example is the hydrolysis of fats by aqueous solutions of alkali. The fatty material must either be in the form of a solid, or a liquid phase immiscible with the aqueous phase. If a fatty substance in the form of a long-chain ester is spread as a monomolecular film on the aqueous alkali as substrate, it is found that the rate of the reaction depends on the state of compression of the film. Thus, ethyl palmitate is hydrolysed about eight times more rapidly when the film is in the expanded state, in which each molecule occupies about 0.8 nm², than when the film is condensed, with an area per molecule of about 0.2 nm². It has been found possible to correlate the reaction rate with the orientation of the ester head group. In the compressed film, the ethyl group is forced below the other atoms of the carboethoxy group and so protects the latter from attack by hydroxyl ions. Another example is the oxidation of oleic acid when spread as a film on an aqueous solution of acidified permanganate. It was mentioned above that oleic acid forms an expanded film. The rate of oxidation of the double bond is found to decrease as the film is compressed. This suggests that at large areas the molecules are lying flat with the double bond close to the liquid surface and that at small areas the double bond is further away from the surface and the molecules are nearly vertical.

Problems

1 At 77 K, unit mass of a microcrystalline sample of sulphadiazine adsorbs the following volumes, v, of nitrogen at the following ratios of pressure p to saturated vapour pressure $p^\ominus$:

$p/p^\ominus$	0.079	0.184	0.281	0.377	0.597	0.807	0.912	0.978
$v/\text{cm}^3\,\text{g}^{-1}$	3.33	3.98	4.56	5.26	7.02	10.2	14.6	21.2

Plot the adsorption isotherm (v versus $p/p^\ominus$) and state the type to which it corresponds in the Brunauer classification.

2 Calculate $\dfrac{p}{v(p^\ominus - p)}\left[= \dfrac{1}{v(p^\ominus/p - 1)}\right]$ for the first four pairs of values given in the previous problem. By plotting the results against $p/p^\ominus$ verify that these data obey the BET adsorption isotherm and calculate v_m and c from the slope and intercept. Comment on the value of c. From the value of v_m calculate the specific surface area of the sample of sulphadiazine, assuming that each adsorbed nitrogen molecule occupies an area of 0.162 nm². Hence calculate the mean particle diameter for the sample given that its bulk density is 1500 kg m⁻³.

3 Identical samples of charcoal each of mass 2.00 g were shaken with 100 cm³ of aqueous solutions of acetone (CH_3COCH_3) until equilibrium had been attained. The concentration, c, of acetone in each solution at equilibrium and the corresponding concentration, d, before addition of the charcoal were as follows:

$d/\text{mol m}^{-3}$	5.993	10.584	19.98	49.93	200.4
$c/\text{mol m}^{-3}$	4.700	8.548	16.62	43.52	183.2

Calculate the mass of acetone adsorbed per unit mass of charcoal (x/m) in each case and plot the composite (or apparent) adsorption isotherm of the solute from solution. Plot the fraction of solute adsorbed, $(d-c)/d$, against d and comment on the form of the graph obtained.

4   Verify that the data in the previous problem conform more closely to the Freundlich isotherm, $x/m = kc^{1/n}$, than to the Langmuir isotherm, $\dfrac{x}{m} = \dfrac{k'kc}{1+kc}$, by making plots of suitable functions of x/m and c which should be linear for the respective types of isotherm. Calculate the constants k and n for the Freundlich isotherm.

5   The surface tension γ, at 293 K, of n-heptyl alcohol solutions of concentrations c, in water, are shown below:

c/per cent (w/v)	0.006	0.017	0.028	0.058	0.069	0.086
γ/N m^{-1}	0.068	0.062	0.057	0.048	0.046	0.042

Plot a graph of γ versus $\log c$ and determine the surface excess of n-heptyl alcohol, and hence the area per molecule, at the concentration 0.025 per cent, (w/v).

6   A dibasic ester giving a 'gaseous' insoluble surface film exerted, at 293 K, the following surface pressures Π at the values of surface area per molecule A quoted:

Π/mN m^{-1}	0.2	0.4	0.6	0.8	1.0	1.2	1.4
$A/10^{-20}$ m^2	1850	834	530	388	323	283	253
Π/mN m^{-1}	1.6	1.8	2.0	4.0	6.0	8.0	
$A/10^{-20}$ m^2	231	215	199	140	120	107	

Plot graphs of Π versus A and ΠA versus Π, and compare the latter with Fig. 2.2, (p. 26), showing pV plotted versus p for gases. Indicate on the second graph the line which would be obtained if the 'gaseous' film obeyed the ideal equation $\Pi A = kT$, where $k = 1.381 \times 10^{-23}$ J K^{-1}.

7   An insoluble film of a long-chain saturated fatty acid spread on dilute HCl gave, at 293 K, the following surface pressures Π at the areas per molecule A indicated:

Π/mN m^{-1}	32	29	26	23	20	18.5	17.5	16	14.5
$A/10^{-20}$ m^2	20.2	20.3	20.4	20.5	20.6	20.7	20.8	21.1	21.4
Π/mN m^{-1}	13	11.5	10	8.5	7	5.5	4	3	2.5	2.2
$A/10^{-20}$ m^2	21.7	22.1	22.4	22.8	23.1	23.5	23.9	24.3	24.7	25.1

Plot the curve of Π versus A, and extrapolate the two linear portions to zero compression to obtain the molecular cross-sectional areas corresponding to the two states of close-packing of the molecules.

12   Colloids

The Nature of Colloidal Particles

The term 'colloid' (derived from the Greek word for glue) was devised by Graham in 1861 to describe substances which diffuse slowly in solution through animal membranes (e.g. gelatin, starch, albumin), as distinct from crystalloids – the easily crystallisable substances – which diffuse rapidly in solution. It was later realised that the term should be used to describe a state of aggregation of matter rather than a particular class of substances, since many crystalline materials can be obtained in a colloidal form under appropriate conditions.

A system is now said to be in the colloidal state when one or more of its components have one or more dimensions within the range of about 1 nm to 1 μm. These components are known as colloids and may often exist in particulate form. A colloidal solution may be regarded as being intermediate between a solution of small molecules and a coarse suspension or emulsion.

Colloidal particles are usually too small to be seen under a normal optical microscope when the sample is illuminated along the axis of the microscope, a procedure known as 'bright field illumination'. Particles of a coarse suspension, e.g. small crystals or bacteria, would then appear as dark specks on a bright background. Colloid particles are, however, usually large enough to scatter light provided that the refractive indices of the particles and the medium differ sufficiently. This scattering of light is known as 'Rayleigh scattering' or as the 'Tyndall effect' and enables the presence of the particles to be detected by means of an optical microscope when the sample is illuminated at right-angles to the axis. This method is known as 'dark field illumination', because each particle gives a bright speck, much larger than itself, superimposed on a dark background. The method may also be called the 'ultramicroscope' technique. It enables the movement of the particles to be observed and their number in a given volume to be counted, even though the particles themselves are not resolved. When viewed in this way, the particles are found to be undergoing a ceaseless, hap-hazard, zig-zag motion. Such movement was first observed by the botanist Brown with pollen grains in water and is therefore known as Brownian movement (or motion). It is due to the unequal and random bombardment of the particles of small inertia by molecules of the liquid.

The mean Brownian displacement, $\bar{x}$, of a particle in a given time interval, t, is related to the diffusion coefficient, D, by equation [12.1] which is due to Einstein and may be derived as follows. Let the displacement, $\bar{x}$, be perpendicular to a plane YZ of area A separating a

Fig. 12.1 Mean displacement, $\bar{x}$, in Brownian motion

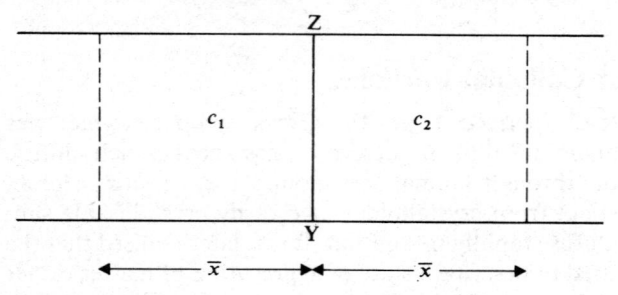

region of momentary higher concentration c_1 from one of lower concentration c_2 as shown in Fig. 12.1. Consider a volume defined by the length $\bar{x}$ and the area A on either side of the plane YZ. Half of the particles undergo net movement to the right and half undergo net movement to the left. Within the time interval t half the particles on the left will cross the plane YZ from the left to the right and half of those on the right will cross the plane from right to left. (Those particles outside the volumes to which c_1 and c_2 refer will not be able to reach the plane.) The net amount, n (mol) of particles displaced from left to right is equal to ($\frac{1}{2}c_1 \cdot$ volume $- \frac{1}{2}c_2 \cdot$ volume). Now the volume is equal to $A\bar{x}$, therefore

$$n = (c_1 - c_2) A\bar{x}/2.$$

Since in Brownian movement, $\bar{x}$, and t are small,

$$\frac{c_1 - c_2}{\bar{x}} = -\frac{dc}{dx}$$

and

$$\frac{n}{t} = \frac{dn}{dt}.$$

Eliminating n and $(c_1 - c_2)$ from the last three equations, we obtain:

$$\frac{dn}{dt} = -A \cdot \frac{\bar{x}^2}{2t} \cdot \frac{dc}{dx}.$$

Comparison with equation [6.58] for Fick's first law of diffusion shows that

$$\bar{x}^2/2t = D \quad \text{or} \quad \bar{x} = \sqrt{2Dt}. \qquad [12.1]$$

The diffusion coefficient, D, of the dispersed substance is related to the frictional coefficient, f, of the particles by equation [12.2] which is known as Einstein's law of diffusion and may be derived as follows. The frictional force, F, acting on a body is a function of its velocity dx/dt, and is given by

$$F = -f \cdot \frac{dx}{dt},$$

where f is the frictional coefficient. The negative sign allows for the fact that the frictional force opposes motion. The infinitesimal net work done in moving a particle through a distance dx in time dt against a frictional resistance is given by

$$dw' = F \cdot dx = -f \cdot \frac{dx}{dt} \cdot dx.$$

This is equal to the infinitesimal change in Gibbs energy of the particle at constant temperature and pressure. For one particle the Gibbs energy, G, is equal to the partial molar Gibbs energy, μ, divided by the Avogadro constant, L, i.e. $G = \mu/L$, whence $dw' = dG = d\mu/L$:

$$\therefore -f \cdot \frac{dx}{dt} \cdot dx = \frac{d\mu}{L}$$

Substituting $RT \, d \ln a$ for $d\mu$ (equation [7.89]) we have, at constant temperature and pressure:

$$-f \cdot \frac{dx}{dt} \cdot dx = \frac{RT}{L} \cdot d \ln a.$$

R/L is equal to k, the Boltzmann constant. If the solutions behave ideally a may be replaced by c expressed in any units, and $d \ln c = dc/c$ (see p. 12). The last equation becomes:

$$-f \cdot \frac{dx}{dt} \cdot dx = kT \cdot \frac{dc}{c}.$$

Multiplying both sides by $-c/f \, dx$, we have

$$c \cdot \frac{dx}{dt} = -\frac{kT}{f} \cdot \frac{dc}{dx}.$$

Equation [6.62] states that $c \cdot \frac{dx}{dt} = J_n$

whence $$J_n = -\frac{kT}{f} \cdot \frac{dc}{dx}.$$

Comparison with equation [6.56] for Fick's first law of diffusion indicates that

$$kT/f = D \quad \text{or} \quad Df = kT. \qquad [12.2]$$

If the particles are spherical, the frictional coefficient is related to the viscosity η of the medium and to the radius r of the particles by Stokes' law, thus:

$$f = 6\pi\eta r. \qquad [12.3]$$

Introducing this expression into equation [12.2]:

$$D = \frac{kT}{6\pi\eta r}. \qquad [12.4]$$

Introducing this expression into equation [12.1] we obtain the following equation for the Brownian displacement:

$$\bar{x} = \sqrt{\frac{kTt}{3\pi\eta r}}. \qquad [12.5]$$

Equations [12.1] to [12.5] are in frequent use in colloid chemistry. Equations [12.3] to [12.5] only apply to spherical particles.

Colloidal particles are sufficiently small to pass through ordinary filter papers capable of retaining coarse suspensions. Thus, normal filtration suffices to separate colloids from coarse suspensions. If the pores of the filter are very fine, they will not allow certain colloidal particles to pass. Such filters are known as ultrafilters or membrane filters and can separate the larger colloidal particles from solutions. Membrane filters are often made of regenerated cellulose materials and and are available in various ranges of pore diameter. Since the pore diameters are small, the flow rate is low, and is usually increased by applying suction or, preferably, pressure. The membrane is supported on a porous disc or grid. Care must be taken not to block the pores of membrane filters.

Colloids and small molecules (or ions) may usually be separated by the differences in their powers of diffusing through membranes which was mentioned above. This is a process known as *dialysis*. The material to be separated is placed in a 'tube' or thimble which consists of a membrane of a regenerated cellulose material, such as collodion, cellophane or Visking, and which is suspended in a beaker of water. Dialysis is speeded up by stirring or replacing the water in the beaker. The small molecules diffuse through the membrane into the external water, but the colloidal material remains within the membrane. Examples of the use of dialysis are the removal of silver nitrate from a silver iodide sol and the removal of loosely bound metal ions, e.g. Mg^{2+}, Mn^{2+}, from enzymes.

Classification of Colloidal Systems

A colloidal system, may be classified under one of the following headings: (*a*) a colloidal dispersion; (*b*) a solution of macromolecules; (*c*) an association colloid. An older classification divides colloids into

Classification of Colloidal Systems

lyophobic (or hydrophobic) and lyophilic (or hydrophilic) colloids (see p. 485).

Colloidal Dispersions

In a colloidal dispersion the colloidal particles are dispersed as a separate phase, known as the disperse (or dispersed) phase, in a continuous medium of different state or chemical nature, known as the dispersion medium. Table 12.1 shows how the nature of the dispersed phase and the dispersion medium govern the type of colloidal dispersion obtained. A definite surface or interface separates each colloidal particle from the dispersion medium. An essential property of all colloidal dispersions is the large ratio of surface area to volume (or mass) for the colloidal particles. If a piece of material is imagined to be successively subdivided, the ratio of surface area to volume (or mass) will increase rapidly as the particles become smaller. When the particles are of colloidal dimensions, their behaviour and that of the system as a whole will be governed predominatly by the properties of the surfaces or interfaces, such as adsorption, discussed in Chapter 11, and electrical effects, discussed later. In contrast the properties of small molecules in solution mainly depend upon 'chemical' forces of attraction and repulsion and electric charge, whereas the large particles in coarse suspensions or emulsions mainly exhibit macroscopic properties, such as density, melting and boiling points of the matter in bulk.

Table 12.1. Classification of colloidal dispersions in terms of the three states of matter

Dispersion medium	Gas	Liquid	Solid
Disperse phase			
Gas	(completely miscible)	foam, e.g. froth on beer	solid foam, e.g. bread, foam plastics
Liquid	liquid aerosol, e.g. mist, sprays	emulsion, e.g. milk	solid emulsion, e.g. pearl, opal
Solid	solid aerosol, e.g. smoke	colloidal suspension, e.g. sol, gel.	solid suspension, e.g. dyed plastics

Colloidal dispersions of solids in liquids are known as sols when they are capable of flowing and as gels when they are fairly rigid.

480 Colloids

The surface free energy of a colloidal dispersion is given by $G = \gamma A$ (equation [11.8]) where γ is the interfacial tension between the dispersed phase and the dispersion medium, and A is the total surface area of the dispersed phase. Since A is large and positive, and γ is positive, G is large and positive. In order to reduce their high surface free energy (and their large interfacial area), colloidal dispersions tend spontaneously to coagulate or separate, and are not easily reconstituted after phase separation. This thermodynamic instability is the most characteristic property of all colloidal dispersions.

Fig. 12.2 Stabilisation of colloidal dispersions by a surface active agent (denoted by ○—; cf. Fig. 11.10):
(a) oil in water emulsion or an aqueous suspension;
(b) water in oil emulsion or a suspension in oil

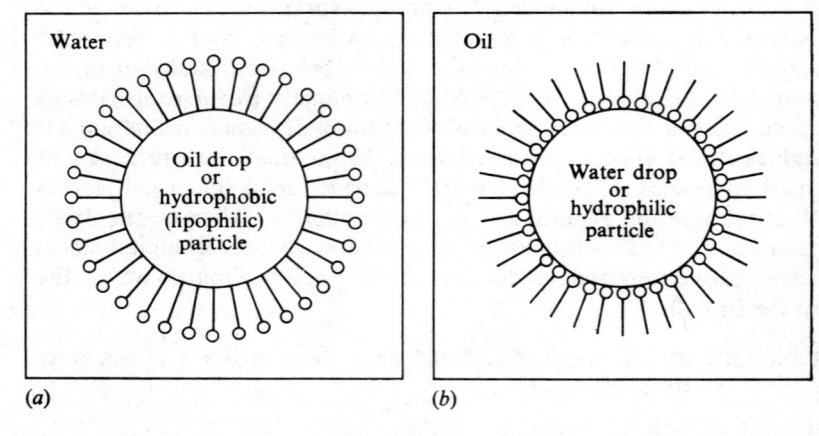

A suspension or an emulsion can be made thermodynamically less unstable by reducing the interfacial tension through the addition of a surface active agent, whereby the surface free energy is reduced. Molecules of surface active agent are positively adsorbed at the interface as discussed on p. 465. The polar head group is either in the aqueous phase or attaches itself to a hydrophilic particle, whereas the hydrocarbon tail is either in the lipid phase or attaches itself to a hydrophobic (or lipophilic) particle. Figure 12.2 illustrates these interactions in colloidal suspensions (sols), in oil in water (O/W) emulsions and in water in oil (W/O) emulsions.

Although colloidal dispersions are thermodynamically unstable, they may be kinetically quite stable, in the sense that coagulation may be slow. The Brownian movement of the colloidal particles, resulting from thermal bombardment by the dispersion medium, may be sufficient to keep them permanently in suspension. Furthermore, a high activation energy may intervene between the dispersed and coagulated states. Electrical forces of repulsion tend to keep the particles apart and are common causes of this activation energy (see p. 516). Sedimentation can usually be brought about by subjecting the particles to centrifugal forces many thousand times the force due to gravity, in an ultracentrifuge (see p. 497).

The distinction between the molecular and colloidal states on the one hand and between colloidal and coarse dispersions on the other hand, is not sharp and, as mentioned before, is merely one of particle size. Some particles may exist as single molecules or as aggregates of small numbers of molecules. At the other end of the size scale, colloidal particles of inorganic origin if allowed to grow, form first a suspension and then a microcrystalline precipitate. X-ray examination of the colloidal particles and the precipitate, in such cases, show that they both have the same structure.

Solutions of Macromolecules

Some macromolecular substances are able to dissolve in certain solvents forming true solutions of single macromolecules. These solutions will be colloidal if at least one molecular dimension is within the colloidal range (1 nm to 1 μm). Important examples are aqueous solutions of the natural macromolecules, namely proteins (e.g. albumin, gelatin, enzymes), polysaccharides (e.g. starch) and nucleic acids (DNA and RNA). Other examples are synthetic macromolecules in organic solvents, e.g. polystyrene in toluene or chloroform. Since these colloids are true solutions, like solutions of small molecules discussed in Chapter 6, they are thermodynamically stable and are easily reconstituted after separation of solute and solvent. Macromolecules, such as proteins and nucleic acids, which contain many ionisable groups are often known as 'polyelectrolytes'. The arrangement and folding of the macromolecular chains in the natural state is altered to a more random conformation by heating, by changes in pH or ionic strength, or by adsorption at an interface. The change is known as *denaturation*, and alters considerably the properties of the colloidal solution. The rate of denaturation is discussed on p. 171.

Association Colloids

Certain types of small molecule are capable of associating in solution as a result of their mutual intermolecular interactions to form clusters or aggregates of colloidal dimensions, known as association colloids or *micelles*. Some important substances which behave in this way are surface active agents (p. 462), certain dyes (e.g. methylene blue) and phospholipids (e.g. lecithin). If the substance is ionic, as is often the case, the association colloid will be ionised in water and may be termed a 'colloidal electrolyte'. The volume and mass of the micelles may be comparable with those of macromolecules. Association colloids, like solutions of both macromolecules and small molecules, are thermodynamically stable and are easily reconstituted after phase separation.

The existence of micelles was first proposed by McBain to explain the unusual physical properties of solutions of surface active agents, here exemplified in Fig. 12.3 by sodium dodecyl sulphate, $C_{12}H_{25}SO_4^-Na^+$. In very dilute solutions anionic and cationic surface active agents behave as normal electrolytes. Over a restricted range of concentration, whose mean value is known as the critical micelle

Fig. 12.3 Physical properties of aqueous solutions of sodium dodecyl sulphate at 25° C
(γ_{air} = surface tension; γ_l = lipid-solution interfacial tension; Λ = molar conductivity; π = osmotic pressure; $\Delta\varrho$ = increment in density; τ = turbidity; c.m.c. = critical micelle concentration)

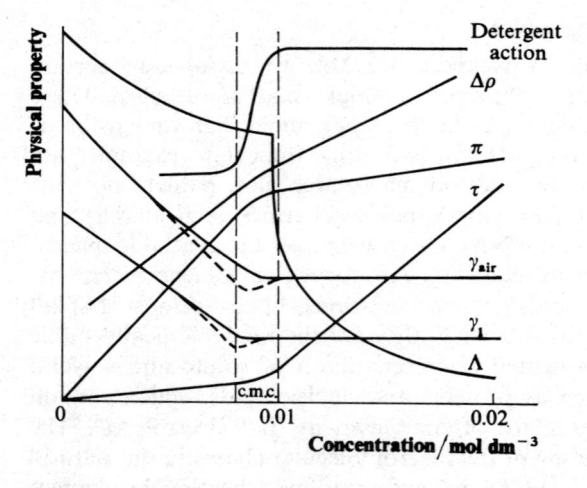

concentration (abbreviated to c.m.c), the trends in certain physical properties undergo abrupt changes. At concentrations above the c.m.c. the rate of increase in osmotic pressure becomes abnormally low which shows that the molecules are associating, whereas the molar conductivity is still appreciable which indicates that the molecules are still undergoing ionic dissociation. These apparently contradictory trends were attributed by McBain to the formation of the organised aggregates known as micelles and illustrated in Fig. 12.4.

The abruptness of the change in physical properties over a very narrow range of concentrations about the c.m.c. can be explained by applying the law of mass action to the equilibrium between unassociated molecules or ions (monomers, indicated by X) and micelles (indicated by X_n). If n is the number of monomers in each micelle, if α is the fraction of monomer units associated, i.e. the degree of association, and if c is the total concentration of surface active agent in solution.

Fig. 12.4 Cross section of a spherical micelle of an anionic surface active agent in an aqueous medium

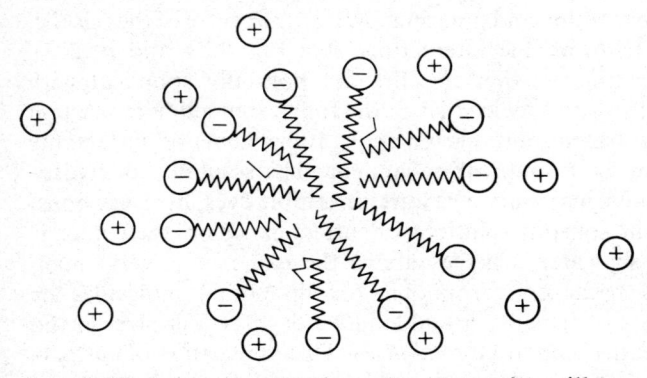

the equilibrium concentration of each species will be:

$$nX \rightleftharpoons X_n$$
$$c(1-\alpha) \quad c(\alpha/n).$$

The equilibrium constant is given by

$$K = \frac{c\alpha/n}{(c(1-\alpha))^n}. \qquad [12.6]$$

For any values of K and for fairly large values of n, as c is increased from zero, α remains very small up to a certain value of c (the c.m.c.) beyond which α rapidly increases. The larger the value of n the sharper will be the discontinuity. An infinite value of n gives a perfect discontinuity which is normally associated with a phase change. The larger the value of K the smaller will be the c.m.c. For sodium dodecyl sulphate in water at 25° C the c.m.c. is about 8 mmol dm^{-3}, and n is about 60.

The c.m.c. is the concentration above which micelle formation (or association, as measured by α) becomes appreciable. Therefore above the c.m.c. addition of more surfactant produces more micelles while the concentration of the monomer increases so slowly as to remain practically constant. Therefore at this stage the turbidity or light scattering increases, since it is produced mainly by the colloidal particles (p. 493), whereas the surface and interfacial tensions remain almost constant, since they are affected only by the monomer adsorbed at the interface (p. 462). As a result of the repulsions between water molecules and the hydrocarbon tail groups the monomer may be adsorbed at the interface or may associate to form micelles. Impurities in the surfactant may cause a minimum in the curves for surface and interfacial tension at the c.m.c. (broken curves in Fig. 12.3).

When micelles are formed in an aqueous medium the ionic or polar head groups are in contact with the medium, where they can remain ionised or become solvated, whereas the lipophilic (hydrocarbon) tail groups are close together and pointing towards the centre of the micelle, as a result of hydrophobic interactions (see Fig. 12.4 and p. 203). Physical studies suggest that micelles are normally spherical with diameters equal to about twice that of the fully extended hydrocarbon chain. In concentrated solutions (> about 10 per cent) of surfactants the micelles may be cylindrical or laminar. The solubility of hydrophobic or lipophilic materials, e.g. water insoluble dyes, hydrocarbons, lipids and dirt, in aqueous solutions of surfactant below the c.m.c., is similar to that in water, whereas above the c.m.c. it is very much higher. This is because hydrophobic (or lipophilic) molecules or particles, which are virtually insoluble in water, may 'dissolve' in the micelles and are then said to be *solubilised*. This application of surfactants is very important in pharmacy and is discussed further in connection with the stabilisation of emulsions (p. 524). Solubilisation only becomes appreciable when the number of micelles is large, that is, above the c.m.c. Solubilisation of dirt or, in other words, detergent action, also increases markedly as the concentration is increased through the c.m.c. (see Fig. 12.3). The hydrophobic interaction is primarily responsible for solubilisation.

The c.m.c. in water usually decreases with increasing hydrocarbon content of the surfactant molecule, with decreasing temperature and with addition of increasing concentrations of simple salts. Increasing ionic strength reduces the repulsive forces between the polar head groups of the monomer units within the micelle. Increasing hydrocarbon content of the surfactant molecule increases the aggregation number, n, of the micelles. Non-ionic and zwitter-ionic surfactants with no net charge have lower c.m.c. values and higher aggregation numbers. These facts suggest that micelle formation, like adsorption of surfactants at surfaces, proceeds because it reduces the interfacial free energy between the hydrocarbon tail group and the water molecules. The above facts also indicate that the formation of micelles is opposed by thermal agitation as well as by electrostatic repulsion between the head groups if charged.

The solubility of substances that form micelles is another property which changes anomolously, because micelles are much more soluble in water than the unassociated molecules. The observed solubility increases rapidly above a certain temperature, known as the *Krafft point*. The reason is as follows. Below the Krafft point the concentration of a saturated solution of unassociated surfactant is less than the c.m.c. With increasing temperature the solubility gradually increases until the c.m.c. is reached at the Krafft point. At higher temperatures appreciable micelle formation can occur so that very much more surfactant can dissolve.

Surfactants which are soluble in lipophilic solvents, e.g. hydrocarbons, form aggregates or micelles in which the hydrocarbon groups are in contact with the solvent and pointing outwards and the polar groups are together in the centre. The micellar structure is now essentially the reverse of that shown in Fig. 12.4 and can solubilise polar molecules which are themselves virtually insoluble in non-polar solvents. The structure and aggregation number of the micelles depend on the degree or polarity and relative permittivity (dielectric constant) of the solvent.

Lyophilic and Lyophobic Colloids

In classical colloid chemistry sols are classified, on the basis of the affinity of the particles for the dispersion medium, as *lyophilic* (solvent liking) or *lyophobic* (solvent hating). If the dispersion medium is aqueous, the terms *hydrophilic* and *hydrophobic* are often used. All solutions of macromolecules and association colloids are lyophilic because they are true solutions, which is a direct result of affinity for the solvent. All sols which are colloidal dispersions are virtually insoluble in the solvent and are classified as lyophobic. However, some colloidal dispersions, particularly aqueous metallic oxide sols, e.g. alumina, are easily wetted by water and have some affinity for it, so that they are strictly speaking lyophilic (or hydrophilic). Nevertheless such sols are in practice always classified as lyophobic (or hydrophobic).

Preparation of Colloidal Solutions

Since solutions of macromolecules and association colloids are true solutions and are thermodynamically stable, they are formed spontaneously on mixing the solute and the solvent and their preparation presents few problems. Examples include the preparation of aqueous solutions of proteins, dextrin, starch and surfactants, and the reconstitution of blood serum and plasma. The preparation of concentrated aqueous solutions of agar and gelatin requires heating and such solutions form gels on cooling. The preparation of lyophobic sols is less simple, since, like all colloidal dispersions, they are thermodynamically unstable.

Lyophobic sols may be produced either by breaking down larger particles, by what is known as a 'dispersion method', or by causing molecules or ions to aggregate, in what may be termed an 'aggregation method'. One of the most obvious of the dispersion methods is to grind the material very finely in a colloid mill. Alternatively, ultrasonic treatment may be used. These methods tend to produce particles having sizes which vary over a considerable range. The mechanical forces and the interparticle attractions tend to cause the smaller particles to coalesce. To reduce this, the liquid medium must wet the surface of the solid and any air on the surface and in the channels and interstices of the particles must be removed. These processes are greatly facilitated by addition of a surface active agent to the dispersion medium. A surface active agent may also stabilise the colloidal particles produced. In some cases, a precipitate may be converted into a sol by providing the particles with the ions necessary to give them stability. Thus, when freshly precipitated ferric hydroxide is treated, in suspension, with small quantities of hydrochloric acid, the ferric chloride produced provides the ions necessary to stabilise ferric hydroxide colloidal particles, and a sol is formed. This process is known as 'peptisation' by analogy with peptic digestion.

Fig. 12.5 The relationship between particle size and reagent concentration for the precipitation of a sparingly soluble material. The concentration scale depends on the solubility of the material. The example given above applies to barium sulphate

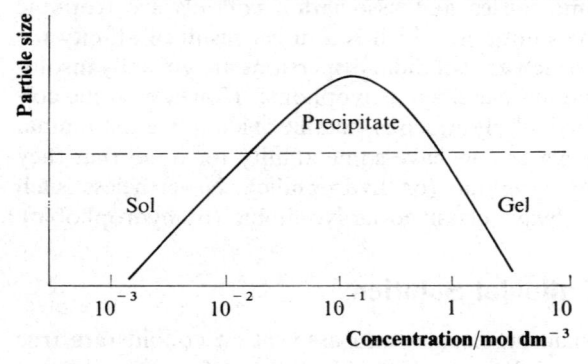

Most chemical or physical processes of aggregation which cause precipitation or crystallisation may be made to give colloidal particles by choosing the experimental conditions such that the initial rate of formation of centres of crystallisation, known as nuclei, is high while the rate of growth of the nuclei is low. Unless the conditions are very carefully controlled, so that all the nuclei are formed simultaneously and grow in step, a polydisperse system is formed.

Von Weimarn in 1908 investigated the effect of reagent concentration on the nature of the barium sulphate precipitates produced in ethanol-water mixtures (see Fig. 12.5) by reactions of the type:

$$Ba(CNS)_2 + SO_4^{2-} = BaSO_4 + 2CNS^-.$$

In order to obtain colloidal particles rather than a precipitate, it is necessary to produce rapidly considerably more of the material than would be required to saturate the dispersion medium. If c is the concentration of the material produced momentarily by chemical reaction, and s is its solubility in the dispersion medium, then the amount of supersaturation is $(c-s)$. The initial rate of formation of nuclei is proportional to the number of times this quantity is larger than the solubility, i.e. proportional to $(c-s)/s$. In order to obtain a high initial rate of formation of nuclei necessary for the formation of colloidal particles, $(c-s)/s$ should be very high, e.g. of the order 10^5. This may be achieved by arranging the conditions such that either s is small or c is large. If the latter method is adopted the total amount of particles formed will also be large and they will tend to link up to form a gel (see Fig. 12.5). For example, on mixing 1.5 mol dm^{-3} aqueous solutions of barium thiocyanate and manganous sulphate a gel of barium sulphate is formed. The best conditions for producing a sol are when s is very small, so that a high value of $(c-s)/s$ can be obtained with a relatively small value of c, and the total number of particles formed is not large (see Fig. 12.5). For example, the solubility of barium sulphate in water is lowered by the presence of ethanol and on mixing 0.01 mol dm^{-3} solutions of barium thiocyanate and cobaltous sulphate each in a 50 per cent (v/v) ethanol–water mixture, a fairly stable sol of barium sulphate is formed. A low rate of growth of the nuclei, which is required for the formation of colloidal particles, is aided by low reactant concentrations, by the presence of impurities which act as growth inhibitors and by a high viscosity of the medium. If c lies between the values appropriate for the formation of a sol and a gel, the nuclei grow more rapidly than for the formation of a sol but are produced less rapidly than for the formation of a gel. As a result a small number of comparatively large, often crystalline, particles are formed and a filterable precipitate results (see Fig. 12.5).

The above considerations also apply to the formation of colloidal particles by rapid precipitation resulting from a sudden reduction in solubility. Thus, if an alcoholic solution of benzoin is poured into a large volume of water, a faintly opalescent colloidal solution of benzoin is formed. This is because benzoin is much less soluble in water than in alcohol, and a high value of $(c-s)/s$ is therefore obtained.

The Size and Molecular Weight of Colloidal Particles

The term particle size is applied mainly to lyophobic sols and other colloidal dispersions, and may be expressed as the mass, volume or linear dimensions of the particles or, if the particles are spherical or approximately spherical, as the radius or diameter. The terms molecular weight (or relative molecular mass, M_r) and molar mass M are applied mainly to macromolecules in solution. We note that M/kg mol^{-1} = $M_r/1000$. For the micelles of association colloids, particle size, molar mass or aggregation number may be used.

If the particle size or molar mass is the same for all the particles, the system is called monodisperse. If these quantities differ from particle to particle, the system is called polydisperse. Methods of determining particle size and molar mass in polydisperse systems give average values, and different methods give different types of average. Since colligative properties (p. 236) depend only on the number of the particles, measurement of these properties lead to the so-called number average molar mass which is defined by:

$$\bar{M}_N = \frac{\Sigma N_i M_i}{\Sigma N_i}, \qquad [12.7]$$

where N_i is the number of particles of a given molar mass M_i. The magnitude of certain properties (symbolized by q_i, e.g. light scattering) is proportional to the size or molar mass of the particles as well as to their number, i.e.

$$q_i \propto M_i N_i.$$

When used to determine the molar mass such methods yield the mass average molar mass, which is defined by

$$\bar{M}_m = \frac{\Sigma q_i M_i}{\Sigma q_i} = \frac{\Sigma N_i M_i^2}{\Sigma N_i M_i}. \qquad [12.8]$$

In polydisperse systems the larger particles exert a greater effect on $\bar{M}_m$ than the smaller particles, whereas $\bar{M}_N$ is affected equally by all particles irrespective of size. Consequently, for all polydisperse systems the mass average molar mass is always greater than the number average molar mass, whereas for all monodisperse systems these values are equal. If equal numbers of molecules with $M = 20$ kg mol^{-1} and $M = 200$ kg mol^{-1} are present (e.g. N molecules of each),

then $\quad \bar{M}_N = \dfrac{(N \times 20)+(N \times 200)}{N+N} = 110$ kg mol^{-1}

and $\quad \bar{M}_m = \dfrac{(N \times 20^2)+(N \times 200^2)}{(N \times 20)+(N \times 200)} = \dfrac{40\,400}{220} = 183.6$ kg mol^{-1}.

The ratio $\bar{M}_m/\bar{M}_N$ increases with polydispersity and is unity for monodisperse systems only.

One simple method of size determination, which is applicable mainly to lyophobic colloids, is that of ultrafiltration. In this method, the dispersion medium is forced, by pressure, through specially prepared filters, which may be of unglazed porcelain with the pore size suitably reduced by impregnating with collodion. Filters of various pore sizes may be produced by altering the extent of the treatment with collodion, and the pore size may be determined by examining the rate at which water may be forced through the pores under pressure. When the sol is filtered under pressure in a number of such filters of different pore sizes, the filter which just retains the colloidal particles of the sol is known to have a pore size a little smaller than the size of the particles.

Another simple method, again applicable mainly to lyophobic colloids, is to count the actual number of particles in a known illuminated volume of the sol using the ultramicroscope technique. The total weight of disperse phase in a given volume of dispersion medium is then found by coagulating and separating the colloidal material. By comparing this with the average number of particles in a given volume, as counted in the ultramicroscope, the average weight of material in one particle may be found. If the density of the particles may then be assumed to be the same as that of the same material in bulk, the average volume of one particle may be found. Finally, from the average volume, the average radius may be calculated, if the particles are assumed to be spherical in shape.

Osmotic Pressure of Lyophilic Colloids

The only method of molecular weight determination by measurement of colligative properties which is applicable to colloidal solutions is the osmotic pressure method (see p. 243). The molecular weights to be determined are so large that in all the other colligative methods (by lowering of vapour pressure, elevation of boiling-point, or depression of freezing-point) the effect of the colloidal particles is too small to be measured. Such methods would be influenced, in any case, by the presence of ionic material which is difficult to remove completely from the colloidal solution. In the osmotic pressure method, however, the membrane can be chosen so as to be permeable to ions, and the removal of such impurity is not necessary. If the ionic material distributes itself equally on both sides of the membrane, it will exert an equal osmotic pressure on both sides and so will not affect the osmotic pressure due to the colloidal particles.

In practice, the ionic material does not distribute itself equally if the colloidal particles bear a resultant charge. Thus, if a protein is present as its potassium salt, and potassium chloride is also present as an impurity, the potassium and chloride ions can diffuse through the membranes but the negatively charged protein molecules are retained on one side. The Donnan membrane equilibrium is established in such cases and has already been discussed on p. 427.

The distribution of ions on both sides of the membrane can be obtained quantitatively as follows. Consider a membrane dividing a vessel into two parts of equal volume. Let there be aqueous solutions of initial concentrations c_1 mol dm^{-3} of potassium proteinate K$^+$P$^-$ and c_2 mol dm^{-3} of potassium chloride on the left hand side of the membrane and pure water on the right hand side before diffusion commences. When equilibrium has been reached there must still be c_1 mol dm^{-3} of P$^-$ on the left of the membrane because this species cannot diffuse. However, the potassium ion concentration will be reduced by an amount a mol dm^{-3} and the chloride ion concentration must also be reduced by a mol dm^{-3} in order to maintain electrical neutrality on each side of the membrane.

The concentrations on the two sides of the membrane will now be:

A	B
P$^-(c_1)$	—
K$^+(c_1+c_2-a)$	K$^+(a)$
Cl$^-(c_2-a)$	Cl$^-(a)$

Substituting these equilibrium concentrations into equation [10.46] on p. 427, we have:

$$1 = \frac{a^2}{(c_1+c_2-a)(c_2-a)}.$$

Multiplying both sides of this equation by the denominator of the fraction and, at the same time, partly multiplying out the latter, we obtain:

$$c_2(c_1+c_2-a) - ac_1 - ac_2 + a^2 = a^2.$$

Subtracting a^2 from both sides and adding ac_1 to both sides this becomes:

$$c_2(c_1+c_2-a) - ac_2 = ac_1$$
$$\therefore c_2(c_1+c_2-2a) = ac_1. \qquad [12.9]$$

Now, (c_1+c_2-2a) is the excess concentration of potassium ions in A over that in B, i.e. [K$^+$]$_A$ − [K$^+$]$_B$. This is therefore equal to ac_1/c_2. The simplest way of appreciating how this excess varies with c_1 and c_2 is to consider the ratio of the excess to a, the concentration of potassium ion in B. This is equal to c_1/c_2 so it is increased by increasing the protein concentration and decreased by increasing the concentration of potassium chloride originally in A. An extreme case of this is mentioned below.

The unequal distribution of ionic material will tend to interfere with the measurement of the osmotic pressure due to the colloid, but there are three ways in which the difficulty may be overcome. These can be illustrated by calculating the ratio of the osmotic pressure π which would be obtained in the presence of potassium chloride to the value π_0 which would be obtained with no potassium chloride present. For each ionic species the contribution to the osmotic pressure is cRT, (cf. p. 245), where c is the concentration in mol dm^{-3}. The total osmotic pressure in compartment A is therefore equal to

$$c_1RT + (c_1+c_2-a)RT + (c_2-a)RT.$$

This simplifies to $\qquad 2RT(c_1+c_2-a).$

The osmotic pressure in compartment B is similarly $2RTa$. The difference in osmotic pressure between A and B, which is what would be actually measured, is therefore

$$2RT(c_1+c_2-a)-2RTa, \quad \text{or} \quad 2RT(c_1+c_2-2a). \qquad [12.10]$$

Substituting ac_1/c_2 for the value of (c_1+c_2-2a), as found above (equation [12.9]), we find that the measured osmotic pressure π would be given by

$$\pi = 2RTac_1/c_2. \qquad [12.11]$$

To find a in terms of c_1 and c_2 we start from equation [12.9]. Dividing both sides by c_2 and adding $2a$ to both sides, this equation becomes:

$$c_1+c_2 = 2a+ac_1/c_2$$
$$\text{or} \quad c_1+c_2 = a(2+c_1/c_2).$$

Dividing both sides by the bracket on the right, we obtain

$$a = \frac{c_1+c_2}{2+c_1/c_2}. \qquad [12.12]$$

The value of π or $2RTac_1/c_2$ is therefore given by

$$\pi = 2RT \times \frac{c_1}{c_2} \times \frac{c_1+c_2}{2+c_1/c_2}$$

$$\therefore \pi = \frac{2RTc_1(c_1+c_2)}{2c_2+c_1}. \qquad [12.13]$$

Now, the value of π_0 in the absence of potassium chloride is c_1RT (for P^-)$+c_1RT$ (for K^+).

Hence
$$\pi_0 = 2c_1RT. \qquad [12.14]$$

Therefore
$$\frac{\pi}{\pi_0} = \frac{c_1+c_2}{2c_2+c_1}. \qquad [12.15]$$

One way of overcoming the difficulty of the unequal distribution of ionic material in measuring osmotic pressures of colloids is to make allowance for the effect by calculation. In the example above one would calculate π_0 from a measured value π and known concentrations c_1 and c_2.

Another method is to minimise the effect by making measurements on the colloid at its isoelectric point if it is stable under these conditions. Since it has no net charge at the isoelectric point the colloid has very little influence on the distribution of ionic material.

The most usual method, however, is to swamp the effect of the colloid by making the measurements in the presence of excess ionic material

This possibility follows from the discussion above on the magnitude of the excess of potassium ions. In the extreme, when c_2 is much greater than c_1, the ratio π/π_0 becomes approximately equal to $c_2/2c_2$, i.e. $\frac{1}{2}$. The observed osmotic pressure π is therefore one-half of that for the dissociated colloid K^+P^- or the same as that for the undissociated colloid.

Measurements of osmotic pressures of solutions of macromolecules require special techniques, because the pressures are low and equilibrium is only attained slowly. A simple method is described in Chapter 6 on p. 244. The semipermeable membrane is commonly made of regenerated cellulose materials, but these are often permeable to molecules of molecular weight, M_r, below about 10^4. The effects of Donnan membrane equilibria must, of course, be allowed for or eliminated. Macromolecules tend to interact with and immobilise solvent molecules, usually water, thereby increasing the observed value of osmotic pressure above the ideal value. Correction for non-ideality is achieved by extrapolation to zero concentration as described on p. 245. The observed osmotic pressure usually becomes too small for accurate measurements, if the molecular weight, M_r, is greater than about 10^6. For larger macromolecules the light scattering method is appropriate.

Light Scattering by Colloidal Particles

When light impinges on a material, some may be absorbed, if its molecules contain appropriate chromophores, some may be scattered, and some may be transmitted unchanged. The ability of colloidal

Fig. 12.6 Nephelometer for studying light-scattering by colloids

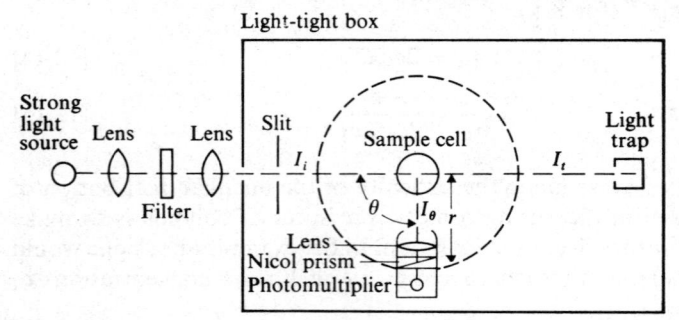

particles to scatter light has already been mentioned in connection with the Tyndall effect. The intensity, polarisation and angular distribution of light scattered from colloids can be used to determine the size, shape and interactions of the colloidal particles.

The measurement of scattered light is carried out using a nephelometer shown in Fig. 12.6. A strong monochromatic light beam is shone through the colloid. The intensity of the scattered light, I_θ, at any given angle θ to the incident beam is measured by means of a photocell on a rotatable arm at a constant distance, r, from the sample. The intensity of the incident light, I_i, is always measured at $\theta = 0$ after removing the sample. The scattering power of the sample at an angle θ is expressed by the Rayleigh ratio which is given by

$$R_\theta = \frac{I_\theta}{I_i} r^2. \qquad [12.16]$$

The turbidity, τ, of the sample is usually calculated from R_{90}, the Rayleigh ratio at right angles to the incident beam, by means of the formula

$$\tau = \frac{16\pi}{3} R_{90}. \qquad [12.17]$$

The turbidity may alternatively be calculated from the intensity, I_t, of the light transmitted by the sample of length l at $\theta = 0$, using the equation:

$$\tau l = \ln\left(\frac{I_i}{I_t}\right), \qquad [12.18]$$

which resembles the Beer–Lambert law for adsorption (p. 124). This method of determining τ is less accurate than the former method (using equation [12.17]), when I_t is not much less than I_i, or, in other words, when the sample is not visibly turbid. The photocell may be preceded by a Nicol prism to study the polarisation of the scattered light (see p. 105). Care must be taken to exclude from the sample particulate impurities, such as dust, which would contribute greatly to the light scattering and introduce serious errors.

Since light is electromagnetic radiation, Rayleigh successfully applied the electromagnetic theory to the light scattered by molecules of a gas. Mie later developed the theory for spherical particles of all sizes and Gans extended it to certain other shapes. The electric field associated with the incident light interacts with the atoms in the light path inducing periodic oscillations in the electrons. The material then behaves as a secondary source of (scattered) light, which has the same wavelength as the incident light. For a spherical, isotropic, non-absorbing particle of small size compared with the light wavelength *in vacuo*, λ, the intensity of scattered light, I_θ, is proportional to the intensity of the incident light, I_i, the square of the volume per particle, V, and increases with increasing difference between the refractive index, n_1, of the particle and of the dispersion medium, n_0. This is shown by the equation:

$$R_\theta = \frac{I_\theta}{I_i} r^2 = \frac{9\pi^2 V^2}{2\lambda^4} \left(\frac{n_1^2 - n_0^2}{n_1^2 + 2n_0^2}\right)^2 (1+\cos^2\theta). \qquad [12.19]$$

Since the scattered intensity is inversely proportional to the fourth power of the wavelength, the shorter wavelengths (bluish) are scattered with greater intensity than the longer wavelengths (reddish). This explains why colloidal dispersions have a blue hue when viewed at right angles to the incident white light, whereas the transmitted light has a red tinge. This also accounts for the blue colour of the sky on a fine day and the red colour of the sun at sunrise and sunset. If both sides of equation [12.19] are divided by $(1+\cos^2 \theta)$, the right hand side

Fig. 12.7 Radiation envelope for light scattered from small spherical isotropic particle(s). Distance from the origin represents the relative intensity of: the vertically polarised component (— —); horizontally polarised component (- - - - - -); and total scattered light (———)

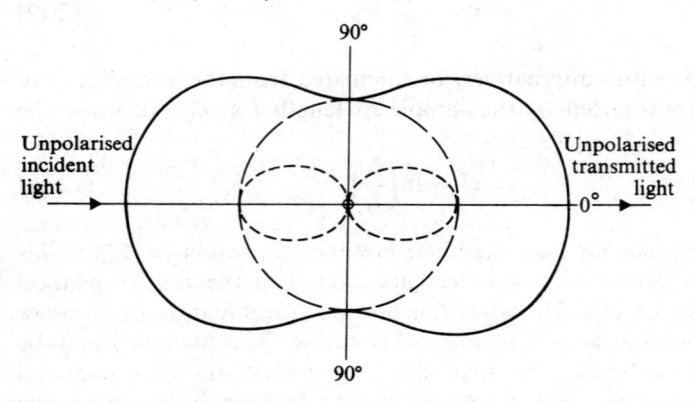

of the equation will now be independent of the angle θ. Therefore $R_\theta/(1+\cos^2 \theta)$ is independent of θ, provided that every dimension of the particle(s) is less than about $\lambda/20$. Under these conditions the light scattering is totally and vertically polarised at $90°$ to the unpolarised incident beam and is partially polarised in all other directions except at $0°$ and $180°$ to the incident beam where it is unpolarised (see Fig. 12.7). In the factor $(1+\cos^2 \theta)$ of equation [12.19] the unity term refers to the vertically polarised component and the $\cos^2 \theta$ term refers to the horizontally polarised component. For N small scattering particles randomly distributed I_θ and R_θ have values N times greater than for a single particle. Putting $\theta = 90°$ in equation [12.19] enables the turbidity to be expressed by the following equation:

$$\tau = \frac{16\pi}{3} R_{90} = \frac{24\pi^3}{\lambda^4} \left(\frac{n_1^2-n_0^2}{n_1^2+2n_0^2}\right)^2 NV^2. \qquad [12.20]$$

If the particles have one or more dimensions greater than about $\lambda/20$, the particles can no longer be considered to be point sources of scattered light. The light waves originating from different parts of the particle may then be out of step. These phase differences cause destructive light interference effects in certain directions. In particular, more light is scattered forwards (at $\theta = 0°$) than backwards (at $\theta = 180°$) and the light scattered at $90°$ to the incident beam is only partially polarised. The radiation envelope remains symmetrical about the $0°$ axis but becomes asymmetrical about the $90°$ axis, the centre of gravity shifting in the forward direction of the incident beam. The ratio $I_\theta/I_{180-\theta}$ (e.g. I_{45}/I_{135}), which is known as the dissymmetry ratio of scattered light, is unity for small particles and increases with increasing interparticle interference and therefore with increasing particle size.

For a spherical particle of diameter $\lambda/4$ destructive interference causes no light to be scattered at $\theta = 180°$. With larger particles maxima and minima of scattering occur at intermediate angles which depend on λ and particle size. With incident white light a monodisperse system gives rise to coloured interference bands known as the 'higher order Tyndall spectrum', but polydisperse systems do not. Calculations from intensity measurements indicate particle size and number.

Macromolecules cause negligible light scattering as individuals, because their refractive index is very similar to that of the dispersion medium (i.e. $n_1 \approx n_0$ in equation [12.19]) and because they are very small. Nevertheless, solutions of macromolecules do scatter light significantly. This is explained by the fluctuation theory which disregards the optical effects of the molecules themselves. According to this theory the random Brownian movement of the molecules causes fluctuations in concentration and therefore in refractive index in each small volume element of the solution. A momentary difference in refractive index between each volume element and its surroundings causes the observed light scattering. The turbidity of the solution depends in the amount of work done by each fluctuation in concentration against the osmotic pressure of the solution and also on the change in refractive index produced by the fluctuation. Thus, in principle, the molecular weights of lyophilic particles can be calculated from light scattering measurements.

For a random colloidal solution containing N particles or macromolecules each of volume, V, and of molecular weight, M_r : $\tau \propto$ total intensity of scattered light $\propto NV^2 \propto NM_r^2$ (since $V \propto M_r$). This relationship for lyophilic colloids is analogous to equation [12.20] for lyophobic colloids. If the turbidity, τ, refers to unit volume of solution $NM_r \propto \varrho$, the mass concentration, and therefore $\tau \propto \varrho M_r$. Consequently

$$\tau = H\varrho M_r, \qquad [12.21]$$

where H is an optical constant of proportionality for the solute–solvent system which is given by

$$H = \frac{32\pi^3 n_0^2}{3\lambda^4 L}\left(\frac{dn}{d\varrho}\right)^2, \qquad [12.22]$$

where L is the Avogadro constant,

λ is the wavelength of the incident light *in vacuo*,
n_0 is the refractive index of the solvent,
n is the refractive index of the solution and
ϱ is the mass concentration.

The rate of change of n with respect to ϱ, $(dn/d\varrho)$, must be determined very accurately by means of a differential refractometer. Equation [12.21] above is applicable only to ideal solutions. For non-ideal solutions of approximately spherical particles of diameter less than $\lambda/20$, the following equation is satisfactory:

$$\frac{H\varrho}{\tau} = \frac{1}{M_r} + B\varrho, \qquad [12.23]$$

where B is a coefficient of non-ideality of the solution. τ is determined at various mass concentrations, ϱ, of lyophilic colloid and $H\varrho/\tau$ is plotted against ϱ. A straight line is usually obtained which is extrapolated to $\varrho = 0$. The intercept on the ordinate, $(H\varrho/\tau)_{\varrho=0}$, is equal to $1/M_r$.

For small particles $R_\theta/(1+\cos^2\theta)$ is independent of θ, therefore we can write:

$$\frac{R_\theta}{1+\cos^2\theta} = \frac{R_{90}}{1+\cos^2 90°} = R_{90}, \qquad [12.24]$$

since $\cos 90° = 0$. Substituting into equation [12.17]:

$$\tau = \frac{16\pi R_\theta}{3(1+\cos^2\theta)}. \qquad [12.25]$$

Thus, for small particles τ can be determined from intensity measurements at any angle using this equation. If the solute macromolecules have diameters greater than $\lambda/20$, the interference effects mentioned above alter the intensity of the scattered light at all angles except at $\theta = 0$. Consequently, the turbidity cannot be measured accurately at $\theta = 90$ using equation [12.17] but can be calculated using equation [12.25] if $R_\theta/(1+\cos^2\theta)$ can be obtained at $\theta = 0$. Since the transmitted light is always present at $\theta = 0$, $R_\theta/(1+\cos^2\theta)$ cannot be obtained by direct measurement at this angle and must therefore be extrapolated to $\theta = 0$ from intensity measurements at small angles.

For a polydisperse system the molecular weight obtained is the mass average value. Light scattering becomes progressively more intense, and in principle more reliable, as the size or molecular weight of the colloidal particles increases. The reverse is the case for the osmotic pressure method.

Sedimentation by Ultracentrifugation

A most important method of measuring the sizes of colloidal particles is that in which they are caused to undergo sedimentation by subjecting them to very high centrifugal forces in an ultracentrifuge. When a body is rotated in a circular path, as is a particle in an ultracentrifuge, the body experiences an inwards acceleration $\omega^2 x$ along the radius, where ω is the angular velocity in radians per unit time and x is the distance of the body from the axis of rotation. This acceleration causes a centrifugal force to act on the body. The magnitude of the force is proportional to the acceleration and the mass of the body. Although its SI unit is rad s^{-1}, angular velocity is frequently stated as the number of revolutions per minute (r.p.m.), u. Since one revolution is equal to 2π radians, u and ω are related as follows:

$$\omega \text{ (in rad s}^{-1}\text{)} = \frac{2\pi}{60} . u \text{ (in r.p.m.)}. \qquad [12.27]$$

The centrifugal acceleration is then given by:

$$a \text{ (SI unit m s}^{-2}\text{)} = \omega^2 x = \frac{4\pi^2 u^2 x}{3600}. \qquad [12.28]$$

Centrifugal accelerations are frequently stated relative to g, the acceleration due to gravity. Since $g = 9.81$ m s^{-2} in most places:

$$a = \omega^2 x = \frac{4\pi^2 u^2 x}{3600 \times 9.81} . g. \qquad [12.29]$$

For example, at a distance 10 cm from the axis of rotation a rotational speed of 100 000 revolutions per minute gives a relative centrifugal acceleration equal to 1.12×10^6 g. Relative centrifugal acceleration is often incorrectly termed relative centrifugal force which is abbreviated to r.c.f.

Acceleration of the order of magnitude 10^5 g to 10^6 g may be achieved by supporting cells of the solution in a rotor driven by a geared electric motor or a turbine. In one form of the instrument, the turbine is air-driven and another jet of air lifts the turbine just clear of its bearings so that it spins on a cushion of air and problems of high-speed lubrication are eliminated (see Fig. 12.8) To minimise frictional resistance which would cause heating of the sample, the chamber in which the rotor is suspended is evacuated or filled with hydrogen gas at low pressure. The sedimentation of the colloidal particles (away from the axis of rotation) may be followed by the change in refractive index, light absorption, or light interference pattern which it produces. Each is observed by a special optical system, and by means of windows in the rotor chamber which permit light to pass through the cells in the

Fig. 12.8 An air-driven ultracentrifuge

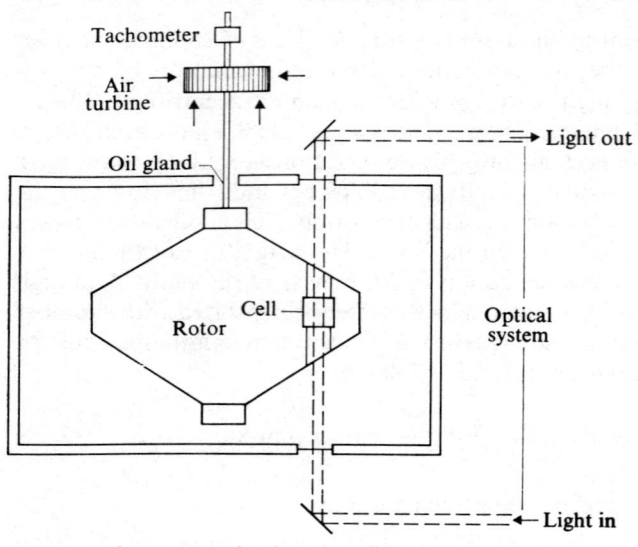

rotor as they are spinning (see Fig. 12.8). A common method is to measure at various distances from the axis across the cell the refractive index gradient (which gives a measure of the concentration gradient) by means of a Schlieren optical system.

In the ultracentrifuge the spontaneous tendency to diffusion which all small particles possess is counteracted by the centrifugal force. If the particles are very small, diffusion will be so rapid and the centrifugal force so small that diffusion will overwhelm sedimentation. The lower size limit of particles which can be sedimented by ultracentrifugation is determined only by the technical difficulties in maintaining very high centrifugal forces without mechanical breakage. Gravity alone is insufficient to cause significant sedimentation of particles with dimensions smaller than about 1 μm. Molar masses and molecular weights of colloidal particles can be determined using the ultracentrifuge either by the sedimentation rate method or by the sedimentation equilibrium method.

In the sedimentation rate method high centrifugal accelerations up to about 10^5 g to 10^6 g are used. These accelerations are capable of causing complete sedimentation of most sizes of particle, leaving behind the pure dispersion medium, and this is best followed by observing the movement of the boundary between the colloidal solution and the dispersion medium. The boundary appears as a peak in the refractive index gradient, as a change in absorbance (of light) or as a discontinuity in the light interference pattern.

When centrifuged, a particle experiences a centrifugal force which is balanced by a force due to the frictional resistance of the medium (see p. 477) thus:

$$m(1 - v\varrho)\omega^2 x - f \cdot \frac{dx}{dt} = 0$$

where m is the mass of the particle, $(1 - v\varrho)$ is a correction for buoyancy, v is the partial specific volume of the disperse phase (equal to the reciprocal of the density of that phase), ϱ is the density of the dispersion medium, ω is the angular velocity of the rotor (measured in radians per unit time), x is the distance of the boundary from the axis of rotation, dx/dt is the sedimentation rate (i.e. the rate of change of the position of the boundary with time) and f is the frictional coefficient. Inserting $f = kT/D$ (equation [12.2], where k is the Boltzmann constant, T is the absolute temperature and D is the coefficient of free diffusion of the particles, we have:

$$m(1 - v\varrho)\omega^2 x = \frac{kT}{D} \cdot \frac{dx}{dt}$$

whence

$$m = \frac{kT}{D(1 - v\varrho)} \cdot \frac{dx/dt}{\omega^2 x}.$$

Since the molar mass of the particles is given by $M = Lm$, where L is the Avogadro constant, and since the gas constant is similarly given by $R = Lk$,

$$M = \frac{RT}{D(1 - v\varrho)} \cdot \frac{dx/dt}{\omega^2 x}. \qquad [12.30]$$

The quantity $\frac{dx/dt}{\omega^2 x}$ is known as the sedimentation coefficient. It is given the symbol s and has the dimension of time. Equation [12.30] may be written:

$$M = \frac{RTs}{D(1 - v\varrho)}. \qquad [12.31]$$

Since for a given colloidal species dispersed in a given medium at a given temperature, M, D, v, ϱ and T are constant, s is a constant. For many proteins and subcellular particles in water at room temperature s often has values between about 10^{-13} and 10^{-10} second. In this connection 10^{-13} second is sometimes known as the Svedberg unit and is given the symbol S. For example, certain particles in bacterial ribosomes for which s is 70×10^{-13} second are said to have a sedimentation coefficient of 70 S. In practice s is found by measuring the distance, x, of the boundary from the axis of rotation at two values of the time, t. For this purpose the equation

$$s = \frac{1}{\omega^2 x} \cdot \frac{dx}{dt} \qquad [12.32]$$

must be integrated. This equation may be written in the form:

$$s \, dt = \frac{1}{\omega^2} \cdot \frac{1}{x} \, dx.$$

For a given type of particle at a constant angular velocity, s and ω are constants and may be placed before the integral signs. We integrate between the limits x_1 at t_1 and x_2 at t_2 as follows:

$$s \int_{t_1}^{t_2} dt = \frac{1}{\omega^2} \int_{x_1}^{x_2} \frac{1}{x} \, dx$$

$$\therefore \quad s\Big[t\Big]_{t_1}^{t_2} = \frac{1}{\omega^2} \Big[\ln x\Big]_{x_1}^{x_2}$$

$$\therefore \quad s(t_2 - t_1) = \frac{1}{\omega^2}(\ln x_2 - \ln x_1)$$

$$\therefore \quad s = \frac{\ln(x_2/x_1)}{\omega^2(t_2 - t_1)}. \qquad [12.33]$$

s is calculated by inserting ω and the two pairs of experimental values x_1, t_1 and x_2, t_2 into this expression. The molar mass, and hence the molecular weight, is calculated by inserting s and the other constants into equation [12.31]. The diffusion coefficient, D, must be determined in a separate and often difficult experiment. Knowledge of D may be avoided by combining the viscosity method, to be discussed in the next section, with the sedimentation rate method and special equations have been derived for this.

The alternative method, the sedimentation equilibrium method, employs a low relative centrifugal acceleration of the order $10^4 \, g$ to $10^5 \, g$ so that the colloidal particles do not undergo complete sedimentation. Instead, after being centrifuged for some time, and equilibrium is set up between the sedimenting tendency and the diffusing tendency of the particles, with the result that they became more densely packed towards the end of the cell remote from the rotation axis. The flux of particles due to sedimentation is given by an equation analogous to [6.62] and that due to diffusion is given by equation [6.56]. When sedimentation equilibrium has been attained, the total flux is zero, so that

$$0 = J_n = c\frac{dx}{dt} - D\frac{dc}{dx},$$

whence

$$\frac{dx}{dt} = \frac{D}{c} \cdot \frac{dc}{dx}.$$

When dx/dt is introduced into equation [12.30], D cancels out and we have

$$M = \frac{RT}{(1 - v\varrho)\omega^2} \cdot \frac{1}{cx} \cdot \frac{dc}{dx}. \qquad [12.34]$$

In order to evaluate M, the concentration gradient dc/dx may be determined directly, for example, from the refractive index gradient at a distance, x, from the axis of rotation. More frequently, the concentration, c, of the colloid at equilibrium is measured at two distances from the axis of rotation. For this purpose equation [12.34] must be written as

$$Mx.dx = \frac{RT}{(1-v\varrho)\omega^2} \cdot \frac{1}{c} \cdot dc$$

and integrated between the limits c_1 at x_1 and c_2 at x_2, as follows:

$$M\int_{x_1}^{x_2} x.dx = \frac{RT}{(1-v\varrho)\omega^2} \int_{c_1}^{c_2} \frac{1}{c} \cdot dc$$

$$\therefore \quad M\left[\tfrac{1}{2}x^2\right]_{x_1}^{x_2} = \frac{RT}{(1-v\varrho)\omega^2} \left[\ln c\right]_{c_1}^{c_2}$$

$$\therefore \quad \frac{M}{2}(x_2^2 - x_1^2) = \frac{RT}{(1-v\varrho)\omega^2} (\ln c_2 - \ln c_1)$$

$$\therefore \quad M = \frac{2RT \ln(c_2/c_1)}{(1-v\varrho)\omega^2(x_2^2 - x_1^2)}. \qquad [12.35]$$

Since the concentrations, c_1 and c_2 of colloid at distances x_1 and x_2 from the axis of rotation appear as a ratio, any quantity or physical property which is proportional to concentration may be inserted into this equation, e.g. mass concentration, absorbance (of light) or difference in refractive index between the solution and the medium. The molar mass, M, and hence the molecular weight, can be calculated from the sedimentation equilibrium without any knowledge of the diffusion coefficient which is required for the sedimentation rate method. A disadvantage of the sedimentation equilibrium method is that a considerable time, sometimes several days, may be required for equilibrium to be established. The method is not much used for colloids with a molecular weight greater than 5000 ($M = 5$ kg mol^{-1}).

If the colloidal solution under examination is polydisperse (i.e. contains particles of more than one molar mass), the value of M determined by the sedimentation equilibrium method will be found to depend upon the position in the cell at which it is determined, and the results are difficult to interpret. Each component of the colloid establishes its own equilibrium distribution of concentration and in any given volume element all components will be present together. In the sedimentation velocity method, however, each different species gives rise to a different boundary (provided there are only a few species and their molecular weights are not too close). It is then possible to determine their separate molar masses and also to obtain an estimate of the relative amounts of the different species present.

The molar masses determined by sedimentation methods usually refer to the unsolvated particle since the partial molar volume usually refers to the unsolvated state.

A useful technique for analysing macromolecules, such as nucleic acids and proteins, is known as *density gradient ultracentrifugation*. If one centrifuges a solution of a substance of low molecular weight, such as an aqueous solution of caesium chloride or sucrose, there will be at equilibrium a density gradient across the cell. The same result can be achieved by layering solutions of decreasing concentration in the cell. If a substance of high molecular weight is added to this solution of graded density, it will float at a definite position at which the density of its molecules equals the density of the solution. If a mixture of macromolecules of different molecular weights is added instead, each component will eventually be present at a band or layer at a particular plane in the cell. The separation can be seen by means of an appropriate optical method.

Viscosity of Colloids

The measurement of viscosity of colloidal solutions may provide some evidence of the molecular weight of the particles. This technique is employed to a considerable extent in the study of colloid systems, because the experimental method is relatively simple. However, the results cannot usually be interpreted with any certainty except with a limited range of materials and molecular weights where check determinations have been made by other methods. Einstein derived an expression for the viscosities of colloidal solutions which is only valid if various assumptions can be justified; in particular, that there is a low concentration of the disperse phase, and that it is present in the form of spherical particles. Under these circumstances, the dynamic viscosity of the sol, η, should be related to that of the dispersion medium, η_0, by the equation:

$$\eta = \eta_0(1+2.5\phi), \qquad [12.36]$$

where ϕ is the volume fraction of the disperse phase (i.e. the volume of the particles divided by the total volume). If both sides of this equation are divided by η_0, an expression for the *viscosity ratio* (or *relative viscosity*), η_r, is obtained:

$$\eta_r = \eta/\eta_0 = 1+2.5\phi. \qquad [12.37]$$

Subtracting 1 from both sides:

$$\eta_r - 1 = 2.5\phi. \qquad [12.38]$$

The term $(\eta_r - 1)$ is called the *viscosity ratio increment* (or *specific viscosity*) for the system, and is represented by η_{sp}. Thus, under the conditions in which Einstein's equation is applicable, the viscosity ratio increment of the sol depends only on the total volume fraction of the disperse phase, and is independent of its nature and of the actual size of the particles. Since ϕ includes any solvent which moves with the particle, lyophilic colloids, being solvated, increase the viscosity of the dispersion medium to a greater extent than do lyophobic colloids. For dilute solutions ϕ is proportional to the mass concentration, ϱ, so the Einstein equation may be written

$$\eta_{sp}/\varrho = \text{constant.} \qquad [12.39]$$

This equation is obeyed by dilute solutions of spherical macromolecules, e.g. globular proteins in water, and by solutions of heavily solvated small molecules, e.g. glycerol in water.

Experiment 12.1 Determination of the volume of the hydrated glycerol molecule.

Let the volume of the hydrated glycerol molecule be v (in m^3) and let the concentration of a given aqueous solution of glycerol be c (in mol m^{-3}). One mole of hydrated glycerol molecules takes up a volume Lv (in m^3 mol^{-1}) within the solution, where L is the Avogadro constant, and is dispersed in a total volume $1/c$ (in m^3 mol^{-1}) of solution. Consequently the volume fraction ϕ, of hydrated glycerol molecules is equal to $Lv \div 1/c$, i.e.

$$\phi = Lvc. \qquad [12.40]$$

Substituting for ϕ in equation [12.37] gives

$$\eta/\eta_0 = 1 + 2.5Lvc, \qquad [12.41]$$

from which v is calculated as we shall see. The viscosity ratio, η/η_0, is most simply determined in a U-tube (or Ostwald) viscometer (see p. 539) and is then given by the following equation which is derived from equation [13.13]:

$$\frac{\eta}{\eta_0} = \frac{\varrho}{\varrho_0} \cdot \frac{t}{t_0}, \qquad [12.42]$$

where η and η_0 are the dynamic viscosities of the solution and pure solvent respectively, ϱ and ϱ_0 are the corresponding densities, t is the time of flow of the solution in the viscometer and t_0 is that for the same volume of solvent.

A solution of glycerol ($M_r = 92.10$) with an accurately known concentration close to 1000 mol m^{-3} is prepared by weighing about 18 cm^3 of dry glycerol in a 100 cm^3 beaker and making it up to 250 cm^3 with distilled water in a volumetric flask. Further solutions of concentration $\frac{4}{5}$, $\frac{3}{5}$, $\frac{2}{5}$ and $\frac{1}{5}$ that of the original solution are prepared by accurate dilution of the original solution using a 25 cm^3

graduated pipette and a 50 cm³ volumetric flask. Using a clean Ostwald viscometer in a thermostat at 25° C the flow time of water and of an equal volume of each glycerol solution in order of increasing concentration is measured in duplicate (see p. 539). The density of each solution may be determined in a density bottle or by accurately weighing a pipetted volume; alternatively the formula $\varrho/\varrho_0 = 1 + 2.10 \times 10^{-5}\ c/\text{mol m}^{-3}$ may be used. The results are recorded in tabular form under the following headings:

$$c \quad t \quad t/t_0 \quad \varrho \quad \varrho/\varrho_0 \quad \eta/\eta_0.$$

A graph of η/η_0 against c is plotted including a point corresponding to the pure solvent, i.e. $c = 0, \eta/\eta_0 = 1.00$. According to equation [12.41] the plot should give a straight line of gradient $2.5Lv$, from which the volume v of the hydrated glycerol molecule is calculated ($L = 6.022 \times 10^{23}\ \text{mol}^{-1}$).

Most lyophilic colloids are not spherical but elongated in the form of long-chain molecules. The ratio η_{sp}/ϱ, which is termed the *viscosity number* (or the *reduced viscosity*) is then not constant but depends on the molar mass, M, shape and mass concentration, ϱ, of the particles. Staudinger proposed the relationship:

$$\eta_{sp}/\varrho \approx K.M, \qquad [12.43]$$

where K is a constant for a given colloid system. The dependence of the viscosity number on mass concentration is eliminated by extrapolation to $\varrho = 0$. The limiting value of η_{sp}/ϱ at zero concentration is called the *limiting viscosity number* (or *intrinsic viscosity*) and is written with the symbol $[\eta]$. The modified form of Staudinger's law is therefore:

$$[\eta] = K.M. \qquad [12.44]$$

Although this law has proved useful in individual colloid systems, it is not now regarded as being generally valid. It has been suggested, for instance, that such a relationship is only valid for colloidal particles which are in the form of long-chain molecules which are randomly kinked. A generalised form of the law which was proposed by Mark and Houwink is:

$$[\eta] = K.M^\alpha, \qquad [12.45]$$

where K and α are constants characteristic of the dispersed phase, the disperion medium and the temperature. Taking logarithms of both sides of equation [12.45], we have

$$\log[\eta] = \log K + \alpha \log M. \qquad [12.46]$$

Thus, at a given temperature and in a given solvent, macromolecules with the same configuration of polymer chains but with different numbers of monomer units, give linear plots of log $[\eta]$ against log M. The constant α is equal to the slope and depends on the configuration of the polymer chains. Although α can have values from 0 to 2, it usually lies between 0.5 and 1.5. For flexible coils $\alpha \approx 2$ and for random coils $\alpha \approx 0.5$. The latter value is found, for example, for proteins in 6 mol dm^{-3} guanidinium chloride with 0.5 mol dm^{-3} 2-mercaptoethanol. For spherical molecules, e.g. globular proteins, the Einstein equation [12.39] holds, $\alpha = 0$ and $[\eta]$ is independent of M.

For polydisperse systems the viscosity method gives an average value of M which is known as the viscosity average molar mass and which is given by:

$$\bar{M}_\eta = \left(\frac{\Sigma N_i M_i^{\alpha+1}}{\Sigma N_i M_i}\right)^{1/\alpha} \qquad [12.47]$$

$\bar{M}_\eta$ lies between $\bar{M}_N$ and $\bar{M}_m$ (cf. equations [12.7] and [12.8]). When the Staudinger equation [12.44] is obeyed, $\alpha = 1$ and $\bar{M}_\eta = \bar{M}_m$.

Concentrated lyophilic colloids exhibit complex rheological phenomena which result from the entanglement of long chain molecules and which are best studied by means of a rotational viscometer (see p. 540, et seq.)

The Electrical Properties of Colloids

Colloidal particles in a polar solvent, e.g. water, often acquire a surface charge either by losing ions from their surface (ionisation or dissolution) or by gaining ions from solution (adsorption). The charged particle attracts ions of unlike charge, called counter-ions, and repels ions of like charge, called co-ions. As a result counterions and molecules of the polar solvent are ordered around the charged particles. The charged surface of the particle and the neighbouring counter ions constitute an electrical double layer. As an example, Fig. 12.9(a) illustrates the double layer formed when a colloidal particle, such as a protein molecule ionises so as to give hydrogen ions in the dispersion medium, becoming negatively charged itself. The negative charges on the particle form one side of the double layer, and the positive ions (counter ions) in the dispersion medium form the other side.

A proportion of the counter ions is firmly attracted electrically by the charged particle and therefore constitute the fixed part of the double layer, which is also known as the *Stern layer*. As a result of the randomising effects of thermal motions, a proportion of the counter ions further from the particle are not firmly held and therefore constitute the *mobile or diffuse part* of the double layer, which is also known as the *Gouy–Chapman layer*. The diffuse part also contains mobile solvent molecules and co-ions, but counter ions predominate over co-ions increasingly as the surface of the particle is approached. The boundary

506   Colloids

Fig. 12.9 (a) Electric double layer at the surface of a colloidal particle
(b) Variation of potential, ψ, with distance in the double layer

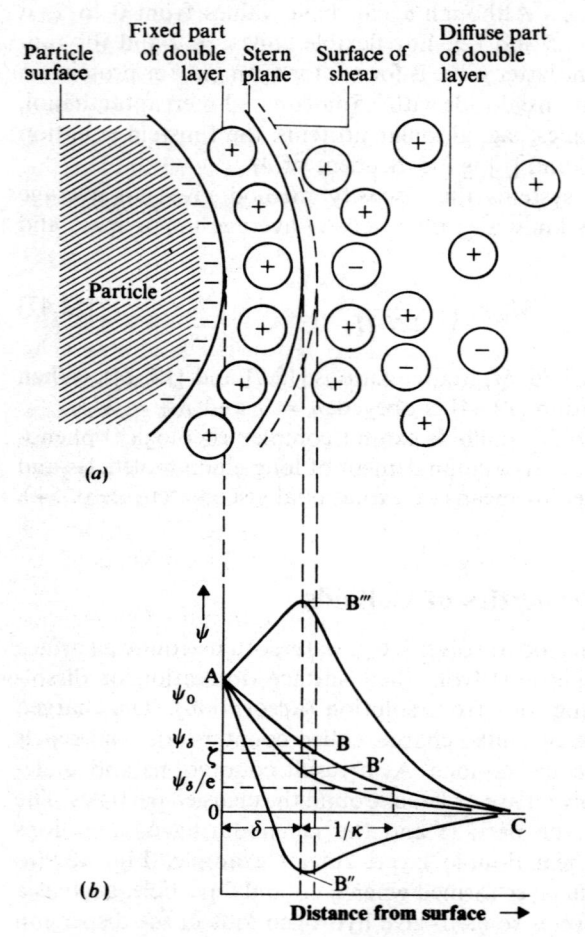

B″ B′ B B‴ between the fixed and diffuse parts of the double layer is often known as the *Stern plane*.

The electric potential, ψ (SI unit the volt, V), is defined as the work which must be done to bring unit electric charge from infinity to that point. Fig. 12.9(b) illustrates some possible examples of the variation in ψ with distance from the surface of a charged colloidal particle. By definition ψ is zero at an infinite distance from the particle. The potential at any point is greater the greater the resultant charge density at that point, since more work is done to bring unit charge from infinity to that point. The value of ψ at the surface of the particle, ψ_0, is therefore large. Curve ABC corresponds to the situation represented by Fig. 12.9(a) and discussed above. The change in potential from A to B takes place within the fixed part of the double layer and is due partly to the normal variation of potential with distance and partly to the adsorbed ions. The change from B to C takes place within the diffuse part of the double layer. As the distance from the particle increases, the more equally distributed are the counter ions and co-ions and the closer to zero is the resultant charge density and the potential. Gouy and Chapman applied the Boltzmann distribution to the ions in the diffuse part of the double layer and derived the following equation for the variation of potential, ψ, with distance x outwards from the Stern plane:

$$\psi = \psi_\delta e^{-\kappa x}, \qquad [12.48]$$

where ψ_δ is the potential at the Stern plane and κ is a constant with the dimensions length^{-1}. Although the diffuse part of the double layer is theoretically limitless, the potential within it tails away rapidly and exponentially with distance. $1/\kappa$ is the distance over which ψ decreases by the ratio $1/e$ (e.g. the value of x when $\psi = \psi_\delta/e$; see Fig. 12.9b) and can be considered to be a measure of the thickness of the diffuse part of the double layer. Figure 12.9(b) is not strictly to scale. The thickness of the fixed part of the double layer, δ, is usually small in comparison with $1/\kappa$, so $1/\kappa$ can be thought of as a measure of the thickness of the whole electrical double layer.

If a colloidal particle moves, for example as a result of an electric field, Brownian motion or mechanical disturbance, the fixed part of the double layer and some solvent molecules in the diffuse part of the double layer move with the particle as a single unit. The outer boundary of this unit is known as the *surface of shear* or *shear plane*. Although the surface of shear is a little further away from the particle than the Stern plane (B″ B′ B B‴ in Fig. 12.9b), the two surfaces are very close and can usually be considered identical. The potential at the surface of shear is called the electrokinetic or ζ (zeta) potential because it determines the rate of motion of the particle in an electric field, discussed in the next section, as well as other electrical properties of the colloid. The ζ potential is approximately equal to ψ_δ in equation [12.48].

The reason why one side of the double layer is thought to include some adsorbed ions is that the ζ potential is strongly influenced by the presence of other ions in the solution. In general, it is made more positive by acids (interpreted as being due to the adsorption of hydrogen ions, and their inclusion within the fixed part of the double layer), and more negative by alkalis (due to hydroxyl ions becoming included in the fixed part of the double layer). At a fixed pH, addition of other ionic materials to the solution modifies the ζ potential. Ions of charge opposite to that of the particles generally become incorporated in the fixed part of the double layer, and the higher their valency the greater is the effect which they have on the ζ potential (see p. 517). Ions of the same sign as the particles also affect the ζ potential, but usually to a smaller extent. In Fig. 12.9(b), the curve AB'C shows how the ζ potential may be reduced by some adsorption of ions opposite in charge to the particles. Strong adsorption of such ions (particularly if they are of high valency) might result in the curve AB''C, in which the ζ potential is actually reversed in sign. The curve AB'''C represents the phenomenon which occurs less frequently when ions of the same sign as the particles are adsorbed and the ζ potential is increased. In each case, the potential difference between the colloidal particle and the bulk of the solution (A to C) remains the same. If the particles are of a material which is capable of setting up a reversible (electrode) potential with its ions in solution, this potential would be equal to the difference in potential between the points A and C, ψ_0.

Another way in which the alteration of ionic concentrations in solution can affect the ζ potential of a colloid is by altering the distribution of charge in the diffuse part of the double layer. Increasing concentration causes a greater accumulation of ions of sign opposite to that of the colloid in the region of the double layer. The potential falls more rapidly with increasing distance from the surface of the particle (i.e. from A of Fig. 12.9(b) and the ζ potential at the surface of the fixed part of the double layer (corresponding to B of Fig. 12.9(b) is therefore less than before. This effect and that of adsorption of ions are difficult to distinguish experimentally and it is probable that they both operate together.

Electrophoresis

Electrophoresis is the term applied to the movement of colloidal particles which takes place under the influence of an electric field, as a consequence of the ζ potential. An electric field strength (or potential gradient), E (SI unit V m^{-1}), is created by applying a potential difference, ΔV, (SI unit V) to two electrodes separated by a distance l (SI unit m) in the colloidal solution, when

$$E = \Delta V/l. \qquad [12.49]$$

If we can assume that the particle is small enough to be treated as a point charge and large enough to obey Stoke's law (equation [12.3]) the electrical force on the particle may be equated with the viscous force which is equal to the frictional coefficient, $6\pi\eta r$, times the speed, thus:

$$QE = 6\pi\eta rv, \qquad [12.50]$$

where Q is the overall charge (SI unit C) on the particle treated as a single electrokinetic unit bounded by the surface of shear, r is the radius of this unit (SI unit m). v is the speed (SI unit m s^{-1}) of migration of the particle as a result of the electric field and η is the dynamic viscosity of the medium (SI unit kg m^{-1}s^{-1}). Since v is proportional to E, the rate of motion of the particles is commonly expressed as the ratio, u, which is known as the electrophoretic (or electric) mobility (SI unit m^2 s^{-1} V^{-1}) and which is defined by

$$u = v/E. \quad [12.51]$$

Equation [12.50] may now be written:

$$u = \frac{Q}{6\pi\eta r}. \quad [12.52]$$

The zeta potential, ζ (SI unit V), of a particle treated as a point charge is related to Q by the equation

$$\zeta = \frac{Q}{4\pi\varepsilon r}, \quad [12.53]$$

where ε is the permittivity of the medium (SI unit C V^{-1} m^{-1}). We note that the relative permittivity (or dielectric constant) of the medium, ε_r, is a dimensionless quantity equal to ε divided by ε_0, the permittivity of a vacuum, where $\varepsilon_0 = 8.8542 \times 10^{-12}$ C V^{-1} m^{-1}, so that

$$\varepsilon = \varepsilon_r \varepsilon_0. \quad [12.54]$$

Introducing the expression for Q from equation [12.53] into equation [12.52] we have:

$$u = \frac{2}{3} \cdot \frac{\zeta\varepsilon}{\eta}. \quad [12.55]$$

Equations [12.53] and [12.55] were derived by assuming that the thickness of the double layer is much less than the radius of the particle (i.e. when $\kappa r \gg 1$) which is very unlikely in aqueous media. The equations may sometimes apply to colloids in non-aqueous media of low conductance. When the thickness of the double layer is much greater than the radius of the particle (i.e. when $\kappa r \ll 1$), equation [12.55] must be replaced by:

$$u = \frac{\zeta\varepsilon}{\eta} \quad \text{or} \quad \zeta = \frac{\eta}{\varepsilon} \cdot u. \quad [12.56]$$

This is usually satisfactory for large non-conducting particles in aqueous media and enables their zeta potential to be calculated from the electrophoretic mobility. For water at 20° C, $\eta = 1.0019 \times 10^{-3}$ kg m^{-1} s^{-1} and $\varepsilon_r = 80.36$ (i.e. $\varepsilon = 80.36 \times 8.8542 \times 10^{-12}$ C V^{-1} m^{-1}, where 1 V = 1 kg m^2 s^{-1}A^{-1}), therefore

$$\zeta \text{ (in V)} = 1.41 \times 10^6 \times u \text{ (in m}^2\text{s}^{-1}\text{V}^{-1}\text{)}. \quad [12.57]$$

The electrophoretic mobility may be measured either microscopically or by a moving-boundary method. The former method is applied mostly to the study of particles rather larger in size than colloidal particles, e.g. bacteria and red blood cells, though by using the ultramicroscope technique the method may be extended to colloidal particles themselves. The colloidal solution is contained in a tube placed horizontally in the field of view of a microscope, and an electric field of known magnitude is applied by means of a reversible

Fig. 12.10 (a) Tiselius electrophoresis apparatus
  (b) Tiselius cell in section and plan view

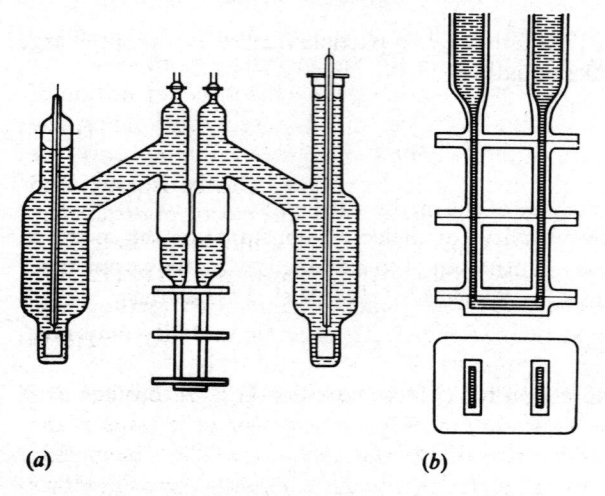

electrode at each end. Individual particles are observed, and their motion across the field of view of the microscope is timed. The mobility is calculated by dividing this rate of motion by the magnitude of the potential gradient (equations [12.49] and [12.51]).

The determination of electrophoretic mobility by the moving boundary method is similar in principle to the corresponding method for the determination of transport numbers (see p. 336). The apparatus most commonly used is that designed by Tiselius and illustrated in Fig. 12.10(a). The cell in which the boundary movement is observed is a U-shaped tube of rectangular cross-section, shown in Fig. 12.10(b). It is built in sections which can slide laterally relative to one another for the purpose of forming the boundary. The colloidal solution is placed in the lower half of the tube, which is then displaced relative to the upper sections. The dispersion medium is then placed in the two limbs of the upper part of the cell. A sharp boundary may then be formed between the colloidal solution and the medium by moving the two sections carefully into alignment. The electrode vessels contain silver-silver chloride electrodes which are covered with saturated potassium chloride solution. Above this is more of the dispersion medium, which fills the rest of the apparatus. During the passage of the current, the boundary between the colloidal solution and the dispersion medium moves towards the electrode opposite in sign to the charge on the particle. The position of the boundary is determined by a similar optical method to that employed in the ultracentrifuge, i.e. one which detects differences in refractive index at the boundary. Again, if more than one type of particle is present, the movement of their separate boundaries may be followed, and an estimate of the relative amount of each present may be made. Thus, one of the most important applications of electrophoresis is in the investigation of the composition of colloidal solutions, particularly those of biological origin, e.g. serum may be shown to consist of albumin and the α, β and γ globulins.

It is possible, by modifying the technique, to effect the separation of colloidal substances which may be chemically inseparable. Thus, the region between the first and second boundaries contains only the component of highest mobility, and this region can be isolated and removed. From a more theoretical point of view, electrophoresis may be used to determine the resultant charge on the colloidal particle if the size and shape may be assumed (employing equations such as [12.52]). Alternatively, if the charge can be found by some other method, e.g. by titration, a measurement of the electrophoretic mobility may give some idea of the size of the particle.

The potential difference which exists between the glass surface and the liquid causes the dispersion medium to flow along the walls of the dispersion medium and return along the centre of the tube. The motion of particles is therefore measured at a point where the opposing flows cancel. (Movement of the dispersion medium relative to the charged particles as a result of an applied potential difference is known as electro-osmosis.) In accurate work the electrophoresis cell should be placed in a thermostat.

Zone electrophoresis is a powerful technique for separating lyophilic colloids, particularly proteins dispersed in aqueous buffers. Zone electrophoresis is related to electrophoresis in the same way as chromatography is to partition, adsorption or ion exchange (pp. 233, 455 and 457). In one method, a column of the dispersion medium is immobilised by the presence of starch gel or polyacrylamide gel. The mixture of proteins is placed on top of the column. Each end of the column is in contact with a buffer solution into which dips a platinum electrode. A potential difference of the order 100 V is connected to the pair of electrodes. The current flowing is of the order 3 mA. The individual colloidal macromolecules move more slowly within the gel network than in the liquid state. Conditions are usually adjusted so that the protein with the highest electrophoretic mobility moves about 3 cm per hour. In this way blood serum may be separated into about

25 components as compared with about 5 using the moving boundary method. The molecular sieving action of the gel probably contributes to the separation. The different types of protein in the mixture become concentrated within separate bands in the column and can be made visible by placing the column in a staining bath. Afterwards the stain may be removed electrolytically or by placing the column in a suitable solution. Before electrophoresis moderate or large concentrations of salt must be removed from the mixture, e.g. by dialysis.

In another method of electrophoretic separation, the medium is held within a thin horizontal layer of gel or strip of filter paper (cellulose). Opposite sides of the layer or strip dip into buffer solutions in contact with suitable electrodes to which the potential difference is connected. This technique resembles thin layer or paper chromatography (pp. 233 and 234). Before electrophoresis the colloidal mixture is placed on a convenient line on the strip. After electrophoresis the individual components appear as separate bands.

Isoelectric Point of Proteins

Each protein molecule behaves as a hydrophilic colloidal particle. Like their constituent amino acids, proteins are amphoteric electrolytes, which are often known as 'ampholytes', since their molecules carry in their side chains both acid (—COOH) and basic (—NH$_2$) groups. These groups are simultaneously involved in acid-base equilibria:

$$-COOH + H_2O \rightleftharpoons -COO^- + H_3O^+$$
$$-NH_2 + H_2O \rightleftharpoons -NH_3^+ + OH^-$$
$$-COOH + OH^- \rightleftharpoons -COO^- + H_2O$$
$$-NH_2 + H_3O^+ \rightleftharpoons -NH_3^+ + H_2O.$$

Since both positively charged groups and negatively charged groups are present in the same molecule of a protein or amino acid, these molecules are known as 'zwitterions', (see p. 377) In the absence of other ions, the resultant charge on a protein molecule, and the pH of its solution, would depend on the numbers of carboxyl and amino groups in the molecule, and their dissociation constants (neglecting effects due to weak dissociation of other groups). If acid is added to the solution, the dissociation of the carboxyl groups is inhibited and the number of —NH$_3^+$ groups is increased. The resultant charge on the molecule therefore becomes more positive. Addition of a base produces the reverse effect, and the charge becomes more negative. For each protein there is one pH at which the resultant charge on the molecule is zero, and the ζ potential is therefore also zero. This is called the isoelectric point (cf. p. 377). At this point the colloidal solution is least stable, and therefore most susceptible to coagulation. (Insulin at its isoelectric point is precipitated spontaneously, and this is used as a method of purification.)

The most direct method of determining isoelectric points of proteins is by studying the electrophoretic mobility at a number of different pH values. A graph is plotted of mobility against pH, and the pH corresponding to zero mobility is the isoelectric point. The study of titration curves of proteins provides a less reliable method of determining the isoelectric point. Suppose a solution containing a known number of moles of a protein were originally at its isoelectric point, and a known quantity of acid were added. The modification of the acid-base equilibria produced could be regarded as resulting in an adsorption of a certain fraction of the hydrogen ions from the solution. The number of moles of hydrogen ion adsorbed per mole of protein would then be equal to the number of positive charges on each protein molecule. Similarly, if alkali were added at the isoelectric point, the number of moles of hydroxyl ion adsorbed per mole of protein would be equal to the number of negative charges on each protein molecule. The number of moles of hydrogen ion or hydroxyl ion adsorbed may be calculated by measuring the pH of the solution, which gives the concentration of free hydrogen or hydroxyl ion, and by comparing this concentration with the number of moles of H^+ or OH^- added. (The difference must be the number of moles adsorbed.) The pH at which this difference is zero is the isoelectric point.

The actual form of the titration curve is of importance, as it may be related to the actual amino-acid composition of the protein. In practice, a curve is drawn simply of amount (moles) of hydrogen ion adsorbed per unit mass of protein against the pH. If it is desired to obtain information about the charge on the protein molecules, the titration curve is compared with the graph of electrophoretic mobility against pH, for the two curves should have the same form.

The protein constituents of bacterial cell walls and of the enzymes essential for metabolism bear charges which depend on pH, for the reasons discussed above. The drugs which operate by being adsorbed by these proteins will therefore have an effectiveness which depends upon pH. Thus, when the pH is low, and the proteins consequently are positively charged, drugs which are effective in cationic form are not easily adsorbed. A larger concentration of drug in the cationic form is therefore required than at high pH to prevent growth of the bacteria. The variation with pH of the bacteriostatic power of the cationic form of the acridines may be attributed to this effect. It should be noted that this variation with pH is additional to that described in Chapter 9 (p. 383), in which the amount of the drug present in the ionic form varies with pH.

Stability of Lyophobic Colloids

Lyophobic colloids, like other colloidal dispersions are thermodynamically unstable owing to the tendency of the surface free energy and the interfacial area to decrease. However, for reasons to be discussed, the rate of particle aggregation may be very low, in which case the colloid may be termed stable from a kinetic point of view. Aggregation of lyophobic sols is brought about by the collisions that result from Brownian motion. The frequency of collision and therefore the rate of aggregation depend on the temperature and viscosity of the medium, the size and concentration of the particles and the forces of attraction and repulsion, which are themselves affected by factors to be discussed.

Since coagulation proceeds through the collison between two particles, the process is of the second order (see p. 146), i.e.

$$-\frac{dv}{dt} = k_2 v^2, \qquad [12.58]$$

where v is the number of particles in unit volume of solution, t is the time, $-dv/dt$ is the rate of coagulation and k_2 is the second order rate constant for coagulation. Smoluchowski in 1916 proposed a theory of coagulation which assumes that Brownian motion is alone responsible for collision and that every collision leads to permanent contact. According to this theory

$$k_2 = 8\pi r D, \qquad [12.59]$$

where r is the radius and D the diffusion coefficient of the particle. Introducing the expression for D from the Einstein equation [12.4]:

$$k_2 = \frac{4kT}{3\eta}. \qquad [12.60]$$

The time $t_{\frac{1}{2}}$ taken for the number of particles per unit volume to decrease from the initial value v_0 to half the initial value $v_0/2$ is given by the following equation which is analogous to equation [5.42], (p. 161):

$$t_{\frac{1}{2}} = \frac{1}{k_2 v_0} \qquad [12.61]$$

$$\therefore t_{\frac{1}{2}} = \frac{3\eta}{4kTv_0}. \qquad [12.62]$$

For water as the dispersion medium at 20° C, $\eta = 1.0019 \times 10^{-3}$ kg m^{-1} s^{-1}, $T = 293.16$ K while $k = 1.3806 \times 10^{-23}$ J K^{-1} (or kg m^2 s^{-2} K^{-1}), so that

$$t_{\frac{1}{2}} = \frac{3 \times 1.0019 \times 10^{-3}}{4 \times 1.3806 \times 10^{-23} \times 293.16} \cdot \frac{1}{v_0}$$

$$t_{\frac{1}{2}} \text{ (in s)} = 1.86 \times 10^{17}/v_0 \text{ (in m}^{-3}\text{)}. \qquad [12.63]$$

For many hydrophobic sols v_0 is of the order 10^{16}–10^{18} particles per m³ so the calculated value of $t_{\frac{1}{2}}$ is small and of the order 10–10^{-1} s. Smoluchowski's theory is therefore called the theory of rapid coagulation. Experimentally determined rate constants agree fairly well with those calculated from Smoluchowski's theory, but many deviate from it, suggesting the existence of attractive and repulsive forces between the particles. Deryaguin, Landau, Verwey and Overbeek have considered the nature of these forces and the remainder of this section is devoted to their (DLVO) theory.

Fig. 12.11 Potential energy curve for the interaction of two colloidal particles in the presence of a low concentration of ionic material

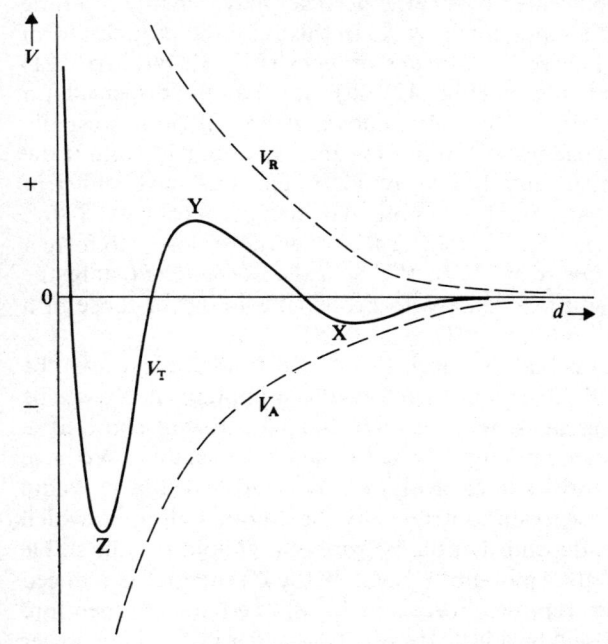

The attractive force between colloidal particles is the van der Waals force (see p. 31), which, of course, increases with decreasing distance between the particles (Fig. 12.11). Between atoms and molecules this force is universely proportional to the sixth power of the distance of separation and hence is of short range comparable to atomic dimensions. Since colloidal particles contain many atoms and molecules, the van der Waals force between two particles is the sum of the attractions between each atom in one particle and each atom in the other particle. The resultant force has a long range comparable with the dimensions of the particles. The repulsive force between colloidal particles is electrical on account of their like charges and increases with increasing charge or ζ potential and with decreasing separation.

The potential energy of interaction, V, which is the integral of the force with respect to the interparticle distance, d, is more commonly considered than the force itself and is plotted against d in Fig. 12.11. The potential energy of attraction, V_A, is approximately proportional to d^{-2}, whereas the potential energy of repulsion, V_R, decreases exponentially with d. The total potential energy, V_T, therefore follows a complex curve (e.g. Fig. 12.11). As the particles approach each other V_T often passes through a secondary minimum at X, a primary maximum at Y and a primary minimum at Z. At smaller interparticle distances the electron clouds of the particles repel each other and V_T increases steeply. If the decrease in V_T at X is large compared with the thermal energy, represented by kT, the particles may remain for a time in a metastable state represented by X. In this state the particles form loose aggregates, known as *floccules* or *flocs* (Fig. 12.14(c), p. 521) which after sedimentation (Fig. 12.14d) are readily dispersed on shaking. The formation of floccules, known as flocculation, is usually only significant for particles of radius in excess of about 100 nm. If the maximum of Y is sufficiently low, or if the particles have sufficient kinetic energy to surmount the potential energy barrier at Y (cf. Fig. 5.2, p. 166), they may approach each other more closely to form a stable state represented by Z. In this so called *coagulated* state the particles form a tight aggregate with properties similar to those of a solid body (Fig. 12.14(b), p. 521).

If the ζ potential is high (see Fig. 12.11), the repulsive force will be strong and V_R will be great; therefore the minimum at X will be shallow, the barrier at Y will be high and colloid will show little tendency to flocculate or coagulate and is said to be 'stable'. We have seen that small quantities of electrolytes may be adsorbed by colloidal particles and give rise to, or contribute to, the ζ potential, upon which the repulsive forces depend. Thus a hydrophobic colloid may be stable in the presence of 10^{-3} mol dm^{-3} NaCl. If the ζ potential is reduced (see Fig. 12.12), the repulsive force and V_R will be reduced; therefore the minimum at X will be a little deeper, the barrier at Y will be lower and the colloid will more readily attain the coagulated or precipitated state at Z. This may be achieved by adding to the hydrophobic colloid a relatively large concentration of an electrolyte, e.g. 10^{-1} mol dm^{-3} NaCl. Ions of charge opposite to that of the colloidal particle are preferentially adsorbed in the fixed part of the double layer and the distribution of ions in the diffuse part of the double layer is altered, with the result that the ζ potential is reduced. The critical electrolyte concentration at which rapid coagulation takes place is that which reduces the ζ potential to a critical value near zero, such that the maximum at Y has the same potential energy as the separated particles, as shown by the dotted curve in Fig. 12.12.

If the precipitating power of ions of different charge (or valency)

Stability of Lyophobic Colloids 517

Fig. 12.12 Potential energy curve for the interaction of two colloidal particles in the presence of sufficient added electrolyte to cause coagulation. (The dotted curve corresponds to the critical electrolyte concentration.)

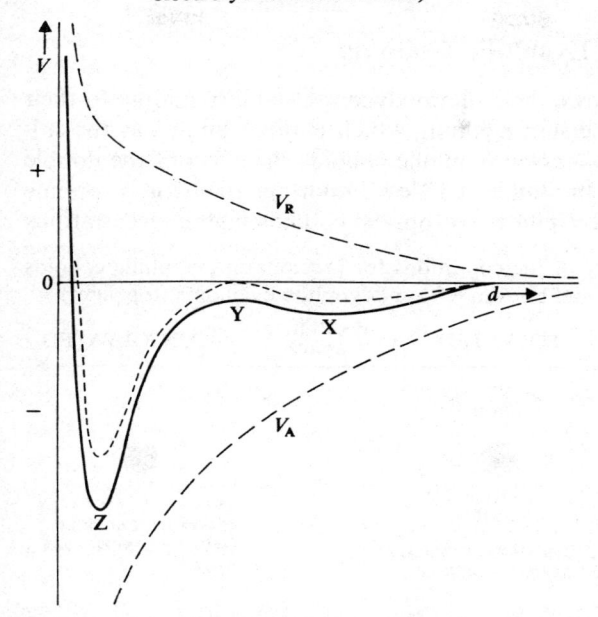

are compared experimentally, the concentrations (moles per unit volume) of monovalent, divalent and trivalent ions required *in solution* to produce the same coagulating effect are in the approximate ratio 1 : 0.02 : 0.001. This is an illustration of the Schulze-Hardy rule, which states that the coagulation is brought about by ions of charge opposite in sign to that of the colloidal particles, and that there is a marked increase in precipitating power with increasing valency of the ions. The ratio of the numbers or moles of monovalent, divalent and trivalent ions actually *adsorbed* by the particles in order to have the same coagulating power are, according to the DLVO theory, in the ratio $1 : \frac{1}{2} : \frac{1}{3}$. According to the theory this is because the greater part of the effect of the ions is due to their being incorporated in the fixed part of the double layer, and this must depend upon their adsorption from solution. Qualitatively, it can be seen that the adsorption of, for example, one trivalent ion from solution will take place for more readily than the adsorption of three monovalent ions. Quantitatively, by applying the Freundlich adsorption isotherm (equation [5.83] and p. 453) to the process and putting, say, $n = 6$, then in order for the ratio of the amounts of monovalent, divalent and trivalent ions *adsorbed* to be in the ratio $1 : \frac{1}{2} : \frac{1}{3}$, their relative concentrations in solution will need to be in the ratio $1^6 : \frac{1}{2}^6 : \frac{1}{3}^6$

i.e. 1 : 0.0156 : 0.00137. If a more refined isotherm is used and if ionic size and other effects of charge are taken into account, better agreement with the experimental ratio can be obtained.

Precipitation of Lyophilic Colloids

Lyophilic colloids owe their thermodynamic stability mainly to their affinity for the dispersion medium, which is often known as the solvent. Although for charged lyophilic colloids the effect of the double layer is essentially the same, it is less important than for lyophobic colloids. Thus, to precipitate hydrophilic colloids high concentrations

Fig. 12.13 Summary of the conditions for precipitating lyophilic colloids (top left and bottom left) and lyophobic colloids (top right)

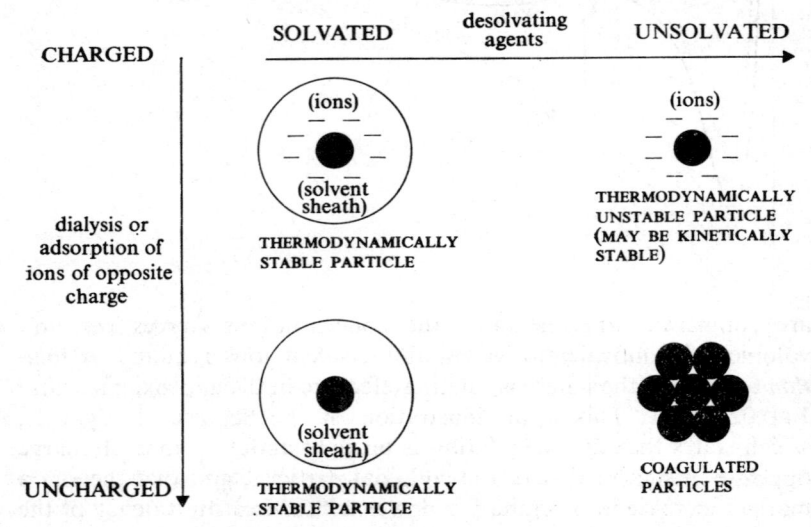

	SOLVATED	desolvating agents	UNSOLVATED
CHARGED	(ions) THERMODYNAMICALLY STABLE PARTICLE (solvent sheath)	→	(ions) THERMODYNAMICALLY UNSTABLE PARTICLE (MAY BE KINETICALLY STABLE)
dialysis or adsorption of ions of opposite charge ↓			
UNCHARGED	(solvent sheath) THERMODYNAMICALLY STABLE PARTICLE		COAGULATED PARTICLES

of electrolytes are generally required, and the process is known as 'salting out'. The effect of the electrolyte is then not mainly one of altering the charge on the particles but one of dehydration of the particles by the fact that the ions take up water molecules. The effectiveness of various ions in precipitating lyophilic colloids is largely independent of the colloid studied, and is governed by the tendencies of the various ions to become hydrated. They can therefore be arranged in a series of decreasing precipitating power, known as the Hofmeister or lyotropic series. Thus anions occur in the series, in the order: citrate > tartrate > sulphate > acetate > chloride > nitrate > iodide > thiocyanate; and the Group IIA and IA cations, which are less effective than the anions, are in the order: magnesium > calcium > strontium > barium > lithium > sodium > potassium. Valency is less important than for lyophobic sols. Ammonium sulphate, which is a highly soluble salt giving ions which have a high affinity for water molecules, is often used to precipitate proteins from colloidal solutions. The dehydration can also be carried out by adding an organic substance with a high affinity for water, such as ethanol or acetone. Precipitation of lyophilic colloids by addition of a liquid which is miscible with the medium but which does not dissolve the colloid is a general phenomenon shared by other solutions of macromolecules. Thus lipophilic plastic materials in hydrocarbon solvents are readily precipitated by alcohols, e.g. rubber from benzene solution by ethanol. If the desolvated particles retain sufficient charge to withstand precipitation by the non-solvent liquid, they behave like lyophobic colloids and are very sensitive to a small quantity of an electrolyte, provided, of course, that it dissolves in the medium (see Fig. 12.13).

Interactions between Colloidal Systems

Addition of one colloidal system to another may cause one or both to aggregate and precipitate or may bring about an increase in stability. Two sols which contain particles of opposite charge usually undergo mutual precipitation when mixed in proportions which cause mutual neutralisation of charge. Thus a mixture of a bismuth subnitrate suspension and tragacanth mucilage forms a firm mass at the bottom of the container. Coagulation arises from the mutual attraction between the positively charged bismuth ions on the bismuth subnitrate particles and the negatively charged carboxylate ions of tragacanth. Coagulation is prevented by the inclusion of sodium citrate or sodium phosphate, presumably because their anions form complexes with the bismuth ions.

Colloidal interactions can explain the stable existence of uncharged lyophobic sols, including some dispersions in non-polar media. Stability obviously cannot be attributed to the electrical double layer effect, since it is absent. These uncharged colloidal particles are invariably dispersed in media containing soluble, non-ionic macromolecules or surface active agents. The stability results from the adsorption of these molecules at the surface of the particles. The nature and thickness of the adsorbed layer is crucial to the stability and depends on the nature and concentration of the particles and the soluble material. Stability may arise from the reduced attractive force between the particles which on collision are separated by two thicknesses of adsorbed layer. If the attractive force is strong enough to allow interpenetration of the adsorbed layers on collision, a decrease in entropy arises. As a result an increase in Gibbs energy occurs on collision which may be sufficient to provide a repulsive force strong enough to act as an energy barrier to coagulation.

Lyophobic colloids may be protected from the precipitating

influence of small quantities of electrolytes by adding an excess of a lyophilic colloid to the sol. The colloids which give the most protection are those which bear a weak charge opposite in sign to that of the lyophobic colloid. They are more easily adsorbed, and the resultant charge which remains helps to prevent coagulation. Gelatin is frequently used to protect hydrophobic sols. In biological systems protective colloids maintain in solution particles of lyophobic colloids which might otherwise be precipitated too easily. Thus, calcium phosphate is maintained in 'solution' in milk by protective colloids such as casein.

In certain cases the added lyophilic colloid makes the lyophobic sol less stable. This may happen on adding a low concentration of an agent which protects the sol at higher concentrations. For example, suspensions of clay are precipitated by solutions of gelatin at 1 to 10 mg dm^{-3}, but 1 g dm^{-3} promotes protective action. When the sol particles and the agent are oppositely charged, adsorption of the agent may neutralise the charge of the sol particles so that the sol becomes unstable, whereas at high concentrations, protection may occur, e.g. silica with polyethylenimine of low molar mass. If a given lyophilic particle, such as a macromolecule, is attached to two or more lyophobic particles, aggregation may take place by a bridging process at low concentrations of the lyophilic colloid. At higher concentrations each lyophilic particle will be adsorbed by only one lyophobic particle, so that bridging does not arise and protective action occurs.

Suspensions

A suspension is a two phase system in which solid particles are dispersed in a liquid medium. The particles are frequently known as the internal, discontinuous or dispersed phase and the medium as the external, continuous or dispersing phase. In pharmacy suspensons are used as orally administered mixtures, as externally applied lotions, aerosols or injections. The size of the particles should be small, but is usually within, or slightly beyond, the upper colloidal range.

Like other dispersions, suspensions are thermodynamically unstable but the rate of aggregation is reduced by dispersion stabilisers, which are also known as suspending agents. These may be surface active agents, deflocculating agents or protective colloids. Stabilisation by surface active agents has been discussed on p. 480. These agents also promote wetting by displacing air from the surface of the particles (see p. 485). Deflocculating (or peptising) agents increase the repulsions between the particles by increasing the ζ potential (p. 516). To do this their molecules must be adsorbed by the particles and possess ionised groups. The action of protective colloids has been discussed above. For aqueous pharmaceutical suspensions these are hydrophilic colloids. Important organic examples include polypeptides (e.g.

Fig. 12.14 Arrangement of particles in various types of suspensions and emulsions
  (a) peptised suspension (discrete particles)
  (b) hard sediment (aggregate)
  (c) flocculated suspension (floccules)
  (d) flocculated sediment (floccules)
  (e) polymeric flocculated suspension (floccules)
  (f) polymeric flocculated sedimented (floccules)

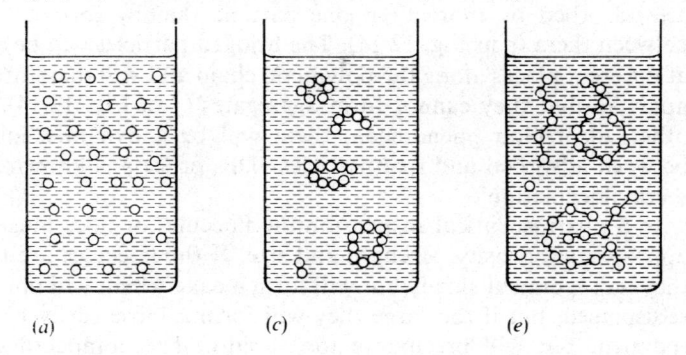

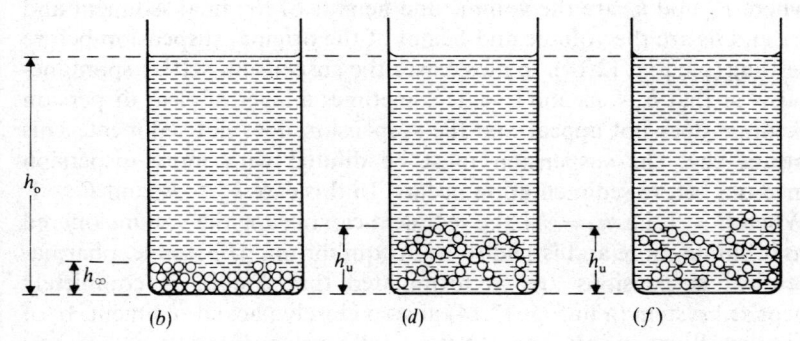

gelatin, casein), polysaccharides and their derivatives (e.g. natural gums, chondrus, sodium alginate, methyl cellulose, sodium carboxymethylcellulose), and synthetic polymers (e.g. carboxypolymethylene). Inorganic examples include powders and clays with an affinity for water (e.g. alumina, magnesia, silica, bentonite).

A pharmaceutical suspension should not settle rapidly, but any particles which do should form loose floccules (d or f in Fig. 12.14, see also p. 516) which are readily redispersed on shaking (c or e in Fig. 12.14) and should not form a compact aggregate or hard cake at the bottom of the container (b in Fig. 12.14). On p. 516 it was mentioned that lyophobic colloids form a compact aggregate (point Z in Fig.

12.11) if the activation energy barrier Y is overcome, but form loose floccules (at point X) if the activation energy is sufficiently high. The coarse suspensions used in pharmacy behave similarly, but if their ζ potential is insufficient to prevent aggreagation of such large particles, some other means must be employed to increase the activation energy and encourage flocculation. Hydrophilic colloids are commonly used to promote the controlled flocculation required. Their long molecules are adsorbed by more than one particle thereby forming a bridge between them (e in Fig. 12.14). The bridged particles will be adsorbed at different places along the polymeric chain and will therefore be kept apart so that they cannot form aggregates (f in Fig. 12.14). As with other adsorption phenomena, there will be a dynamic equilibrium between adsorbed and free particles. This process is known as 'polymeric flocculation'.

In a pharmaceutical suspension the floccules or flocs must have an appropriate porosity, strength and size. If the floccules are too small they will sediment slowly but will form a cake which will not easily be redispersed, but if too large they will form a loose easily redispersed sediment but will precipitate too rapidly. The compactness of the sediment is measured by the relative sedimentation volume which is given by

$$F = V_u/V_0 = h_u/h_0, \qquad [12.64]$$

where V_u and h_u are the volume and heights of the final sediment and V_0 and h_0 are the volume and height of the original suspension before settling (see Fig. 12.14). In most cases the suspension settles spontaneously so that $h_u < h_0$ and $F < 1$. Sometimes a layer of clear dispersion medium does not appear and the suspension does not sediment. This means that the suspension must be diluted with more dispersion medium before sedimentation occurs. In this case $h_u > h_0$ and $F > 1$. When $F = 1$, i.e. $h_u = h_0$, the sediment can occupy the volume offered to it but no more and is said to be in equilibrium. If possible, pharmaceutical suspensions are so formulated that $F \geqslant 1$. A completely peptised system (a in Fig. 12.14) gives a closely packed sediment, b, of very small volume, V_∞, height, h_∞, and F value, F_∞, where

$$F_\infty = V_\infty/V_0 = h_\infty/h_0. \qquad [12.65]$$

The degree of flocculation, β, is defined by:

$$\beta = F/F_\infty \qquad [12.66]$$

and is unity for a close packed sediment and increases with increasing porosity of the flocculated sediment, and with decreasing co-ordination number. The latter is the number of particles linked to a single particle, i.e. the number of nearest neighbours. For a compact sediment the co-ordination number is large, e.g. 12 for close-packed spheres. A loose floc has an open, three-dimensional structure, with many large empty spaces, like a scaffold, and the co-ordination number will be small, e.g. 2 to 4.

Since the particles (peptised particles or flocs) of a suspension are relatively large, gravity alone is sufficient to cause sedimentation (Fig. 12.14) and an expression for its rate may be derived as follows. If the forces between the individual particles are negligible and if the system is left undisturbed, the downward force on the particle exerted by gravity is balanced by the upward force of buoyancy due to the displaced dispersion medium (by Archimedes' principle) plus the frictional resistance of the medium. Stated mathematically:

$$mg = m'g + fv,$$

where m is the mass of the particle, m' is the mass of an equal volume of medium, g is the acceleration due to gravity, f is the frictional coefficient and v is the rate of sedimentation.

$$\therefore v = (m-m')g/f. \qquad [12.67]$$

The mass in each case is equal to the product of the volume and the density. If the particles are spherical and uniform:

$$m = \tfrac{4}{3}\pi r^3 \varrho_i, \quad m' = \tfrac{4}{3}\pi r^3 \varrho_e,$$

where r is the radius of a particle, ϱ_i is the density of the internal phase (particle) and ϱ_e is the density of the external phase (medium) and f is given by Stokes' law (equation [12.3]), thus $f = 6\pi r\eta$. Substituting into equation [12.67] gives

$$v = \tfrac{4}{3}\pi r^3(\varrho_i - \varrho_e)g/6\pi\eta r$$

$$v = \frac{2r^2(\varrho_i - \varrho_e)g}{9\eta}. \qquad [12.68]$$

This is the equation required and is another form of Stokes' law. To account for the non-uniformity of size and shape of the particles a correction factor must be introduced which can be determined by experiment. The individual particles noticeably affect each other by mutual interaction and by introducing turbulence, or in other words, hindered settling occurs, if the volume of the internal phase is greater than about 5 per cent of the total volume. Such mutual interferences are negligible, i.e. free settling occurs, if the volume fraction of the internal phase is less than about 5 per cent. In general Stokes' law is obeyed by dilute suspensions which contain less than about 1 per cent (w/v) of solids.

Equation [12.68] shows that the rate of sedimentation increases with increasing gravitational field and with increasing difference in density between the particles and medium. Sedimentation will not occur when the two densities are equal. The rate decreases with decreasing particle radius and with increasing viscosity of the medium (p. 537).

Emulsions

An emulsion is a dispersion in which the two phases are immiscible or partially miscible liquids. One liquid is dispersed in the other in the form of fine droplets. The terminology applied to the phases is the same as for suspensions. In pharmacy the two liquids are usually water and an oil. The size of the droplets should be small but is usually within or slightly beyond the upper colloidal range.

The use of emulsions in pharmacy is so widespread that it is impossible to attempt to discuss specific examples here. However, a few of the reasons for their use might be mentioned. Oily substances become much more palatable, and are often much more easily absorbed, when they are administered in the form of emulsions. Drugs which are insoluble in water but soluble in an oil can be presented in an emulsion as a fluid dosage form. Where it is necessary to prolong the action of a drug which would be dispersed too quickly in aqueous solution, it may be caused to dissolve more slowly if it can be made oil-soluble and administered as an emulsion. By forming an appropriate emulsion, a pharmaceutical preparation may be given the correct physical properties (varying from the highly viscous and plastic ointments to the emulsions of low viscosity for internal use).

Like other dispersions, emulsions are thermodynamically unstable but the rate of aggregation is reduced by the presence of an emulsifying agent. This substance is adsorbed at the surface of the droplets (the oil–water interface), thereby lowering the interfacial free energy (see p. 480) and increasing the ζ potential. Thus some surface active agents are emulsifying agents.

Emulsifying agents may be classified according to whether they produce an oil-in-water (abbreviated to O/W), or a water-in-oil (abbreviated to W/O) emulsion. The type of emulsion formed depends on the relative degrees of hydrophilic and lipophilic character of the emulsifying agent. Agents which are predominantly hydrophilic form O/W emulsions. The agents include the alkali metal salts of fatty acids and most of the hydrophilic organic and inorganic colloids listed on p. 521 as stabilising aqueous suspensions. For example, milk, an O/W emulsion, is stabilised by the polypeptide, casein. On the other hand, agents which are predominantly hydrophobic form W/O emulsions. These agents include divalent and trivalent metal salts of fatty acids (e.g. oleates and stearates of Mg^{2+}, Al^{3+} and Ca^{2+}), certain non-ionic surface active agents such as long chain alcohols and esters, certain fats such as lanolin, and hydrophobic powders such as carbon black.

Factors in addition to the lowering of interfacial free energy and the existence of a ζ potential are important to the stability of emulsions. For the stabilisation of an O/W emulsion by a surface active agent, the interfacial film must be of the condensed liquid type (p. 471). The film must be condensed so as to surround the droplets completely without any gaps and must be liquid so that it can easily be reformed on distortion. The stabilisation of an O/W emulsion by a surfactant is very similar to the solubilisation of a non-polar material in the interior of a micelle. The oil may be imagined to form a spherical droplet with the film of surfactant molecules adsorbed on its surface (p. 480, Fig. 12.2a). The hydrocarbon chains are directed towards the oil droplet and the polar groups are directed outwards towards the water. The stability of an emulsion is greatly increased and the interfacial free energy further reduced, if the emulsifying agent is not a single substance, but consists of at least two components, one of which is appreciably water soluble (e.g. sodium cetyl sulphate or a sodium soap) and the other appreciably oil soluble (e.g. cholesterol or cetyl alcohol). To produce the most stable emulsions, the two substances together must be able to form a stable complex condensed film with the molecules of both types at the interface (see p. 470) and with a negligible interfacial free energy. The relative amounts of each component are important for this. For example, the quantities of sodium cetyl sulphate and cholesterol required to form the most stable emulsion of liquid paraffin in water are equimolecular in the actual interfacial film but with some additional sodium cetyl sulphate in the aqueous phase.

For the stabilisation of W/O emulsions by a surface active agent, the same conditions apply except that the interfacial film must be solid, or at least very viscous, rather than liquid. In this case the water may be imagined to form a spherical droplet with a film of surfactant molecules adsorbed on its surface, (p. 480, Fig. 12.2b). The polar groups are directed towards the water droplet and the hydrocarbon chain are directed outwards towards the oil. A soap derived from a divalent or trivalent metal (e.g. calcium or barium cetyl sulphate) mixed with a second substance (e.g. cholesterol) stabilises W/O emulsions. In cases where a surface active agent appears to be effective alone as an emulsifying agent, it is possible that an impurity is acting as the second component.

The hydrophilic colloids, such as polypeptides and polysaccharides, which stabilise O/W emulsions form visible, coherent, polymolecular films on the droplets. The observed properties of the droplets are often those of the emulsifying agent rather than of the internal phase. For example, droplets of mineral oil emulsified in dilute gelatin solution move in on electric field as gelatin does and show an isoelectric point at pH 4.7 corresponding to that of gelatin itself.

Solids may also act as emulsifying agents, provided that they are finely divided, remain at the interface and form a closely packed film. Solid particles which are too large will be pulled through the interface by gravity.

The relationship between the nature of the emulsifying agent and the type of emulsion formed is explained by the theory of preferential wetting which was originally proposed for finely divided solids. When the emulsifying agent is preferentially wetted by one phase, more particles or molecules of emulsifying agent can be accommodated at the interface if the interface is convex towards the side of the preferred phase, or, in other words, if the preferentially wetted phase is the dispersion medium. Agents which are more hydrophilic will be preferentially wetted by the water phase which therefore becomes the dispersion medium. Conversely, agents which are more lipophilic will be preferentially wetted by the oil phase which therefore becomes the dispersion medium. If the hydrophilic and lipophilic strengths are too similar, the agent will not be selective enough to form an emulsion of any type and will act as a wetting or spreading agent. If the hydrophilic and lipophilic strengths are too dissimilar, the agent will migrate to the phase for which it has the greater affinity, or, in other words, it will be negatively adsorbed at the interface, so emulsification will not occur.

There are four methods which are commonly used for determining whether a given emulsion is of the O/W or the W/O type:

(*a*) O/W emulsions have a higher electrical conductivity than W/O emulsions, since oils are normally good insulators;

(*b*) Dyestuffs which are oil-soluble will only spread through the liquid, when dusted on the surface, if the emulsion is of the W/O type. Conversely, O/W emulsions are coloured by water-soluble dyes.

(*c*) Water is only rapidly miscible with O/W emulsions, and oils are only rapidly miscible with W/O emulsions.

(*d*) If the oil is fluorescent, O/W emulsions have a speckled appearance during exposure to ultra-violet light while W/O emulsions exhibit a fairly uniform fluorescence.

Emulsions may be made by agitating the two liquid phases in the presence of the emulsifying agent. Special machines capable of generating high shearing forces are used to produce a smaller and more uniform particle size than is possible by normal shaking. Examples of such machinery are stirrers, vibrating mixers, colloid mills and orifice homogenisers. In the latter, a coarse emulsion is forced under high pressure through a small adjustable orifice. In the colloid mill the mixture is forced between a plate rotating at a high speed and a stationary plate.

Continued agitation may cause an emulsion to separate into its component liquids, a process known as breaking, cracking or demulsification, or to undergo phase inversion or phase reversal. In the latter process the roles of the internal and external phases are interchanged, as for example when the O/W emulsion, cream, is churned to produce butter, a W/O emulsion.

Breaking of an emulsion may be brought about by undue physical agitation as mentioned above, by extremes of temperature or by addition of a substance which precipitates or interacts with the emulsifying agent. Examples of the latter are addition of acid or metal ions of different valency to an emulsion stabilised by a soap or by addition of a second emulsifying agent or electrolyte capable of neutralising the charge on the droplets. The latter effect suggests that electrical double layer repulsions contribute to the stability of emulsions. Addition of an electrolyte reduces the ζ potential of the particles, thereby causing them to coagulate.

Demulsification is often accompanied by phase inversion or creaming. The breakdown of an emulsion proceeds in two stages. The first is flocculation in which the droplets come together to form aggregates but in which each droplet retains its identity. Flocculation is often reversible. In the second stage groups of aggregates unite to form a single larger drop. This process is called coalescence, is irreversible and may continue until demulsification is complete. The onset of demulsification may be detected by microscopic determination of the particle size distribution, which is not an easy task. Incipient breaking of an emulsion shows itself in the formation of droplets in the upper size range.

Phase inversion may be brought about by undue physical agitation as mentioned above, by a change in temperature, by an alteration of the volume fractions of the phases or by the addition of a suitable electrolyte. An example of the latter is the addition of a calcium salt to a O/W emulsion stabilised by a sodium soap. Phase reversal occurs because a calcium soap is formed which stabilises a W/O emulsion. A possible mechanism of this process is as follows. Addition of the electrolyte reduces the ζ potential whereupon the droplets come together in a closely packed structure. The new film of emulsifying agent breaks in various places around each droplet and reforms around the interstices (spaces) between them. The liquid in the interstices which was formerly the external phase therefore becomes the internal phase. The rupture of the film around the original internal phase causes the release of this liquid so that it becomes the external phase.

The phase-volume theory of emulsification is based upon the fact that when uniform spheres are closely packed together they cannot occupy more than 74.02 per cent of the total volume. Ostwald postulated that an emulsion prepared from more than 74 per cent (v/v) of the internal phase would either invert or break. The fact that some stable emulsions contain more than 74 per cent (v/v) of internal phase (occasionally as much as 99 per cent) can be explained by variations in size and irregularities in the shape of the droplets. Thus the phase present in the higher concentration may not necessarily be the external phase.

In addition to breaking and inversion some emulsions can undergo creaming which may be defined as the separation of an emulsion into two or more layers which differ in the relative proportions of the internal and external phase. Creaming is exemplified by the spontaneous separation of milk into cream and skimmed milk on standing. With shaking or agitation the creamed emulsion always returns to the original state, because in creaming the droplets of the creamed emulsion are still dispersed and covered by an interfacial film of the emulsifying agent. Equation [12.68] applies to a single droplet involved in creaming as well as in sedimentation and with the same reservations. It is found that the rate of creaming, v, is inversely proportional to the concentration of internal phase, ϕ, so that equation [12.68] may be written

$$v = \frac{kr^2(\varrho_i - \varrho_e)g}{\phi \eta}, \qquad [12.69]$$

where k is a constant which depends on the system. The other quantities have the same significance as before. This equation shows that the relative magnitude of the densities of the internal and external phases, ϱ_i and ϱ_e respectively, is crucial to the rate and course of creaming. If $\varrho_i = \varrho_e$, $v = 0$ and creaming does not occur. If $\varrho_i > \varrho_e$, v is positive, which means that the layer richer in the internal phase falls to the bottom, a process known as downward creaming. If $\varrho_i < \varrho_e$, v is negative, which means that the layer richer in the internal phase rises to the top, a process known as upward creaming. The separation of cream from milk is an example of upward creaming.

In pharmacy the rate of creaming of emulsions like the rate of sedimentation of suspensions (p. 523) must be reduced as much as possible. One method is to reduce the radius, r, of the droplets or particles by further dispersion (see equations [12.68] and [12.69]). An additional method is simply to increase the viscosity, η, of the medium through the addition of 'thickeners' such as glucose, sucrose or glycerol. Alternatively, the rheological properties may be altered by adding lyophilic colloids to the system so that it becomes pseudoplastic as described in the next chapter. Viscosity is insufficient to describe the flow properties in systems as complex as these.

Problems

1 In water at 20° C the diffusion coefficients of tobacco mosaic virus and beef insulin are 5.3×10^{-8} cm^2 s^{-1} and 7.53×10^{-7} cm^2 s^{-1} respectively. Calculate the mean time for each of these molecules to diffuse 10 μm, which approximates to the diameter of a typical living cell. Assuming that the viscosity of water is 1.002 g m^{-1} s^{-1} at 20° C and that the insulin molecule is spherical, calculate the mean diameter of this molecule.

2 Calculate the number average and mass average molar masses for a colloidal solution which contains equal *numbers* of particles of molar masses 10 kg mol^{-1} and 50 kg mol^{-1} and for one which contains equal *masses* of these particles.

3 A membrane impermeable to carboxymethylcellulose anions but permeable to sodium ions and choloride ions separates initially pure water from an aqueous solution of sodium carboxymethylcellulose containing 0.01 mol dm^{-3} sodium chloride. At equilibrium the former liquid is found to contain 0.0065 mol dm^{-3} chloride ion. By application of the Donnan membrane equilibrium (e.g. equation [12.9]) calculate the concentration of carboxymethylcellulose present, expressed as moles of electronic charge per dm^3. If the sodium chloride had been initially present only on the opposite side of the membrane from the sodium carboxymethylcellulose, what would be the concentration of chloride ions on that side at equilibrium?

4 Aqueous solutions of a polypyrrolidone at various mass concentrations, ϱ, have the following values of osmotic pressure, π, at 298 K:

ϱ/g dm^{-3}	2.5	5.0	7.5	10.0
π/Pa	289	590	904	1230

By plotting a suitable graph and extrapolating it to $\varrho = 0$, calculate the molar mass of the polypyrrolidone. If the polypyrrolidone were polydisperse, what type of average molar mass would be obtained from these data?

5 Calculate the molar mass of a sample of pyroxylin (nitrated cellulose) from the following data. For solutions in acetone, of refractive index $n_0 = 1.3589$ at 589.3 nm and at 298 K, the rate of change of refractive index with respect to mass concentration, $dn/d\varrho = 0.105$ cm^3 g^{-1}. The value of the ratio of mass concentration to turbidity, ϱ/τ, extrapolated to zero concentration is 77.05 g cm^{-3}. If the pyroxylin were polydisperse, what type of average molar mass would be obtained from these data?

6 Determine the molar mass of myoglobin from the following ultracentrifuge sedimentation rate data for the colloidal solution at 293 K. The sedimentation coefficient, $\dfrac{dx/dt}{\omega^2 x}$, has the value 2.04×10^{-13} s. The partial specific volume of the disperse phase, v, is 7.41×10^{-4} m^3 kg^{-1} and the diffusion coefficient, D, is 1.13×10^{-10} m^2 s^{-1}. The density of the dispersion medium may be taken as 1000 kg m^{-3}.

7 A colloidal solution of congo-red in sodium chloride solution as dispersion medium was rotated in a low-speed ultracentrifuge at

293 K until sedimentation equilibrium was established. The speed of rotation was 1883 radians per sec, the density of the dispersion medium was 1002.3 kg m^{-3} and the partial specific volume of the disperse phase, v, was 6.00×10^{-4} m^3 kg^{-1}. The concentrations c_1 and c_2 of congo red were measured in arbitrary units at distances x_1 and x_2 respectively, from the axis of rotation. The values were: $x_1 = 5.72$ cm, $x_2 = 5.75$ cm; $c_1 = 39.76$, $c_2 = 42.18$. Calculate the molar mass of the congo-red colloidal particles.

8 The molar masses of various polymer fractions of polystyrene are found to be related to the limiting viscosity numbers, in benzene solution, by Staudinger's law, with the constant K equal to 5.2×10^{-3} m^3 kg^{-2} mol^{-1}. The following table gives the viscosity ratios of one such fraction at various mass concentrations ϱ:

ϱ/kg m^{-3}	10.4	14.0	20.8	26.0	31.0	36.8
η_r	1.098	1.133	1.200	1.253	1.304	1.365

At each concentration calculate the viscosity ratio increment and the viscosity number and plot values of the latter versus ϱ. Hence obtain the limiting viscosity number for this fraction and its viscosity average and mass average molar masses.

9 The electrophoretic mobility of horse serum albumin in water at pH 4.0 is 9.3×10^{-5} cm^2 s^{-1} V^{-1} at 20° C. Calculate the ζ-potential of the protein under these conditions.

10 A sample of magnesium oxide powder which has a mean particle diameter of 1.5 μm and true density 3580 kg m^{-3} is suspended in an aqueous solution of density 1080 kg m^{-3} and viscosity 1.12 g m^{-1} s^{-1}. Calculate the rate of sedimentation of the particles under gravity ($g = 9.81$ m s^{-2}) assuming negligible interference between the particles. What distance would a particle fall in one hour?

13 Rheology

Introduction

Rheology is the science of flow and deformation of matter. It is important in pharmacy because the texture, consistency and physical properties of most types of formulation are largely governed by the properties of flow and deformation of the constituent solutions, colloids and dispersions (suspensions, emulsions and powders). The types of formulation of interest are diverse and include 'mixtures', lotions, creams, pastes, ointments, suppositories and tablets. In order to make preparations more acceptable to the patient, pharmaceutical technologists are studying the relationships between the physical properties of preparations and subjective sensory assessments, such as consistency and texture, mentioned above. This branch of rheology is often known as psychorheology. This chapter, however, is concerned only with physical properties and attempts to give molecular explanations whenever possible.

Deformation, Elasticity and Flow

When acted on by a force, a body will be deformed. The force per unit area (F/A) producing a deformation is known as the *stress*. If the force is applied in a direction perpendicular to the area, the stress is known as the normal stress, σ, or sometimes as the tensile stress in the case of a solid.

Deformation may be defined as a change in the relative positions of different parts of the body. If the body tends to return to its original shape after the force has been removed, the body is said to be elastic (or to exhibit elasticity). Many solids, such as metals, are elastic. Elastic deformation is usually expressed as *strain* which is defined as the *relative* change in a dimension, namely a length, an area or the volume of the body. For example, the linear strain (or relative elongation), ε, of a solid is the change in length, Δl, in the direction of the force divided by the length, l_0, in the absence of the force. Strain is always dimensionless.

Fig. 13.1 Stress-strain relationship of an elastic solid. The point L represents the elastic limit. The line OL represents Hooke's law

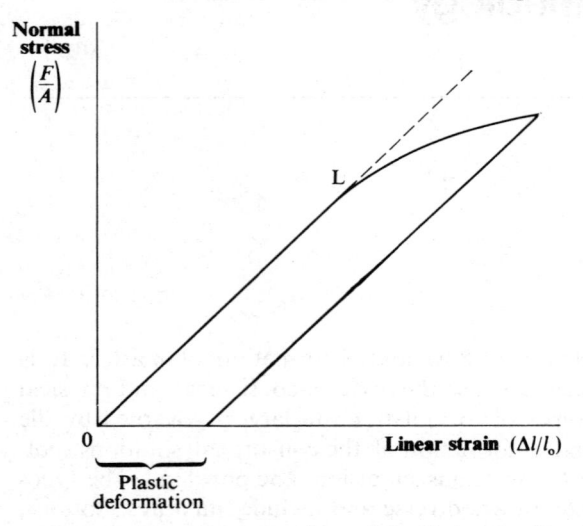

If the body returns to its original shape after the force has been removed, deformation is said to be reversible and the body is referred to as *perfectly elastic* or as an *ideal elastic* body. Such bodies are also known as *Hookean* (or Hooke) bodies since they usually obey Hooke's law which states that the normal stress, σ, is proportional to the linear strain, ε, i.e.

$$\sigma = E\varepsilon \qquad [13.1]$$

or
$$\frac{F}{A} = E\frac{\Delta l}{l_0}, \qquad [13.2]$$

where E is a proportionality constant known as the modulus of elasticity or Young's modulus. This behaviour is illustrated by the line OL in Fig. 13.1. For the compression of a Hookean body on all sides in a confined space the pressure ($p = F/A$) is the stress and this is proportional to the volume strain, θ, which is equal to $\Delta V/V_0$, thus:

$$\frac{F}{A} = -K\frac{\Delta V}{V_0} \qquad [13.3]$$

or
$$p = -K\theta, \qquad [13.4]$$

where K is a proportionality constant known as the compression modulus or bulk modulus. This is the only elastic constant for gases and pure liquids. The minus sign allows for the fact that under compression the volume decreases, so that p is positive and ΔV and θ are negative. Since strain is dimensionless, E and K have the same dimension as stress and pressure and all have the SI unit Pa or $N\,m^{-2}$ or kg m^{-1} s^{-2}. For steel $E = 2.5 \times 10^{11}$ N m^{-2} and $K = 1.6 \times 10^{11}$ N m^{-2}; for rubber $E = 8 \times 10^5$ N m^{-2} and $K = 1.9 \times 10^7$ N m^{-2}.

Stretching of a body is nearly always accompanied by a decrease in the dimensions at right angles to the force. Conversely, longitudinal compression is accompanied by an increase in the width. For a perfectly elastic solid the ratio of lateral strain (i.e. change in width ÷ original width) to longitudinal strain (i.e. change in length ÷ original length) is a constant known as the Poisson ratio. This quantity is given the symbol μ and is dimensionless. For steel, $\mu = 0.30$ and for rubber $\mu = 0.49$. If the properties of a material are the same in different directions, it is said to be isotropic. For an isotropic and perfectly elastic material, the Poisson ratio is related to Young's modulus and to the compression modulus by the equation:

$$\mu = \frac{1}{2} - \frac{E}{6K} \qquad [13.5]$$

Many solids, especially metals, are perfectly elastic at low to moderate tensile stresses (below L in Fig. 13.1). A metal spring is a typical Hookean body and is often chosen as a model for elastic behaviour as shown in Fig. 13.2.

Fig. 13.2 The three basic elements in models of rheological behaviour. The Hooke body exhibits perfectly elastic behaviour and is represented by a spring. The Newton body gives ideal viscous behaviour and is represented by a dashpot. The Saint-Vernant element is a block that is not pulled along until friction is overcome and represents a yield value.

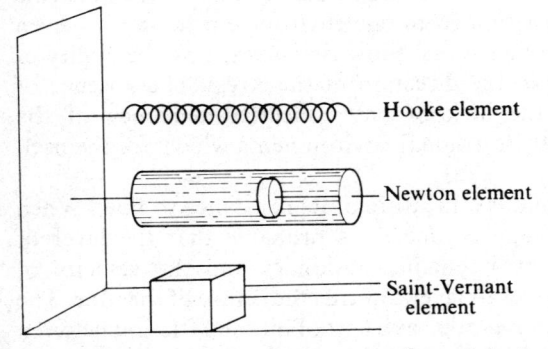

If the body does not return to its original shape when the force is removed, deformation is irreversible and flow takes place. For example, if the normal stress on a solid exceeds a certain value, corresponding to the *elastic limit* which is indicated by L in Fig. 13.1, Δl does not return to zero when the stress is removed. The solid is permanently deformed at zero stress by an amount which is known as the *plastic deformation* and which gives a measure of the flow that occurred when the elastic limit was exceeded. A molecular explanation is as follows.

Below the elastic limit the body obeys Hooke's law because the stress is insufficient to move two neighbouring molecules beyond the range of their mutual intermolecular forces in the crystal lattice, so that the forces can restore the relative positions of the molecules when the stress is removed. Beyond the elastic limit the stress is sufficient to move two molecules beyond the range of their mutual intermolecular forces, and their new neighbours exert intermolecular forces which restore them to new equilibrium positions when the stress is removed. The observed elastic limit for crystals, such as metals, is only $1/10^4$ to $1/10^2$ of that calculated for a perfect crystal lattice. The explanation is that actual crystals contain dislocations (described on p. 36), which facilitate the movement of planes of molecules in the crystal lattice thereby lowering the elastic limit. This movement of planes of molecules constitutes flow and is equivalent to the movement of the dislocations in the opposite direction.

A body which has no tendency to return to its original shape after the stress has been removed exhibits flow but no elasticity and may be termed an *ideal viscous* body. Such a body is also known as a *Newtonian* body because it obeys Newton's law of viscous flow which is discussed in the section on p. 537. Liquids and gases are typical Newtonian bodies. A dashpot, which consists of a cylinder containing a Newtonian liquid and a loosely fitting piston as shown in Fig. 13.2, is often chosen as a model for ideal viscous behaviour. A molecular explanation of the phenomenon is based on the disordered structure of the liquid and gaseous states which can be thought of as a lattice with many vacant positions and with a very large number of dislocations. Each molecule can undergo rapid translation from one environment to another even in the absence of an applied stress. Flow is explained by the ability of the molecules to move in the direction of the stress. The absence of elasticity arises from the lack of any concerted tendency of the molecules to return to their original environments which are themselves constantly changing.

The viscosity of a liquid is a measure of its resistance to flow. When a liquid is flowing through a tube, it is probable that the layer of molecules next to the wall remains stationary and the velocity of motion increases from layer to layer towards the centre of the tube. The viscosity can therefore be regarded as a sort of internal friction between adjacent layers and is a result of molecular collisions many of which are caused by that component of the molecular velocities which is at right angles to flow. A similar effect occurs in problems of flow of gases, but since the molecules of a gas move about far more freely than those of a liquid, the effect is not as noticeable.

Shear

Shear is an important type of deformation applicable to flow as well as to certain types of elasticity. In its simplest form shear is a process whereby parallel planes of infinitesimal thickness, dz, and of equal area, A_{xy}, slide over each other in the direction of flow, x, as do playing cards in a pack (Fig. 13.3). If δx is the displacement along the x axis of a plane at a perpendicular distance z from the origin, the *shear strain* is $\delta x/z$ which is equal to tan α in Fig. 13.3. In this discussion x, y or z represent the distance measured parallel to the respective axis, but if used as a subscript each indicates that the subscripted quantity is measured parallel to the corresponding axis. Shear is also possible in other geometrical arrangements some of which are shown in Fig. 13.4.

The force per unit area (F_x/A_{xy} in Fig. 13.3) producing shear is known as the *shear stress* and is given the symbol τ. In this type of stress the force producing deformation is parallel to the area over which it is applied as well as to the direction of deformation.

When a Hookean body is sheared, the shear stress, τ, is proportional to the shear strain which, for simple shear shown in Fig. 13.3, is equal to tan α. Stated mathematically

$$\frac{F_x}{A_{xy}} = G \frac{\delta x}{z} \qquad [13.6]$$

or
$$\tau = G \tan \alpha, \qquad [13.7]$$

where G is a proportionality constant known as the shear modulus or rigidity modulus, which like the other moduli, has the same dimensions and units as stress. $G/\text{N m}^{-2}$ is 8×10^{10} for steel, 2.9×10^5 for rubber.

Fig. 13.3 Simple shear

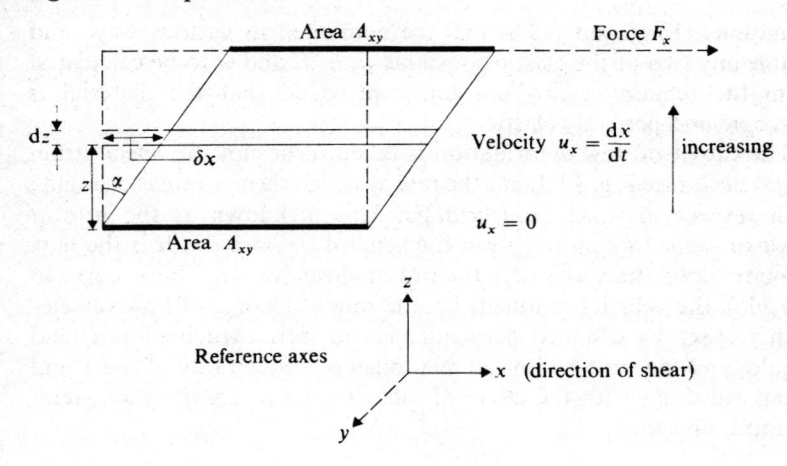

Fig. 13.4 Some more complex types of shear
  (a) Rotational shear. The cylindrical elements rotate with different angular velocities
  (b) Capillary shear. The cylindrical elements move parallel with the axis at different linear velocities.
  (c) Circular shear. The disc elements rotate with different angular velocities.

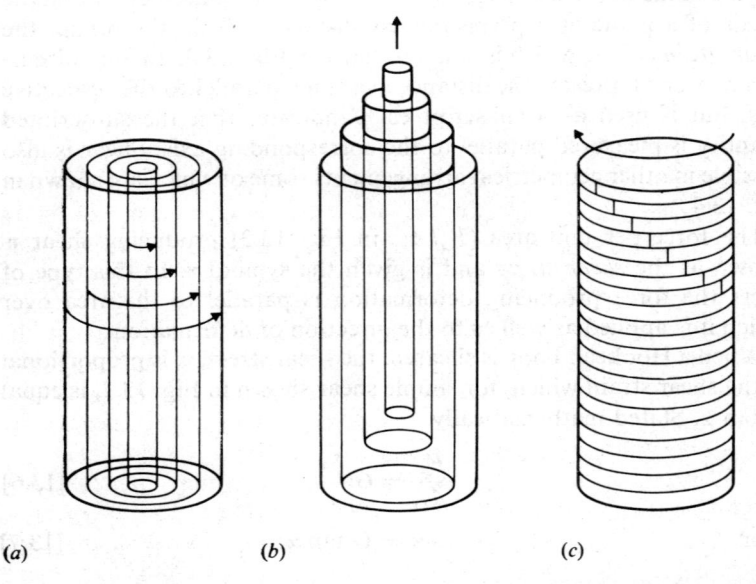

(a)   (b)   (c)

For an isotropic and perfectly elastic solid, the three moduli are related by the equation:

$$\frac{1}{E} = \frac{1}{9K} + \frac{1}{3G}. \quad [13.8]$$

Equations [13.5] and [13.8] may be combined in various ways and enable any two of the elastic constants μ, E, K and G to be calculated from the remaining two constants, provided that the material is isotropic and perfectly elastic.

The extent of flow deformation is taken to be not the shear strain ($\delta x/z$) itself (see Fig. 13.3) but the rate at which shear strain is changing with respect to time, i.e. $(dx/dt)/z$. This is known as the *rate of shear* or *shear rate* and is given the symbol D. Since dx/dt is the flow velocity along the x axis (u_x), the rate of shear is u_x/z. This is equal to (du_x/dz), the velocity gradient, i.e. the rate of change of flow velocity with respect to distance perpendicular to flow. Much of practical rheology is concerned with the relationship between rate of shear and shear stress and with the effect of factors such as composition, temperature and time.

Newtonian Flow

All gases and all pure liquids obey Newton's law of viscous flow which states that the shear stress, τ (SI unit $N\,m^{-2}$ or $kg\,m^{-1}s^{-2}$), is proportional to the rate of shear, D, (equal to du_x/dz and having the SI unit s^{-1}), i.e.

$$\tau = \eta D \qquad [13.9]$$

or

$$\frac{F_x}{A_{xy}} = \eta \frac{du_x}{dz}. \qquad [13.10]$$

The proportionality constant, η, is called the *dynamic viscosity*, coefficient of viscosity, or simply the viscosity, and its SI unit is $N\,s\,m^{-2}$ or $kg\,m^{-1}\,s^{-1}$. At 20° C $\eta/kg\,m^{-1}\,s^{-1}$ is 1.49 for glycerol, 1.00×10^{-3} for water and 2.42×10^{-4} for diethyl ether. The older c.g.s unit was the poise (P) which was $1\,g\,cm^{-1}\,s^{-1}$ and therefore one-tenth of the SI unit. The old practical unit, the centipoise (cP), is therefore $10^{-3}\,kg\,m^{-1}\,s^{-1}$. The reciprocal of the dynamic viscosity is known as the *fluidity*, ϕ, i.e. $\phi = 1/\eta$.

Newton's law (equations [13.9] and [13.10]) breaks down if the flow of the liquid or gas is not streamline (i.e. if it is turbulent) and η apparently increases. If flow is streamline, the law applies quite well to solutions of substances of low molecular weight (macromolecules of chain length less than 1000 atoms) and to dilute dispersions.

If a Newtonian liquid is made to flow in a streamline manner along a cylindrical tube of length l and radius r by virtue of a difference in pressure Δp between its ends, the volume V of liquid flowing in a given time t is given by

$$V = \frac{\pi r^4 t \, \Delta p}{8 l \eta}. \qquad [13.11]$$

This equation is named after Poiseuille, a pioneer in the study of viscosity, and enables the viscosity, η, of a Newtonian liquid to be determined by absolute measurement of the flow of the liquid along a tube. It is, however, much easier to obtain an accurate *comparison* of two viscosities than an absolute measure of viscosities and special capillary tube viscometers have been devised for such comparisons. The most common of these is the Ostwald or U-tube viscometer shown in Fig. 13.5. It is the usual practice to measure the time taken for a given volume of liquid (between mark B and mark C) to flow through the capillary tube as described in experiment 13.1. The quantities V, r, and l in equation [13.11] are constant. Although the pressure difference Δp is continually decreasing as the liquid flows through the U-tube, its mean value is proportional to the density, ϱ, of the liquid. Poiseuille's equation [13.11] then reduces to

$$\eta = \frac{\varrho t}{k} \quad \text{or} \quad \varrho = \frac{k\eta}{t} \quad \text{or} \quad t = \frac{k\eta}{\varrho}, \qquad [13.12]$$

Fig. 13.5 A viscometer of the Ostwald type

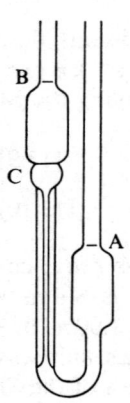

where k is a constant of proportionality characteristic of the viscometer. This equation can be derived directly from Newton's law (equation [13.9]) as follows. At any instant the shear stress is proportional to the pressure, Δp, of the head of liquid which forces the liquid through the tube. Thus, the mean values of Δp and of the shear stress are proportional to the density ϱ of the liquid. The mean value of the rate of shear is inversely proportional to the time, t, taken for the liquid to flow from mark B to mark C in Fig. 13.5. Introducing these proportionalities into equation [13.9] shows that ϱ is proportional to η/t, as in equation [13.12]. The equation is applied to the liquid of unknown viscosity η_u and to a standard liquid of known viscosity η_s. The viscometer constant k cancels on division giving:

$$\frac{t_u}{t_s} = \frac{\eta_u}{\varrho_u} \cdot \frac{\varrho_s}{\eta_s} \qquad [13.13]$$

where t_s is the time of flow and ϱ_s is the density of the standard liquid and t_u and ϱ_u are the corresponding values for the liquid whose viscosity is to be determined. The quantities t_u and t_s are measured, ϱ_u, ϱ_s and η_s are known and η_u is calculated from equation [13.13]. The precision of capillary tube viscometers when used in this way is better than c. 0.1 per cent, provided that the times of flow are not too short.

The kinematic viscosity, ν, is obtained by dividing the dynamic viscosity, η, by the density, ϱ, of the liquid, i.e.

$$\nu = \eta/\varrho. \qquad [13.14]$$

In SI units, therefore, kinematic viscosity is expressed in $m^2\ s^{-1}$. The c.g.s. unit was $cm^2\ s^{-1}$ and was called the stokes (St). The stokes is therefore $10^{-4}\ m^2\ s^{-1}$ and the old practical unit the centistokes (cSt) is $10^{-6}\ m^2\ s^{-1}$. Equation [13.13] may be rewritten in terms of the kinematic viscosities ν_u and ν_s of the liquid under investigation and of the standard respectively, thus:

$$\frac{t_u}{t_s} = \frac{v_u}{v_s}.$$ [13.15]

The value of the kinematic viscosity is used to characterise liquid paraffin, light liquid paraffin and the macrogols (polyethylene glycols). Liquid paraffin should have v not less than 6.4×10^{-5} m² s⁻¹ (64 cSt) at 37.8° C and light liquid paraffin should have v not greater than 3.0×10^{-5} m² s⁻¹ (30 cSt) at this temperature.

Experiment 13.1 The determination of the kinematic viscosity of light liquid paraffin at room temperature.

A viscometer of the Ostwald type is used as illustrated in Fig. 13.5 (The British Standards specification No. 188–1957 should be consulted for details of dimensions.) It is first cleaned with chromic acid, thoroughly washed with distilled water, dried with ethanol and ether, and then filled with light liquid paraffin at room temperature up to the mark A. (To characterise the light liquid paraffin, the liquid and viscometer would have to stand for some time in a bath thermostatically controlled to 37.8° C, but this is avoided in the present determination by carrying out the experiment at room temperature.) The liquid is then blown by means of a rubber bulb, or sucked up the left-hand limb, until the meniscus is 1 cm above the mark B. As the liquid is allowed to fall back again, a stop-watch is started as the meniscus passes B, and stopped as it passes the mark C. The determination is repeated twice, and an average time found for the fall from B to C. The paraffin is then removed from the apparatus, and, after cleaning, washing, and drying once more, it is filled with a standard sucrose solution of known viscosity. A suitable solution for the purposes of this experiment is a 60 per cent (w/w) sucrose solution which has a kinematic viscosity of 4.39×10^{-5} m² s⁻¹ at 20° C. More accurate and comprehensive data on solutions for calibration purposes are given in the British Standards specification mentioned above.

The kinematic viscosity is calculated by means of equation [13.15]. If the dynamic viscosity were required this would be obtained by multiplying the kinematic viscosity by the density of the liquid determined at the same temperature in a separate experiment.

Another useful viscometer is the falling body type which is based on Stoke's law ($f = 6\pi\eta r$, equation [12.3], p. 478). One form of this instrument is the falling sphere viscometer. A metal sphere of radius r (c. 0.795 mm) and mass m is dropped into a cylinder of the liquid and the terminal velocity u is measured by carefully timing it between two marks. The force causing motion is equal to $(m-m')g$, where m' is the mass of liquid displaced and g is the acceleration due to gravity. ($m' = \frac{4}{3}\pi r^3 \varrho$, where ϱ is the density of the liquid.) The force resisting motion is equal to $-fu$. When the terminal velocity is reached, the opposing forces balance and sum to zero, i.e.

$$(m-m')g - 6\pi\eta r u = 0$$

Whence
$$\eta = \frac{(m-m')g}{6\pi r u}. \qquad [13.16]$$

The viscosity is calculated by substituting the measured quantities into this equation. This method is used to characterise pyroxylin.

During measurements of viscosity, accurate temperature control is necessary because viscosity usually decreases markedly with increasing temperature. For liquids without hydrogen bonds log η is usually a linear function of $1/T$ (cf. equation [5.55], p. 165). Thus, dynamic viscosity, like reaction rates, obeys a law similar to the Arrhenius equation ([5.52] p. 164) thus:

$$\eta = A e^{E_{\text{vis}}/RT}, \qquad [13.17]$$

where E_{vis} is the activation energy barrier per mole which a molecule must overcome before it can take part in the process of flow. E_{vis} is usually a factor of about 0.3 of the molar latent heat of vaporisation of the liquid. For liquids with hydrogen bonded structures increasing temperature causes η to decrease more markedly than expected from the above linear relationship or in other words E_{vis} decreases with increasing temperature. Presumably the reason is that the hydrogen bonds between adjacent molecules in these liquids are progressively broken as the temperature increases.

The viscosity of a liquid or solution affects the rate at which molecules can diffuse through the solution. This, in turn, affects the rate at which sensitive components can suffer oxidation at the liquid surface. For example, ascorbic acid in aqueous solution is decomposed in a few hours on exposure to air, but in the form of syrup it may be preserved for some time, because the increased viscosity makes access to air at the surface more difficult. Where it is necessary to increase the viscosity of a solution, this can be done by addition of such substances as sugar, gum, or sodium alginate.

Rotational Viscometers

The U-tube and falling body viscometers discussed above operate over a wide range of shear stresses and rates of shear during a measurement. The range and the mean values depend on the constants for the U-tube or the sphere as well as on the viscosity of the liquid. The measured time corresponds to a mean value of the rate of shear and a mean value of the shear stress, i.e. to a single point on the graph of τ versus D shown on Fig. 13.7. These viscometers are therefore often known as 'single point viscometers'. Their measurements define completely the relationship between τ and D and the viscosity of a Newtonian liquid, because the relationship is a linear one. Single point viscometers are quite inadequate to study non-Newtonian flow to be described in the next section. Non-Newtonian flow must be studied by measuring the shear stress at several rates of shear and for this purpose

Fig. 13.6 Rotational viscometers
(a) concentric cylinder viscometer (b) cone and plate viscometer

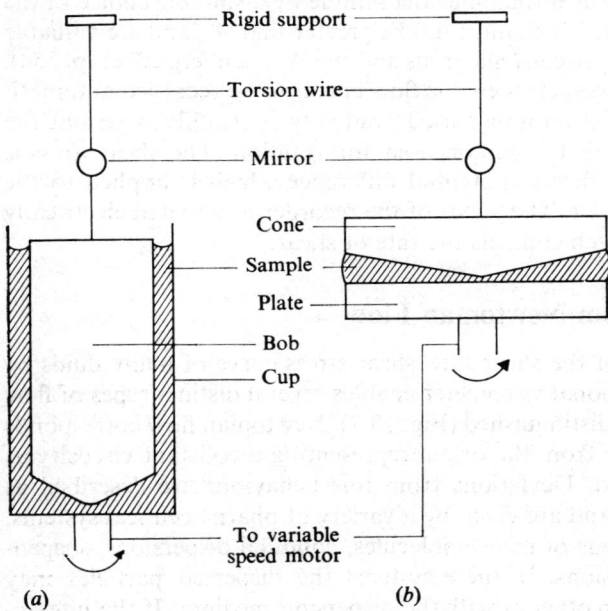

rotational viscometers have been devised. Two important types of rotational viscometer are the concentric cylinder and the cone and plate instruments shown in Fig. 13.6. These instruments also enable continuous measurements at a given rate of shear to be made for extended periods of time, so that time-dependent phenomena, such as thixotropy, may be studied. Although rotational viscometers are much more versatile than U-tube viscometers, they are not so precise.

In the concentric cylinder viscometer the material under investigation is placed in the space between the two concentric cylinders. In some versions of this instrument shown in Fig. 13.6(a), the outer cylinder, called the cup, is rotated and the rate of rotation determines the rate of shear. The inner cylinder, called the bob, is attached to a mechanism, such as a torsion wire, which measures the twisting force, F. The shear stress is equal to F divided by the mean area, A, of the inner and outer cylindrical surfaces of the liquid. In other forms of concentric cylinder viscometer the cup is stationary and the bob is rotated by a motor or falling weight. The shear stress is calculated from the weight or from the resistance offered to the motor, while the rate of shear is measured by the rate of rotation. With the concentric cylinder viscometers an end correction (for the effect of the base of the cylinders) must be applied or in some way eliminated. Another disadvantage of this type of viscometer is the variation of the rate of shear across the sample in the gap.

Cone and plate viscometers (see Fig. 13.6b) enable the rate of shear to be kept constant throughout the sample by a suitable choice of the angle of the cone, which must not be greater than 4°, and are valuable for studying very viscous materials and the Weissenberg effect (p. 554).

Rotational viscometers enable flow curves to be recorded automatically. The rate of shear is increased gradually by steadily increasing the voltage applied to the motor used for rotation. The shear stress is measured as an electric potential difference which is applied to the x axis of a recorder. The y axis of the recorder is actuated electrically by the motor which controls the rate of shear.

Deviations from Newtonian Flow

Determination of the shear rate–shear stress curve of many fluids by means of a rotational viscometer enables several distinct types of flow behaviour to be distinguished (Fig. 13.7). Newtonian flow corresponds to a straight line from the origin representing a constant viscosity as already discussed. Deviations from this behaviour are described as non-Newtonian and are given by a variety of pharmaceutical systems, especially solutions of macromolecules, colloidal dispersions, suspensions, and emulsions. In these systems the dispersed particles may interact with each other or with the dispersion medium. If the interaction depends on the flow rate, the viscosity is no longer constant and a curve results. Deviations from Newtonian behaviour become pronounced if the disperse phase is concentrated or if its particles are asymmetric. Solutions of long chain polymers are more viscous than those of branched chain polymers of the same molecular weight. This is because the linear molecules sweep out a larger volume on rotation and interfere more with each others motion and with the flow of solvent.

For non-newtonian systems it is useful to define a quantity known as 'apparent viscosity' which is simply the shear stress divided by the rate of shear at a given point on the flow curve. It is therefore equal to the reciprocal of the gradient of the tangent of the line drawn from the origin to the point in question on the graph of shear rate versus shear stress. Single point viscometers give a measure of the apparent viscosity.

In the case of dilatant flow (Fig. 13.7) increasing shear stress and shear rate gives rise to an increase in the apparent viscosity. Dilatant materials are therefore said to be 'shear thickening'. They are rather rare and comprise thick suspensions containing high concentrations of closely packed particles of approximately the same size, e.g. a slurry of sand and water and concentrated starch pastes. At rest the densely packed deflocculated particles are surrounded by just enough liquid to

Fig. 13.7 Various types of flow curve (independent of time)

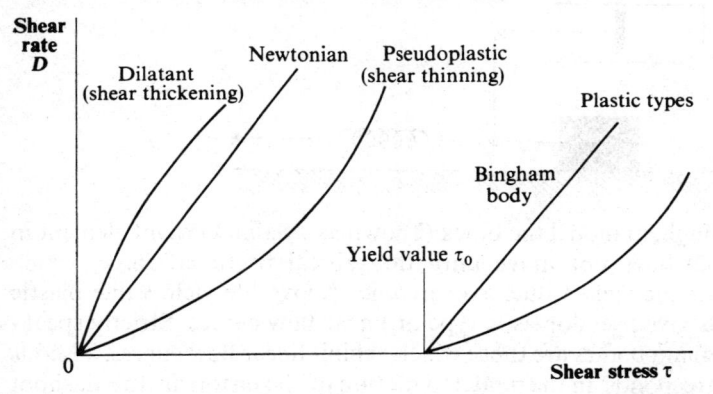

[Graph showing Shear rate D on vertical axis and Shear stress τ on horizontal axis, with curves for: Dilatant (shear thickening), Newtonian, Pseudoplastic (shear thinning), Bingham body, and Plastic types. Yield value τ_0 marked on horizontal axis.]

fill the spaces between them. It is believed that, as the shear stress and shear rate are increased, the particles rearrange to occupy a larger volume so that there is insufficient liquid to fill the voids. As a result the friction between the particles increases and hence the viscosity increases. At rest dilatant materials appear wet, but when sheared appear dry. This effect is often noticeable when one walks on wet sand on the sea shore.

Pseudoplastic materials exhibit decreased apparent and differential viscosities with increasing shear stress and shear rate and are therefore said to be 'shear thinning' (Fig. 13.7). Pseudoplasticity is evidently the opposite of dilatancy, and is shown by solutions of long chain molecules (e.g. nucleic acids and polysaccharides, especially gums, such as tragacanth and sodium carboxymethylcellulose) and by dispersions of small particles (e.g. pigments, inks and blood). At rest the long chain molecules in the gums are randomly arranged through the dispersion medium. At low shear stresses and shear rates the flow lines are disturbed by the randomly orientated macromolecules so that the apparent viscosity is high. At high shear stresses and shear rates the long chain molecules align themselves in the direction of flow, thus facilitating shear and reducing the apparent viscosity. In the case of the pseudoplastic dispersions the particles form aggregates at rest but under shear the aggregates break down releasing immobilised solvent which acts as a lubricant. As a result the viscosity is reduced with increasing shear.

Plastic materials do not shear or flow until a certain shearing stress, known as the 'yield value', is exceeded (Fig. 13.7). These materials are often referred to as Bingham bodies and their behaviour is usually represented by the mechanical model shown in Fig. 13.8. At shear stresses below the yield value the body behaves as a solid elastic material, namely a Hookean body, which is represented by a spring.

Fig. 13.8 Rheological model for a plastic material (a Bingham body)

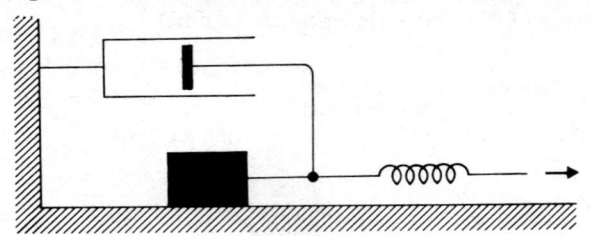

In the Bingham model the block (known as a Saint-Vernant element in Fig. 13.2) does not move until the critical frictional force, which represents the yield value, is overcome. Above the yield value plastic materials give pseudoplastic type or linear flow curves. Strictly speaking, Bingham bodies are those which exhibit linear flow curves. Viscous flow corresponds in the model to motion of the piston in the dashpot which cannot occur until the critical frictional force of the block is overcome. The dashpot then prevents the block from gaining speed. Similarly, models of other rheological phenomena can be set up by coupling together the basic elements shown in Fig. 13.2 in suitable ways and more examples will be discussed later (Fig. 13.12).

Examples of plastic materials include paint, modelling clay and certain dispersions. The yield value of paint should be such that it is exceeded only by brushing and not by its own weight, so that it flows only during application and not down the surface afterwards.

The molecular explanation of plastic behaviour in dispersions and polymer solutions is as follows. At low shear stresses the dispersed particles tend to aggregate and entrap the dispersion medium thus forming a continuous rigid network which does not flow. The yield value is the minimum shearing stress necessary to break down the aggregate or network and thereby permit flow. The yield value of dispersions is altered by changes in the ratio of the surface area to the volume of the particles and is increased by increasing concentration of the disperse phase and by increasing particle aggregation. The latter effect may be brought about by addition of a flocculating agent or a surface active agent.

The yield values of plastic dispersions and solutions of polymers are relatively small (< 100 N m^{-2}) because flow proceeds merely by displacement of particles in a solvent. On the other hand, the yield value of metals, which corresponds to the elastic limit L in Fig. 13.1, is comparatively large. (c. 70–400 MN m^{-2}) because flow proceeds by the displacement of the atoms in a sturdy crystal lattice.

Each flow curve shown in Fig. 13.7 can often be represented fairly well by an empirical equation of the type

$$\tau = \tau_0 + KD^n, \qquad [13.18]$$

where τ is the shear stress, D is the rate of shear and K and n are empirical constants. τ_0 represents the yield value and is zero for all but plastic materials. For a Newtonian material equation [13.9] shows that $\tau_0 = 0$, $n = 1$ and $K = \eta$. For dilatant materials $\tau_0 = 0$ and $n > 1$. For pseudoplastic materials $\tau_0 = 0$ and $n < 1$.

Thixotropy

The types of system discussed above and illustrated in Fig. 13.7 are independent of time. In practice this means that for a given system there is always a fixed shear rate corresponding to a given shear stress and vice versa and this correspondence remains unaffected by time or by the past history of the sample. When flow behaviour depends on either of these variables it is said to be time-dependent. Two such time dependent phenomena are known as *thixotropy* and *rheopexy*.

The word thixotropy is derived from the Greek meaning 'change by touch', and was originally defined as the isothermal transformation of a gel to a sol on agitation and back to a gel again on standing. Examples include some flocculated hydrophobic sols, such as ferric oxide and aluminium oxide, and clays. The definition has since been expanded. Thixotropy may now be defined as the property of a material to become less viscous the longer it flows. At a constant rate of shear the shear stress in thixotropic systems decreases with time to a constant equilibrium value (Fig. 13.9a), and as a result the apparent viscosity decreases. Furthermore, when the stress is removed the system takes time to recover its original apparent viscosity. Thixotropy is common among plastic and pseudoplastic materials, e.g. bentonite suspensions and petroleum jelly. It is only detected if the decrease in shear stress with time (Fig. 13.9a) is slow compared with the response of the instrument. If this is the case, an increase in shear rate in a rotational viscometer will give a non-equilibrium value of the shear stress ('up curve' in Fig. 13.9b). When the shear rate is reduced after a high value has been reached, the material will usually have been flowing for a sufficient time for the shear stress to be close to the equilibrium value (corresponding to a point on the 'down curve' in Fig. 13.9b).

Fig. 13.9 Flow behaviour of a thixotropic system
  (a) fall in shear stress with time at a constant rate of shear
  (b) flow curve of a thixotropic pseudoplastic material

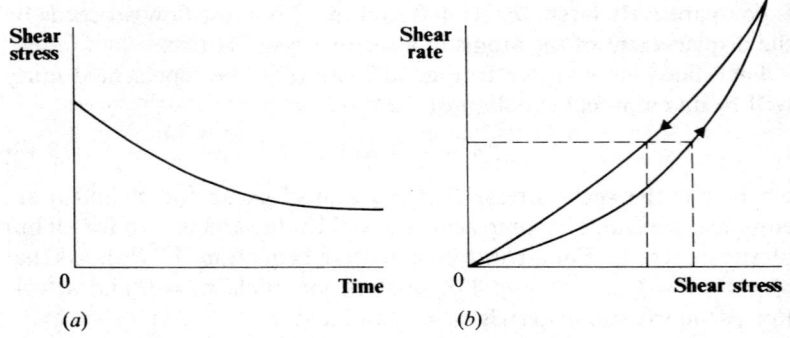

Thixotropy is not easy to quantify. Probably the best measure of the extent of thixotropy is the area of the hysteresis loop in the flow curve, (Fig. 13.9b), determined under controlled conditions. Another measure of time dependence is the rate of decay of shear stress at constant shear rate (Fig. 13.9a).

Thixotropy is associated with the breakdown of certain structures in the material on shear to an extent which determines the equilibrium value of the shear stress. When the stress is removed the structures re-form slowly. Solutions of high polymers are often thixotropic. On shearing these materials the intermolecular entanglements and attractions and the immobilisation of the solvent are reduced.

Pseudoplastic and plastic materials are usually in some degree thixotropic, although the extent of thixotropy may be very small. Sometimes the accompanying thixotropy is very pronounced and quite complex flow behaviour may be observed as exemplified by Fig. 13.10.

Fig. 13.10 Complex flow curves
  (a) flow curve of an aqueous suspension of bentonite (12 % w/v)
  (b) flow curves of thick aqueous solutions of procaine penicillin G

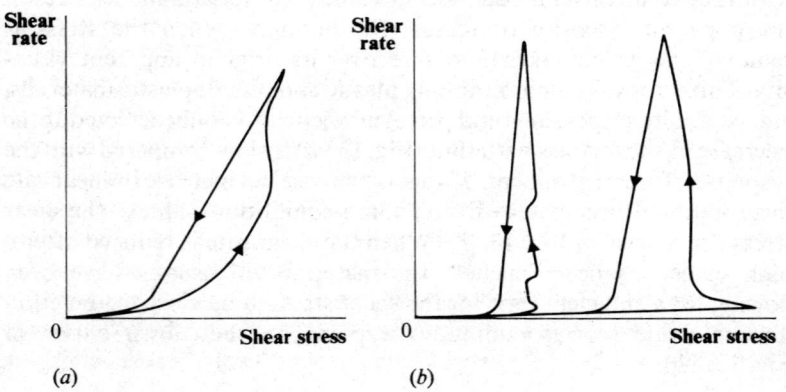

Complex flow curves are often found among suspensions and gels and sometimes among solutions of macromolecules, such as carboxymethylcellulose in water. The bulges, spurs and kinks in the flow curves are attributed to the orientation of the particles. The type and extent of orientation depends on the immediate history of the sample as well as on the rate of shear.

Rheopexy is sometimes known as anti-thixotropy and is the phenomenon in which a material becomes *more* viscous the longer it flows. When observed, it is usually accompanied by dilatancy.

Elastomers

Elastomers are elastic polymers of which rubber is a typical example. Elastomers consist of polymer chains cross-linked at various points. The chains in natural rubber consist of repeating *trans*-isoprene units:

$$\begin{array}{c} -CH_2 \diagdown \qquad \diagup H \\ \qquad C=C \\ CH_3 \diagup \qquad \diagdown CH_2- \end{array}$$

Heating natural rubber with sulphur, a process known as vulcanisation, increases the amount of cross-linking. On stretching an elastomer, the polymer chains are extended in the direction of stress and compressed in perpendicular directions as shown in Fig. 13.11. Stretching

Fig. 13.11 Stretching an elastomer such as rubber

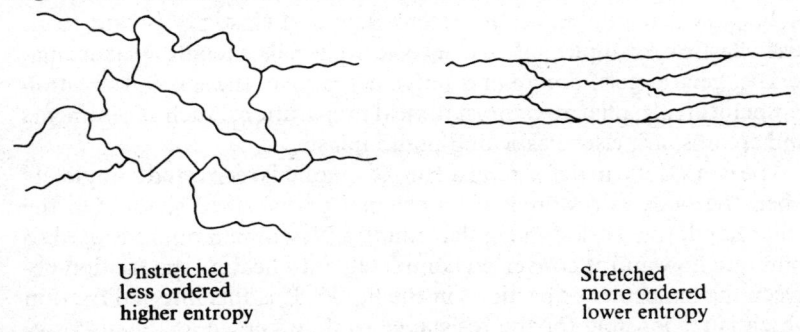

Unstretched	Stretched
less ordered	more ordered
higher entropy	lower entropy

therefore introduces into the random or amorphous structure of the elastomer some order in the form of some crystallinity, and therefore decreases the entropy. On the other hand the total volume and the enthalpy are little changed, because no bonds are broken and because the interactions between the chains are weak. Since the spontaneous tendency is for the entropy to increase, the elastomer returns rapidly to its original shape on releasing the stress. On increasing the temperature of rubber under constant stress the thermal motions increase, the entropy increases and so deformation decreases. On the other hand, when Hookean bodies, such as metals, are heated under constant stress, their molecules become more displaced and the deformation increases.

Elastomers and synthetic polymers can exhibit viscous flow as well as elasticity. This property, known as viscoelasticity, is discussed in the next section. If there is little cross-linking, as in crude rubber, the chains can move relative to each other and the material flows visibly under stress. Considerable cross-linking hampers movement of the polymer chains, preventing flow, reducing the elasticity and hardening the material. Introduction of crystalline regions into the amorphous structure of the polymer by stretching or cooling has the same effects.

The amplitudes of the molecular motions and vibrations depend greatly on the temperature. There is for each amorphous polymer a 'glass transition temperature' above which the polymer is ductile and rubber-like, because the macromolecules have a certain freedom of motion, and below which the polymer is hard, brittle and glass-like, because the attractive forces between the macromolecules are able to fix their positions. The glass transition temperature, T_g, for polyethylene is $-125°$ C and for poly(methyl methacrylate) or 'Perspex' is $+105°$ C. The temperature curve for many physical properties, including thermodynamic functions, exhibits a discontinuity (or kink) at T_g.

Viscoelasticity

A body which exhibits both viscous flow and elasticity is said to be viscoelastic. Examples of viscoelastic materials include plastic materials, gelatin gels, dough and polymers above the glass transition temperature. In pharmacy most topical preparations, such as ointments and creams, are viscoelastic and pseudoplastic.

The work done in deforming a Hookean solid is recovered completely when the body is restored to its original undeformed shape. On the other hand, the work done in deforming a Newtonian liquid is used to maintain flow and is converted completely into heat by the friction between the molecules or particles in the liquid. It is this internal friction which is responsible for the resistance to flow known as the viscosity (see p. 534). When a viscoelastic material is stressed some of the work done is stored elastically and can be recovered and the rest is dissipated as heat in maintaining flow.

In the case of plastic materials, elasticity predominates at low stresses and viscous flow at high stresses, the two types of behaviour being separated by the yield value. However, this is only an approximation. With all viscoelastic materials the elasticity and flow occur simultaneously. Their rheological models will therefore contain springs and dashpots, which may be connected in series or in parallel, as shown in Fig. 13.12.

Fig. 13.12 Some rheological models for viscoelastic behaviour
(a) Maxwell model (b) Kelvin–Voigt model (c) Burgers model

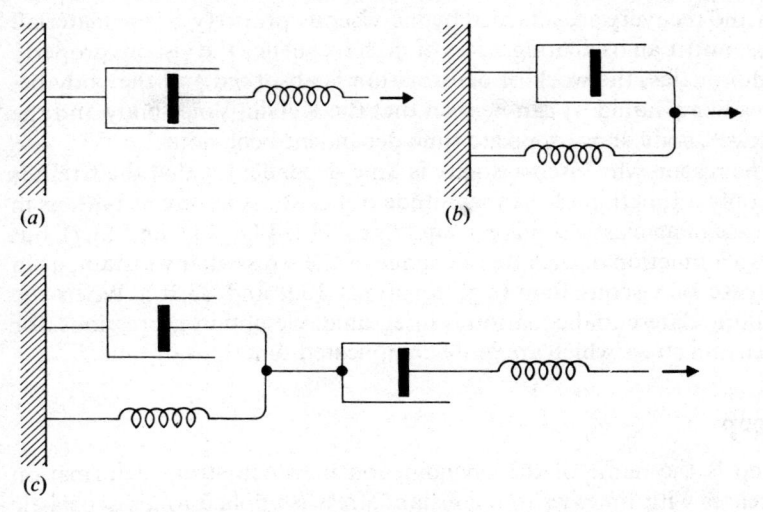

A viscoelastic liquid is one for which the slightest stress causes a permanent deformation in the form of viscous flow. The simplest rheological model for this consists of a dashpot and spring in series and known as the Maxwell model (Fig. 13.12a). If the spring is imagined to be rather strong, this model explains the elasticity of fluids, such as liquids and gases. If the dashpot is imagined to contain an extremely viscous oil, the model represents the slow creep of an elastic solid, such as metals and elastomers below the elastic limit. We can see that the times over which the stress is applied and the observations made are most important when considering the behaviour of viscoelastic materials. The Maxwell body also explains the strange behaviour of 'bouncing putty', a material which exhibits elasticity under rapidly changing stress, so that it bounces when dropped on to a surface, but which exhibits viscous flow under slow sustained stress, so that it can be moulded like putty.

A viscoelastic solid will eventually return to its original shape after a small deforming stress has been applied and removed. The simplest rheological model for this behaviour consists of a spring and a dashpot in parallel and known as a Kelvin or Voigt body (Fig. 13.12b). This body explains the behaviour of a gel consisting of gelatin and water or agar and water. The hydrophilic macromolecules are linked together by electrical forces and by hydrogen bonding to form a solid three-dimensional gel network. An excess of relatively free water molecules fills the spaces. The structure approximates to a sponge of perfectly elastic material whose holes are filled with a viscous liquid. The elastic property predominates if the stress is slowly increased or decreases or oscillates at a low frequency and the body behaves like a solid. When the body is subjected to a rapidly changing stress, the elastic deformation and recovery are retarded by the viscous property of the material. Thus, under an oscillating stress of high frequency the viscous property predominates, the work of deformation is absorbed and the body behaves like a liquid. It can be seen that the Kelvin–Voigt body and the Maxwell body show opposite time-dependent behaviour.

The reason why viscoelasticity is time-dependent is that the strain is not only a function of the magnitude of the stress at any instant, as in the case of an elastic solid (e.g. equations 13.1–13.4, 13.6 and 13.7), but is also a function of the rate of change of the stress at any instant, as in the case of viscous flow (e.g. equations 13.9 and 13.10). When the resulting differential equation is integrated, we obtain expressions for strain and stress which are quite complicated functions of time.

Creep

Creep is the name of the phenomenon in which strain deformation increases with time, when a constant stress is applied to a viscoelastic material. When the stress is removed, the deformation will usually decrease, a process which is known as recovery. During recovery the

Fig. 13.13 Creep and recovery of a viscoelastic material, - - - - - creep compliance (= strain/constant stress)

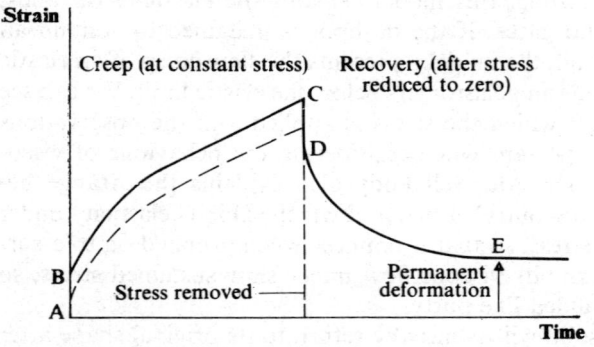

strain falls to an equilibrium value which is usually higher than the initial value before deformation, owing to viscous flow. Typical creep and recovery curves are shown in Fig. 13.13 and are exemplified by many polymers and by a mixture of cetyl alcohol and sodium dodecyl sulphate in a mole ratio of 8 : 1. This behaviour may be represented approximately by the Burgers model which consists of a Maxwell model and a Kelvin–Voigt model connected in series as shown in Fig. 13.12(c). The instantaneous elasticity (AB in the creep curve, Fig. 13.13) and the instantaneous recovery (CD) are attributed to the spring in the Maxwell model. The delayed (or retarded) elasticity (BC) and the delayed (or retarded) recovery (DE) are due to the Kelvin-Voigt model. The dashpot in the Maxwell model explains the viscous deformation (E) remaining after recovery. In practice AB and BC usually form a continuous curve as do CD and DE, because elasticity and recovery are never really instantaneous.

The ratio of the strain (due to creep) at a given time to the constant stress is known as the *creep compliance* at that time. Since strain is dimensionless, creep compliance has the same dimension as stress^{-1} and its SI unit is $m^2\ N^{-1}$ or Pa^{-1} or $m\ kg^{-1}\ s^2$. The creep compliance may refer to any type of deformation, e.g. elongation, compression or shear, but is most frequently applied to shear, when it may be called shear compliance.

Fig. 13.14 Logarithmic creep compliance—time curves for various types of material

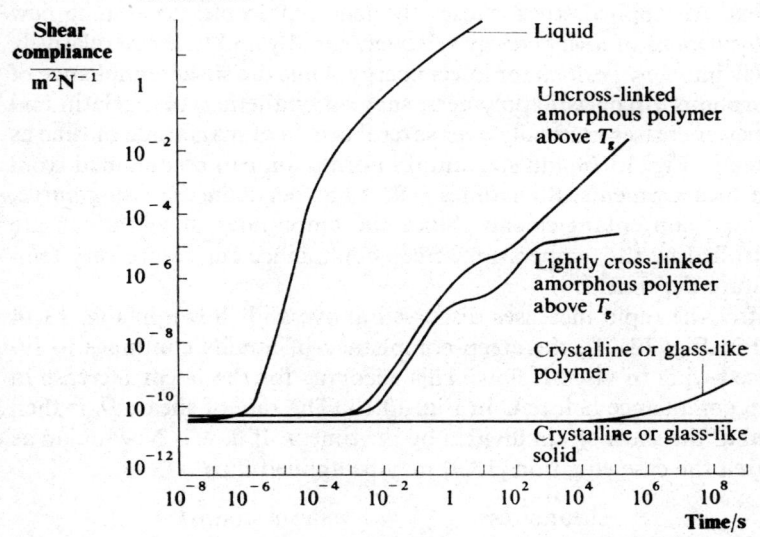

Figures 13.13 and 13.14 show typical creep compliance-time curves. The scales in Fig. 13.13 are linear while those in Fig. 13.14 are logarithmic. The horizontal portion in the initial part of the curves in Fig. 13.14 is too short to appear in Fig. 13.13. At times less than about 10^{-6} s all materials, including liquids, have shear compliances of the same small order (10^{-11} to $10^{-9}\ m^2\ N^{-1}$). The reason for this is that only deformations within molecules, such as stretching of valency bonds or bending of bond angles, can take place within the time available, and all materials respond similarly in this respect.

A steep increase in the shear compliance shown in Fig. 13.14 occurs when the molecules in the material move into new portions relative to each other. The time at which this arises depends on the intermolecular forces and on the size and shape of the molecules. In the case of liquids which have relatively weak intermolecular forces, this happens after the shear stress has been applied for only about 10^{-6} s, whereas for crystals, e.g. sodium chloride and diamond and for glasses, which have relatively strong intermolecular or interionic forces, this does not happen for a very considerable time. The time taken for the molecules in viscoelastic polymers to move into new positions relative to each other lies between these two extremes.

An increase in creep compliance, indicating some relative molecular movement, occurs after appreciable times with highly crystalline polymers, such as linear polyethylene of high density and with amorphous polymers, such as poly(methyl methacrylate) below T_g.

In amorphous viscoelastic polymers above T_g the macromolecules are continually undergoing translations and vibrations, including stretching, compression, bending, twisting and entanglement of their chains. An applied stress causes the macromolecules to assume new *conformations* of lower energy relatively rapidly and to move relatively slowly into new *positions* of lower energy. Thus the shear compliance of amorphous viscoelastic polymers, such as synthetic types, gelatin and rubber, increases gradually over several orders of magnitude of time as shown in Fig. 13.14 and structural information can be obtained from such measurements. Sometimes inflections occur in the rising curve, due to chain entanglements. Since the amplitudes of vibrations are controlled by the temperature, creep compliance curves are very temperature-dependent.

After the rapid increases discussed above and shown in Fig. 13.14 (and in Fig. 13.13), the creep compliance of liquids continues to increase owing to viscous flow. This accounts for the linear increase in creep compliance before C in Fig. 13.13. The rate of shear, D, is then equal to the shear strain divided by the time, t. If flow is Newtonian as is often the case equation [13.9] may be applied thus:

$$\text{shear stress} = \eta \times \text{shear strain}/t$$

$$\therefore \quad \text{shear compliance} = \frac{\text{shear strain}}{\text{shear stress}} = \frac{t}{\eta}. \quad [13.19]$$

Thus during Newtonian flow the creep compliance of liquids increases by an amount equal to the time divided by the dynamic viscosity. This also applies to uncross-linked polymers, e.g. poly(methyl methacrylate), above T_g. Any cross-linking, as in elastomers, greatly reduces this effect and causes a levelling of the creep compliance curve in Fig. 13.14 and before point C in Fig. 13.13 (not shown in this diagram).

When the creep compliance–time curves at different stresses all coincide, the viscoelasticity of the material concerned is said to be linear, and when not, it is said to be non-linear. Truly linear viscoelasticity is exceptional. The range of shear stresses over which viscoelasticity is approximately linear usually decreases in the order : elastomers > amorphous polymers > fats. Typical non-linear behaviour occurs when the stress is such that the rate at which the intermolecular forces take effect after deformation is less than the rate at which they are overcome during deformation. The structure is weakened and in parts the remaining intermolecular forces are overcome more easily under stress, so the material may develop cracks.

Other Methods of Studying Viscoelasticity

Instead of measuring the creep at constant stress, viscoelasticity may be studied by alternative time-dependent techniques. In the stress-relaxation method the sample is forced to undergo an instantaneous strain and the decay of stress within the sample is measured as a function of time. A concentric cylinder viscometer shown in Fig. 13.6(a) may be used for this. Fig. 13.15 shows a typical result. If the material

Fig. 13.15 Decay of stress at constant strain

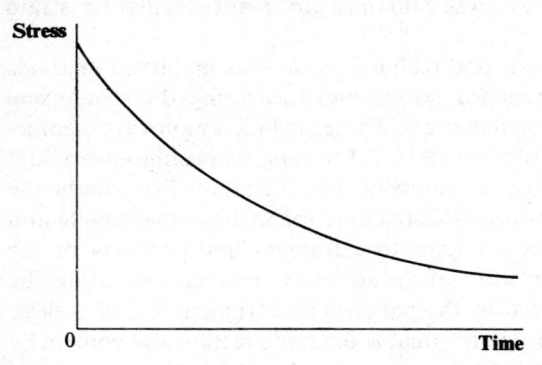

behaves like a Maxwell body shown in Fig 13.12(a), the decay of stress within the spring takes place through the dashpot and is a first order process, i.e. the logarithm of the stress decays linearly with time (cf. equation [5.8] p. 151). The reciprocal of the rate constant is known as the relaxation time and is equal to the viscosity (represented by the dashpot) divided by the shear modulus (represented by the spring). Usually the logarithm of the stress does not decay linearly with time. The system can then be represented by several Maxwell models in parallel, if the viscoelasticity is linear. There is then a variety of relaxation times and a mean value can be calculated.

A further technique for studying time-dependent deformation of viscoelastic materials is to apply an oscillating stress to the sample and to measure the oscillating strain produced. Figure 13.16 illustrates this. For Hookean solids the strain is in phase with the stress but is damped. With Newtonian liquids the strain lags a quarter of a wave

Fig. 13.16 Oscillation of a viscoelastic material

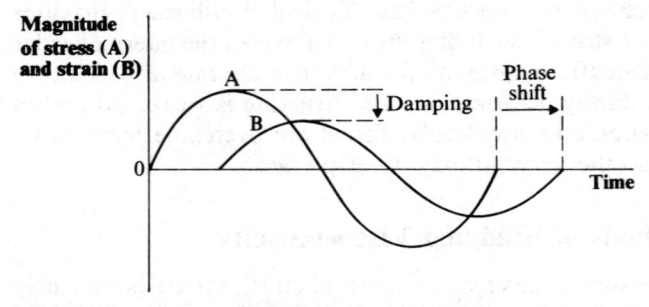

behind the stress and is not damped. For a viscoelastic material the phase difference and the damping have intermediate values depending on its rheological properties. The alternating stress may be applied to the cone of a cone and plate viscometer shown in Fig. 13.6(b) and the alternating strain on the plate may be measured. Transducers which interconvert electrical and mechanical vibrations may be used for applying the oscillating stress and measuring the oscillating strain produced.

In pharmaceutical and food technology various empirical methods and instruments are used for testing and measuring the rheological behaviour of viscoelastic materials. These include empirical viscometers, penetrenometers, plastometers, gel testers, recording mixers and apparatus for stretching, compression and bending. Frequently the applied stress is not uniformly distributed throughout the sample and the results obtained do not refer to a fundamental property of the material but are in the form of an arbitrary number measuring the behaviour of the material to the particular instrument. Nevertheless, these empirical methods are invaluable for rapid testing and control by the manufacturer and the user.

Viscoelasticity may also be studied during rotational shear by measuring the normal stress component which is responsible for the Weissenberg effect to be discussed in the next section.

The Weissenberg Effect

If a rotating rod is lowered into a Newtonian liquid, the circular motion causes the liquid to move outwards leaving a depression around the rod as shown in Fig. 13.17(a). If the same experiment is performed on a viscoelastic liquid, the liquid may actually climb up the rod as shown in Fig. 13.17(b). This phenomenon is known as the 'Weissenberg effect' and is observed with polymer solutions, malt and many other, but not all, viscoelastic materials. The explanation is as follows. The angular velocity at the bottom of the liquid is lower than at the top, so that the liquid is experiencing circular shear as shown in Fig. 13.4(c) on

Fig. 13.17 Effects of rotating a rod in a liquid
(a) formation of a vortex by a Newtonian liquid
(b) the Weissenberg effect for a viscoelastic liquid

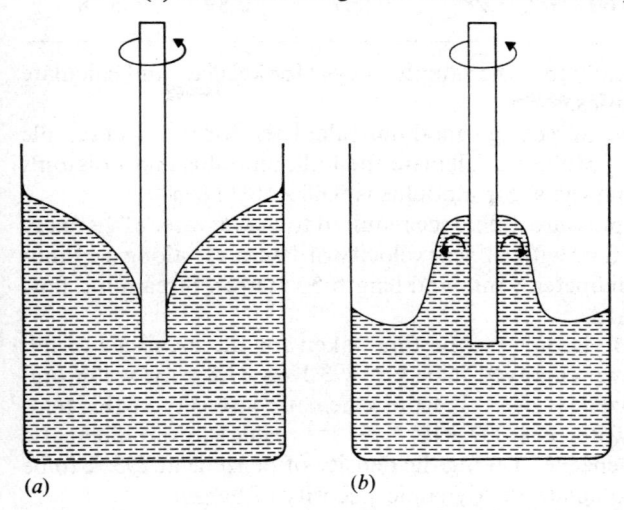

p. 536. If the liquid has elasticity a normal stress component will arise which is greater at the centre than the edges and will force the liquid up the rod as shown in Fig. 13.4(b). The origin of this component can be visualised by imagining a rubber cylinder which is held at the bottom and twisted at the top. The outer elements shown in Fig. 13.4(a) will be stretched more than the inner elements, so the top surface will be depressed at the edges and raised at the centre as in the Weissenberg effect.

The effect causes a viscoelastic liquid in a concentric cylinder viscometer, shown in Fig. 13.6(a), to climb up the inner cylinder and to be depressed near the outer cylinder, thus making viscosity measurements extremely difficult. The cone and plate viscometer, shown in Fig. 13.6(b), does not suffer from this disadvantage. Indeed, the normal stress component can actually be determined by measuring the upward force on the cone. This feature and several others, including facilities for oscillation measurements, are incorporated in a versatile cone and plate viscometer, known as the Weissenberg rheogoniometer.

Problems

1 A sample of steel of length 15.0 cm and of cross sectional area 2.0 cm^2 was tested, with a view to its use in a tabletting machine, by applying various longitudinal stretching forces. The measured elongations, Δl, and the corresponding longitudinal forces, F, were as follows:

F/MN	0.197	0.422	0.684	0.954	1.248
$\Delta l/\mu\mathrm{m}$	0.60	1.28	2.07	2.89	3.78

Verify graphically that the sample obeys Hooke's law and calculate Young's modulus.

2 From the value of Young's modulus calculated for the steel sample in the previous problem, calculate the bulk modulus and Poisson's ratio, given that the shear modulus is 9.68×10^{10} N m^{-2}.

3 Calculate the pressure difference required to make water of viscosity 1.0 g m^{-1} s^{-1} flow with a linear velocity of 10 cm s^{-1} along a cylindrical pipe of diameter 8 mm and length 50 metres. Streamline flow may be assumed.

4 In an Ostwald viscometer, the time taken for the meniscus to fall between the two engraved marks at 293 K is 87.8 s for water and 64.8 s for benzene. Given that the kinematic viscosity at this temperature for water is 1.007×10^{-6} m^2 s^{-1}, calculate the kinematic viscosity for benzene. Taking the density of benzene at 293 K to be 879 kg m^{-3}, calculate the dynamic viscosity of benzene.

5 In the British Pharmacopœia (1973) the kinematic viscosity (v in centistokes) of pyroxylin is determined by means of a falling sphere viscometer and the equation for the calculation is given as follows:

$$v = \frac{d^2 g(\delta - \varrho) 0.867}{0.18 v \varrho}$$

where d = diameter of the sphere in cm
       δ = density of the sphere in g per ml
       ϱ = density of the liquid being tested in g per ml
       v = velocity of fall in cm per sec
       g = local acceleration due to gravity in cm per sec per sec.

Verify this equation assuming that the factor 0.867 is necessary to correct for 'end effects' in the method of measuring the velocity.

6 The dynamic viscosity, η, and the saturated vapour pressure, p, of ethanol have the following values at the stated temperatures, T:

T/K	273.15	283.15	293.15	303.15	313.15
$\eta/\mathrm{mN\ s\ m^{-2}}$	1.773	1.466	1.200	1.003	0.834
p/kPa	1.56	3.04	5.67	10.07	17.33

From the gradients of suitable plots, which should be approximately linear, calculate E_{vis}, the activation energy for viscous flow, and

ΔH_{vap}, the molar enthalpy of vapourisation. Comment on the ratio $E_{vis}/\Delta H_{vap}$ and on the linearity of your plots.

7 When the flow of a commercial suspension was studied in a concentric cylinder viscometer, the shear stresses, τ, at various rates of shear, D, were as follows:

D/r.p.m.	6.5	24	55	99	154	216	300
τ/Nm^{-2}	8.25	16.5	24.8	33.0	41.3	49.5	57.8

After expressing D in the SI units radians per second, plot the flow curve and comment on its shape. Plot a suitable log-log graph to test whether the data conform to the equation $\tau = KD^n$. If so, determine the constants K and n.

8 A gel prepared from 1.5 per cent (w/w) gelatin gave the following values of shear strain, G, at various times, t, when subjected to constant shear stresses, τ:

	t/min	0	2	4	10	20	30
$\tau = 1.5$ N m^{-2},	G	0.141	0.177	0.192	0.217	0.244	0.265
$\tau = 3.0$ N m^{-2},	G	0.282	0.354	0.384	0.435	0.489	0.531
$\tau = 5.0$ N m^{-2},	G	0.470	0.590	0.640	0.725	0.815	0.885
$\tau = 7.0$ N m^{-2},	G	0.693	0.840	0.924	1.057	1.204	1.309

Plot the shear compliance against time for each shear stress and deduce the value of the stress above which the viscoelasticity of the gel becomes non-linear.

Answers to Problems

Chapter 1

1 $-273°$ C; 22.4 dm^3; 0.082 dm^3 deg^{-1}

2 $\ln N_1 = \ln N - E/RT$; plot $\ln N_1$ versus $1/T$; or $\log N_1 = \log N - E/2.303\ RT$; plot $\log N_1$ versus $1/T$

3 $\alpha = 4.56 \times 10^{-3}$

4 $N_1/N = 1.18 \times 10^{-14}$

5 $l = 0.023$ min^{-1}; $T = 30.0$ min

6 $d \ln k/dT = E/RT^2$; 0.107 K^{-1}

7 (a) $dy/dx = -1/(a-x)$; (b) $dy/dx = 1/(a-x)^2$

8 (a) $\left(\dfrac{\partial H}{\partial p}\right)_V = \left(\dfrac{\partial U}{\partial p}\right)_V + V$; (b) $\left(\dfrac{\partial H}{\partial V}\right)_p = \left(\dfrac{\partial U}{\partial V}\right) + p$

9 $M = \dfrac{2.\ \text{constant} . \ln(c_2/c_1)}{(x_2^2 - x_1^2)}$

Chapter 2

1 $p = 1.18 \times 10^7$ Pa; $V = 426$ m^3

2 diffusion rates, oxygen : ether $= 1.52 : 1$

3 N_2 79.9 per cent, 8.09×10^4 Pa; O_2 14.1 per cent, 1.42×10^4 Pa; CO_2 6.0 per cent, 6.1×10^3 Pa

4

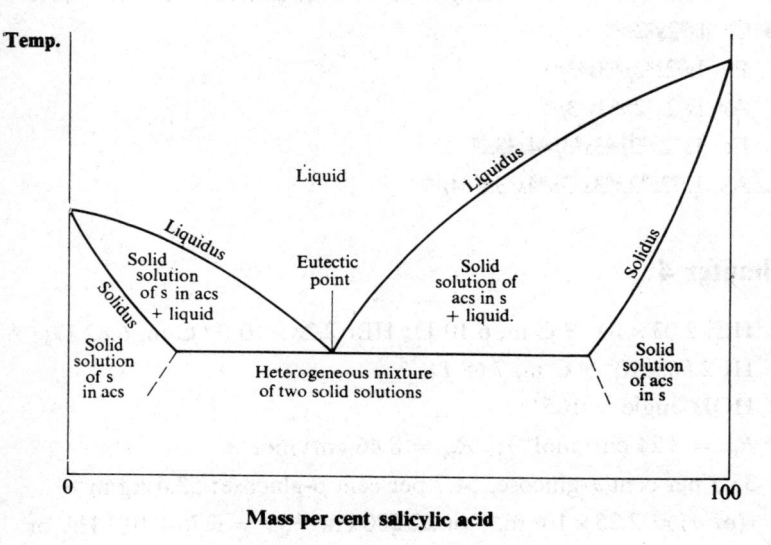

Temp.

Liquid

Liquidus

Liquidus

Solidus

Solid solution of s in acs + liquid

Eutectic point

Solid solution of acs in s + liquid.

Solidus

Solid solution of s in acs

Heterogeneous mixture of two solid solutions

Solid solution of acs in s

0   100

Mass per cent salicylic acid

Cooling curves

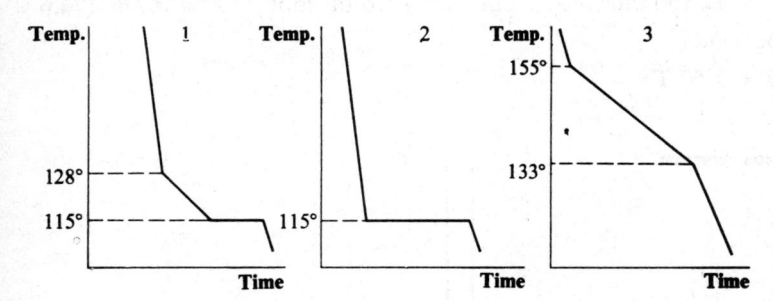

Temp.   1   Temp.   2   Temp.   3
                                   155°
128°
                                   133°
115°   115°

Time   Time   Time

Chapter 3

1 $^{208}_{82}$Pb

2 mass defect = 0.528 46 m_u; 8.79 MeV

3 $^{6}_{3}\text{Li} + ^{1}_{0}\text{n} = ^{4}_{2}\text{He} + ^{3}_{1}\text{H}$;

$^{9}_{4}\text{Be} + ^{4}_{2}\text{He} = ^{1}_{0}\text{n} + ^{12}_{6}\text{C}$;

$^{130}_{52}\text{Te} + ^{1}_{0}\text{n} = \gamma + ^{131}_{52}\text{Te}$, $^{131}_{52}\text{Te} = ^{131}_{53}\text{I} + ^{0}_{-1}\beta^{-}$;

$^{59}_{27}\text{Co} + ^{1}_{0}\text{n} = \gamma + ^{60}_{27}\text{Co}$, $^{60}_{27}\text{Co} = ^{60}_{28}\text{Ni} + ^{0}_{-1}\beta^{-} + \gamma$

4 5.10×10^3, 8.08×10^3, 1.02×10^4, 1.32×10^4 years

5 $\lambda/\text{min}^{-1} = 0.227$, $0.025\ 9$, $0.035\ 2$; abundances $= 1 : 8.79 : 6.46$
6 C $1s^2 2s^2 2p^2$
  P $1s^2 2s^2 2p^6 3s^2 3p^3$
  Ar $1s^2 2s^2 2p^6 3s^2 3p^6$
  Fe $1s^2 2s^2 2p^6 3s^2 3p^6 4s^2 3d^6$
  As $1s^2 2s^2 2p^6 3s^2 3p^6 4s^2 3d^{10} 4p^3$

Chapter 4

1 HCl 2.03×10^{-29} C m, 6.10 D; HBr 2.28×10^{-29} C m, 6.82 D; HI 2.56×10^{-29} C m, 7.68 D
2 HOH angle $= 105°$
3 $R_\text{m} = 8.24 \text{ cm}^3 \text{ mol}^{-1}$; $R_\text{m} = 8.46 \text{ cm}^3 \text{ mol}^{-1}$
4 35.3 per cent α-glucose, 64.7 per cent β-glucose; 53.4 kg m^{-3}
5 $\bar{\nu}(\text{or } \sigma) = 2.25 \times 10^6 \text{ m}^{-1}$ or $22\ 500 \text{ cm}^{-1}$; $\nu = 6.75 + 10^{14}$ Hz or 675 THz; $\varepsilon = 4.47 \times 10^{-19}$ J; $E = 270 \text{ kJ mol}^{-1}$
6 $c = 9.83 \times 10^{-5}$ g per 100 cm^3 $= 2.61 \times 10^{-3}$ mol m^{-3}; $\varepsilon = 12\ 160 \text{ dm}^3 \text{ mol}^{-1} \text{ cm}^{-1} = 1216 \text{ m}^2 \text{ mol}^{-1}$; $T = 0.746$ (74.6 per cent)
7 $B = 1.50$ T
8

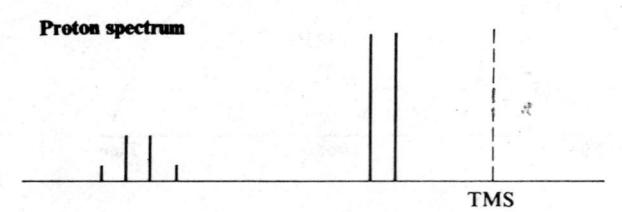

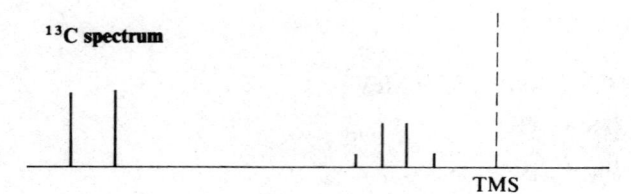

Proton spectrum

TMS

^{13}C spectrum

TMS

9 C_2H_5I
$27 = C_2H_3^+$; $29 = C_2H_5^+$; $127 = I^+$; $128 = HI^+$; $141 = CH_2I^+$; $156 = C_2H_5I^+$; $157 = {}^{13}C^{12}CH_5I^+$.

Chapter 5

1 $k_1 = 1.61 \times 10^{-3}$ s^{-1}; $t_{\frac{1}{2}} = 430$ s
2 $k_1 = 1.50 \times 10^{-2}$ min^{-1}; $t_{\frac{1}{2}} = 46.2$ min; 80.0 min
3 $k_1 = 4.94 \times 10^{-2}$ min^{-1}, 1.01×10^{-2} min^{-1}; see pp.
4 $K = 1.83$; $k_1 = 0.022\,3$ min^{-1}; $k_{-1} = 0.012\,2$ min^{-1}
5 $k_2 = 5.70$ dm^3 mol^{-1} min^{-1}; $t_{\frac{1}{2}} = 17.3$ min
6 $k_2 = 8.70 \times 10^{-8}$ Pa^{-1} s^{-1}
7 $E = 95.3$ kJ mol^{-1}
8 $E = 113$ kJ mol^{-1}; $A = 1.4 \times 10^{15}$ s^{-1};
  $k_1(37.0°$ C$) = 12 \times 10^{-5}$ s^{-1}; $k_1(18.0°$ C$) = 0.70 \times 10^{-5}$ s^{-1}
9 $\phi = 0.110$
10 $V = 7.5$ (arbitrary units); $K_m = 3.25 \times 10^{-4}$ mol dm^{-3}
11 $K_m = 24$ mmol dm^{-3} or 24 mol m^{-3};
  $V = 200$ nmol min^{-1}; non-competitive inhibition;
  $K_i = 2.7$ nmol dm^{-3} at 1.61 mmol dm^{-3},
      2.7 nmol dm^{-3} at 5.08 mmol dm^{-3}

Chapter 6

1 (a) $p = 38.0$ kPa; (b) $x_A = 0.526$, (c) $p = 40.5$ kPa

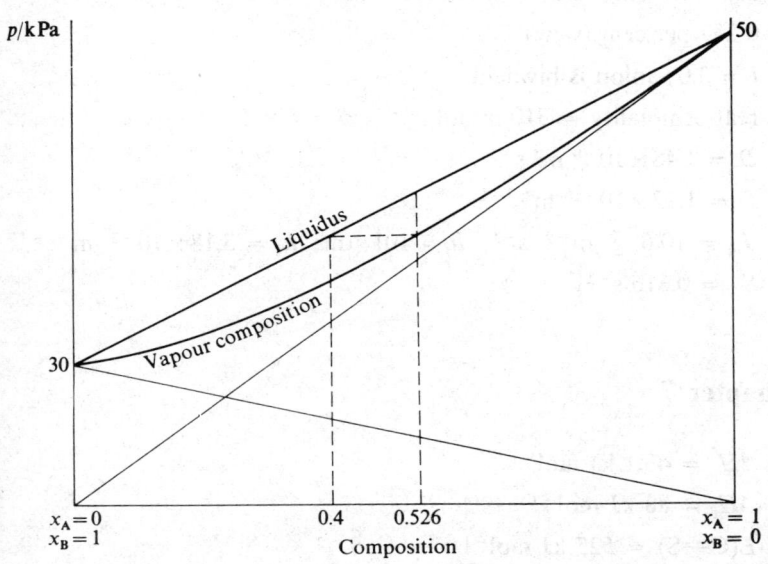

2 20 per cent nitrobenzene, 80 per cent water
3 p/kPa = 101.3,   5.88,   5.07,   3.44,   0.963,   0.51;
   a = 0.0128,   0.0232,   0.0213,   0.0311,   0.0160,   0.012;
   a is approximately constant, in accordance with Ferguson's principle.
4 Volume fractions, ϕs, and concentrations, c, in cell lipid are as follows:
   ϕs 1.4, 1.80, 2.03, 1.24, 1.33;
   c/mol m^{-3} 55, 71,   80, 49,   52;
   Both ϕs and c are approximately constant, in accordance with the Overton–Meyer theory.
5 Concentrations in cell lipid (cP) are as follows:
   cP/mol dm^{-3} 0.033, 0.020, 0.021, 0.041, 0.047, 0.045
   These are approximately constant, in accordance with the Overton–Meyer theory.
6 Linear plot confirms double molecules.
7 $p^{\ominus} - p$ = 12.3 Pa
8 M_r = 156
9 K_f = 5.12 K kg mol^{-1}
10 π = 517 kPa = 5.10 atmospheres
11 M_r = 49 100; ΔT_f = 1.99 × 10^{-3} K
12 0.257 per cent (w/w)
13 i = 3.0; anion is bivalent
14 real osmolality = 310 mmol kg^{-1}; ϕ = 0.905
15 D = 1.48 × 10^{-9} m^2 s^{-1}
16 D = 1.12 × 10^{-10} m^2 s^{-1}
17 J_m = 10.6 g m^{-2} s^{-1}; h = 101 μm; k_i = 3.18 × 10^{-5} m s^{-1};
    S = 0.316 s^{-1}

Chapter 7

1 ΔH = 43.1 kJ mol^{-1}
2 ΔH_f = 88 kJ mol^{-1}
3 E(C=S) = 525 kJ mol^{-1}

4 $E(C—H) = 416$ kJ mol^{-1}; $E(C—C) = 337$ kJ mol^{-1}; $E(C=C) = 595$ kJ mol^{-1}; $E(C—Cl) = 342$ kJ mol^{-1}

5 E_r/kJ mol^{-1} = 13.4, -2.2, -1.1, 20.5, 39.3; positive E_r occurs where there is conjugation between the two double bonds.

6 maximum work done = (a) -3.10 kJ, (b) = -2.19 kJ

7 (a) $\Delta S = 5.08$ J K^{-1} (b) $\Delta G = -1.89$ kJ

8 $\Delta S = 8.34$ J K^{-1} mol^{-1}

9 ΔS/J K^{-1} mol^{-1} = 109, 93.8, 83.7, 85.8, 104, 60.3; water and methanol are hydrogen-bonded in the liquid state, ethanoic acid forms dimers in the vapour state.

10 ΔH = (a) 30.8 kJ mol^{-1}, (b) 29.1 kJ mol^{-1}

11 $V(H_2O) = 1.63 \times 10^{-5}$ m^3 mol^{-1}

12 $\Delta G^\ominus = -4.82$ kJ

13 $\Delta S^\ominus = -21.5$ J K^{-1}

14 $\Delta G = -11.9$ kJ

5 (a) $\Delta G^\ominus = 0.244$ kJ mol^{-1} (b) ratio = 0.194

16 $\Delta H^\ominus = -16$ kJ; $\Delta G^\ominus = -3.43$ kJ; $\Delta S^\ominus = -42$ J K^{-1}

Chapter 8

1 $K_s = 1.85 \times 10^{-10}$; $c(Ag^+) = 9.26 \times 10^{-11}$ mol dm^{-3}

2 number of faradays = 0.003 08

3 $t(Ag^+) = 0.467$; $t(NO_3^-) = 0.533$; $\lambda(Ag^+) = 6.22 \times 10^{-3}$ Ω^{-1} m^2 mol^{-1}; $\lambda(NO_3^-) = 7.11 \times 10^{-3}$ Ω^{-1} m^2 mol^{-1}

4 (a) (i) $I = 0.003$ mol kg^{-1}; (ii) $\gamma_\pm = 0.880$
(iii) $m_\pm = 0.001\ 59$ mol kg^{-1}; (iv) $a_\pm = 0.001\ 40$
(v) $a = 2.72 \times 10^{-9}$

(b) (i) $I = 0.001\ 2$ mol kg^{-1}; (ii) $\gamma_\pm = 0.850$
(iii) $m_\pm = 0.000\ 3$ mol kg^{-1}; (iv) $a_\pm = 0.000\ 255$
(v) $a = 6.50 \times 10^{-8}$

5 Λ/Ω^{-1} m^2 mol^{-1} = 4.12×10^{-2}, 1.41×10^{-2}, 1.62×10^{-3}

6 $\Lambda_0(HCl) \approx 4.22 \times 10^{-2}$ Ω^{-1} m^2 mol^{-1}

7 $\Lambda_0(\text{HCl}) = 4.26 \times 10^{-2}\,\Omega^{-1}\,\text{m}^2\,\text{mol}^{-1}$; $a = 4.3 \times 10^{-4}\,\Omega^{-1}\,\text{m}^{3.5}\,\text{mol}^{-1.5}$; a (Onsager) $= 4.98 \times 10^{-4}\,\Omega^{-1}\,\text{m}^{3.5}\,\text{mol}^{-1.5}$

8 $10^2\alpha = 1.33, 1.88, 2.96, 4.14, 5.83, 9.11, 12.43$; plot α versus $1/\sqrt{c}$ and slope is $\sqrt{K}$; $K = 0.017\,\text{mol}\,\text{m}^{-3}$

9 $S = 16.9\,\text{mg dm}^{-3} = 16.9\,\text{g m}^{-3}$; $K_s = 4.05 \times 10^{-11}$

10 $t(\text{Na}^+) = 0.393$, $t(\text{Cl}^-) = 0.607$; $u(\text{Na}^+) = 4.44 \times 10^{-8}\,\text{m}^2\,\text{s}^{-1}\,\text{V}^{-1}$, $u(\text{Cl}^-) = 6.84 \times 10^{-8}\,\text{m}^2\,\text{s}^{-1}\,\text{V}^{-1}$

Chapter 9

1 pH $= 1.00, 2.28, 2.44, 4.96, 10.80, 13.00$

2 $k_a/\text{mol dm}^{-3} = 1.79 \times 10^{-4}, 1.40 \times 10^{-3}, 1.21 \times 10^{-10}$; p$k_a = 3.75, 2.85, 9.92$

3 $0.055\,5\,\text{mol}$

4 $0.007\,5\,\text{mol dm}^{-3}$; $0.001\,1\,\text{mol dm}^{-3}$

5 p$k_a = 9.24$; $k_h = 5.75 \times 10^{-10}\,\text{mol dm}^{-3}$; pH $= 11.12$; pH $= 5.12$

6 acid is dibasic; $M_r = 90$

7 isoelectric pH $= 6.07$

8 c (neutral molecules)/mol dm$^{-3} = 2.0 \times 10^{-4}, 1.6 \times 10^{-4}, 1.6 \times 10^{-4}, 2.0 \times 10^{-4}, 1.6 \times 10^{-4}, 1.4 \times 10^{-4}, 1.3 \times 10^{-4}, 1.6 \times 10^{-4}$

9 $[\text{BH}^+]_{\text{acidic}}/[\text{BH}^+]_{\text{basic}} = 1.12 \times 10^3$, ratio of total concentrations $= 234$

10 $k_{\text{acid}} = 1.10 \times 10^{-4}\,\text{mol}^{-1}\,\text{dm}^3\,\text{s}^{-1}$; $k_{\text{ion}} = 1.95 \times 10^{-4}\,\text{mol}^{-1}\,\text{dm}^3\,\text{s}^{-1}$; $k_0 = 0.867 \times 10^{-4}\,\text{s}^{-1}$

Chapter 10

1 $\Delta G = -4390\,\text{J}$; $\Delta S = 32.6\,\text{J K}^{-1}$; $\Delta H = 5330\,\text{J}$. For the doubled reaction these values are all doubled

2 $E = 1.133\,\text{V}$

3 $E = -0.018\,\text{V}$

4 $E_h^\ominus/\text{V} = -0.761, -0.761, -0.759, -0.761$, mean $= -0.760$

5 $E_h/\text{V} = -0.058, -0.132, -0.141, -0.286, -0.624, -0.751$

6 $E_h/\text{V} = 0.122, 0.132, 0.145, 0.150, 0.155, 0.168, 0.178, 0.711, 0.747, 0.764, 0.770, 0.775, 0.779$

7 $E_h/V = -0.148, -0.314$

8 $E^\ominus = 1.14$ V; $\Delta G^{\ominus\prime} = -440$ kJ mol^{-1}; K $= 1.2 \times 10^{77}$

9 The 0.05 mol dm^{-3} solution is in contact with the positive pole of the cell; $n = 2.06$ so M is bivalent; n is not a whole number because activity coefficients are not unity as assumed and the assumption that there is no liquid junction potential may not be valid.

10 $t(K^+) = 0.374$

11 $E = -0.112\ 0$ V

12 $c = 0.079\ 7$ mol dm^{-3}; $E_m = 98.5$ mV

Chapter 11

1 Type II

2 $v_m = 3.29$ cm^3 g^{-1}; c is very large, which indicates that ΔH_l is much more negative than ΔH_g^1; $a = 14.3$ m^2 g^{-1}; diameter $= 0.279$ μm

3 $10^3\ x/m = 3.754, 5.913, 9.757, 18.61, 49.95$

4 $k = 1.3 \times 10^{-3}$ (when concentrations are expressed in mol m^{-3}); $n = 1.4$

5 $\Gamma \approx 4.25 \times 10^{-6}$ mol m^{-2}, depending on the gradient; $A \approx 39 \times 10^{-20}$ m^2 (per molecule), depending on the gradient

6 Ideal $\Pi A = 4.05 \times 10^{-21}$ N m (or J)

7 $A \approx 21.3 \times 10^{-20}$ m^2 (per molecule); $A \approx 24.8 \times 10^{-20}$ m^2 (per molecule)

Chapter 12

1 $t = 9.43$ s; $t = 0.664$ s; diameter $= 5.70$ nm

2 For equal numbers $\overline{M}_N = 30$ kg mol^{-1}, $\overline{M}_m = 43.3$ kg mol^{-1}; for equal masses $\overline{M}_N = 16.7$ kg mol^{-1}, $\overline{M}_m = 30$ kg mol^{-1}

3 $c = 0.008\ 57$ mol dm^{-3}; $c = 0.006\ 5$ mol dm^{-3}

4 $M_N = 21.9$ kg mol^{-1}

5 $M_m = 140$ kg mol^{-1}

6 $M = 17.0$ kg mol^{-1}

7 $M = 5.97$ kg mol^{-1}

8 $M_\eta = M_m = 1.77$ kg mol^{-1}

9 $\zeta = 13.1$ mV
10 $v = 2.74$ μm s^{-1}; distance = 9.85 mm

Chapter 13

1 $E = 2.48 \times 10^{11}$ N m^{-2}
2 $K = 1.89 \times 10^{11}$ N m^{-2}; $\mu = 0.281$
3 $\Delta p = 2.50$ kPa
4 $v = 7.43 \times 10^{-7}$ m^2 s^{-1}; $\eta = 6.53 \times 10^{-4}$ kg m^{-1} s^{-1}
6 $E_{vis} = 13.6$ kJ mol^{-1}; $\Delta H_{vap} = 42.6$ kJ mol^{-1}; $E_{vis}/\Delta H_{vap}$ (=0.32) ≈ 0.3 and the log p versus $1/T$ plot is closely linear, as for most liquids, but the log η versus $1/T$ plot is slightly curved due to hydrogen bonding
7 Pseudoplastic type; $n = 0.508$, $K = 6.15$
8 Non-linear at $\tau > 5.0$ N m^{-2}

Appendices

Appendix I Physical Quantities and Units

A *physical quantity* is a product of a *numerical value* (a pure number) and a *unit*. In this text the various physical quantities are, wherever possible, represented by symbols recommended by the International Union of Pure and Applied Chemistry (IUPAC) and expressed in the International System of Units (SI).

Table A.1. Basic Physical Quantities and Units in SI

Basic physical quantity	symbol	Basic unit	symbol
length	l	metre	m
mass	m	kilogram(me)	kg
time	t	second	s
electric current	I	ampere	A
thermodynamic temperature	T	kelvin	K
luminous intensity	I_v	candela	cd
amount of substance	n	mole	mol

The SI *base* units and their corresponding basic physical quantities are quoted in Table A.1. The *kilogram* is defined by a very carefully preserved international prototype, whereas the *metre* and the *second* are now defined in terms of very accurate measurements of certain spectroscopic transitions (p. 80). The *ampere* is that constant current, which, if maintained in two straight parallel conductors of infinite length, of negligible cross-section and placed 1 metre apart in vacuum, would produce between these conductors a force equal to 2×10^{-7} newton per metre of length (p. 331). The SI unit of thermodynamic temperature, the kelvin, is the fraction 1/273.16 of the thermodynamic temperature of the triple point of water (p. 40). The *candela* will not be defined, since it is seldom needed in physical chemistry. The *mole* is the amount of substance of a system which contains as many elementary entities as there are atoms in 0.012 kilogram of carbon 12 (p. 59). When the mole is used, the elementary entities must be specified and may be atoms, molecules, ions, electrons, photons, other particles or a specified group of such particles. The number of elementary entities or elementary particles in a mole is known as Avogadro's constant L, whose value is $6.022\,2 \times 10^{23}$ mol^{-1}. Values of other fundamental constants which are important in physical chemistry are given in Table A.5.

The *derived* SI units and quantities are obtained from the base units and quantities by processes of multiplication and division. When two or more basic quantities are multiplied or divided in a certain way to obtain a derived quantity, the units of the basic quantities have to be treated in the same way to obtain the derived unit. When this is done with SI units, no constant factors are introduced; SI is therefore said to be a *coherent* system of units. Certain derived units have special names and symbols and are shown in Table A.2. For example, unit mass, kg, multiplied by unit acceleration, m s^{-2}, gives unit force, m kg s^{-2}, which is given the special name, newton, whose symbol is N.

Table A.2. SI derived units with special names

Quantity		SI unit		Expression in terms of other units	Expression in terms of SI base units
Name	Symbol	Name	Symbol		
frequency	v	hertz	Hz		s^{-1}
force	F	newton	N		m kg s^{-2}
pressure	p	pascal	Pa	N m^{-2}	m^{-1} kg s^{-2}
energy, work, quantity of heat		joule	J	N m	m^2 kg s^{-2}
power	P	watt	W	J s^{-1}	m^2 kg s^{-3}
quantity of electricity, electric charge	Q	coulomb	C	A s	s A
electric tension, electric potential, electromotive force	E	volt	V	W A^{-1}	m^2 kg s^{-3} A^{-1}
electric capacitance	C	farad	F	C V^{-1}	m^{-2} kg^{-1} s^4 A^2
electric resistance	R	ohm	Ω	V A^{-1}	m^2 kg s^{-3} A^{-2}
electric conductance	G	siemens	S	A V^{-1}	m^{-2} kg^{-1} s^3 A^2
magnetic flux		weber	Wb	V s	m^2 kg s^{-2} A^{-1}
magnetic flux density		tesla	T	Wb m^{-2}	kg s^{-2} A^{-1}

Table A.3 shows some units which are not SI but which are still in use in physical chemistry for reasons of convenience and habit. The less common ones are defined in terms of SI units, whenever they are used in the book.

Table A.3. Other Units

Name	Symbol	Value in SI units
Units in use with the International System		
degree	°	$1° = (\pi/180)$ rad
litre	l	$1\,l = 1\,dm^3 = 10^{-3}\,m^3$
electronvolt	eV	$1\,eV = 1.602\,19 \times 10^{-19}$ J
unified atomic mass unit	m_u	$1\,m_u = 1.660\,53 \times 10^{-27}$ kg
Units to be used with the International System for a limited time		
ångström	Å	$1\,Å = 0.1\,nm = 10^{-10}\,m$
bar	bar	$1\,bar = 0.1\,MPa = 10^5\,Pa$
standard atmosphere	atm	$1\,atm = 101\,325\,Pa$
curie	Ci	$1\,Ci = 3.7 \times 10^{10}\,s^{-1}$
röntgen	R	$1\,R = 2.58 \times 10^{-4}\,C\,kg^{-1}$
rad	rad	$1\,rad = 10^{-2}\,J\,kg^{-1}$
c.g.s. units with special names		
erg	erg	$1\,erg = 10^{-7}$ J
dyne	dyn	$1\,dyn = 10^{-5}$ N
poise	P	$1\,P = 1\,dyn\,s\,cm^{-2} = 0.1\,Pa\,s$
stokes	St	$1\,St = 1\,cm^2\,s^{-1} = 10^{-4}\,m^2\,s^{-1}$
gauss	Gs, G	$1\,Gs = 10^{-4}\,T$
oersted	Oe	$1\,Oe = 10^3/4\pi\,A\,m^{-1}$
maxwell	Mx	$1\,Mx$ corresponds to 10^{-8} Wb
Other units generally deprecated		
calorie (thermochemical)	cal_{th}	$1\,cal_{th} = 4.184$ J
molar electronvolt	LeV	$1\,LeV = 96\,487$ J
micron	μ	$1\,\mu = 1\,\mu m = 10^{-6}\,m$
molal	m	$1\,m = 1\,mol\,kg^{-1}$
molar	M	$1\,M = 1\,mol\,dm^{-3} = 1000\,mol\,m^{-3}$
mm of Hg	mmHg	$1\,mmHg = 13.595\,1 \times 9.806\,65\,Pa$ $= 133.322\,39\,Pa$
torr	Torr	$1\,Torr = (101\,325/760)\,Pa$ $= 133.322\,37\,Pa$

To avoid numbers involving large positive or negative powers of ten, certain decimal multiples and submultiples of SI units may be represented by the prefixes shown in Table A.4, e.g. 5×10^{-2} m = 5 cm; 10^{-10} m = 100 pm = 0.1 nm. An exponent affixed to a symbol containing a prefix indicates that the multiple or submultiple of the unit is raised to the power expressed by the exponent, e.g. 1 cm³ = $(10^{-2}\,\text{m})^3 = 10^{-6}\,\text{m}^3$; 1 cm⁻¹ = $(10^{-2}\,\text{m})^{-1} = 10^2\,\text{m}^{-1}$. The use of more than one prefix is avoided. Since the kg is the only basic unit which includes a prefix, we note that 10^{-3} kg may be written as 1 g, but not as 1 mkg, and 2×10^{-5} kg may be written as 20 mg, but not as 20 μkg.

Table A.4. SI Prefixes

Factor	Prefix	Symbol	Factor	Prefix	Symbol
10^{12}	tera	T	10^{-1}	deci	d
10^{9}	giga	G	10^{-2}	centi	c
10^{6}	mega	M	10^{-3}	milli	m
10^{3}	kilo	k	10^{-6}	micro	μ
10^{2}	hecto	h	10^{-9}	nano	n
10^{1}	deca	da	10^{-12}	pico	p
			10^{-15}	femto	f
			10^{-18}	atto	a

Table A.5. SI Values of the Fundamental Constants

Quantity	Symbol	Value
'ice-point' temperature	T_{ice}	273.150 K
gas constant	R	8.314 3 J K⁻¹ mol⁻¹
molar volume of an ideal gas at NTP	$\dfrac{RT_{\text{ice}}}{101\,325\,\text{Pa}}$	0.022 414 m³ mol⁻¹
Avogadro constant	L, N_A	$6.022\,2 \times 10^{23}$ mol⁻¹
Boltzmann constant	$k\,(= R/L)$	$1.380\,6 \times 10^{-23}$ J K⁻¹
Faraday constant	F	$9.648\,7 \times 10^{4}$ C mol⁻¹
charge of proton	$e\,(= F/L)$	$1.602\,2 \times 10^{-19}$ C
unified atomic mass constant	m_{u}	$1.660\,53 \times 10^{-27}$ kg
speed of light in a vacuum	c	$2.997\,93 \times 10^{8}$ m s⁻¹
Planck constant	h	$6.626\,2 \times 10^{-34}$ J s
constant relating wavenumber and energy	$Z\,(= Lhc)$	$1.196\,6 \times 10^{-1}$ J m mol⁻¹
Bohr magneton	μ_B (or β)	$9.274\,1 \times 10^{-24}$ m² A
nuclear magneton	β_N	$5.050\,4 \times 10^{27}$ m² A = 6.347×10^{-3} Wb m

Appendix II. Periodic Table

Periods	Main groups 1	2				Transition block										3	4	5	6	7	8
1	$_1$H																				$_2$He
2	$_3$Li	$_4$Be														$_5$B	$_6$C	$_7$N	$_8$O	$_9$F	$_{10}$Ne
3	$_{11}$Na	$_{12}$Mg														$_{13}$Al	$_{14}$Si	$_{15}$P	$_{16}$S	$_{17}$Cl	$_{18}$Ar
4	$_{19}$K	$_{20}$Ca	$_{21}$Sc	$_{22}$Ti	$_{23}$V	$_{24}$Cr	$_{25}$Mn	$_{26}$Fe	$_{27}$Co	$_{28}$Ni	$_{29}$Cu	$_{30}$Zn				$_{31}$Ga	$_{32}$Ge	$_{33}$As	$_{34}$Se	$_{35}$Br	$_{36}$Kr
5	$_{37}$Rb	$_{38}$Sr	$_{39}$Y	$_{40}$Zr	$_{41}$Nb	$_{42}$Mo	$_{43}$Tc	$_{44}$Ru	$_{45}$Rh	$_{46}$Pd	$_{47}$Ag	$_{48}$Cd				$_{49}$In	$_{50}$Sn	$_{51}$Sb	$_{52}$Te	$_{53}$I	$_{54}$Xe
6	$_{55}$Cs	$_{56}$Ba	$_{57}$La and 58–71	$_{72}$Hf	$_{73}$Ta	$_{74}$W	$_{75}$Re	$_{76}$Os	$_{77}$Ir	$_{78}$Pt	$_{79}$Au	$_{80}$Hg				$_{81}$Tl	$_{82}$Pb	$_{83}$Bi	$_{84}$Po	$_{85}$At	$_{86}$Rn
7	$_{87}$Fr	$_{88}$Ra	$_{89}$Ac and 90–103																		

Lanthanides: $_{58}$Ce $_{59}$Pr $_{60}$Nd $_{61}$Pm $_{62}$Sm $_{63}$Eu $_{64}$Gd $_{65}$Tb $_{66}$Dy $_{67}$Ho $_{68}$Er $_{69}$Tm $_{70}$Yb $_{71}$Lu

Actinides: $_{90}$Th $_{91}$Pa $_{92}$U $_{93}$Np $_{94}$Pu $_{95}$Am $_{96}$Cm $_{97}$Bk $_{98}$Cf $_{99}$Es $_{100}$Fm $_{101}$Md $_{102}$No $_{103}$Lr

Appendix III-Atomic Weights, 1975

The values of relative atomic masses ($A_r(E)$) of the element E are given based on the relative atomic mass of $^{12}C = 12$ (i.e. $A_r(^{12}C) = 12$). Recommended by the International Union of Pure and Applied Chemistry Commission on Atomic Weights (1975). The values in italics are the mass numbers of the longest-lived isotopes.

Element	Symbol	Atomic No.	Atomic weight
Actinium	Ac	89	*227*
Aluminium	Al	13	26.981 54[a]
Americium	Am	95	*243*
Antimony	Sb	51	121.7_5
Argon	Ar	18	39.94_8[b, c, d, g]
Arsenic	As	33	74.921 6[a]
Astatine	At	85	*210*
Barium	Ba	56	137.33
Berkelium	Bk	97	*247*
Beryllium	Be	4	9.012 18[a]
Bismuth	Bi	83	208.980 4[a]
Boron	B	5	10.81[c, d, e]
Bromine	Br	35	79.904[c]
Cadmium	Cd	48	112.41
Caesium	Cs	55	132.905 4[a]
Calcium	Ca	20	40.08[g]
Californium	Cf	98	*251*
Carbon	C	6	12.011[b, d]
Cerium	Ce	58	140.12
Chlorine	Cl	17	35.453[c]
Chromium	Cr	24	51.996[c]
Cobalt	Co	27	58.933 2[a]
Copper	Cu	29	63.54_6[c, d]
Curium	Cm	96	*247*
Dysprosium	Dy	66	162.5_0
Einsteinium	Es	99	*254*
Erbium	Er	68	167.2_6
Europium	Eu	63	151.96
Fermium	Fm	100	*257*
Fluorine	F	9	18.998 403[a]
Francium	Fr	87	*223*
Gadolinium	Gd	64	157.2_5
Gallium	Ga	31	69.72
Germanium	Ge	32	72.5_9
Gold	Au	79	196.966 5[a]
Hafnium	Hf	72	178.4_9
Helium	He	2	4.002 60[b, c]

Appendix III Atomic Weights, 1975

Element	Symbol	Atomic No.	Atomic weight
Holmium	Ho	67	164.930 4[a]
Hydrogen	H	1	1.007 9[b, d]
Indium	In	49	114.82
Iodine	I	53	126.904 5[a]
Iridium	Ir	77	192.2$_2$
Iron	Fe	26	55.84$_7$
Krypton	Kr	36	83.80[e]
Lanthanum	La	57	138.905$_5$[b]
Lawrencium	Lr	103	*260*
Lead	Pb	82	207.2[d, g]
Lithium	Li	3	6.94$_1$[c, d, e, g]
Lutetium	Lu	71	174.97
Magnesium	Mg	12	24.305[c, g]
Manganese	Mn	25	54.938 0[a]
Mendelevium	Md	101	*258*
Mercury	Hg	80	200.5$_9$
Molybdenum	Mo	42	95.94
Neodymium	Nd	60	144.2$_4$
Neon	Ne	10	20.17$_9$[c, e]
Neptunium	Np	93	237.048 2[f]
Nickel	Ni	28	58.70
Niobium	Nb	41	92.906 4[a]
Nitrogen	N	7	14.006 7[b, c]
Nobelium	No	102	*255*
Osmium	Os	76	190.2[g]
Oxygen	O	8	15.999$_4$[b, c, d]
Palladium	Pd	46	106.4
Phosphorus	P	15	30.973 76[a]
Platinum	Pt	78	195.0$_9$
Plutonium	Pu	94	*244*
Polonium	Po	84	*209*
Potassium	K	19	39.098$_3$[c]
Praseodymium	Pr	59	140.907 7[a]
Promethium	Pm	61	*145*
Protactinium	Pa	91	231.035 9[f]
Radium	Ra	88	226.025 4[f, g]
Radon	Rn	86	*222*
Rhenium	Re	75	186.207[c]
Rhodium	Rh	45	102.905 5[a]
Rubidium	Rb	37	85.467$_8$[c]
Ruthenium	Ru	44	101.0$_7$
Samarium	Sm	62	150.4
Scandium	Sc	21	44.955 9[a]
Selinium	Se	34	78.9$_6$
Silicon	Si	14	28.085$_5$[c, d]
Silver	Ag	47	107.868[c]

Element	Symbol	Atomic No.	Atomic weight
Sodium	Na	11	22.989 77[a]
Strontium	Sr	38	87.62[g]
Sulphur	S	16	32.06[d]
Tantalum	Ta	73	180.947_9[b]
Technetium	Tc	43	97
Tellurium	Te	52	127.6_0
Terbium	Tb	65	158.925 4[a]
Thallium	Tl	81	204.3_7
Thorium	Th	90	232.038 1[f, g]
Thulium	Tm	69	168.934 2[a]
Tin	Sn	50	118.6_9
Titanium	Ti	22	47.9_0
Tungsten	W	74	183.8_5
Uranium	U	92	238.029[b, c, e, g]
Vanadium	V	23	50.941_4[b, c]
Xenon	Xe	54	131.30[e]
Ytterbium	Yb	70	173.0_4
Yttrium	Y	39	88.905 9[a]
Zinc	Zn	30	65.38
Zirconium	Zr	40	91.22

Notes:

Atomic weights are considered reliable to ± 1 in the last digit, or ± 3 when the last digit is shown as a subscript.

[a] Element with only one stable nuclide.

[b] Element with one predominant isotope (about 99 to 100 per cent abundance); errors or variations in the isotopic composition have a corresponding small effect on the value of $A_r(E)$.

[c] Element for which the value of $A_r(E)$ derives its reliability from calibrated measurements (i.e. from comparisons with synthetic mixtures of known iosotopic composition).

[d] Element for which known variations in isotopic abundance in terrestrial material prevent a more precise atomic weight being given; $A_r(E)$ values should be applicable to any 'normal' material.

[e] Element for which values of A_r may be found in commercially available products that differ from the tabulated value of $A_r(E)$ because of inadvertent or undisclosed changes of isotopic composition.

[f] Element for which the value of A_r is that of the most commonly available long-lived nuclide.

[g] Element for which geological specimens are known in which the element has an anomalous isotopic composition.

Appendices
Logarithms

	0	1	2	3	4	5	6	7	8	9	1 2 3 4	5	6 7 8 9
10	0000	0043	0086	0128	0170	0212	0253	0294	0334	0374	4 9 13 17 4 8 12 16	21 20	26 30 34 38 24 28 32 37
11	0414	0453	0492	0531	0569	0607	0645	0682	0719	0755	4 8 12 15 4 7 11 15	19 19	23 27 31 35 22 26 30 33
12	0792	0828	0864	0899	0934	0969	1004	1038	1072	1106	3 7 11 14 3 7 10 14	18 17	21 25 28 32 20 24 27 31
13	1139	1173	1206	1239	1271	1303	1335	1367	1399	1430	3 7 10 13 3 7 10 12	16 16	20 23 26 30 19 22 25 29
14	1461	1492	1523	1553	1584	1614	1644	1673	1703	1732	3 6 9 12 3 6 9 12	15 15	18 21 24 28 17 20 23 26
15	1761	1790	1818	1847	1875	1903	1931	1959	1987	2014	3 6 9 11 3 5 8 11	14 14	17 20 23 26 16 19 22 25
16	2041	2068	2095	2122	2148	2175	2201	2227	2253	2279	3 5 8 11 3 5 8 10	14 13	16 19 22 24 15 18 21 23
17	2304	2330	2355	2380	2405	2430	2455	2480	2504	2529	3 5 8 10 2 5 7 10	13 12	15 18 20 23 15 17 19 22
18	2553	2577	2601	2625	2648	2672	2695	2718	2742	2765	2 5 7 9 2 5 7 9	12 11	14 16 19 21 14 16 18 21
19	2788	2810	2833	2856	2878	2900	2923	2945	2967	2989	2 4 7 9 2 4 6 8	11 11	13 16 18 20 13 15 17 19
20	3010	3032	3054	3075	3096	3118	3139	3160	3181	3201	2 4 6 8	11	13 15 17 19
21 22 23 24	3222 3424 3617 3802	3243 3444 3636 3820	3263 3464 3655 3838	3284 3483 3674<>3856	3304 3502 3692 3874	3324 3522 3711 3892	3345 3541 3729 3900	3365 3560 3747 3927	3385 3579 3760 3945	3404 3598 3784 3962	2 4 6 8 2 4 6 8 2 4 6 7 2 4 5 7	10 10 9 9	12 14 16 18 12 14 15 17 11 13 15 17 11 12 14 16
25	3979	3997	4014	4031	4048	4065	4082	4099	4116	4133	2 3 5 7	9	10 12 14 15
26 27 28 29	4150 4314 4472 4624	4166 4330 4487 4639	4183 4346 4502 4654	4200 4362 4518 4669	4216 4378 4533 4683	4232 4393 4548 4698	4249 4409 4564 4713	4265 4425 4579 4728	4281 4440 4594 4742	4298 4456 4609 4757	2 3 5 7 2 3 5 6 2 3 5 6 1 3 4 6	8 8 8 7	10 11 13 15 9 11 13 14 9 11 12 14 9 10 12 13
30	4771	4786	4800	4814	4829	4843	4857	4871	4886	4900	1 3 4 6	7	9 10 11 13
31 32 33 34	4914 5051 5185 5315	4928 5065 5198 5328	4942 5079 5211 5340	4955 5092 5224 5353	4969 5105 5237 5366	4983 5119 5250 5378	4997 5132 5263 5391	5011 5145 5276 5403	5024 5159 5289 5416	5038 5172 5302 5428	1 3 4 6 1 3 4 5 1 3 4 5 1 3 4 5	7 7 6 6	8 10 11 12 8 9 11 12 8 9 10 12 8 9 10 11
35	5441	5453	5465	5478	5490	5502	5514	5527	5539	5551	1 2 4 5	6	7 9 10 11
36 37 38 39	5563 5682 5798 5911	5575 5694 5809 5922	5587 5705 5821 5933	5599 5717 5832 5944	5611 5729 5843 5955	5623 5740 5855 5966	5635 5752 5866 5977	5647 5763 5877 5988	5658 5775 5888 5999	5670 5780 5899 6010	1 2 4 5 1 2 3 5 1 2 3 5 1 2 3 4	6 6 6 5	7 8 10 11 7 8 9 10 7 8 9 10 7 8 9 10
40	6021	6031	6042	6053	6064	6075	6085	6096	6107	6117	1 2 3 4	5	6 8 9 10
41 42 43 44	6128 6232 6335 6435	6138 6243 6345 6444	6149 6253 6355 6454	6160 6263 6365 6464	6170 6274 6375 6474	6180 6284 6385 6484	6191 6294 6395 6493	6201 6304 6405 6503	6212 6314 6415 6513	6222 6325 6425 6522	1 2 3 4 1 2 3 4 1 2 3 4 1 2 3 4	5 5 5 5	6 7 8 9 6 7 8 9 6 7 8 9 6 7 8 9
45	6532	6542	6551	6561	6571	6580	6590	6599	6609	6618	1 2 3 4	5	6 7 8 9
46 47 48 49	6628 6721 6812 6902	6637 6730 6821 6911	6646 6739 6830 6920	6656 6749 6839 6928	6665 6758 6848 6937	6675 6767 6857 6946	6684 6776 6866 6955	6693 6785 6875 6964	6702 6794 6884 6972	6712 6803 6893 6981	1 2 3 4 1 2 3 4 1 2 3 4 1 2 3 4	5 5 4 4	6 7 7 8 5 6 7 8 5 6 7 8 5 6 7 8
50	6990	6993	7007	7016	7024	7033	7042	7050	7059	7067	1 2 3 3	4	5 6 7 8

Logarithms

	0	1	2	3	4	5	6	7	8	9	1	2	3	4	5	6	7	8	9
51	7076	7084	7093	7101	7110	7118	7126	7135	7143	7152	1	2	3	4	5	6	7	8
52	7160	7168	7177	7185	7193	7202	7210	7218	7226	7235	1	2	3	4	5	6	7	7
53	7243	7251	7259	7267	7275	7284	7292	7300	7308	7316	1	2	2	3	4	5	6	6	7
54	7324	7332	7340	7348	7356	7364	7372	7380	7388	7396	1	2	2	3	4	5	6	6	7
55	7404	7412	7419	7427	7435	7443	7451	7459	7466	7474	1	2	2	3	4	5	5	6	7
56	7482	7490	7497	7505	7513	7520	7528	7536	7543	7551	1	2	2	3	4	5	5	6	7
57	7559	7566	7574	7582	7589	7597	7604	7612	7619	7627	1	2	2	3	4	5	5	6	7
58	7634	7642	7649	7657	7664	7672	7679	7686	7694	7701	1	1	2	3	4	4	5	6	7
59	7709	7716	7723	7731	7738	7745	7752	7760	7767	7774	1	1	2	3	4	4	5	6	7
60	7782	7789	7796	7803	7810	7818	7825	7832	7839	7846	1	1	2	3	4	4	5	6	6
61	7853	7860	7868	7875	7882	7889	7896	7903	7910	7917	1	1	2	3	4	4	5	6	6
62	7924	7931	7938	7945	7952	7959	7966	7973	7980	7987	1	1	2	3	3	4	5	6	6
63	7993	8000	8007	8014	8021	8028	8035	8041	8048	8055	1	1	2	3	3	4	5	5	6
64	8062	8069	8075	8082	8089	8096	8102	8109	8116	8122	1	1	2	3	3	4	5	5	6
65	8129	8136	8142	8149	8156	8162	8169	8176	8182	8189	1	1	2	3	3	4	5	5	6
66	8195	8202	8209	8215	8222	8228	8235	8241	8248	8254	1	1	2	3	3	4	5	5	6
67	8261	8267	8274	8280	8287	8293	8299	8306	8312	8319	1	1	2	3	3	4	5	5	6
68	8325	8331	8338	8344	8351	8357	8363	8370	8376	8382	1	1	2	3	3	4	4	5	6
69	8388	8395	8401	8407	8414	8420	8426	8432	8439	8445	1	1	2	2	3	4	4	5	6
70	8451	8457	8463	8470	8476	8482	8488	8494	8500	8506	1	1	2	2	3	4	4	5	6
71	8513	8519	8525	8531	8537	8543	8549	8555	8561	8567	1	1	2	2	3	4	4	5	5
72	8573	8579	8585	8591	8597	8603	8609	8615	8621	8627	1	1	2	2	3	4	4	5	5
73	8633	8639	8645	8651	8657	8663	8669	8675	8681	8686	1	1	2	2	3	4	4	5	5
74	8692	8698	8704	8710	8716	8722	8727	8733	8739	8745	1	1	2	2	3	4	4	5	5
75	8751	8756	8762	8768	8774	8779	8785	8791	8797	8802	1	1	2	2	3	3	4	5	5
76	8808	8814	8820	8825	8831	8837	8842	8848	8854	8859	1	1	2	2	3	3	4	5	5
77	8865	8871	8876	8882	8887	8893	8899	8904	8910	8915	1	1	2	2	3	3	4	4	5
78	8921	8927	8932	8938	8943	8949	8954	8960	8965	8971	1	1	2	2	3	3	4	4	5
79	8976	8982	8987	8993	8998	9004	9009	9015	9020	9025	1	1	2	2	3	3	4	4	5
80	9031	9036	9042	9047	9053	9058	9063	9069	9074	9079	1	1	2	2	3	3	4	4	5
81	9085	9090	9096	9101	9106	9112	9117	9122	9128	9133	1	1	2	2	3	3	4	4	5
82	9138	9143	9149	9154	9159	9165	9170	9175	9180	9186	1	1	2	2	3	3	4	4	5
83	9191	9196	9201	9206	9212	9217	9222	9227	9232	9238	1	1	2	2	3	3	4	4	5
84	9243	9248	9253	9258	9263	9269	9274	9279	9284	9289	1	1	2	2	3	3	4	4	5
85	9294	9299	9304	9309	9315	9320	9325	9330	9335	9340	1	1	2	2	3	3	4	4	5
86	9345	9350	9355	9360	9365	9370	9375	9380	9385	9390	1	1	2	2	3	3	4	4	5
87	9395	9400	9405	9410	9415	9420	9425	9430	9435	9440	0	1	1	2	2	3	3	4	4
88	9445	9450	9455	9460	9465	9469	9474	9479	9484	9489	0	1	1	2	2	3	3	4	4
89	9494	9499	9504	9509	9513	9518	9523	9528	9533	9538	0	1	1	2	2	3	3	4	4
90	9542	9547	9552	9557	9562	9566	9571	9576	9581	9586	0	1	1	2	2	3	3	4	4
91	9590	9595	9600	9605	9609	9614	9619	9624	9628	9633	0	1	1	2	2	3	3	4	4
92	9638	9643	9647	9652	9657	9661	9666	9671	9675	9680	0	1	1	2	2	3	3	4	4
93	9685	9689	9694	9699	9703	9708	9713	9717	9722	9727	0	1	1	2	2	3	3	4	4
94	9731	9736	9741	9745	9750	9754	9759	9763	9768	9773	0	1	1	2	2	3	3	4	4
95	9777	9782	9786	9791	9795	9800	9805	9809	9814	9818	0	1	1	2	2	3	3	4	4
96	9823	9827	9832	9836	9841	9845	9850	9854	9859	9863	0	1	1	2	2	3	3	4	4
97	9868	9872	9877	9881	9886	9890	9894	9899	9903	9908	0	1	1	2	2	3	3	4	4
98	9912	9917	9921	9926	9930	9934	9939	9943	9948	9952	0	1	1	2	2	3	3	4	4
99	9956	9961	9965	9969	9974	9978	9983	9987	9991	9996	0	1	1	2	2	3	3	3	4

Antilogarithms

	0	1	2	3	4	5	6	7	8	9	1	2	3	4	5	6	7	8	9
.00	1000	1002	1005	1007	1009	1012	1014	1016	1019	1021	0	0	1	1	1	1	2	2	2
.01	1023	1026	1028	1030	1033	1035	1038	1040	1042	1045	0	0	1	1	1	1	2	2	2
.02	1047	1050	1052	1054	1057	1059	1062	1064	1067	1069	0	0	1	1	1	1	2	2	2
.03	1072	1074	1076	1079	1081	1084	1086	1089	1091	1094	0	0	1	1	1	1	2	2	2
.04	1096	1099	1102	1104	1107	1109	1112	1114	1117	1119	0	1	1	1	1	2	2	2	2
.05	1122	1125	1127	1130	1132	1135	1138	1140	1143	1146	0	1	1	1	1	2	2	2	2
.06	1148	1151	1153	1156	1159	1161	1164	1167	1169	1172	0	1	1	1	1	2	2	2	2
.07	1175	1178	1180	1183	1186	1189	1191	1194	1197	1199	0	1	1	1	1	2	2	2	2
.08	1202	1205	1208	1211	1213	1216	1219	1222	1225	1227	0	1	1	1	1	2	2	2	3
.09	1230	1233	1236	1239	1242	1245	1247	1250	1253	1256	0	1	1	1	1	2	2	2	3
.10	1259	1262	1265	1268	1271	1274	1276	1279	1282	1285	0	1	1	1	1	2	2	2	3
.11	1288	1291	1294	1297	1300	1303	1306	1309	1312	1315	0	1	1	1	2	2	2	2	3
.12	1318	1321	1324	1327	1330	1334	1337	1340	1343	1346	0	1	1	1	2	2	2	2	3
.13	1349	1352	1355	1358	1361	1365	1368	1371	1374	1377	0	1	1	1	2	2	2	2	3
.14	1380	1384	1387	1390	1393	1396	1400	1403	1406	1409	0	1	1	1	2	2	2	3	3
.15	1413	1416	1419	1422	1426	1429	1432	1435	1439	1442	0	1	1	1	2	2	2	3	3
.16	1445	1449	1452	1455	1459	1462	1466	1469	1472	1476	0	1	1	1	2	2	2	3	3
.17	1479	1483	1486	1489	1493	1496	1500	1503	1507	1510	0	1	1	1	2	2	2	3	3
.18	1514	1517	1521	1524	1528	1531	1535	1538	1542	1545	0	1	1	1	2	2	2	3	3
.19	1549	1552	1556	1560	1563	1567	1570	1574	1578	1581	0	1	1	1	2	2	3	3	3
.20	1585	1589	1592	1596	1600	1603	1607	1611	1614	1618	0	1	1	1	2	2	3	3	3
.21	1622	1626	1629	1633	1637	1641	1644	1648	1652	1656	0	1	1	2	2	2	3	3	3
.22	1660	1663	1667	1671	1675	1679	1683	1687	1690	1694	0	1	1	2	2	2	3	3	3
.23	1698	1702	1706	1710	1714	1718	1723	1726	1730	1734	0	1	1	2	2	2	3	3	4
.24	1738	1742	1746	1750	1754	1758	1762	1766	1770	1774	0	1	1	2	2	2	3	3	4
.25	1778	1782	1786	1791	1795	1799	1803	1807	1811	1816	0	1	1	2	2	2	3	3	4
.26	1820	1824	1828	1832	1837	1841	1845	1849	1854	1858	0	1	1	2	2	3	3	3	4
.27	1862	1866	1871	1875	1879	1884	1888	1892	1897	1901	0	1	1	2	2	3	3	3	4
.28	1905	1910	1914	1919	1923	1928	1932	1936	1941	1945	0	1	1	2	2	3	3	4	4
.29	1950	1954	1959	1963	1968	1972	1977	1982	1986	1991	0	1	1	2	2	3	3	4	4
.30	1995	2000	2004	2009	2014	2018	2023	2028	2032	2037	0	1	1	2	2	3	3	4	4
.31	2042	2046	2051	2056	2061	2065	2070	2075	2080	2084	0	1	1	2	2	3	3	4	4
.32	2089	2094	2099	2104	2109	2113	2118	2123	2128	2133	0	1	1	2	2	3	3	4	4
.33	2138	2143	2148	2153	2158	2163	2168	2173	2178	2183	0	1	1	2	2	3	3	4	4
.34	2188	2193	2198	2203	2208	2213	2218	2223	2228	2234	1	1	2	2	3	3	4	4	5
.35	2239	2244	2249	2254	2259	2265	2270	2275	2280	2286	1	1	2	2	3	3	4	4	5
.36	2291	2296	2301	2307	2312	2317	2323	2328	2333	2339	1	1	2	2	3	3	4	4	5
.37	2344	2350	2355	2360	2366	2371	2377	2382	2388	2393	1	1	2	2	3	3	4	4	5
.38	2399	2404	2410	2415	2421	2427	2432	2438	2443	2449	1	1	2	2	3	3	4	4	5
.39	2455	2460	2466	2472	2477	2483	2489	2495	2500	2506	1	1	2	2	3	3	4	5	5
.40	2512	2518	2523	2529	2535	2541	2547	2553	2559	2564	1	1	2	2	3	4	4	5	5
.41	2570	2576	2582	2588	2594	2600	2606	2612	2618	2624	1	1	2	2	3	4	4	5	5
.42	2630	2636	2642	2649	2655	2661	2667	2673	2679	2685	1	1	2	2	3	4	4	5	6
.43	2692	2698	2704	2710	2716	2723	2729	2735	2742	2748	1	1	2	3	3	4	4	5	6
.44	2754	2761	2767	2773	2780	2786	2793	2799	2805	2812	1	1	2	3	3	4	4	5	6
.45	2818	2825	2831	2838	2844	2851	2858	2864	2871	2877	1	1	2	3	3	4	5	5	6
.46	2884	2891	2897	2904	2911	2917	2924	2931	2938	2944	1	1	2	3	3	4	5	5	6
.47	2951	2958	2965	2972	2979	2985	2992	2999	3006	3013	1	1	2	3	3	4	5	5	6
.48	3020	3027	3034	3041	3048	3055	3062	3069	3076	3083	1	1	2	3	4	4	5	6	6
.49	3090	3097	3105	3112	3119	3126	3133	3141	3148	3155	1	1	2	3	4	4	5	6	6

Antilogarithms

	0	1	2	3	4	5	6	7	8	9	1	2	3	4	5	6	7	8	9
.50	3162	3170	3177	3184	3192	3199	3206	3214	3221	3228	1	1	2	3	4	4	5	6	7
.51	3236	3243	3251	3258	3266	3273	3281	3289	3296	3304	1	2	2	3	4	5	5	6	7
.52	3311	3319	3327	3334	3342	3350	3357	3365	3373	3381	1	2	2	3	4	5	5	6	7
.53	3388	3396	3404	3412	3420	3428	3436	3443	3451	3459	1	2	2	3	4	5	6	6	7
.54	3467	3475	3483	3491	3499	3508	3516	3524	3532	3540	1	2	2	3	4	5	6	6	7
.55	3548	3556	3565	3573	3581	3589	3597	3606	3614	3622	1	2	2	3	4	5	6	7	7
.56	3631	3639	3648	3656	3664	3673	3681	3690	3698	3707	1	2	3	3	4	5	6	7	8
.57	3715	3724	3733	3741	3750	3758	3767	3776	3784	3793	1	2	3	3	4	5	6	7	8
.58	3802	3811	3819	3828	3837	3846	3855	3864	3873	3882	1	2	3	4	4	5	6	7	8
.59	3890	3899	3908	3917	3926	3936	3945	3954	3963	3972	1	2	3	4	5	5	6	7	8
.60	3981	3990	3999	4009	4018	4027	4036	4046	4055	4064	1	2	3	4	5	6	6	7	9
.61	4074	4083	4093	4102	4111	4121	4130	4140	4150	4159	1	2	3	4	5	6	7	8	9
.62	4169	4178	4188	4198	4207	4217	4227	4236	4246	4256	1	2	3	4	5	6	7	8	9
.63	4266	4276	4285	4295	4305	4315	4325	4335	4345	4355	1	2	3	4	5	6	7	8	9
.64	4365	4375	4385	4395	4406	4416	4426	4436	4446	4457	1	2	3	4	5	6	7	8	9
.65	4467	4477	4487	4498	4508	4519	4529	4539	4550	4560	1	2	3	4	5	6	7	8	9
.66	4571	4581	4592	4603	4613	4624	4634	4645	4656	4667	1	2	3	4	5	6	7	9	10
.67	4677	4688	4699	4710	4721	4732	4742	4753	4764	4775	1	2	3	4	5	7	8	9	10
.68	4786	4797	4808	4819	4831	4842	4853	4864	4875	4887	1	2	3	4	6	7	8	9	10
.69	4898	4909	4920	4932	4943	4955	4966	4977	4989	5000	1	2	3	5	6	7	8	9	10
.70	5012	5023	5035	5047	5058	5070	5082	5093	5105	5117	1	2	3	5	6	7	8	9	11
.71	5129	5140	5152	5164	5176	5188	5200	5212	5224	5236	1	2	4	5	6	7	8	10	11
.72	5248	5260	5272	5284	5297	5309	5321	5333	5346	5358	1	2	4	5	6	7	9	10	11
.73	5370	5383	5395	5408	5420	5433	5445	5458	5470	5483	1	3	4	5	6	8	9	10	11
.74	5495	5508	5521	5534	5546	5559	5572	5585	5598	5610	1	3	4	5	6	8	9	10	12
.75	5623	5636	5649	5662	5675	5689	5702	5715	5728	5741	1	3	4	5	7	8	9	10	12
.76	5754	5768	5781	5794	5808	5821	5834	5848	5861	5875	1	3	4	5	7	8	9	11	12
.77	5888	5902	5916	5929	5943	5957	5970	5984	5998	6012	1	3	4	5	7	8	10	11	12
.78	6026	6039	6053	6067	6081	6095	6109	6124	6138	6152	1	3	4	6	7	8	10	11	13
.79	6166	6180	6194	6209	6223	6237	6252	6266	6281	6295	1	3	4	6	7	9	10	11	13
.80	6310	6324	6339	6353	6368	6383	6397	6412	6427	6442	1	3	4	6	7	9	10	12	13
.81	6457	6471	6486	6501	6516	6531	6546	6561	6577	6592	2	3	5	6	8	9	11	12	14
.82	6607	6622	6637	6653	6668	6683	6699	6714	6730	6745	2	3	5	6	8	9	11	12	14
.83	6761	6776	6792	6808	6823	6839	6855	6871	6887	6902	2	3	5	6	8	9	11	13	14
.84	6918	6934	6950	6966	6982	6998	7015	7031	7047	7063	2	3	5	6	8	10	11	13	15
.85	7079	7096	7112	7129	7145	7161	7178	7194	7211	7228	2	3	5	7	8	10	12	13	15
.86	7244	7261	7278	7295	7311	7328	7345	7362	7379	7396	2	3	5	7	8	10	12	13	15
.87	7413	7430	7447	7464	7482	7499	7516	7534	7551	7568	2	3	5	7	9	10	12	14	16
.88	7586	7603	7621	7638	7656	7674	7691	7709	7727	7745	2	4	5	7	9	11	12	14	16
.89	7762	7780	7798	7816	7834	7852	7870	7889	7907	7925	2	4	5	7	9	11	13	14	16
.90	7943	7962	7980	7998	8017	8035	8054	8072	8091	8110	2	4	6	7	9	11	13	15	17
.91	8128	8147	8166	8185	8204	8222	8241	8260	8279	8299	2	4	6	8	9	11	13	15	17
.92	8318	8337	8356	8375	8395	8414	8433	8453	8472	8492	2	4	6	8	10	12	14	15	17
.93	8511	8531	8551	8570	8590	8610	8630	8650	8670	8690	2	4	6	8	10	12	14	16	18
.94	8710	8730	8750	8770	8790	8810	8831	8851	8872	8892	2	4	6	8	10	12	14	16	18
.95	8913	8933	8954	8974	8995	9016	9036	9057	9078	9099	2	4	6	8	10	12	15	17	19
.96	9120	9141	9162	9183	9204	9226	9247	9268	9290	9311	2	4	6	8	11	13	15	17	19
.97	9333	9354	9376	9397	9419	9441	9462	9484	9506	9528	2	4	7	9	11	13	15	17	20
.98	9550	9572	9594	9616	9638	9661	9683	9705	9727	9750	2	4	7	9	11	13	16	18	20
.99	9772	9795	9817	9840	9862	9886	9908	9931	9954	9977	2	5	7	9	11	14	16	18	20

Index

abscissa, 2
absolute
  activity, 308, 309
  temperature (*see also* thermodynamic temperature), 21, 40, 287, 291
  zero of temperature, 21, 291
absorbance, 115
absorption
  coefficient, 210
  spectra of atoms, 81, 82
  of molecules, 110–21
absorptivity, 116
abundance, isotropic, 58, 574
acetylene (ethyne), bonding in, 98, 99
acid
  definition of, 356–8, 389
  dissociation constant, 359–62, 377–82
  solution, pH of, 362–4, 366–9
  strength of (*see also* dissociation constant), 358–60
acids and bases
  Brönsted-Lowry theory, 182, 357–61, 387, 388
  dissociation theory, 356, 357
  Lewis theory, 389
acid-base
  catalysis, 182, 387–9
  titration, 352–4, 373–5, 413, 414
actinides, 87, 571
actinometer, 173, 174
activated
  adsorption, 444, 445
  complex, 166, 169–71, 179–81
activation energy, 17, 18, 164–71, 181, 190, 444, 445, 516, 522, 540

active transport, 428
activity
  absolute, 308, 309
  biological, 224, 225
  coefficient, 311, 312, 338–40, 407, 408
  mean ionic, 337–40
  radioactive, 73–5
  thermodynamic, 213–15, 222–5, 243, 247, 307–14, 403–8
adiabatic expansion, 29
adsorbate, 442
adsorbent, 442
adsorption, 442–73, 505–9, 516, 517, 522, 524–7
  chromatography, 455, 456
  isobar, 444
  isotherm, 184–6, 446–54, 464–6
affinity, 313
aggregation
  number, 482–5
  of particles, 475, 481–5, 513–23
air–liquid interface, 457–73
alkali metals, 84–9, 469
allotropy, 41–3
alpha (α)
  helix, 109, 139–42
  particles, 61–3
    emission of, 62, 63, 68
  rays, 61
amino acids, 377, 378
amorphous
  polymers, 547, 548, 551–3
  solids, 54
ampere, 331, 567
amperometric titration, 436, 437
amphiphile, 462, 463, 480–5
ampholyte, 377, 378, 512, 513

amphoteric electrolyte, 377, 378, 512, 513
anaesthesia, Pauling theory of, 450, 451
anaesthetic, 29, 224–6, 450, 451
Andrews' isothermals for carbon dioxide, 27, 28
angular
  momentum of nucleus, 126–9
  of electron, 132, 133
  velocity, 497–501
ångström unit, 569
anion exchangers, 456, 457
anisotropic medium, 34
antilogarithms, 4, 6, 578, 579
anti-parallel β-pleated sheet, 109, 139, 140, 142
anti-thixotropy, 547
apparent
  adsorption isotherm, 451, 452
  transport number, 332–7
  viscosity, 542, 543
appearance potential, 138
area per molecule (on a surface), 467–73
association
  and the distribution law, 226, 227
  colloids, 480–5
  of molecules, 31, 321
  reactions, 321
Aston, mass spectrograph, 65, 66
atmosphere, 569
atom
  definition of, 57
  structure of, 57–94
atomic
  absorption spectroscopy, 82
  emission spectra, 81, 82
  energy, 68
  mass constant (unified), 58, 66, 569, 570
  nucleus, 65–70, 126–32
  number, 59, 61–5, 571–74
  orbital, 85–94, 97–9
    hybridised, 93, 94, 97–9
    shapes of, 92, 93, 98, 99
  positions, 37, 38
  spectra, 79–82
  structure, 82–7

atomic
  structure Bohr theory of, 82, 83
  and periodic table, 85–7
  theory, 57–62, 82
    Dalton's, 57, 60
    Thomson's, 60
    Rutherford's, 61, 62, 82
  weight, 57–9, 61, 62, 64–6, 138, 572–4
  definition of, 57–9
atto, 570
attractive forces, 30–6
autoxidation, 177
auxochromes, 117, 118
available energy, 298–300, 393
Avogadro
  constant, 568, 570
  hypothesis, 22
azeotropic mixture, 215, 218, 219

bacteria
  oxidation-reduction potentials of, 421
  water activity, 223, 224
bacteriostatic activity of drugs, 224, 225, 383–7
bar, 569
bar notation, 6
base
  Brönsted-Lowry theory, 182, 357–61, 387, 388
  dissociation theory, 356, 357
  dissociation constant of, 360–2, 364–7, 370–3, 377–8
  Lewis concept, 178, 179, 389
  pH of solutions of, 364–6
  strength of, 360–2
base, of exponentials and logarithms, 4, 5
bathochromic effect, 117, 118
Beckmann
  freezing point method, 241, 242
  thermometer, 241, 242
bending vibrations, 113–15
Beer–Lambert law, 115–17, 124
benzene
  Kekule formulae of, 96, 97
  molecular orbitals of, 99
  resonance energy of, 282

benzene
  ultraviolet spectrum of, 117
Berthelot, reversible reactions, 316
beta (β)
  -particles, 61–3
    emission of, 62, 63, 67–9
  -pleated sheet, 139, 140, 142
  -rays, 61
BET equation, 447, 448
bicarbonate buffer in biological systems, 369
bimolecular reaction, 146, 147, 167–71
binary liquid mixtures, 204–6, 212–22
  boiling point of, 213–22
  vapour pressure of, 212–22
binding energy, nuclear, 66, 67
Bingham body, 543–5
biological
  activity of drugs, 224, 225, 383–7
  effect of drugs, 224, 225, 383–7
  effect of radiation, 76–9
  half life of radioactive substances, 78
  importance of colligative properties, 248–51
birefringence, circular, 108
blending agent, 206–10
Bohr
  energy states of atoms, 80–3
  magneton, 132, 570
  theory of atomic structure, 82, 83
boiling point
  composition diagram, 213–22
  elevation of, 237–9, 245–7
  of liquid mixtures, 213–22
Boltzmann
  constant, 290, 291, 570
  distribution, 24, 289, 290
  equation, 291
  factor, 24, 166
bond
  co-ordinate, 88, 91, 203
  covalent, 87–94, 96–100
  directional, 91–4, 97–9
  dissociation energy, 138, 280, 281
  electrovalent, 53, 54, 88
  energy, 279–81

bond
  energy table, 280
  hydrogen, 31, 32, 54, 93, 139–42
  ionic, 53, 54, 88
  molar refraction of, 104
  moments, 101–3
  refractions, 104
bonding, non-localised, 99, 100
Boyle's law, 21
  deviations from, 26
branching chain reactions, 177
bright field illumination, 475
Brönsted-Lowry theory of acids and bases, 182, 357–61, 387, 388
Brownian
  displacement, 476, 478
  motion, 475–8, 495, 514
Brunauer, Emmett and Teller adsorption isotherm, 447, 448
buffer
  capacity, 367–9, 374
  solution, 367–9, 374, 379–82
    biological, 191, 369
bulk modulus, 532, 533
Bunsen absorption coefficient, 210
Burgers model, 549–51
butadiene, delocalisation in, 99, 100

calculus notation, 9, 10
calomel electrode, 409, 410
calorie, 275, 569
capacitance, electric, 568
capacity factor, 272, 273
capillary, 443
  adsorption, 443, 446, 447
  rise, 459, 460
capillator, 379, 380
carbon
  dioxide, Andrews' isothermals for, 27, 28
  radioactive, 69
carbonium ions, 135, 136, 179, 183
cascade process, 29, 30
CAT, 131
catalysis, 181–95
  by acids and bases, 182, 387–9
  enzyme, 186–95
  heterogeneous, 182–6

catalysis
  homogeneous, 181, 182, 387–9
catalytic coefficient, 388
catalyst, 181–95
  active centres in, 183, 191–5
cathode
  lamp, 82
  rays, 60, 64
cation exchanger, 456, 457
CD, 108–10
cell
  constant, 257, 258, 346
  Daniell, 393, 397, 400, 401, 411
  diagram, 399–402, 415, 421–4, 429
  electrochemical, 392–439
  membrane, 225, 226, 248–51, 384–7
  standard, 399
  wall, in osmosis, 248
  Weston, 399
Celsius, 21
centrigrade, 21
centipoise, 537
centistokes, 538
centrifugal
  acceleration, 497–500
  force, 497, 498
Chadwick, α-particle scattering, 65
c.g.s. units, 569
chain
  carrier, 176, 177
  reaction, 175–7
  rule, 12, 13
characteristic
  frequencies, 113–15
  of logarithms, 5, 6
charge
  electric, 134–7, 330–2, 505–9, 568
  of colloidal particle, 505–9, 512, 513, 515–20
  of electron, 60, 61
  of gaseous ions, 134–7
  of nucleus, 64, 65
  of proton, 64, 570
  transfer interaction, 54
Charles' law, 21, 22
chemical
  adsorption, 443–6

chemical
  bond, 87–94, 96–100
  combination, laws of, 57
  potential, 306–10, 313
  shift, 128–31
chemisorption, 443–6
chlorophyll, 174
chromophore, 108–10, 117–19
chromatography
  adsorption, 455, 456
  gas-liquid, 235, 236
  ion exchange, 457
  partition, 233–6
  thin layer, 233–5
  vapour phase, 235, 236
circular
  birefringence, 108
  dichroism, 108–10
  polarisation, 107, 108
cis-trans isomerisation, 174, 175
cis-trans isomerism, 103
Clapeyron–Clausius equation, 40, 303–5
Clark oxygen cell, 434, 435
clathrate, 204, 450, 451
Claude's liquefaction method, 29
Clausius–Clapeyron equation, 40, 303–5
close packing of atoms and ions, 53, 54
c.m.c., 482–4
^{13}C n.m.r., 131, 132
coagulation of colloids, 513–23
coalescence in emulstions, 526–8
coefficient
  of diffusion, 250–9, 476–8
  of viscosity, 537–40
co-enzymes, 419, 420
coherent system of units, 568
co-ions, 505–8
colligative properties, 236–51, 489–92
collision theory of reaction rates, 165–9
colloid, 475–528, 542–7, 548
  adsorption of ions, 505–8, 513, 516–18, 520, 524, 527
  aggregation, 475, 481–5, 513–23
  charge on particles, 505–9, 512, 513, 515–20

Index 585

colloid
  classification, 478–85
  dehydration, 518, 519
  diffusion, 476–8, 499, 500
  electric mobility, 509–13
  electrical properties, 505–13, 515–18
  electrophoretic mobility, 509–13
  light scattering, 492–6
  molar mass, 487–505
  molecular weight, 469, 470, 487–505
  mutual precipitation, 519, 520
  osmotic pressure, 489–92
  particle dimensions, 475, 481, 487–9, 503, 504
  precipitation, 514–23, 526–8
  preparation, 485–7
  protection, 519, 520
  'salting out' process, 518, 519
  sedimentation, 497–502
  separation by electrophoresis, 511, 512
    ultracentrifugation, 497–502
  size and molecular weight, 475, 481, 487–505
  stability, 514–18, 520–8
  ultrafiltration, 478, 488
  ultramicroscopic examination, 475, 489
  viscosity, 502–5, 537–45
  zeta (ζ) potential, 507–9, 512, 515, 516, 520, 522, 524, 527
colloidal
  dispersion, 479–81, 485–9, 492–5, 505–11, 514–28
  electrolyte, 481–4
common ion effect, 328, 329, 364
competitive inhibition, 191–3
complex formation in phase equilibria, 46–8, 205, 215, 218, 219
complete differential, 15
compliance
  creep, 551, 552
  shear, 551, 552
composite isotherm, 451–4
compound
  definition of, 57
compound
  formations in phase equilibria, 44–8, 205, 215, 218, 219
compression modulus, 532, 533
computer averaging of transients, 131
concentration
  cell, 421–5
    with transport, 422–4
    without transport, 424, 425
  definition of, 200
  gradient, 252–9, 500
concentric cylinder viscometer, 541, 542, 555
condensed surface film, 470, 471, 525
conductance, 342–6, 568
  definition, 342
  measurement, 344–6
conductimetric titrations, 352–4
conduction band, 343, 344
conductivity
  definition, 343
  equivalent, 349
  measurement, 344–6
  molar, 346–52
  of association colloids, 482
  of colloidal electrolytes, 482
  of electrolytes, 344–54
  of micelles, 482
  of solids and gases, 343, 344
conductors, 329, 343
cone and plate viscometer, 541, 542, 555
configurational entropy, 288, 292–4
congruent melting point, 46, 47
conjugate
  acid-base pair, 359–62, 370, 371
  solutions, binary, 205, 208
  solutions, ternary, 208, 209
conjugation
  band, 118, 119
  of single and multiple bonds, 99, 100
constant boiling mixture, 218, 219
contact
  angle, 459, 460
  catalysis, 182–6
continuity of states, 28, 32, 33

586 Index

continuous phase, 479, 520
cooling curve, 51–3
co-ordinate bond, 88, 91
co-ordination
   compound, 91
   number, 522, 523
Cotton effect, 107–10
   curve, 107, 109
coulomb, definition of, 331, 568
coulometer, 332, 333, 335
counter ion, 505–8
covalent bond, 87–94, 96–100
cracking
   of emulsions, 526–8
   pattern, in mass spectra, 136, 137
creaming of emulsions, 526–8
creep, 550–2
   compliance, 551, 552
criteria for equilibrium and spontaneity, 296–301
critical
   electrolyte concentration, 516, 517
   micelle concentration, 482–4
   point, 28, 40
   solution temperature, 205, 206
   temperature, 28
cross-linking of polymers, 547, 548, 551, 552
crystal
   creep compliance of, 551, 552
   lattice, 35, 36, 44
   liquid, 34, 35
   shear compliance of, 551, 552
   structure, 35–9, 41–4, 53, 54
   determination, 36–9
crystalline solid, 35–9, 41–4, 53–5, 551, 552
crystallisation, 54, 55, 202
curie, 76, 79, 569
current, electric, 331, 332, 567
cytochromes, 420

Dalton
   atomic theory, 57, 60
   law of partial pressures, 24–6
Danckwerts' theory of dissolution rates, 261, 262
Daniell cell, 393, 397, 400, 401, 411

dark field illumination, 475
dative bond, 81, 91, 203
dead-stop end-point titration, 437–9
dead time of counter, 72
Debye
   and Hückel theory of strong electrolytes, 340, 347, 348
   units, 100, 101
deca, 570
decay law, 8, 9
   radioactive, 73–6
defects in crystals, 35, 36
definite integral, 17, 18
deformation, 531–6, 548–54
degree of
   association of colloids, 482, 483
   dissociation, 322, 323, 341, 342, 348, 362–4
   flocculation, 522
   freedom, 274
   hydrolysis, 372, 373
dehydration of colloidal particles, 518, 519
delocalisation of molecular orbitals, 99, 100
demulsification, 526–8
denaturation (denaturing), 171, 172, 191, 470, 481
density
   gradient, 502
   of gas, 23, 28
   of liquid, 28, 504
deoxyribonucleic acid (DNA), 140, 141
depolariser, 432–9
depressant drug, 224–6
depression of freezing point, 238–43, 245–50
derived SI units, 568
Deryaguin, Landau, Verwey and Overbeek (DLVO) theory, 515–18
desalting of macromolecular solutions, 457, 511, 512
desolvation of colloids, 518, 519
detachment method, 460, 461
detergent action, 482, 484
deuterons, 69
deviations from
   gas law, 26–8

deviations from
  Raoult's law, 214–16, 218, 219, 301, 311, 312
dialysis, 478
dichroism, circular, 108–10
dielectric, 344
  breakdown, 344
  constant, 178, 203, 509
differential, 10–15
  coefficient, 10
  complete, 15
  enthalpy of solution, 202, 203, 301
  partial, 14, 15
  standard, 11–13
  thermal analysis, 51
differentiation, 10–15
diffraction
  gratings, 80, 121
  of light, 80
  of X-rays, 36–9
diffuse part of double layer, 505–8, 516
diffusion, 251–9, 260–3, 285, 310, 432–5
  coefficient, 252–9, 260–3, 476–8, 499–501, 514
    determination, 254–9
  current, 432–5
  definition, 251, 252
  Graham's law, 23
  in solids, 36
  layer model, 260–3
  of gases, 23
  through cell membranes, 248–51
dilatancy, 543, 547
dilalant flow, 543
dilution
  definition, 347, 348
  law (Ostwald's), 341, 342
dipole, 30, 31, 100–3, 203
  interactions, 30, 31, 203
  moment, 100–3
    induced, 30, 31, 103
directional bonds, 91–4, 97–9
discontinuous phase, 520
disintegration constant, radioactive, 9, 73–6
dislocations in solids, 35, 36, 534

disperse phase, 479, 520
dispersion, 479–81, 485–9, 492–5, 505–11, 514–28
  medium, 479
  method for colloid preparation, 485, 486
  stabilizer, 519–22
dissociation
  constant, 342, 388
    acid, 359–62, 377–82
    and catalytic coefficient, 388
    base, 360–2, 364–7, 370–3, 377, 378
    measurement, 342, 362
  degree of, 322, 323, 341, 342, 348, 362–4
  equilibrium, 321–3, 341, 342
  ionic, 245–7, 326–9, 341, 342
  of electrolytes, 245–7, 326–9, 341, 342
  of molecules, 227, 321–3
  theory of acids and bases, 356, 357
dissolution of solids in liquids, 200–4, 259–64
  rate, 259–64
dissymmetry ratio, 495
distribution
  between immiscible solvents, 225–33
  coefficient, 225–8
  law, 225–7
  of molecular energies, 24, 289, 290
  of molecular velocities, 24
DLVO theory, 515–18
DNA, 140, 141
Donnan membrane equilibrium, 426–8, 489–92
d-orbital, 85–7, 91, 92, 203
double helix (in DNA), 140, 141
  layer of colloidal particle, 505–8, 516, 517
  resonance technique, 131, 132
drop
  volume method, 461, 462
  weight method, 461, 462
drug
  action, Ferguson principle, 224, 225

drug
  action, Overton-Meyer rule, 226
  action, and pH, 383–7
  activity, 224–6
  antagonism, 317
  ionised and non-ionised forms, 383–7
  solubility, 224–6
drying agent, 202, 203
Du Noüy tensiometer, 461
dynamic
  equilibrium, 315–17, 153–5
  viscosity, 478, 502–4, 508, 509, 537–40, 552
dyne, 269, 569
dynode, 73

e, 4, 5
edge dislocation, 36
effective half life, 78
Einstein
  equivalence of mass and energy, 66
  law of diffusion, 477
  law of photochemical equivalence, 172–4
  viscosity equation, 502
einstein, 173
elastic
  collision, 23, 166
  limit, 532–4, 545
  polymers, 547, 548
elasticity, 531–3, 548–55
elastomers, 547, 548
electrical
  double layer, 505–8, 515–18
  energy, 268, 392–8
electric
  capacitance, 568
  charge, 134–7, 330–2, 505–9, 568
  conductance, 342, 344–6, 568
  current, 331, 332, 567
  field strength, 508, 509
  mobility, 350–2, 509–11, 513
  potential (or tension), 350, 506, 508, 568
  resistance, 342–6, 568
electricity, quantity of, 330–2, 568

electrochemical
  cells, 392–439
  series, 405, 406, 410–12
electrode
  calomel, 409, 410
  glass, 414–17
  hydrogen, 396
  irreversible, 429–31
  metal, 394, 395, 399–402, 405–7
  oxidation-reduction, 395, 402–8
  polarisation of, 345, 430–9
  potential, 395, 396, 403–12
    equation, 403–7
    formal, 407, 408
    measurement, 408–10
    sign convention, 401, 402
    significance, 410–12
    standard, 403–7
  processes, 329, 330, 394–6, 429–31
  redox, 395, 402–8
  reference, 396, 407, 409, 410, 415
  reversible, 393–6
  silver, silver halide, 407, 415
electrokinetic potential, 507–9, 512, 515, 516, 520, 522, 524, 527
electrolysis, 329–32
electrolyte
  amphoteric, 377, 378, 512, 513
  colloidal, 481–4
  conductance of, 344–54
  dissociation of, 245–7, 326–9, 341, 342
  strong, 246, 326, 337–40, 347, 348
  weak, 246, 341, 342, 347, 348
electromagnetic radiation, 61, 64, 65, 79–82, 103–33, 172–5, 271, 492–96
electromotive
  force, 395–402, 568
  series, 405, 406, 410–12
electron, 60–4, 71–3, 79–94, 329–31
  affinity, 90, 91
  and electrolysis, 60, 329–31
  density, see also probability distribution, 31, 36–9, 88
  distribution, see probability distribution
  energy levels, 82–7

electron
  orbital, 84–94, 96–9
  probability distribution, 30, 83, 91, 92
  shell, 83–7
  spin resonance, 132, 133
  transport chain, 419, 420
  wave nature, 83, 84
electronegativity, 31, 101
electronic
  energy levels, 79–94, 96–100, 115–19, 121–3, 172
  theory of valency, 87–94, 96–100
electronvolt, 66, 344, 569
electro-osmosis, 511
electrophilic reagent, 178
electrophoresis, 508–12
electrophoretic mobility, 509–11, 513
electrostatic attraction, 30–2, 88, 203
electrovalent bond, 53, 54, 88
elements, list and atomic properties, 571–74
elevation of boiling point, 237–9, 245–7
elution volume, 455, 456
e.m.f., 395–402, 568
emulsifying agent, 524–8
emission
  alpha (α) particles, 61–3, 68
  beta (β) particles, 61–3, 67–70
  negatron, 61–63, 67–70
  positron, 67, 68
  spectra, see also fluorescence, 78–82
emulsions, 479, 524–8
enantiotropic system, 42, 43
end absorption, 118, 119
endothermic
  compounds, 279
  dissolution, 202
  reaction, 275, 279, 320
energy, 268–71, 466
  activation, 17, 18, 164–6, 169–72, 181, 183, 190, 444, 445, 516, 540
  bond, 138, 279–81
  conservation of, 270, 271
  electrical, 268, 392–8

energy
  electronic, 79–94, 96–100, 115, 118, 119, 121–3, 172
  free, 297–303, 313–15, 393, 396, 397
  Gibbs, 297–303, 313–15, 393, 396, 397
  internal, 270–4, 290
  kinetic, 23, 24, 166, 270, 274, 290
  levels, see also –, electronic, 110, 111
  molecular, 110, 111, 115, 118, 119, 121–3, 289, 290
  potential, 166, 270, 274, 290, 444, 445, 515–17
  thermodynamic, 270–4, 290
enthalpy, 271–82, 296–305
  change, 40, 272, 276–82
  and equilibrium constant, 320
  measurement, 271, 272, 320, 375–7, 398
  lattice, 202
  of activation, 169–71
  of adsorption, 442–5
  of atomisation, 279–82
  of combustion, 278
  of condensation, 443
  of desorption, 443
  of evaporation, 31, 33, 303–5
  of formation, 278, 279
  of fusion, 36, 51, 303, 304
  of hydrogenation, 282
  of mixing, 204, 213–15, 301
  of neutralisation, 375–7
  of phase changes, 295, 296, 302–5
  of reaction, 272, 275–7, 320
  of solution, 202–4
  of solvation, 202
  of sublimation, 40, 279, 295, 296, 303–5
  of vaporisation, 31, 33, 295, 296, 303–5
entropy, 286–303, 547
  change, in life processes, 288, 289
  change, measurement, 294–6, 397, 398
  of activation, 169–71
  of adsorption, 442
  of mixing, 204, 213, 214, 292–4, 301

entropy
  of phase changes, 295, 296, 303
  of solution, 204
enzyme reactions, 186–95
enzyme-inhibitor, 191–5
enzyme–substrate complex, 188–90, 192, 193
equation
  of state, 22, 27, 270
  of straight line, 2, 3
equilibrium, 282–5, 315–17
  constant, 181, 314–23
  diagram, 39–53, 213–22
equivalent, 330, 331, 349, 375–7
  conductivity, 349
  weight, 330, 331
erg, 269, 569
e.s.r., 132, 133
ethylene (ethene), bonding in, 96–8
ethyne, bonding in, 98, 99
eutectic, 45–8, 50–3
evaporation, enthalpy (latent heat) of, 31, 33, 303–5
excitation spectrum, 125
exclusion principle, 85
exothermic reaction, 275, 320
expanded surface films, 471, 472
expansion of gases, 28–30, 269, 270
exponential, 6, 7
  factor, 164
  series, 11
extensive quantity, 272, 273
external phase, 520–8
extinction, 115
  coefficient, 115, 116
extraction by solvents, 228–33

falling body viscometer 539, 540
farad, 568
faraday, 330
Faraday constant, 331, 397, 570
  laws of electrolysis, 330, 331, 397
femto, 570
Ferguson's principle, 224, 225
fibrous proteins, structure, 138–40
Fick's first law of diffusion, 252, 253
  second law of diffusion, 255
fingerprint region in spectra, 113

first law of thermodynamics, 270, 271
first-order rate constant, 150–6
first-order reaction, 150–6
  time of half-change, 155, 156
fission, nuclear, 67, 68
fixed part of double layer, 505–8, 516, 517
flame photometry, 81, 82
flocculation of colloids, 516–18, 520–2
flow, 531, 533, 534
  curves, 542–7
  Newtonian, 537–40
    deviations from, 542–5
  viscoelastic, 548–55
fluctuation theory, 495, 496
fluidity, 537
fluorescence, 121–6
  efficiency, 123, 124
  intensity, 124–6
  spectrum, 125, 126
fluorimetry, 125, 126
fluorophor, 126
flux, 252–5
force, 269, 568
formation, enthalpy of, 278, 279
Fourier
  synthesis, 36, 37, 142
  transform n.m.r., 131
fractional
  crystallisation, 55
  distillation, 216–19
fractionating column, 216–18
fragmentation in mass spectra, 135, 136
free
  atoms, 175, 177, 279–81
  energy (see also, Gibbs energy), 297–303
  radicals, 133, 135, 136, 175–8, 280, 281
freeze drying, 41
freezing, 298
  point, 43, 45–53
    depression, 238–43, 245–7, 250
    of blood serum, 250
frequency
  factor, 164, 165, 168–70

frequency
  of radiation, 11, 112, 568
Freundlich adsorption isotherm, 184, 453, 517, 518
frictional
  coefficient, 477, 478, 499, 508, 523
  resistance, 477, 498, 499, 523
function, 2
  of a function, 12, 13
fundamental
  constants, 468
  vibration frequencies, 111, 113–15
fusion
  enthalpy (latent heat) of, 36, 51, 303, 304
  nuclear, 67

g (splitting factor), 132, 133
  nuclear, 126–8
gamma (γ) rays, 61, 62, 71–3, 76–9
gas
  constant, 22, 570
  diffusion of, 23, 251, 252
  equation, 21, 22, 27
  kinetic theory, 23, 24
  law, ideal, 22
    deviations from, 26–31
  liquefaction, 27–30, 40
  multiplication, 71, 72
  noble, 89, 91
  non-ideal behaviour, 26–31
  partial pressures, 24–6
  solubility, 210–12
  velocities, distribution, 24
gaseous state, 21–31, 39–40, 274, 275
  films, on surfaces, 469, 470, 472
gas–liquid
  chromatography (GLC), 137, 235, 236
  interface, 457–73
gas–solid interface, 182–6, 442–51
gauss, 569
Gay-Lussac's law, 21, 22
Geiger–Müller (G–M) counter, 72
gel, 455, 479, 486, 487, 548–50
  electrophoresis, 511, 512
  generation time, 8

geometrical isomerism, 103
Gibbs
  adsorption isotherm, 463–6
  energy (or Gibbs free energy), 204, 297–303
    in chemical reactions, 313–15
    in electrochemical cells, 393, 396, 397
    of activation, 169
    of mixing, 204, 301
Gibbs–Helmholtz equation, 302
giga, 570
glass, 54, 551, 552
  electrode, 414–17
  transition temperature, 552
globular proteins, structure, 141, 142
Golay cell, 121
Gouy–Chapman layer, 505–8
gradient, 3, 9, 10
Graham
  definition of colloids, 475
  law of diffusion, 23
graphs, 1, 2
gravimetric method for adsorption, 445, 446
growth
  law, 7, 8
  rate constant, 8
Guldberg and Waage, law of mass action, 145, 146

haemolysis, 248
half-life
  of lyophobic colloid, 514, 515
  of reaction, 155, 156, 161–3
  of radioactive decay, 8, 75
half-wave potential, 433
heat, 268, 270–2, 568
  capacity, 273–7, 294, 295
  change at constant pressure, 272
    volume, 271, 272
  latent, see enthalpy
  of, see enthalpy of
heavy atom method, 38
hecto, 570
Heisenberg's uncertainty principle 83

592 Index

Henderson equation, 364–8, 379‘ 382–6
Henry's law, 211, 212
hertz, 11, 112, 568
Hess's law, 278
heterogeneous catalysis (or reaction), 182–6
heterolysis, 178
Hofmeister (or lyotropic) series, 518
homogeneous catalysis, 181, 182, 387–9
homolysis, 178
Hooke (or Hookean) body, 532–6, 547, 548, 553
Hooke's law, 532
Hund's rule, 85–7, 90, 122
hybridisation of orbitals, 93, 94, 97–9, 203
hydrated salt, 201, 202
hydration of ions, 30, 202, 203, 337
  enthalpy of, 202, 203
hydrogen
  bond, 31, 32, 54, 93, 139–41, 171, 203, 205, 450, 540
  electrode, 396
  ion, 88, 89, 356–62
    activity, *see* pH
    adsorption on proteins, 512, 513
    mobility, 351, 352
    molar conductivity, 351, 352
    overpotential or overvoltage, 430, 431
hydrogenation
  catalytic, 183, 282
  enthalpy of, 282
hydrolysis
  constant, 360, 371, 372, 381, 382
  degree of, 372, 373, 382
  of fats, 472, 473
  of salts, 370–74, 381, 382
hydrophilic, *see also* lyophilic
  colloid, 485, 518, 519, 549, 550
  group, 462
hydrophobic, *see also* lyophobic
  colloids, 485
  group, 462
  interaction, 203, 204, 484

hyperfine splitting
  in e.s.r., 133
  in n.m.r., 129–31
hypertonic solution, 248–50
hypotonic solution, 248–50
hypsochromic shift, 118
hysteresis loop
  in adsorption, 449, 450
  in flow curves, 546

ice
  point temperature, 570
  structure of, 32
ideal
  elastic body, 532, 533, 536
  gas, 22–6, 274, 275, 292–4
  mixture (or solution), 213, 214, 292–4, 301
  osmolality, 246, 247
  viscous body, 534
Ilkovic equation, 433
immiscible liquids
  distillation, 219–22
  distribution between, 225–33
incongruent melting point, 47, 48
index (or power), 3, 4
indicator
  acid-base, 379–82, 386
  oxidation-reduction, 417–20
  table of pk values, 380
individual (or true) adsorption isotherms, 451, 452
induced dipoles, 30
inductive effect, 103
infrared spectra, 111–15
ingestion of radioactive substances, 78, 79
inhibition
  of enzyme reactions, 191–5
  of heterogeneous reactions, 183
initial rate method for reaction order, 162, 163
injection, osmotic pressure, 248–51
inner transition series, 60, 87
insoluble surface films, 467–73
insulators, 344
integration, 15–18

intensity
  factors, 273
  of scattered light, 492, 493
  of spectra, 115–17, 124, 125
intensive quantities, 273
interaction of molecules, 30–2, 203, 204, 213–15
interatomic vectors in X-ray structure determination, 38
interfacial
  barrier model for dissolution rates, 261
  film in emulsions, 525
  free energy, 458–66, 482–4
  tension, 458–66, 482–4
interference of scattered light, 495
intermediate compound in catalysis, 181, 182, 188–90
intermolecular
  forces, 30–2, 203, 204, 213–15
  hydrogen bond, 31, 32, 54, 93, 140, 141, 203, 205, 450, 540
internal
  conversions in photochemistry, 122, 123
  energy (or thermodynamic energy), 270–4, 290
  phase in dispersions, 520–8
international system of units (SI), 567–70
interstices, 44, 525
interstitial
  sites, 36, 44
  solid solution, 44
intersystem crossing, 121–3
intramolecular hydrogen bond, 31, 32
intrinsic
  dissolution rate, 260–3
  viscosity, 504, 505
inversion temperature, 29
ion
  current, 134, 135
  exchange, 456, 457
ion-dipole interaction, 30, 203
ionic
  bond, 53, 54, 88
  compounds, 53, 54, 88
  product for water, 361, 362

ionic
  strength, 340
ionisation
  chamber, 72
  energy, 89
  of solutes and colligative properties, 245–7
  potential, see also — energy, 138, 344
ions
  activity, 337–9, 406, 407, 416
  adsorption on colloids, 505–8, 512, 513, 516–18
  discharge in electrolysis, 329–31
  hydration, 30, 202, 203, 337
  migration during electrolysis, 329–37, 348–52
  mobility, 350–2
irreversibility, 282–6, 315
irreversible,
  change, 284–9, 296–8, 300, 301
  electrode, 429–31
  inhibition, 195
  path, 284–6
  process, 284–9, 296, 297, 300, 301
  reaction, 314, 315
  thermdynamics, 285
isoelectric point, 377, 379, 491, 512, 513
isomerisation, cis-trans, 174, 175
isomerism
  geometrical (cis-trans), 103
  structural in disubstituted benzenes, 32, 96, 103
isomorphism, molecular, 44
iso-osmotic solutions, 223, 237, 244, 248–51
isopiestic
  method for molecular weight determination, 237
  solutions, 223, 237, 244, 248–51
isotherm
  adsorption, 184–6, 446–54, 517, 518
    apparent, 451–4
    BET, 447, 448
    composite, 451–4
    experimental, 446–54
    Freundlich, 184, 452, 453, 517, 518

isotherm
  adsorption individual, 451
  Langmuir, 184–7, 446–8, 452, 454
  true, 451
isothermal
  Andrews for carbon dioxide, 27, 28
  process, 283–7
isotonic solutions, 248–50
isotope, 58, 59, 63, 65, 66, 68–71 572–4
isotopic
  abundance, 58, 65, 68, 138, 574
  composition, 58, 574
  mass, 58, 65, 66
isotropic
  medium, definition of, 34
  solids, deformation of, 533

joule, 22, 269, 275, 397, 568
Joule–Thomson effect, 29
junction potential, 408, 409, 421–5

K-capture, 68
Kekulé, bond structures, 96, 97
kelvin, 21, 40, 567
Kelvin body or model, 549–51
kinematic viscosity, 538, 539
kinetic
  energy, 23, 24, 166, 270, 274, 290
  equation (or expression), 149, 151, 154, 158–60, 162, 185, 187, 189, 190, 192, 194
  theory of gases, 23, 24
Kirchhoff equation, 277
Kohlrausch's law, 348, 349
Krafft point, 484

labelling, radioactive, 69–71
Langmuir
  adsorption isotherm, 184–7, 446–8, 452, 454
  trough for surface studies, 468, 469
lanthanides, 60, 87, 571

latent heat, *see also* enthalpy of appropriate phase change, 40, 295, 296, 302–5
lattice
  enthalpy, 202, 203
  sites, defects in, 35, 36
law of
  conservation of energy, 270, 271
  mass action, 145, 146, 314–23
  osmosis, 244, 245
  photochemical equivalence or of quantum activation, 172–4
least-squares structure refinement, 37
Le Chatelier's principle, 39, 202, 203, 305, 316, 317, 320–3
Lewis
  acid, 178, 183, 389
  base, 179, 389
  theory of acids and bases, 389
ligand field, 133
litre, 569
light
  interference, 495
  scattering, 492–6
limiting
  current, in polarography, 432, 433, 435–7
  viscosity number, 504
linear strain, 531, 532
Lineweaver–Burk plot, 189, 190, 193, 195
lipophilic
  group, 462, 463, 484
  material, precipitation of, 519
  particle, 480
liquefaction of gases, 27–30
liquid
  creep compliance of, 551, 552
  crystal, 34, 35
  expanded film, 471, 472
  junction potential, 408, 409, 421–5
  Newtonian, 537–40
  state, 28, 32, 33, 39, 40
liquid–gas interface, 457–73
liquid–liquid interface, 457–66, 524–8
liquid–solid interface, 451–7, 520–3

liquidus curve or line, 45–53, 215–19
Littrow mirror, 120, 121
logarithms, 4–6, 576, 577
London dispersion forces, 30
lone pair
  donation, 203, 389
  repulsion, 93, 94
lowering of vapour pressure, 236–8, 245–7, 213–15
lubrication, 35
lyophilic colloids, 485, 487–92, 495–508, 511–13, 518–20
lyophobic colloids, 485–9, 492–5, 505–11, 514–20
lyotropic (or Hofmeister) series, 518
lysozyme, structure, 142

macromolecules, *see also* lyophilic colloids, 138–42, 481, 485
magnetic
  flux, 568
    density, 127–9, 132, 134, 135, 568
  moment of electrons, 132
    nuclei, 126, 127
  quantum number, 84, 85, 132, 133
mantissa of logarithms, 5
Mark and Howink viscosity equation, 504
mass, 465
  action law, 145, 146, 314–23
  average molar mass, 488, 496
  concentration, 200
  defect, 66
  fraction, 199
  isotopic, 58, 65, 66
  number, 66
  spectra, 65, 66, 70, 134–8
maximum
  boiling mixture, 218, 219
  radiation dose, 79
  work, 285, 286, 299
maxwell, 569
Maxwell–Boltzmann distribution, 24, 289, 290
Maxwell body or model, 549–51, 553
McBain–Bakr balance, 445, 446
McBain, micelle formation, 482

mean
  activity, 337–9
    coefficient, 338–40
  bond energy, 280–2
  life, radioactive, 75
  particle radius, 449
mechanical
  systems and thermodynamics, 297
  work, 268–70, 284–6, 299
mechanisms of organic reactions, 178–80
mega, 570
melting, 39–53, 295, 296, 302–4
membrane
  biological, 226, 248–51, 383–6, 426–8
  filters, 478, 488
  glass, 415–17
  potential, 415–17, 425–9
  semi-permeable, 243, 244, 282–6, 489–92
Mendeléeff, periodic table, 59
meritectic point and reaction, 47, 48
metal electrodes, 329, 330, 393–5, 399–402, 405, 407
metallic conductors, 329, 343
metastable species or form, 42, 43
micelle, 481–5
Michaelis
  constant, 188–95
  enzyme-substrate complex, 188
Michaelis–Menten treatment of enzyme reactions, 187–90
micro, 570
micron, 569
microscopic electrophoresis method, 510
milk
  protection of colloids, 520
minimum boiling mixture, 218, 219
miscible liquid, 43–53, 213–19
  boiling point of, 213–19
  distillation of, 214–19
  freezing of, 43–53
  vapour pressure of, 213–19
miscibility of liquids, 203, 204, 301
mixing
  enthalpy of, *see also* enthalpy of solution, 204, 213–15, 292, 301

596 Index

mixing
  entropy of, 204, 213, 214, 292–4, 301
  Gibbs free energy of, 204, 301
mixture
  definition, 199
  ideal, 213, 214, 292–4, 301
  non-ideal, 214–19, 301
  of liquids, 203–10, 212–22
  partial molar quantities in, 305–10
  thermodynamics of, 203, 204, 213–15, 292–4, 301, 305–12
mmHg, 569
mobile or diffuse part of double layer, 505–8, 516
mobility of ions, 348–52
modulus of elasticity, 532, 533
molal, 200, 569
  depression constant, 240
  elevation constant, 239
molality, 200, 239, 240
  mean of electrolytes, 339
molar, 200, 569
  absorption coefficient, 115
  conductivity, 346–52, 482
  electronvolt, 569
  entropy, see also entropy, 296, 303
  enthalpy, see also enthalpy, 273, 303, 304
  Gibbs free energy, see also Gibbs free energy, 303
  heat, see also enthalpy, 303, 304
  latent heat, see also enthalpy, 303, 304
  mass, see also molecular weight, 59
  optical rotatory power, 106
  quantity, 272, 273
  refraction, 104, 105, 109
  volume, 273
    of an ideal gas at N.T.P., 22, 570
mole, 59, 567, 568
  fraction, 25, 199, 213
molecular
  energies, distribution, 24, 290
  isomorphism, degree of, 44

molecular
  orbitals, 97–9
  rotation, 106
  sieve, 450, 455
  spectra, 111–26
  velocities, distribution, 24
  weight, 58, 59
    determination by steam distillation, 220, 221
    determination by colligative properties, 237, 239–47
    determination by mass spectrometry, 137, 138
    of colloids, 455, 456, 469, 470, 487–505
molecularity, 146, 147, 179, 180
molecules
  definition, 57
  structure, 96–142
monomolecular adsorbed layer, 443, 446–8, 462–73
monatomic solid, heat capacity of, 274
monochromator, 125
monodisperse colloid, 487, 488
monolayer, 443, 446–8, 462–73
monotropic system, 42, 43
Moore and Stein, amino acid separation, 457
Moseley, 64
moving-boundary method, 336, 410–12
multilayer, 443, 446–50
multiple fractional extraction, 230–2
multiplicity of energy state, 121, 122
mutarotation, 153–5, 182, 387

nano, 570
negative deviations from Raoult's law, 215, 218, 219, 301, 311, 312
negatron, 67–9
nematic phase, 34
nephelometer, 492, 493
Nernst
  equation, 404–8
  filament, 121
net work, 299

Index 597

neutralisation
  curve, 373–5, 412–14, 513
  enthalpy of, 375–7
neutron, 65, 68, 69, 76, 77
newton, 22, 269, 568
Newton, law of viscous flow, 537
Newtonian
  body, 534, 537
  liquid, 534, 537–40, 548, 554, 555
Nicol prism, 105, 493
noble gas, 89, 91
non-competitive inhibition, 193–5
non-equilibrium or irreversible thermodynamics, 285
non-ideal behaviour
  of gases, 26–31
  of liquid mixtures, 214–19, 301, 311, 312
non-Newtonian flow, 542–9, 554, 555
non-polar
  adsorbents, 454, 455
  molecules, 30, 179–80, 203, 204, 454, 455
  solvents, 30, 178–80, 203, 204
non-specific drugs, 224–6
normal stress, 531–3, 555
Noyes–Whitney equation, 259, 260
N.T.P. (S.T.P.), 77, 210, 212
nuclear
  binding energy, 66, 67
  fission, 67, 68
  fusion, 67
  g factor, 126
  magnetic resonance, 126–32
  magneton, 126, 570
  reaction, 69, 70
  stability, 66–8
nucleic acids, 140, 141, 481
nucleon, 67
nucleophilic
  reagent, 179, 180
  substitution, 179, 180
nucleus, 65–76
nuclide, 58, 59, 67, 68, 574
number average molar mass and molecular weight, 488

ohm, 342, 568
Ohm's law, 342
oil-in-water (O/W) emulsion, 524–7
oil/water partition coefficient 226
Onsager equation, 347, 348
open system, 309, 310
optical
  activity, 105
  density, 116
  microscopy, in colloid chemistry, 475, 489
  rotation, 105–9, 153–5
  rotatory dispersion, 106–10
optimum temperature of enzyme reactions, 191
orbit, of electron, 82, 83
orbital
  angular momentum, 132, 133
  atomic, 85–94, 97–9
  hydridised, 93, 94, 97–9
  molecular, 97–9
  momentum quantum number, 132
O.R.D., 106–10
order of reaction, 146, 147
  determination, 162–5
ordinate, 2
oscillating stress and strain, 553, 554
osmolality, 246, 247
osmole, 246, 247
osmometer
  freezing point, 247
  membrane, 244
  vapour pressure, 237, 247
osmosis, 243–5, 248–51, 282–6
osmotic
  coefficient, 247, 337
  pressure, 243–51, 282–6, 489–92, 496
  in cells, 248–51
  of blood, 248–51
  of colloids, 489–92, 496
  of injections, 248–51
Ostwald
  dilution law, 342
  phase-volume theory, 527
  viscometer (U-tube), 503, 504, 537–9
overlap of atomic orbitals, 93, 97–9
Overton and Meyer theory, 226

overvoltage, 429–31
O/W
  emulsion, 524–7
  partition coefficient, 226
oxidation-reduction, 395, 402, 417–21
  electrode, 395, 402, 407
  indicator, 417–19
  potential, 402–8, 417–21
    measurement, 408–10, 418, 419
    standard, 403–10, 418–21
    standard table, 406
  processes in biology, 419–21
  titration, 412–14, 419

paper chromatograophy, 233
paralysis time, 72
paratonic solution, 248
parent ion peak, 135–7
partial
  differential, 14, 15
  miscibility of liquids, 203–10, 215, 301
  miscibility of solids, 44, 50, 51
  molar Gibbs free energy, 306–10
    quantities, 305, 306
    volume, 305, 306
  pressure, 24–6, 318, 322
    Dalton's law of, 25, 26
  rotation, 108
  vapour pressure, 212–15, 222, 223, 225, 236–40
particle
  alpha (α), 61–3
  beta (β), 61–3
  size in colloids, 485–9
partition, 225–36
  chromatography, 233–6
  coefficient, 225–30, 233–5
    determination, 227, 228
  law, 225–7
pascal, 22, 270, 568
Patterson synthesis, 38
Pauli, exclusion principle, 85
Pauling
  hybridisation of orbitals, 93, 94
  theory of anaesthesia, 450, 451
peptisation of colloids, 486, 520

perfect gas, *see* ideal gas
perfectly elastic body, 532, 533, 536
periodic table, 59, 60, 63, 85–7, 571
  anomalies in, 59
  atomic structure and, 85–7
  electronegativity variation in, 101
permanent dipole, 30, 31, 100–3, 203
permittivity, 178, 203, 485, 509
  of vacuum, 509
  relative, 178, 203, 485, 509
Pfeffer, osmotic pressure relationship, 244, 245
pH, 361–75, 383–7, 413–17
  and drug action, 383–7
  determination, 379–81, 386, 413–17
  indicators, 379–81
  intracellular, 386, 387
  isoelectric, 377, 378, 512, 513
  of acid and base solutions, 362–7
  of ampholytes and proteins, 377, 378, 513
  of buffer solutions, 367–9
  of salt solutions, 370–3
  titration, 373–5, 413, 414, 513
phase
  boundary, 39, 40, 42, 43, 45–51, 213–22
  change, 28, 40–55, 216–22, 285, 295, 296, 302–5
    effect of external conditions on, 302–5
  diagram, 39–53, 204–22, 237–40
  differences in light scattering, 495
  inversion of emulsions, 527
  of X-ray diffracted beam, 37–9
  problem in X-ray crystallography, 37–9
  reversal of emulsions, 527
  volume theory, 527
phosphate bond energy, 281, 419, 420
phospholipid, 481
phosphor, 73
phosphorescence, 121–3
  spectrometry, 126
photocell or photoelectric cell, 81, 493

photochemical reaction, 172–5, 271
photomultiplier, 73, 82
photosensitiser, 174
photosynthesis, 174
physical
  adsorption, 443–50
  quantity, 567
  states, 277
physisorption, 443–50
pi (π) bond, 97, 98
pico, 570
pK, 361
  of acids and bases, 361, 362
  of indicators (and table), 380
plain curve, 107
plait point, 209
Planck
  constant, 80, 570
  equation, 80
  quantum theory, 80, 110, 111, 172, 173
plane polarised light, 34, 105
plasmolysis, 248
plastic
  behaviour, 544, 545, 458
  deformation, 36, 533, 534
  materials, 544, 545, 548
plasticity, 36, 533, 534, 544, 545, 548
plateau region in counters, 71
point defects, 35, 36
poise, 537, 569
Poiseuille's equation, 537
Poisson ratio, 533
polar
  adsorbents, 454, 455
  molecules, 30, 31, 100–3, 203, 204, 454
  solvents, 178, 180, 203
polarimeter, 105
polarisability, 103, 104, 178
polarisation
  of electrodes, 345, 430–9
  of light, 34, 105–8, 494, 495
    circular, 107, 108
    plane, 34, 105–7
  of molecules, 103–5, 178
polariser, 34, 105
polarity
  of adsorbents, 454, 455

polarity
  of electrochemical cell, 399–401
  of electrodes, 401, 402
  of molecules, 30, 31, 100–3, 203, 204, 454, 455
  of solvents, 178–80, 203, 204, 454, 455
polarography, 431–4
polarometric (or amperometric) titration, 436, 437
polybasic acids, dissociation, 360, 375
polydisperse colloids, 487, 488, 496, 501, 505
polyelectrolytes, 481
polymer, *see also* macromolecules, 542–55
polymeric flocculation, 521, 522
polymorphism, 41–3
polysaccharide, 481
p-orbital, 92, 97–9
positive
  deviations from Raoult's law, 214–19, 301, 311, 312
  rays, 63, 64
positron, 67, 68
potential
  difference, 342, 344, 395, 396, 508
  double layer, 505–8
  electric, 395, 568
  energy, 166, 270, 274, 290, 444, 445, 515–17
    barrier, 166, 169, 170, 444, 445, 516
    of attraction, 515–17
    of interaction, 515–17
    of repulsion, 515–17
  gradient, 252, 508, 509
potentiometer, 398, 399
potentiometric titration, 412–14
power, 568
  (in exponential), 3, 4
practical osmotic coefficient, 247
precipitation of colloids, 518–20
pre-exponential (or frequency) factor, 164, 165, 168–70
preparation of colloidal solutions, 485–7

pressure, definition and units, 269, 270, 568, 569
 partial, 24–6, 318, 322
 osmotic, 243–51, 282–6, 489–92, 496
 surface, 467–73
 vapour, 28, 31, 33, 41, 42, 212–22, 225, 236–40, 304, 305, 446–9
primary structure of proteins, 141
principal quantum number, 83, 84
probability factor, 168–71
progress of chemical reaction, 147–9
proportional
 counter, 72
 saturation, 224, 225
protective colloids, 519, 520
protein, 481, 512, 513
 denaturation, 171, 172, 191, 470, 481
 electrophoresis, 512, 513
 in surface films, 470
 isoelectric point, 512, 513
 light scattering, 495, 496
 osmotic pressure, 489–92
 structure, 109, 110, 138–42
 titration curve, 512, 513
 ultracentrifugation, 497–502
 viscosity, 502–5
proton, 88, 89
 in acid-base equilibria, 357, 358
 magnetic resonance, 126–31
protoplast, 248
pseudoplasticity, 543–7
pyrolysis, 176, 177

quantisation of energy levels, 83–5
quantity
 of electricity, 134–7, 330–2, 505–9, 568
 of heat, 268–72, 275–9, 568
quantum, 80, 173
 efficiency, 173–5
 number, 83, 84
 theory, 80
 yield, 173–5
quaternary structure of proteins, 141

quenching
 of counter, 72
 of fluorescence, 123, 125
 of orbital motion, 133

rad (radiation absorbed dose), 77, 78, 569
radian (rad), 497
radiation
 absorbed dose (rad), 77, 78, 569
 biological effect, 76–9
 electromagnetic, 61, 64, 65, 79–82, 103–33, 172–5, 271, 492–6
 emitted in atomic spectra, 79, 80
 envelope, in light scattering, 494, 495
 maximum dose, 79
radiationless transition, 120–3
radioactive
 counter, 71–3
 decay, 9, 73–6
  constant, 9, 73–5
 disintegration, 9, 73–6
  constant, 9, 73–5
 equilibrium, 76
 isotope, 69, 70
 series, 62, 63, 76
radioactivity, 61–3, 66–79
 measurement, 71–3
 units, 76–8, 569
radiocarbon dating, 69
Radiological Protection (International Commission on), 79
random coil, 109, 505
Raoult's law, 213, 214, 236
 deviations from, 214–19, 301, 311, 312
rare earths (lanthanides), 60, 87, 571
Rast's freezing point method, 242, 243
rate
 constant, 17, 18, 150–65, 315, 316
  determination, 152–5, 160, 161, 163, 164
  variation with temperature, 17, 18, 164–72
 equation, 150–63, 514

rate
  of aggregation of particles, 514, 515
  of change, 9, 10
  of coagulation of colloids, 514, 515
  of creaming of emulsions, 528
  of diffusion, 251–9
  of dissolution (or solution), 259–64
  of reaction, 145–95
    determination, 147–9
  of sedimentation, 523
  of shear, 536, 537, 541–7
ray
  alpha (α), 61
  beta (β), 61
  cathode, 60
  gamma (γ), 61, 62
  positive, 63, 64
Rayleigh
  ratio, 493, 494
  scattering, 475
RBE, 77
reaction
  chain, 175–7
  determination,
    of order, 162–4
    of rate, 147–9
    of rate constant, 152–5, 160, 161, 163, 164
  endothermic, 275, 279, 320
  enthalpy of, 275–82, 320
  enzyme, 186–92
  exothermic, 275, 320
  explosive, 177
  heat of, 272, 275–7, 320
  heterogeneous, 182–6
  in solution, 168
  in surface films, 472, 473
  irreversible, 315
  molecularity, 146, 147, 179, 180
  nuclear, 69, 70
  order, 146, 147, 150–64
    determination, 162–4
  photochemical, 172–5
  rate, 145–95
    determination, 147–9
  reversible, 153–5, 315–17

real
  gas, 26–9
  osmolality, 246, 247
receptor-drug complex, 317
reciprocal ohm, 342
recovery
  in rheology, 550
  time of counter, 72
redox, see oxidation-reduction
recrystallisation, 55, 202
red blood cells, 248
reduced viscosity, 504
refinement of crystal structure, 37
refraction
  molar, 104, 105, 109
  of light, 80, 103, 104
refractive index, 103, 104, 108, 493, 494
  gradients, 254, 497, 498
relative
  activity, 309
  atomic mass, 58, 137, 138, 572-4
  biological effectiveness, 77
  centrifugal acceleration, 497
    force, 497
  molecular mass, see also molecular weight, 58, 59, 137, 236–47, 487, 488
  permittivity, 178, 203, 485, 509
  sedimentation volume, 522
  viscosity, 502, 503
relaxation time, 553
Rem, 77–9
Rep, 77
repulsive force, 33, 515, 516
resistance
  electrical, 343, 568
  frictional, 477, 478, 498, 499, 508
resistivity, 343
resonance, 96, 97
  criteria for, 97
  electron spin, 132, 133
  energy, 281, 282
  nuclear magnetic, 126–32
respiration, 23, 26, 419, 420, 435
respiratory chain, 419, 420
retention time, 235
reversibility, 282–6, 315
reversible

reversible
  change, 282–9, 296–300, 302–5
  electrode, 394–6
  inhibition of enzymes, 191–5
  path, 282–6
  process, 282–9, 296–300, 302–5
  reaction, 153–5, 315–17
R_F value, 234, 235
rheogoniometer, 555
rheology, 531–55
rheopexy, 547
rigidity modulus, 535, 536
ring detachment method, 461
roentgen, 76, 77
  equivalent man, 77–9
  physical, 77
röntgen, see roentgen
rotation
  about bonds, 174
  of polarised light, 105–10
rotational
  energy, 111
  spectra, 111
  viscometer, 540–2
rubber, 547, 548
Rutherford
  alpha (α) particle scattering experiment, 61
  atomic theory, 61, 62, 79, 82

Saint-Vernant element, 533
salicylic acid
  hydrogen bonding, 32
  infra-red spectrum, 113–15
salicylaldehyde, hydrogen bonding, 32
salt
  bridge, 408, 409, 424
  hydrolysis, 370–4, 381, 382
salting out of lyophilic colloids, 518, 519
saturated
  solution, 200, 201
  vapour pressure, 28, 33, 213, 214, 219–23, 225, 236–40, 446–9
Schlieren optical system, 498
Schulze–Hardy rule for colloid precipitation, 517

scintillation counter, 73
screening
  by electron shell, 101
  constant, in n.m.r., 128, 129
screw dislocation, 36
second law of thermodynamics, 287–9
second order rate constant, 156–61
  reaction, 156–61
  time of half-change, 161
secondary structure of proteins, 141
sedimentation, 497–502
  coefficient, 499, 500
  equilibrium, 500, 501
  rate, 498–500, 523
  volume, 522
semiconductors, 343, 344
semipermeable membrane, 243
Sephadex (molecular sieve), 455
settling of suspensions, 520–3
shear, 535, 536
  compliance, 551
  modulus, 535, 536
  plane, 506, 507
  rate, 536
  shear stress curve, 546
  strain, 535
  stress, 535, 536
  thickening, 547
  thinning, 545, 546
shell of electrons, 83–91
SI system of units, 567–70
  base unit, 567
  derived unit, 568
  prefix, 570
siemens, 342, 568
sigma (σ) bond, 98, 99
sigmoid isotherm, 447–9
sign convention
  electrochemical cells, 399–401
  electrode potential, 401, 402
  energy, 270, 271
  enthalpy, 275
  heat, 268, 271, 275
  work, 269, 270
silver-silver halide electrode, 407, 415, 424
single point viscometer, 537–41
singlet energy state, 121–3

Index 603

sink conditions, 262
slope of graph, 2, 3
smectic phase, 34
Smoluchowski equation, 514
S_N1 and S_N2 reactions, 179, 180
sodium D light, 106
sol, *see also* colloid, 479
solid
   film, 525
   -liquid phase diagram, 43–53
   nuclei, 201
   solution, 44, 49, 50
   state, 35–43
   viscoelastic, 549–54
solidus curve or line, 45–53
solubilisation, 484
solubility, 200–12, 327–9
   and drug action, 224, 225
   curve, 201–3
   determination, 201
   of colloidal electrolytes, 484
   of gases, 210–12
   of liquids, 204–10
   of solids, 200–4
   product, 327–9, 339, 340
solutes, 199
solution, *see* solubility
   definition, 199
   of macromolecules, 481
   rate, 259–64
solvation, 202, 203
solvent, 199
   extraction, 228–33
s-orbital, 91, 92
sorbitan esters, 463
specific, 273
   conductance, 343
   heat capacity, 273
   optical rotatory power, 106
   quantity, 273
   reaction rate, *see* rate constant
   resistance, 343
   rotation, 106
   supersaturation, 487
   surface area, 448, 449
   viscosity, 502, 503
spectra
   atomic, 79–82
   molecular, 110–26

spectrograph, 80, 81
spectrometer, 80, 81, 120, 121
spectrophotofluorimeter, 125
spectroscope, 80
speed of light, 104, 570
spin
   angular momentum, 126
   quantum number, 83, 84, 125, 132
   -spin splitting, 129–31
splitting factor
   electronic, 132
   nuclear, 126
spontaneous (or natural) process, 285
spreading, 467
square root by logarithms, 6
stability, nuclear, 66–8
stable phase, 41–3
standard, 277
   atmosphere, 569
   electrode potential, 403–7
      table, 405
   oxidation-reduction potential, 403–7
      table, 406
   redox potential, 403–7
      table, 406
   state, 199, 214, 277, 307, 311, 312
state
   equation of, 22, 27, 270
   function, 271
   of system, 270
   standard, 199, 214, 277, 307, 311, 312
stationary chain reaction, 175–7
statistical mechanics or thermodynamics, 289–94
Staudinger viscosity equation, 504
steady state
   approximation in chain reactions, 176
   treatment in enzyme kinetics, 188, 189
steam distillation, 220–2
Stern
   layer, 505–8
   plane, 506, 507
Stirling approximation, 294
stoichiometric number, 146, 275, 317, 318

stokes, 538, 569
Stokes' law, 478, 508, 523
S.T.P. (N.T.P.), 77, 210, 212
straight line, equation of, 2, 3
strain, 531
streamline flow, 537
stress
  definition, 531
  normal, 531
  relaxation, 553
  shear, 535, 537, 543–6, 552
  tensile, 531
stretching
  of elastomers (rubber), 533, 547
  vibrations in spectra, 111–15
strong electrolytes, 246, 326, 337–40, 347, 348
structural chemistry basis, 96, 97
structure
  of atoms, 57–94
  of liquids, 33–5
  of molecules, 96–142
  of solids, 35–9, 53, 54
sublimation, 41, 285
  enthalpy of, 295, 296, 302–5
subsidiary quantum number, 83, 84
substitutional solid solution, 44
substitution reaction, 178–80
substrate in enzyme reactions, 186, 188
supersaturation, 201, 486, 487
  specific, 487
suppository, 42
surface
  active agent, 462–5, 480–5, 519–21, 524–7
  area per molecule, 448, 449, 466–73
    specific, 448, 449
  catalysis, 182–7
  chemistry, see also colloids, 442–73
  excess concentration, 463–7
  film, 467–73
    condensed, 470, 471
    expanded, 471, 472
    gaseous, 469, 470
    insoluble, 467–73
    reactions, 472, 473

film soluble, 467
surface
  film vapour, 469, 470
  free (or Gibbs) energy, 458–68, 480, 482–4
  of shear, 506, 507
  pressure, 467–73
  reaction, 182–7
  tension, 33, 457–68, 480, 482–4
    definition, 33, 457
    determination, 459–62
    in emulsion formation, 524–7
    lowering by adsorbed films, 467–72
  of surfactant solutions, 462–7, 482–4
surfactant (or tenside), 462–5, 480–5, 519–21, 524–7
suspending agents, 520–2
suspension, 479–81, 520–3
Svedberg unit, 499
system, thermodynamic, 268

tau (τ) value in n.m.r., 129
tautomerism, catalysis of, 388
temperature
  absolute scale, 21, 40, 287, 291
  Celsius or centigrade, 21
  coefficient
    of electrochemical cells, 297, 298
    of enzyme reactions, 190, 191
    of equilibrium constants, 319, 320
    of reaction rates, 164–72, 190, 191
    of vapour pressure, 39–42, 304, 305
    of viscous flow, 540
  Kelvin scale, 21, 40, 287, 291, 567
  thermodynamic, see also absolute temperature, 40, 291, 567
tenside, see surfactant
tensile stress, 531
tension
  electric, 568
  surface, see surface tension
tera, 570

ternary mixtures, 206–10
tertiary structure of proteins, 141
tesla, 568
tetramethylsilane, standard in n.m.r., 129
theoretical plate, 217, 218
theory
 of absolute reaction rates, 169–72
 of preferential wetting, 526
 of rapid coagulation, 514, 515
thermal
 analysis, 51
 conduction, 285
 entropy, 286–9
 motion of molecules, 33, 35, 36
thermochemical
 calorie, 275, 569
 equation, 277–81
thermochemistry, 275–82, 375–7
thermocouple, 51
thermodynamic
 activity, 213–15, 222–5, 243, 247, 307–14, 403–8
 energy, 270–7, 290
 probability, 290–3
 temperature, see also absolute temperature, 40, 291, 567
thermodynamics, 268–323
thickener, 528
thin layer chromatography (TLC), 233–5, 455
third law of thermodynamics, 291
thixotropy, 545–7
Thomson
 atomic theory, 60
 discovery of the electron, 60
tie line, 45, 49, 50, 209, 216, 217
time of half change, see also half-life, 155
 determination of order of reaction, 163
 for first order reactions, 155, 156
 for second order reactions, 161
 for zero order reactions, 162
Tiselius, electrophoresis apparatus, 510, 511
titration
 acid-base, 352–4, 373–5, 413, 414
 amperometric, 436, 437

titration
 conductimetric, 352–4
 curve, 374, 414
 of acids and bases, 373–5
 of oxidation-reduction reactions, 412–14
 of proteins, 514
oxidation-reduction, 412–14, 419
polarometric (or amperometric), 436, 437
potentiometric, 412–14
tonicity, 248–50
torr, 569
torsion balance, 461
tracer
 non-radioactive, 70, 71
 radioactive, 69, 70
transfer of matter, thermodynamics of, 285, 309, 310
transference number, see transport number
transition
 $\pi \to \pi^*$, 118
 of polymorphs, 42, 43, 285, 295, 296, 302–4
 of solvates, 201, 202
 series, 59, 87
 state, 166, 169–71, 179–81
 theory, 169–71
transmittance, 117
transport or flow of matter, see also diffusion, 285, 309, 310
transport number, 332–7, 352, 422–5
 and molar conductivity, 351
 apparent, 337
 determination by movement of ions, 335, 336
 determination by e.m.f. measurements, 423, 424
 factors influencing, 336, 337
 relationship to e.m.f. and liquid junction potential of cells, 423–6
 true, 337
triangular coordinates, 207
triple point, 39–41, 239, 240
triplet energy state, 122, 123
true
 adsorption isotherm, 451, 452
 dead time of counter, 72

true
  mass defect, 66
  transport number, 337
turbidity, 34, 35, 493–6
turgidity, 248
Tyndall effect, 475, 492

ultracentrifuge, 497–502
ultrafiltration, 478, 488
ultramicroscope, 475, 489, 510
ultraviolet spectra, 115–19
unavailable energy, 300, 392
uncertainty principle, 83
unified atomic mass constant, 58, 66, 569
unimolecular reaction, 172
unit, 567–9
  cell, 35
unpaired electron, 133, 176–8
U-tube viscometer, 503, 504, 537–40
uranium radioactive series, 63

vacant site, 36
valency, 87–91
van der Waals
  adsorption, 443–50
  equation, 27, 32
  forces, 30–3, 53, 54, 443, 515
  packing, 53
van't Hoff
  equation, 320, 321
  factor (i), 246, 247, 341
  isochore, 320, 321
  isotherm, 313–15
vaporisation
  enthalpy, 31, 33, 295, 296, 303, 305
  entropy, 259, 296, 303
vapour
  phase chromatography, 235, 236
  pressure, 28, 31, 33, 41, 42, 212–22, 225, 236–40, 304, 305, 446–9
    effect of solute on, 236–40, 245–7
    effect of temperature on, 304, 305
    lowering of, 236–40, 245–7
    of liquid mixtures, 212–22, 225

  of single substance, 28, 31, 33, 446–9
  pressure relative lowering of, 236–8, 245–7
  state, 28, 33, 39–42
vector
  addition for dipole moments, 101–3
  in Patterson function, 38
velocity
  constant, see rate constant,
  gradient, 536, 537
  of light, 104, 570
  of molecules, 24
vibration, molecular, 111–15
vibrational
  energy, 111, 115
  spectra, 111–15
viscoelasticity, 548–55
viscosity, 502–5, 534, 537–47
  average molar mass, 505
  dynamic, 537–40
  Einstein equation, 502–5
  effect of temperature on, 540
  intrinsic, 504, 505
  kinematic, 538, 539
  limiting, 504, 505
  Mark and Houwink equation, 504, 505
  measurement, 503, 504, 537–42
  Newtonian, 537–40
  non-Newtonian, 541–7
  number, 504
  of colloids, 502–5
  of emulsions, 528, 542–7
  of liquids, 534, 537–40
  of suspensions, 523, 542–7
  of viscoelastic materials, 548–55
  ratio, 502
    increment, 502–4
  reduced, 504
  relative, 502
  specific, 502–4
  Staudinger equation, 504, 505
viscous flow, 534, 537–50
visible spectra, 80–2, 115–17
Voigt body, 549–51
volt, 397, 568

volume
  fraction, 199, 502
  strain, 532
volumetric method for gas adsorption, 445, 446
vulcanisation, 547

wall of cell in osmosis, 248
water
  activity, 223
  -in-oil (W/O) emulsions, 524–7
  ionic product, 361
  phase diagram, 39, 40
watt, 568
wavelength
  in light scattering, 493–6
  in spectra, 80, 111, 112
  in wave mechanics, 83, 84
wave mechanics, 83–5, 91–4, 97–100
wavenumber, 111, 112, 570
weak
  acid, pH of, 362, 363, 366
  acid and salt, pH of, 363, 364
  base, pH of, 364–6
  base and salt, pH of, 365
  electrolyte, 246, 341, 342
weber, 568
Weissenberg effect, 554, 555
Weston cell, 399
wetting, 526
Wheatstone bridge, 345
Wilhelmy plate method, 460, 461
W/O emulsions, 524–7
work, 268–70, 284–6, 299
  electrical, 268, 397

work
  in dilution of a solution, 284–6
  in expansion of a gas, 28, 29, 284–6
  maximum, *see also* Gibbs free energy, 285, 286, 299
  mechanical, 268–70, 284–6, 299
  net, 299
  osmotic, 284–6
  reversible, *see also* Gibbs free energy, 285, 286, 299
  useful, 299

X-ray
  biological effect of, 76–9
  crystallography, 36–9
  frequencies of elements, 64
  in K-capture, 68
  production of, 64

yield value, 544, 545, 548
Young's modulus, 532, 533

zeolite, 450, 456
zero
  order rate constant, 162
  reaction, 161, 162, 186, 187
  reaction, time of half-change, 162
  point energy, 291
zeta (ζ) potential, 507–9, 512, 515, 516, 520, 522, 524, 527
zwitterion, 377, 378, 512, 513